The Mitchell Beazley Joy of Knowledge Library

Man and Society

Scientiam non dedit natura semina scientiae nobis dedit
"Nature has given us not knowledge itself, but the seeds thereof."
Seneca

The Joy of Knowledge Encyclopaedia is affectionately
dedicated to the memory of John Beazley 1932–1977,
Book Designer, Publisher and Co-Founder of the
publishing house of Mitchell Beazley Limited, by all
his many friends and colleagues in the company.

The Joy of Knowledge Library

General Editor: James Mitchell
With an overall preface by Lord Butler, Master of Trinity College,
University of Cambridge

Science and The Universe   Introduced by Sir Alan Cottrell,
Master of Jesus College,
University of Cambridge; and
Sir Bernard Lovell, Professor of
Radio Astronomy, University of
Manchester

The Physical Earth   Introduced by Dr William
Nierenberg, Director, Scripps
Institution of Oceanography,
University of California

The Natural World   Introduced by Dr Gwynne
Vevers, Assistant Director of
Science, the Zoological Society
of London

Man and Society   Introduced by Dr Alex Comfort,
Professor of Pathology, Irvine
Medical School, University
of California, Santa Barbara

History and Culture 1   Introduced by Dr Christopher
History and Culture 2   Hill, Master of Balliol College,
University of Oxford

Man and Machines   Introduced by Sir Jack Callard,
former Chairman of Imperial
Chemical Industries Limited

The Modern World

Fact Index   A–K

Fact Index   L–Z

The Mitchell Beazley Joy of Knowledge Library

Man and Society

Introduced by Alex Comfort, MB, Dsc

Professor of Pathology, Irvine Medical School, University of California, Santa Barbara

MITCHELL BEAZLEY

The Joy of Knowledge Encyclopaedia
© Mitchell Beazley Encyclopaedias Limited 1976

The Joy of Knowledge Man and Society
© Mitchell Beazley Encyclopaedias Limited 1977

Artwork © Mitchell Beazley Publishers Limited
1970, 1971, 1972, 1973, 1974, 1975 and 1976
© Mitchell Beazley Encyclopaedias Limited 1976
© International Visual Resource 1972

ISBN 0 85533 108 9

Typesetting by Filmtype Services Limited, England
Photoprint Plates Ltd, Rayleigh, Essex, England

Printed in England by Balding + Mansell

The Joy of Knowledge Library

Editorial Director	**Frank Wallis**
Creative Director	**Ed Day**
Project Director	**Harold Bull**

Volume editors
Science and The Universe	John Clark
	Lawrence Clarke
The Natural World	Ruth Binney
The Physical Earth	Erik Abranson
	Dougal Dixon
Man and Society	Max Monsarrat
History and Culture 1 & 2	John Tusa
	Roger Hearn
Time Chart	Jane Kenrick
Man and Machines	John Clark
Fact Index	Stephen Elliott
	Stanley Schindler
	John Clark

Art Director	Rod Stribley
Production Editor	Helen Yeomans
Assistant to the Project Director	Graham Darlow
Associate Art Director	Anthony Cobb
Art Buyer	Ted McCausland
Co-editions Manager	Averil Macintyre
Printing Manager	Bob Towell
Information Consultant	Jeremy Weston
Sub-Editors	Don Binney
	Arthur Butterfield
	Charyn Jones
	Jenny Mulherin
	Shiva Naipaul
	David Sharpe
	Jack Tresidder
Proof-Readers	Jeff Groman
	Anthony Livesey
Researchers	Peter Furtado
	Malcolm Hart
	Peter Kilkenny
	Ann Kramer
	Lloyd Lindo
	Heather Maisner
	Valerie Nicholson
	Elizabeth Peadon
	John Smallwood
	Jim Somerville
Senior Designer	Sally Smallwood
Designers	Rosamund Briggs
	Mike Brown
	Lynn Cawley
	Nigel Chapman
	Pauline Faulks
	Nicole Fothergill
	Juanita Grout
	Ingrid Jacob
	Carole Johnson
	Chrissie Lloyd
	Aean Pinheiro
	Andrew Sutterby
Senior Picture Researchers	Jenny Golden
	Kate Parish
Picture Researchers	Phyllida Holbeach
	Philippa Lewis
	Caroline Lucas
	Ann Usborne
Assistant to the Editorial Director	Judy Garlick
Assistant to the Section Editors	Sandra Creese
Editorial Assistants	Joyce Evison
	Miranda Grinling
Production Controllers	Jeremy Albutt
	John Olive
	Anthony Bonsels
Production Assistants	Nick Rochez
	John Swan

Major contributors and advisers to The Joy of Knowledge Library

Fabian Acker CEng, MIEE, MIMarE; Professor H.C. Allen MC; Leonard Amey OBE; Neil Ardley BSc; Professor H.R.V. Arnstein DSc, PhD, FIBiol; Russell Ash BA(Dunelm), FRAI; Norman Ashford PhD, CEng, MICE, MASCE, MCIT; Professor Robert Ashton; B.W. Atkinson BSc, PhD; Anthony Atmore BA; Professor Philip S. Bagwell BSc(Econ), PhD; Peter Ball MA; Edwin Banks MIOP; Professor Michael Banton; Dulan Barber; Harry Barrett; Professor J.P. Barron MA, DPhil, FSA; Professor W.G. Beasley FBA; Alan Bender PhD, MSc, DIC, ARCS; Lionel Bender BSc; Israel Berkovitch PhD, FRIC, MIChemE; David Berry MA; M.L. Bierbrier PhD; A.T.E. Binsted FBBI (Dipl); David Black; Maurice E.F. Block BA, PhD(Cantab); Richard H. Bomback BSc (London), FRPS; Basil Booth BSc(Hons), PhD, FGS, FRGS; J. Harry Bowen MA(Cantab), PhD(London); Mary Briggs MPS, FLS; John Brodrick BSc (Econ); J.M. Bruce ISO, MA, FRHistS, MRAeS; Professor D.A. Bullough MA, FSA, FRHistS; Tony Buzan BA(Hons) UBC; Dr Alan R. Cane; Dr J.G. de Casparis; Dr Jeremy Catto MA; Denis Chamberlain; E.W. Chanter MA; Professor Colin Cherry DSc(Eng), MIEE; A.H. Christie MA, FRAI, FRAS; Dr Anthony W. Clare MPhil(London), MB, BCh, MRCPI, MRCPsych; Sonia Cole; John R. Collis MA, PhD; Professor Gordon Connell-Smith BA, PhD, FRHistS; Dr A.H. Cook FRS; Professor A.H. Cook FRS; J.A.L. Cooke MA, DPhil; R.W. Cooke BSc, CEng, MICE; B.K. Cooper; Penelope J. Corfield MA; Robin Cormack MA, PhD, FSA; Nona Coxhead; Patricia Crone BA, PhD; Geoffrey P. Crow BSc(Eng), MICE, MIMunE, MInstHE, DIPTE; J.G. Crowther; Professor R.B. Cundall FRIC; Noel Currer-Briggs MA, FSG; Christopher Cviic BA(Zagreb), BSc(Econ, London); Gordon Daniels BSc(Econ, London), DPhil(Oxon); George Darby BA; G.J. Darwin; Dr David Delvin; Robin Denselow BA; Professor Bernard L. Diamond; John Dickson; Paul Dinnage MA; M.L. Dockrill BSc(Econ), MA, PhD; Patricia Dodd BA; James Dowdall; Anne Dowson MA(Cantab); Peter M. Driver BSc, PhD, MIBiol; Rev Professor C.W. Dugmore DD; Herbert L. Edlin BSc, Dip in Forestry; Pamela Egan MA(Oxon); Major S.R. Elliot CD, BComm; Professor H.J. Eysenck PhD, DSc; Dr Peter Fenwick BA, MB, BChir, DPM, MRCPsych; Jim Flegg BSc, PhD, ARCS, MBOU; Andrew M. Fleming MA; Professor Antony Flew MA(Oxon), DLitt(Keele); Wyn K. Ford FRHistS; Paul Freeman DSc(London); G.E. Fussell DLitt, FRHistS; Kenneth W. Gatland FRAS, FBIS; Norman Gelb BA; John Gilbert BA(Hons, London); Professor A.C. Gimson; John Glaves-Smith BA; David Glen; Professor S.J. Goldsack BSc, PhD, FINSTP, FBCS; Richard Gombrich MA, DPhil; A.F. Gomm; Professor A. Goodwin MA; William Gould BA(Wales); Professor J.R. Gray; Christopher Green PhD; Bill Gunston; Professor A. Rupert Hall LittD; Richard Halsey BA(Hons, UEA); Lynette K. Hamblin BSc; Norman Hammond; Professor Thomas G. Harding PhD; Richard Harris; Dr Randall P. Harrison; Cyril Hart MA, PhD, FRICS, FIFor; Anthony P. Harvey; Nigel Hawkes BA(Oxon); F.P. Heath; Peter Hebblethwaite MA(Oxon), LicTheol; Frances Mary Heidensohn BA; Dr Alan Hill MC, FRCP; Robert Hillenbrand MA, DPhil; Professor F.H. Hinsley; Dr Richard Hitchcock; Dorothy Hollingsworth OBE, BSc, FRIC, FIBiol, FIFST, SRD; H.P. Hope BSc (Hons, Agric); Antony Hopkins CBE, FRCM, LRAM, FRSA; Brian Hook; Peter

Howell BPhil, MA(Oxon); Brigadier K. Hunt; Peter Hurst BDS, FDS, LDS, RSCEd, MSc(London); Anthony Hyman MA, PhD; Professor R.S. Illingworth MD, FRCP, DPH, DCH; Oliver Impey MA, DPhil; D.E.G. Irvine PhD; L.M. Irvine BSc; Anne Jamieson cand mag(Copenhagen), MSc(London); Michael A. Janson BSc; Professor P.A. Jewell BSc(Agric), MA, PhD, FIBiol; Hugh Johnson; Commander I.E. Johnston RN; I.P. Jolliffe BSc, MSc, PhD, CompICE, FGS; Dr D.E.H. Jones ARCS, FCS; R.H. Jones PhD, BSc, CEng, MICE, FGS, MASCE; Hugh Kay; Dr Janet Kear; Sam Keen; D.R.C. Kempe BSc, DPhil, FGS; Alan Kendall MA(Cantab); Michael Kenward; John R. King BSc(Eng), DIC, CEng, MIProdE; D.G. King-Hele FRS; Professor J.F. Kirkaldy DSc; Malcolm Kitch; Michael Kitson MA; B.C. Lamb BSc, PhD; Nick Landon; Major J.C. Larminie QDG, Retd; Diana Leat BSc(Econ), PhD; Roger Lewin BSc, PhD; Harold K. Lipset; Norman Longmate MA(Oxon); John Lowry; Kenneth E. Lowther MA; Diana Lucas BA(Hons); Keith Lye BA, FRGS; Dr Peter Lyon; Dr Martin McCauley; Sean McConville BSc; D.F.M. McGregor BSc, PhD(Edin); Jean Macqueen PhD; William Baird MacQuitty MA(Hons), FRGS, FRPS; Jonathan Martin MA; Rev Canon E.L. Mascall DD; Christopher Maynard MSc, DTh; Professor A.J. Meadows; J.S.G. Miller MA, DPhil, BM, BCh; Alaric Millington BSc, DipEd, FIMA; Peter L. Moldon; Patrick Moore OBE; Robin Mowat MA, DPhil; J. Michael Mullin BSc; Alistair Munroe BSc, ARCS; Professor Jacob Needleman; Professor Donald M. Nicol MA, PhD; Gerald Norris; Caroline E. Oakman BA(Hons, Chinese); S. O'Connell MA(Cantab), MInstP; Michael Overman; Di Owen BSc; A.R.D. Pagden MA, FRHistS; Professor E.J. Pagel PhD; Carol Parker BA(Econ), MA(Internat. Aff.); Derek Parker; Julia Parker DFAstrolS; Dr Stanley Parker; Dr Colin Murray Parkes MD, FRC(Psych), DPM; Professor Geoffrey Parrinder MA, PhD, DD(London), DLitt(Lancaster); Moira Paterson; Walter C. Patterson MSc; Sir John H. Peel KCVO, MA, DM, FRCP, FRCS, FRCOG; D.J. Penn; Basil Peters MA. MInstP, FBIS; D.L. Phillips FRCR, MRCOG; B.T. Pickering PhD, DSc; John Picton; Susan Pinkus; Dr C.S. Pitcher MA, DM, FRCPath; Alfred Plaut FRCPsych; A.S. Playfair MRCS, LRCP, DObstRCOG; Dr Antony Polonsky; Joyce Pope BA; B.L. Potter NDA, MRAC, CertEd; Paulette Pratt; Antony Preston; Frank J. Pycroft; Margaret Quass; Dr John Reckless; Trevor Reese BA, PhD, FRHistS; Derek A. Reid BSc, PhD; Clyde Reynolds BSc; John Rivers; Peter Roberts; Colin A. Ronan MSc, FRAS; Professor Richard Rose BA(Johns Hopkins), DPhil(Oxon); Harold Rosenthal; T.G. Rosenthal MA(Cantab); Anne Ross MA, MA(Hons, Celtic Studies), PhD(Archaeol and Celtic Studies, Edin); Georgina Russell MA; Dr Charles Rycroft BA(Cantab), MB(London), FRCPsych; Susan Saunders MSc(Econ); Robert Schell PhD; Anil Seal MA, PhD(Cantab); Michael Sedgwick MA(Oxon); Martin Seymour-Smith BA(Oxon), MA(Oxon); Professor John Shearman; Dr Martin Sherwood; A.C. Simpson BSc; Nigel Sitwell; Dr Alan Sked; Julie and Kenneth Slavin FRGS, FRAI; Alec Xavier Snobel BSc(Econ); Terry Snow BA, ATCL; Rodney Steel; Charles S. Steinger MA, PhD; Geoffrey Stern BSc(Econ); Maryanne Stevens BA(Cantab), MA(London); John Stevenson DPhil, MA; J. Stidworthy MA; D. Michael Stoddart BSc, PhD; Bernard Stonehouse DPhil, MA, BSc, MInstBiol; Anthony Storr FRCP, FRCPsych; Richard Storry; Professor John Taylor; John W.R. Taylor FRHistS, MRAeS, FSLAET; R.B. Taylor BSc(Hons, Microbiol); J. David Thomas MA, PhD; Harvey Tilker PhD; Don Tills PhD, MPhil, MIBiol, FIMLS; Jon Tinker; M. Tregear MA; R.W. Trender; David

Trump MA, PhD, FSA; M.F. Tuke PhD; Christopher Tunney MA; Laurence Urdang Associates (authentication and fact check); Sally Walters BSc; Christopher Wardle; Dr D. Washbrook; David Watkins; George Watkins MSc; J.W.N. Watkins; Anthony J. Watts; Dr Geoff Watts; Melvyn Westlake; Anthony White MA(Oxon), MAPhil(Columbia); P.J.S. Whitmore MBE, PhD; Professor G.R. Wilkinson; Rev H.A. Williams CR; Christopher Wilson BA; Professor David M. Wilson; John B. Wilson BSc, PhD, FGS, FLS; Philip Windsor BA, DPhil(Oxon); Professor M.J. Wise; Roy Wolfe BSc(Econ), MSc; Dr David Woodings MA, MRCP, MRCPath; Bernard Yallop PhD, BSc, ARCS, FRAS Professor John Yudkin MA, MD, PhD(Cantab), FRIC, FIBiol, FRCP.

The General Editor wishes particularly to thank the following for all their support:
Nicolas Bentley
Bill Borchard
Adrianne Bowles
Yves Boisseau
Irv Braun
Theo Bremer
the late Dr Jacob Bronowski
Sir Humphrey Browne
Barry and Helen Cayne
Peter Chubb
William Clark
Sanford and Dorothy Cobb
Alex and Jane Comfort
Jack and Sharlie Davison
Manfred Denneler
Stephen Elliott
Stephen Feldman
Orsola Fenghi
Dr Leo van Grunsven
Jan van Gulden
Graham Hearn
the late Raimund von Hofmansthal
Dr Antonio Houaiss
the late Sir Julian Huxley
Alan Isaacs
Julie Lansdowne
Andrew Leithead
Richard Levin
Oscar Lewenstein
The Rt Hon Selwyn Lloyd
Warren Lynch
Simon macLachlan
George Manina
Stuart Marks
Bruce Marshall
Francis Mildner
Bill and Christine Mitchell
Janice Mitchell
Patrick Moore
Mari Pijnenborg
the late Donna Dorita de Sa Putch
Tony Ruth
Dr Jonas Salk
Stanley Schindler
Guy Schoeller
Tony Schulte
Dr E. F. Schumacher
Christopher Scott
Anthony Storr
Hannu Tarmio
Ludovico Terzi
Ion Trewin
Egil Tveteras
Russ Voisin
Nat Wartels
Hiroshi Watanabe
Adrian Webster
Jeremy Westwood
Harry Williams
the dedicated staff of MB Encyclopaedias who created this *Library* and of MB Multimedia who made the IVR Artwork Bank.

Man and Society/Contents

Keystone

Lord Butler, Master of Trinity College,
Cambridge, knocks on the great door of
the college during his installation
ceremony on October 7, 1965

Preface

I do not think any other group of publishers could be credited with producing so comprehensive and modern an encyclopaedia as this. It is quite original in form and content. A fine team of writers has been enlisted to provide the contents. No library or place of reference would be complete without this modern encyclopaedia, which should also be a treasure in private hands.

The production of an encyclopaedia is often an example that a particular literary, scientific and philosophic civilization is thriving and groping towards further knowledge. This was certainly so when Diderot published his famous encyclopaedia in the eighteenth century. Since science and technology were then not so far developed, his is a very different production from this. It depended to a certain extent on contributions from Rousseau and Voltaire and its publication created a school of adherents known as the encyclopaedists.

In modern times excellent encyclopaedias have been produced, but I think there is none which has the wealth of illustrations which is such a feature of these volumes. I was particularly struck by the section on astronomy, where the illustrations are vivid and unusual. This is only one example of illustrations in the work being, I would almost say, staggering in their originality.

I think it is probable that many responsible schools will have sets, since the publishers have carefully related much of the contents of the encyclopaedia to school and college courses. Parents on occasion feel that it is necessary to supplement school teaching at home, and this encyclopaedia would be invaluable in replying to the queries of adolescents which parents often find awkward to answer. The "two-page-spread" system, where text and explanatory diagrams are integrated into attractive units which relate to one another, makes this encyclopaedia different from others and all the more easy to study.

The whole encyclopaedia will literally be a revelation in the sphere of human and humane knowledge.

Butler

Master of Trinity College,
Cambridge

General Editor's Introduction

The Structure of the Library

 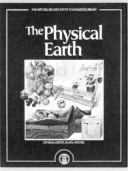   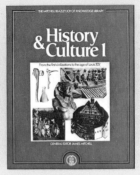

Science and The Universe	The Physical Earth	The Natural World	Man and Society	History and Culture
The growth of science	Structure of the Earth	How life began	Evolution of man	Volume 1 From the first
Mathematics	The Earth in perspective	Plants	How your body works	civilizations to the age of
Atomic theory	Weather	Animals	Illness and health	Louis XIV
Statics and dynamics	Seas and oceans	Insects	Mental health	
Heat, light and sound	Geology	Fish	Human development	The art of prehistory
Electricity	Earth's resources	Amphibians and reptiles	Man and his gods	Classical Greece
Chemistry	Agriculture	Birds	Communications	India, China and Japan
Techniques of astronomy	Cultivated plants	Mammals	Politics	Barbarian invasions
The Solar System	Flesh, fish and fowl	Prehistoric animals and	Law	The crusades
Stars and star maps		plants	Work and play	Age of exploration
Galaxies		Animals and their habitats	Economics	The Renaissance
Man in space		Conservation		The English revolution

Man and Society is a book of popular general knowledge about ourselves. It is a self-contained book with its own index and its own internal system of cross-references to help you to build up a rounded picture of the way we live.

It is one volume in Mitchell Beazley's intended ten-volume library of individual books we have entitled *The Joy of Knowledge Library*—a library which, when complete, will form a comprehensive encyclopaedia.

For a new generation brought up with television, words alone are no longer enough—and so we intend to make the *Library* a new sort of pictorial encyclopaedia for a visually oriented age, a new "family bible" of knowledge which will find acceptance in every home.

Seven other colour volumes in the *Library* are planned to be *Science and The Universe, The Physical Earth, The Natural World, History and Culture (two volumes), Man and Machines,* and *The Modern World. The Modern World* will be arranged alphabetically: the other volumes will be organized by topic and will provide a comprehensive store of general knowledge rather than isolated facts.

The last two volumes in the *Library* will provide a different service. Split up for convenience into A-K and L-Z references, these volumes will be a fact index to the whole work. They will provide factual information of all kinds on peoples, places and things through approximately 25,000 mostly short entries listed in alphabetical order. The entries in the A-Z volumes also act as a comprehensive index to the other eight volumes, thus turning the whole *Library* into a rounded *Encyclopaedia*, which is not only a comprehensive guide to general knowledge in volumes 1–7 but which now also provides access to specific information as well in *The Modern World* and the fact index volumes.

Access to knowledge

Whether you are a systematic reader or an unrepentant browser, my aim as General Editor has been to assemble all the facts you really ought to know into a coherent and logical plan that makes it possible to build up a comprehensive general knowledge of the subject.

Depending on your needs or motives as a reader in search of knowledge, you can find things out from *Man and Society* in four or more ways: for example, you can simply browse pleasurably about in its pages haphazardly (and

that's my way!) or you can browse in a more organized fashion if you use our "See Also" treasure hunt system of connections referring you from spread to spread. Or you can gather specific facts by using the index. Yet again, you can set yourself the solid task of finding out literally everything in the book in logical order by reading it from cover to cover: in this the Contents List (page 6) is there to guide you.

Our basic purpose in organizing the volumes in *The Joy of Knowledge Library* into two elements—the three volumes of A-Z factual information and the seven volumes of general knowledge—was functional. We devised it this way to make it easier to gather the two different sorts of information—simple facts and wider general knowledge, respectively—in appropriate ways.

The functions of an encyclopaedia

An encyclopaedia (the Greek word means "teaching in a circle" or, as we might say, the provision of a *rounded* picture of knowledge) has to perform these two distinct functions for two sorts of users, each seeking information of different sorts.

First, many readers want simple factual answers to straightforward questions like "What is Epicureanism?". They may be intrigued to learn that it is a philosophy founded by the Greek Epicurus in the late 4th century BC and that it held that pleasure and pain are the ultimate good and evil. Such direct and simple facts are best supplied by a short entry and in the *Library* they will be found in the two A-Z *Fact Index* volumes.

But secondly, for the user looking for in-depth knowledge on a subject or on a series of subjects—such as "How did our legal system arise?"—short alphabetical entries alone are inevitably bitty and disjointed. What do you look up first—"law"? "legal systems"? "British law"? "jurisprudence"? "crime"? "courts"? and do you have to read all the entries or only some of them? You normally have to look up *lots* of entries in a purely alphabetical encyclopaedia to get a comprehensive answer to such wide-ranging questions. Yet comprehensive answers are what general knowledge is all about.

A long article or linked series of longer articles, organized by related subjects, is clearly much more helpful to the

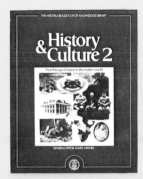

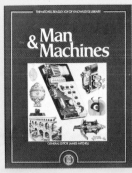

History and Culture

Volume 2 From the Age
of Reason to the
modern world

Neoclassicism
Colonizing Australasia
World War I
Ireland and independence
Twenties and the
 depression
World War II
Hollywood

Man and Machines

The growth of
 technology
Materials and techniques
Power
Machines
Transport
Weapons
Engineering
Communications
Industrial chemistry
Domestic engineering

The Modern World

Almanack
Countries of the world
Atlas
Gazetteer

Fact Index A-K

The first of two volumes
containing 25,000 mostly
short factual entries
on people, places and
things in A-Z order. The
Fact Index also acts as
an index to the eight
colour volumes. In
this volume, everything
from Aachen to Kyzyl.

Fact Index L-Z

The second of the A-Z
volumes that turn the
Library into a complete
encyclopaedia. Like the
first, it acts as an
index to the eight
colour volumes. In this
volume, everything from
Ernest Laas to Zyrardow.

person wanting such comprehensive answers. That is why
we have adopted a logical, so-called *thematic* organization
of knowledge, with a clear system of connections relating
topics to one another, for teaching general knowledge in
Man and Society and the six other general knowledge
volumes in the *Library*.

The spread system
The basic unit of all the general knowledge books is the
"spread"—a nickname for the two-page units that
comprise the working contents of all these books. The
spread is the heart of our approach to explaining things.

Every spread in *Man and Society* tells a story—almost
always a self-contained story—a story on how people
differ, for example (pages 34 to 35) or the way the nervous
system works (pages 38 to 39) or on the occult (pages 196
to 197) or on the meaning of myth (pages 206 to 207). The
spreads on these subjects all work to the same discipline,
which is to tell you all you need to know in two facing
pages of text and pictures. The discipline of having to get in
all the essential and relevant facts in this comparatively
short space actually makes for better results—text that has
to get to the point without any waffle, pictures and
diagrams that illustrate the essential points in a clear and
coherent fashion, captions that really work and explain the
point of the pictures.

The spread system is a strict discipline but once you get
used to it, I hope you'll ask yourself why you ever thought
general knowledge could be communicated in any other way.

The structure of the spread system will also, I hope
prove reassuring when you venture out from the things you
do know about into the unknown areas you don't know,
but want to find out about. There are many virtues in
being systematic. You will start to feel at home in all sorts
of unlikely areas of knowledge with the spread system to
guide you. The spreads are, in a sense, the building blocks
of knowledge. Like living cells which are the building
blocks of plants and animals, they are systematically
"programmed" to help you to learn more easily and to
remember better. Each spread has a main article of 850
words summarising the subject. The article is illustrated
by an average of ten pictures and diagrams, the captions
of which both complement *and* supplement the

information in the article (so please read the captions,
incidentally, or you may miss something!). Each spread,
too, has a "key" picture or diagram in the top right-hand
corner. The purpose of the key picture is twofold: it
summarises the story of the spread visually and it is
intended to act as a memory stimulator to help you to
recall all the integrated facts and pictures on a given
subject.

Finally, each spread has a box of connections headed
"See Also" and, sometimes, "Read First". These are
cross-reference suggestions to other connecting spreads.
The "Read Firsts" normally appear only on spreads with
particularly complicated subjects and indicate that you
might like to learn to swim a little in the elementary
principles of a subject before being dropped in the deep
end of its complexities.

The "See Alsos" are the treasure hunt features of *The
Joy of Knowledge* system and I hope you'll find them
helpful and, indeed, fun to use. They are also essential if
you want to build up a comprehensive general knowledge.
If the spreads are individual living cells, the "See Alsos"
are the secret code that tells you how to fit the cells
together into an organic whole which is the body of
general knowledge.

Level of readership
The level for which we have created *The Joy of Knowledge
Library* is intended to be a universal one. Some aspects of
knowledge are more complicated than others and so readers
will find that the level varies in different parts of the
Library and indeed in different parts of this volume,
Man and Society. This is quite deliberate: *The Joy of
Knowledge Library* is a library for all the family.

Some younger people should be able to enjoy and to
absorb most of the pages in this volume on the human
body, for example, from as young as ten or eleven
onwards—but the level has been set primarily for adults
and older children who will need some basic knowledge to
makes sense of the pages on evolution or international
economics, for example.

Whatever their level, the greatest and the bestselling
popular encyclopaedias of the past have always had one
thing in common—simplicity. The ability to make even

complicated subjects clear, to distil, to extract the simple principles from behind the complicated formulae, the gift of getting to the heart of things: these are the elements that make popular encyclopaedias really useful to the people who read them. I hope we have followed these precepts throughout the *Library*: if so our level will be found to be truly universal.

Philosophy of the Library

The aim of *all* the books—general knowledge and *Fact Index* volumes—in the *Library* is to make knowledge more readily available to everyone, and to make it fun. This is not new in encyclopaedias. The great classics enlightened whole generations of readers with essential information, popularly presented and positively inspired. Equally, some works in the past seem to have been extensions of an educational system that believed that unless knowledge was painfully acquired it couldn't be good for you, would be inevitably superficial, and wouldn't stick. Many of us know in our own lives the boredom and disinterest generated by such an approach at school, and most of us have seen it too in certain types of adult books. Such an approach locks up knowledge, not liberates it.

The great educators have been the men and women who have enthralled their listeners or readers by the self-evident passion they themselves have felt for their subjects. Their joy is natural and infectious. We remember what they say and cherish it for ever. The philosophy of *The Joy of Knowledge Library* is one that precisely mirrors that enthusiasm. We aim to seduce you with our pictures, absorb you with our text, entertain you with the multitude of facts we have marshalled for your pleasure—yes, *pleasure*. Why not pleasure?

There are three uses of knowledge: education (things you ought to know because they are important); pleasure (things which are intriguing or entertaining in themselves); application (things we can do with our knowledge for the world at large).

As far as education is concerned there are certain elementary facts we need to learn in our schooldays. The *Library*, with its vast store of information, is primarily designed to have an educational function—to inform, to be a constant companion and to guide everyone through school, college and other forms of higher education.

But most facts, except to the student or specialist (and these books are not only for students and specialists, they are for everyone) aren't vital to know at all. You don't *need* to know them. But discovering them can be a source of endless pleasure and delight, nonetheless, like learning the pleasures of food or wine or love or travel. Who wouldn't give a king's ransom to know when man really became man and stopped being an ape? Who wouldn't have loved to have spent a day at the feet of Leonardo or to have met the historical Jesus or to have been there when Stephenson's *Rocket* first moved? The excitement of discovering new things is like meeting new people—it is one of the great pleasures of life.

There is always the chance, too, that some of the things you find out in these pages may inspire you with a lifelong passion to apply your knowledge in an area which really interests you. My friend Patrick Moore, the astronomer, who first suggested we publish this *Library* and wrote much of the astronomy section in our volume on *Science and The Universe*, once told me that he became an astronomer through the thrill he experienced on first reading an encyclopaedia of astronomy called *The Splendour of the Heavens*, published when he was a boy. Revelation is the reward of encyclopaedists. Our job, my job, is to remind you always that the joy of knowledge knows no boundaries and can work untold miracles.

In an age when we are increasingly creators (and less creatures) of our world, the people who *know*, who have a sense of proportion, a sense of balance, above all perhaps a sense of insight (the inner as well as the outer eye) in the application of their knowledge, are the most valuable people on earth. They, and they alone, will have the capacity to save this earth as a happy and a habitable planet for all its creatures. For the true joy of knowledge lies not only in its acquisition and its enjoyment, but in its wise and loving application in the service of the world.

Thus the Latin tag "Scientiam non dedit natura, semina scientiae nobis dedit" on the first page of this book. It translates as "Nature has given us not knowledge itself, but the seeds thereof."

It is, in the end, up to each of us to make the most of what we find in these pages.

General Editor's Introduction
The Structure of this Book

Man and Society is a book of general knowledge containing all the information that we think is most interesting and relevant about ourselves and the way we live. In its 336 pages it covers man's evolution, his physical and mental make-up, his beliefs and dreams, and the systems he has created to enable him to understand and exploit the environment he lives in. It has been our intention to present that mass of facts in such a way that they make sense and tell a logical, comprehensible and coherent story rather than appear in a meaningless jumble.

One of the more interesting problems we found ourselves confronted with when we were planning the book was that of the sheer diversity of man's endeavour. Broadly speaking, science is universal—Italian astronomers use the same equipment and apply the same scientific truths in their work as do American astronomers or British astronomers. The same is true, again broadly speaking, of other fields of applied science such as technology and medicine. But in many human pursuits that closely affect our daily lives there are no international rules—or if there are, they are so general that particular applications of them can differ widely. In philosophy, religion, living standards, economics and attitudes to law, work, leisure and politics, there are not only wide differences between one nation and the next, but also between one man and his neighbour.

Though a pessimist might claim that a good deal of man's history could be seen as the direct expression of those differences, an optimist might see it as an attempt to reconcile them. We are frankly, on the side of the optimists—in planning *Man and Society* we set out to record the differences while at the same time trying to show what men, whatever race they might be and whatever systems they might operate, had in common.

Where to start
Before outlining the plan of the contents of *Man and Society* I'm going to assume for a moment that you are coming to this subject, just as I came to it when planning the book, as a "know-nothing" rather than as a "know-all". Knowing nothing, incidentally, can be a great advantage as a reader—or as an editor, as I discovered in making this book. If you know nothing, but want to find things out,

you ask difficult questions all the time. I spent much of my time as General Editor of this *Library* asking experts difficult questions and refusing to be fobbed off with complicated answers I couldn't understand. *Man and Society*, like every other book in this *Library*, has thus been through the sieve of my personal ignorance in its attempt to re-state things simply and understandably.

If you know nothing, my suggestion is that you start with Alex Comfort's introduction to the subject (pages 16 to 19). He sets out the factors that produced modern man, examines why we think and behave the way we do, and—because he is an optimist also—delivers a rebuttal of certain views that serve to divide mankind rather than unite it. His views form a useful framework for the facts that follow. If, however, you prefer to plunge straight into the book, but don't have much basic knowledge, I suggest you study eight spreads in the book before anything else (see panel on page 14). These spreads are the "Read First" spreads. They will give you the basic facts about man's evolution, the structure of his body and mind, the importance of ritual, and the way in which he studies and governs himself and his fellows. Once you have digested these spreads you can build up a more comprehensive general knowledge by exploring the rest of the book.

Plan of the book
There are broadly ten sections, or blocks of spreads, in *Man and Society*. The divisions between them are not marked in the text because we thought that would spoil the continuity of the book. They are:
Evolution of Man
There are three parts to the story of how modern man came to live in today's nation states. First, there is the story of how he evolved from his ape-like ancestors. Then there is the story of how he spread across the world. Finally, there is today's story—who the people of the world are, where they live and how they differ.
How your body works
Before man's intellectual, social and emotional development can be described, it is necessary to understand the mental and physical structure that produced—and limits—that development. Having established who man is, *Man and Society* looks at what he is—a collection of cells with many

Man and Society, like most volumes in *The Joy of Knowledge Library*, tackles its subject topically on a two-page spread basis. Though the spreads are self-contained, you may find some of them easier to understand if you read certain basic spreads first. Those spreads are illustrated here. They are "scene-setters" that will give you an understanding of the fundamentals of life—its origins and continuity—and of how botanists and zoologists classify plants and animals. With them as background, the rest of the spreads in *Man and Society* can be more readily understood. The eight spreads are:

The first hominids

Human development

specialized functions, some responsible for memory, some for sight, some for touch, some for perpetuating his own kind. The marvellous machine, even now imperfectly understood, that is a man is broken down into its component parts—and there are spreads on how to keep those parts in first-class working order.

Illness and health
Machines break down—and so does man. One of the most creative of his endeavours throughout history has been to prevent breakdown or to treat it when it occurs. In 68 pages, *Man and Society* looks at the causes of both physical and mental illness, at their manifestations, and at the ways in which they can be cured. Increasingly, the problem of illness and disease has expanded from one that concerned a local community to one that concerns a nation and, today, to one that concerns the whole world. In an age of rapid transport and world-wide movements of people, disease can no longer be localized. In an age of increasing interdependence between nations, the ill-health of one may dramatically affect the economic viability of another. The section ends by examining how illness and disease have become international issues.

Human development
The route from the cradle to the grave is one of constant learning and re-adjustment. From the moment, as newly born babies, they first become aware of the world, to the moment they accept that life is over, human beings have lessons to learn. Social, emotional, intellectual and moral development are all essential if we are to cope with the business of living and so is the ability to handle the crises that develop through childhood, adolescence and adulthood. This section of the book concludes with an examination of the most elusive of concepts, those of human personality and individuality.

Man and his gods
Man does not live by bread alone—even in the most secular of ages there have been men who have sought a meaning to life that lies outside themselves, an explanation of those aspects of life for which science offered no laws. This section of *Man and Society* opens with spreads on the occult, parapsychology and astrology and then proceeds to discuss what lies behind the rituals that even the most sophisticated of us follow. It

considers myth and religion before concluding with five spreads on the nature and functions of philosophy.

Communications
Speech is one of the attributes that separates man from his animal background—but he still retains the ability to communicate without speech. Facial expressions, gestures and stance reflect effective but wordless ways of conveying messages, ways habitually used by animals. This section ranges from the openness of a smile to the secretiveness of a cypher.

Politics
Some kind of political organization has been a feature of human society since the first true men appeared. There is evidence that *Homo erectus*—upright man—returned year after year to the same settled sites and by the time of Neanderthal man, quarter of a million years ago, it is likely that society had developed rigid rules regarding the behaviour of individuals. In historical terms, sophisticated civilizations with complex political structures have existed since the time of the Sumerians, more than 5,000 years ago. Since that time, Western man has evolved his present democratic system of government. That system, which is by no means universal, functions in different ways in different countries and is enormously complex. *Man and Society* traces the story of political evolution from its beginnings to the present day, looks at the systems man has tried, abandoned or refined, and examines the ways in which governments work.

Law
The need for ways of regulating conduct probably arose simultaneously with the need for political organization, though some kind of primitive system of justice doubtless existed earlier than Neanderthal times, even if only on "an eye for an eye" basis. Modern man has increasingly sought to regulate the conduct of individuals, groups and nations by extending and refining principles laid down in earlier times—notably, in the West, by the Romans. Like political systems, legal systems differ widely from country to country. This section outlines the principles of law and how it works, and examines crime and the ways in which it is detected and prevented.

Work and Play
The thought that leisure might one day become a problem

An introduction to body and mind

An introduction to illness and health

Introduction to mental health

The meaning of ritual

The study of man

The rule of law

would, even thirty years ago, have seemed bizarre. But modern industrial methods have created a situation, in the Western world at least, in which workers may have to face enforced leisure. They have also created a situation in which the worker feels less and less that he is taking part in a creative process. What motivates us to work, how we perform, and how production is organized and managed are among today's vital issues. This section discusses the problems of both work and leisure.

Economics

The concluding section of *Man and Society* considers the economic basis of our lives and the questions of international trade, finance and co-operation. Help at the international level, whether in money or goods, is one of the major ways in which mankind is attempting to break down the barriers that exist between countries and races. It is a practical way of ironing out those differences between men which, as I suggested at the beginning of this introduction, may overshadow the things they have in common.

Man and Society

Dr Alex Comfort

Professor of Pathology, Irvine Medical School, University of California

Imagine that the names of a page of a city telephone directory are those of your ancestors in chronological order. The first name is your own, the second is your father's, the third is your father's father and so on. If your ancestors could be traced back over the whole of human history (say 750,000 years) their names would occupy about 100 such pages. But all those who lived within the period of written history would be listed on the first page alone. Most of modern science would effectively be covered by the life-spans of the first seven names – about 200 years. At the bottom of the first column would appear the name of your ancestor who lived in the Iron Age (about 720 BC – the time that Shalmaneser V deported the Hebrews from Israel). The horse would have come into service some way down the second column. The name at the bottom of the first page would be that of an ancestor who lived about the time the first sizeable city was founded (approximately 5200 BC). Around the bottom of the second page, the dog would have been domesticated (about 15,000 BC). All the names occupying the following 99 pages would be those of ancestors who lived in small bands. Such is the time scale of human social history.

What are the characteristics of man and how would we tell him from pre-man or not-man? He is a tool-using, speaking walker, with a highly developed capacity for conceptual thought, probably related to his capacity for language. Other animals use rudimentary natural tools, but man can invent them to suit his purpose. Language may have appeared in pre-men; the conceptual machinery to understand simple language exists not only in apes, which can be taught deaf-and-dumb language, but also in dogs, which cannot. No other contemporary hominid is a fully bipedal walker.

The enlargement of man's brain into the conceptual "computer" (now able to infer the existence of black holes or pi-mesons, neither of which can be seen) increased the learning period of childhood. Growth in humans is rapid between birth and the age of five but it then reaches a plateau, or slows down, to be resumed at puberty. Man is "altricial", born non-self-supporting – unlike a piglet, which can walk as soon as it is born – and he depends largely on parental protection during the whole birth-to-puberty period. Relationships with both parents are usually intense and play a major part in development and socialization. Childhood and the family are to man what the pouch is to the infant wallaby – a continuation of the womb.

With the division of growth into two phases there is a significant division of another process, psychosexual development. Most infant animals play tentatively "sexual" games that model later mating behaviour. These are usually connected with dominance (the finding of social place in the group). True sexual behaviour in animals begins at puberty, with the appearance of sex hormones and secondary sexual characteristics. At this time, in species where males are competitive, competition and aggression against other males appear and in these species newly mature males leave or are driven out of the family by the more dominant male parent.

In humans, psychosexual development, although continuous, has two major phases. The first of these is in early childhood just before growth slows down. The drives that will later be used in mating are directed towards the most important objects close at hand, the parents. This appearance of "infantile sexuality", the activation of parts of the later mating-associated apparatus at an age when mating is impossible, was first studied by Sigmund Freud (1856–1939). One feature of it is that the fear or avoidance of a more dominant male (in this case the father) surfaces in humans about the age of four or five as the "Oedipal reaction" and appears to play a crucial part in character formation. In some individuals, it may persist throughout the plateau phase of growth and adversely pattern adult sexual behaviour.

This rather extraordinary phenomenon, in which a reaction is displaced from the phase where it originally evolved to a much earlier period and acquires a new function, is called "anticipation". It may have risen in man in the following way. If we imagine a primate with an increasing brain, an increasing period of maternal dependence that kept the infant male close to the mother and on the father's "patch" or territory, and then imagine that the signal by which maleness is recognized shifts from the secondary sex characteristics (beard, voice and behaviour) to the genitals themselves, we have a biological crisis. The young male needs to stay close to his mother to survive and yet at the same time to avoid conflict with his father, for whom a penis of any size means a possibly competing male. This shift from secondary to primary sex characteristics as dominance signals explains the focus on the penis (which is unusually large in man compared with other primates and is almost certainly a "dominance signal" as well as a reproductive organ) rather than on, say, the beard as the equivalent of the adult robin's red breast.

The switchback behaviour of human psychosexual development has had other effects. One is that all manner of unlikely things and situations can, in man, be made erotic by learning – another important effect of big-brain plasticity. There may be instincts in man but we can never identify them, because they can always be overridden by learning.

Man at war –
A South Vietnamese
soldier holds his son's
body, tagged for burial.

Much blood and ink has been expended in the attempt to decide which human characteristics or abilities are learned or cultivated and which innate. The search in man for instincts (prepatterned and built-in patterns of behaviour such as those that instruct an aviary-reared bird to fear anything resembling a hawk, though it has never seen a hawk) occupied much of human psychology up to and including the time of Freud. The truth is that man is an animal programmed to be modified by social learning. There are no intelligibly identifiable human instincts, if we exclude simple reactions such as those of a baby to heartbeat or to a face; even the mothering response can be overridden by learning. Male and female chromosomes, and the hormones they induce, are about as innate as any biological predisposition could be, yet social "maleness" and "femaleness" are almost totally learned reactions and the predispositions of biology can be overridden by experience or upbringing.

Of all animals, man is quintessentially the "learning" animal and his most obvious human characteristic in all contexts where nature and nurture are discussed is the size and power of the social experience. What is "built into" man is less a series of rigid instincts than a series of spaces, like the blanks in a student's copybook, "intended" to be filled in by social experience. The copybook itself is not, however, blank; in contrast to behaviouristic models that regard all behaviours as written on a virgin page by learning, social animals generally have a capacity for social response. Thus, during behaviouristic training, a dog comes to co-operate, while a cat is simply conditioned. Ethical sense, one of the great standbys of past philosophers in distinguishing man from beast, is a name for the response of the individual to specific demands and is not a human preserve. Its activity in man, and its overactivity as guilt, are certainly prominent human characteristics but what is "built in" is a space for ethics; what ethics are written into the space depends on learning and may range from the unassertiveness of some cultures to the extreme aggression of others (including at times our own).

Another consequence or function of man's big brain is the growth in him of two different systems of data processing, one of them logical and verbal and the other analogic. The first is familiar and you are reading this page with it. The other follows different rules. You may have been jolted into it by the description of early psychosexual development, for the patterns of thought derived from that period bulk large in it. It is the older human implement and for the first 75 per cent of human history the analysis of experience by patterns, and the

verbal and ceremonial expression of these in myths, ceremonies and so on, and later in literature, was the predominant activity of people whose chief technology was an emotional one. They could not change the world greatly, so they tended to treat it as a person. More recently, the development of language and logic has given us formidable control and we tend by contrast to treat people as things, rather than the other way around.

An almost even mix between these two systems of thought was achieved by the old alchemists, who set out to manufacture gold by what we now call chemistry, but steadily adopted patterns drawn from the intuitive mode, very like those studied in human myth by the psychiatrist Carl Jung (1875–1961). Such patterns are called "archetypes" and are less patterns with content than ways of treating incoming information non-logically. They almost certainly reflect features of our nervous system, but they shape what and how we see. Similarly, the circular pattern (mandala) that some people see if they close their eyes and meditate is both a picture of how the retina is divided up and at the same time a kind of map of inner experience which we readily adopt when a structure (atom, solar system, universe) happens to fit it objectively.

The history of science is the history of trying to keep intuitive patterns out of places where they do not belong. But since intuitive patterns are also the source of innovative ideas the task is a difficult one and in a culture with no mythology they appear unwanted: at best irrelevant, as when an adult thinks of a locomotive or a ship as a person; at worst noxious, as when many adults took Adolf Hitler for a divine hero. The analog side of human thinking, submerged as it is in our awareness by the logical, is, however, an integral part of humanness. It has had its explorers – yogis, shamans and artists. Today our attempt to design thinking-machines has made us realize that we can apply the logical mode to analog thinking, and should do so, to see what it is for rather than to dismiss it as primitive nonsense.

Another important and general human experience which is almost certainly unique to man (though there is no way of knowing this) is the objective sense of "I" – the awareness that there is a little man or woman inside, driving, who is "I". This experience develops during childhood and can be suspended by some mystics and others who have vivid impressions of being boundless and of one substance with all things. The actual sensation probably results from the fact that our pattern-system of thought can somehow observe our logic-

Man the social animal – Pilgrims congregate in Jerusalem for an Easter service.

system of thought working. But its importance is that states in which awareness of the "I" is suspended are extremely compelling to the experiencer and have played a large part in the appearance of a typically human behaviour, namely religion. For in religion the "I" attempts to address or be addressed by "That" – That being the universe, spirits, ancestors, a God or gods according to culture.

The analog system of thought expresses itself in ritual. Primitives and others have been adept at cultivating social or biofeedback techniques (dancing, drumming, meditation, fasting, yoga, the eating of psychedelic mushrooms) by which "I-ness" can be altered and the response of That more clearly heard. If we wanted one marker in the remains of a very early culture to settle whether it was human or nonhuman, tools or fire would be inconclusive. Art, ritual and practices such as burial, prayer or circumcision would fix it as human.

Humans are polymorphic and polytypic (they differ from race to race, from place to place and within a population) and intellectual and physical capacities are included in this variation. Intelligence (if we define it by certain specified performances) obviously differs between Isaac Newton and the average citizen even though the latter may perform far beyond his present level. No dispute arises when the interplay of nature and nurture is applied to athletics. Applied to intelligence this interplay has engendered an envenomed dispute conducted, at times, by persons bent on justifying political conclusions.

Instead of following the ancient rabbis in praising God who made his creatures of diverse forms, discussions of the genetic and ethnic distribution of aptitudes (particularly intellectual aptitudes or disabilities of the type that are measured by the misnamed "intelligence quotient test") have led to much pseudo-science. Given an observation that African, Chinese and Anglo-Saxon inheritances in persons living nominally in the same culture lead to different kinds of mental performance, there is no way to ascertain that this difference is "genetic".

For a start, these persons do not live in the same culture, but often in quite discrepant cultures existing alongside one another, with attitudes of prejudice or superiority and greater or lesser opportunity as parts of the learning landscape. Were it to turn out, indeed, that African peoples with non-discursive cultures function as "right brain" or non-discursive people, and Europeans and Chinese from discursive cultures as "left brain" or discursive people (or vice versa), it would still be impossible to demonstrate how much of this difference was innate. All that such an observation would show would be that Western discursive and test-based culture grossly undervalues an important aptitude in its educational reckoning.

The writing of racist nonsense about human differences, and even of less motivated and would-be scientific studies attempting to unravel "nature" from what-is-learned, both imply an ignorance of modern human biology. Genetic differences, for instance in appearance, which we have made socially prestigious or detrimental, attract far more attention than the vast range of invisible genetic differences that attract no such fallout. Adolf Hitler no doubt differed in nearly as many respects from a red-haired Irishman as from, say, a gypsy, whom he regarded as "inferior". All too often, writing about human diversity that contains the word "race" is likely to be divisive nonsense. There are very few human "races" in the biological sense. If we want a word to convey that Chinese people, for instance, tend to have features or hair within a common range, the expression "ethnic" group is about as far as it is safe to go.

All classification of man must start from two aspects most characteristic of him – diversity and modification by social experience. It is the second of these to which the term "intelligence" should really be applied and the ability so to be modified is innate in the species man.

Primate to hominid

The primates (the first or highest order of mammals) originated 70 million years ago towards the end of the Cretaceous period. The earliest members of the group were clawed quadrupeds very similar in appearance to tree shrews. They remained extremely primitive until, with climatic changes in the Eocene period about 55 million years ago, they increased in number.

How the prosimians developed

Eocene primates were very similar to the prosimians (bush babies, lorises, lemurs and tarsiers) of today [1]. Some of them lost their four-footed form of movement and became vertical climbers capable of leaping from tree to tree. This different method of getting about required important physical changes. Although they still had the characteristic five digit hand and foot, the thumb and the big toe gradually turned in to face the other digits. This greatly increased their ability to grasp things [2, 3]. The claws were replaced by nails and sensitive touch pads developed on the ends of the digits.

The eyes of these early primates were also positioned differently from those of other mammals of the time. Their eyes were larger and more forward-looking because accident-free life in the trees depended on better three-dimensional vision and so needed overlapping, rather than sideways, vision. This greater dependence on sight reduced the importance of the sense of smell so the nose and snout became smaller. Greater use of hands, feet and eyes was probably responsible for the increased size of the brain, which became the largest among the mammals of that period. Animals that live and climb trees, such as the prosimians, sometimes hop or walk upright when they come down to the ground. This behaviour perhaps signals the origins of bipedalism (moving on two legs).

Towards the end of the Eocene period the fossil record shows a great reduction in the number and variety of prosimians. The survivors lingered on in a few isolated places such as Madagascar, Sri Lanka and South-East Asia and became nocturnal to avoid daytime predators and competitors.

Fossils of animals very similar to present-day monkeys date from about 35 to 45 million years ago. Two main subdivisions are recognized: the Catarrhini (Old World monkeys including the apes and man) and the Platyrrhini (New World monkeys). New World monkeys are not very important in the evolutionary history of man, although they do provide instructive comparisons in methods of locomotion.

The Old World monkey group is still further divided into the monkeys (Cercopithecoidea) and the apes and man (Hominoidea). All living monkey species are agile four-footed climbers that clutch branches with their well-developed hands and feet. Fossil finds from many countries show that monkeys have been extremely successful, but their range has now been reduced by climatic changes and, apart from some species in Japan, all live in the tropics.

Evolution of the ape

The next evolutionary stage found in the fossil records occurred about 30 million years ago with the evolution of early apes (dryopithecines). *Aegytopithecus zeuxis* is one of the earliest known and was found in

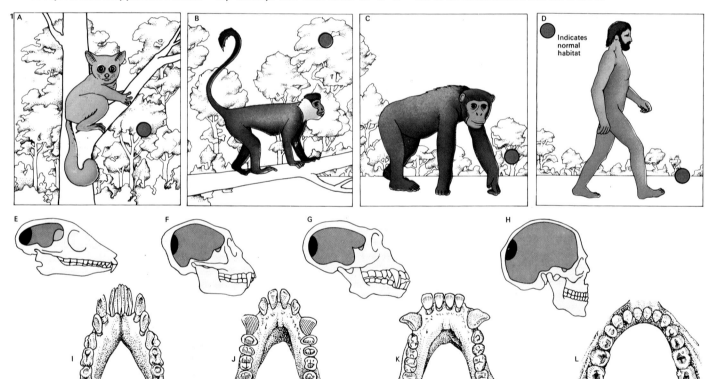

1 The primates have evolved through many stages, from the bush baby (*Galago*) [A], guenon (*Cercopithecus*) [B] and chimpanzee (*Pan*) [C] to man (*Homo*) [D]. This gradual evolution has involved changes to many parts of the body. The skulls E, F, G, H, for instance, show the gradual shortening of the snout and reduction in the size of the teeth (especially the canines) and jaws. Brain size increases and the change in the relative importance of the olfactory centres [blue] and the visual centres [black] is noticeable. The shortening of the snout and changes in the arrangement of the teeth [I, J, K, L] have affected the shape of the chin. The dental arcade has altered from a rectangular to a parabolic shape. Locomotion has progressed from vertical clinging and leaping and walking on all fours in the trees (arboreal quadrupedalism) to swinging (brachiation) that required an upright posture. Eventually, the striding walk of man (bipedalism) on the ground developed. The latter was probably successful because once man was upright his visual field was much greater and his hands were freed to carry food and to use tools. The feet [M, N, O, P] also changed shape and function. In man the ability to grasp with the feet has been lost. The big toe is in line with the others and is large because it takes most of the force of each stride. Man's foot is unique in being sprung on two arches, one running lengthways, the other across.

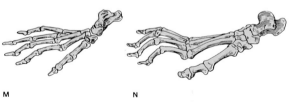

rocks of Oligocene age at Fayum in the Egyptian desert. It was a fairly small animal with a tail and the characteristic limbs of a tree-dwelling quadruped. The snout was long and the teeth were well developed.

A somewhat controversial fossil is that of the *Propliopithecus*, specimens of which have been found at the same location in Fayum, Egypt. The fossil was so named because, at the time of its discovery early in the 1900s, it was considered to be an ancestor of *Pliopithecus*, which is also on the evolutionary line leading to the gibbon. *Propliopithecus* may be an ancestor of *Aegyptopithecus* or it may be at the beginning of the evolutionary line that leads to man.

Knowledge accumulated to date shows that few of the fossil prosimians, monkeys and apes have left living descendants. Most of them became extinct and disappeared. Two of these extinct lines whose remains have been discovered are *Oreopithecus* and *Gigantopithecus*. The best-preserved skeletons of *Oreopithecus* have been found in the brown coal (lignite) of the Pliocene age of about seven million years ago. This animal had long

arms and was about the size of a chimpanzee. It was originally considered to be on the evolutionary line leading to man and became known as "the abominable coalman". *Gigantopithecus* also lived in the Pliocene and was the largest of the primates. This huge animal was nearly 3m (9ft) tall and weighed 300kg (660lb). Remains have also been found in Pleistocene deposits and some people believe that *Gigantopithecus* is not extinct but has taken refuge in mountain ranges such as the Himalayas and is the "abominable snowman" [6].

Lines of human descent

The different species of apes have their ancestry within the dryopithecine group so it would seem likely that the immediate forerunner of man was a similar animal. At present the best candidate for the role would appear to be a small ape-like creature called *Ramapithecus* that lived between 12 and 14 million years ago. Although only the teeth and jaws are known they have features strongly suggestive of the earliest stage on the direct line of descent to man [Key].

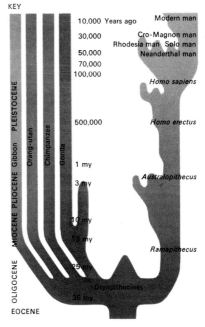

KEY

An important evolutionary step on the way to man was taken 30 million years ago with the evolution of the apes (present-day forms include orang-utan, chimpanzee and gorilla). The earliest undisputed member of the group is *Aegyptopithecus*, first discovered in 1964. This is probably one of the early apes called dryopithecines and is ancestral to the hominid line. The first dryopithecine was found in France in 1856 but specimens have also been found in Africa and Asia. Some of the African dryopithecines are believed to be ancestors of the chimpanzee and gorilla while the Asian finds are regarded as early forms of orang-utan and gibbon.

2 Primates evolved from a stock that resembles present-day tree shrews. The hand is intermediate in structure between a simple five-digit hand and that of a typical primate. It shows the beginning of a specialized thumb although this does not function very differently from the other four fingers. Sensitive finger bulbs are beginning to develop underneath the long claws.

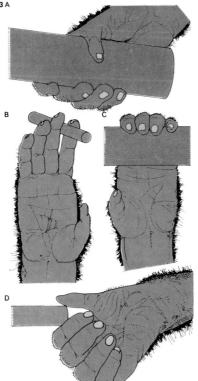

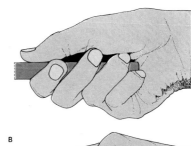

3 The ape's hand has nails instead of claws, and fingertips that are delicate instruments of touch. Although the hand resembles man's it cannot perform delicate operations. The power grip [A] is similar to that of man but there is no real precision grip, for the ape's hand lacks a fully opposable thumb. The ape's equivalent is shown [B and D]. Tree swingers use the hook grip [C].

4 The two basic human grips are the power grip [A] and the precision grip [B]. It is the latter that sets man apart from the apes. The fully developed opposable thumb has greatly increased the accuracy of man's touch. Once this stage had been reached he was able to manufacture and use tools with accuracy [C]. The continuing interaction between hand and brain has been a major influence in human evolution.

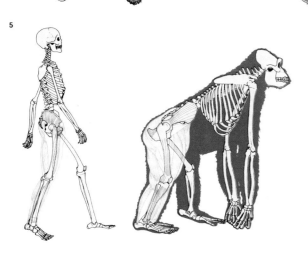

5 The long pelvis of the gorilla is necessary to accommodate the special muscles needed for quadrupedal walking. The big toe is able to grasp branches and the fingers are curved inwards for knuckle walking. In man the pelvis is short and adapted to a striding gait. The skull, unlike that of the gorilla, is balanced on top of the spinal column.

6 Reconstructions of *Gigantopithecus*, *Oreopithecus* and a dryopithecine.

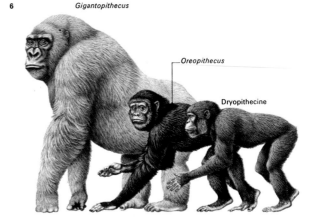

Gigantopithecus

Oreopithecus

Dryopithecine

The first hominids

Tantalizing gaps remain in our knowledge of early man, even though they have been narrowed dramatically by recent fossil discoveries. Detailed examination of bones, weapons and tools has resulted in interpretations of where and when man first appeared and who his ancestors were. But the remains are usually fragmentary – rarely are whole bones found, let alone a complete skeleton – which makes the assignment of any find within the family of man problematic.

To this difficulty must be added another. Collecting prehistoric remains seems to be a highly personalized activity and there has been a strong tendency for each find to receive a new name – perhaps understandably, after possibly years of digging and the sifting of millions of fragments. The resulting confusion in naming prehistoric man has considerably hindered the interpretation of the finds on a worldwide scale.

The process of evolution

Every find can, theoretically, be positioned on the evolutionary ladder from the structure of the skeleton. The brain has gradually increased in size, shape and complexity, and thus the skull has also changed. Stance has advanced from a bent-knee walk to the striding gait of modern man. To make this possible the shape of the pelvis has altered, and the foot and big toe have undergone modification. The hand has changed too, so that man can grip with finger and thumb.

The upright stance and the various stages to this are well shown by the position of the foramen magnum, the hole at the base of the skull through which the spine passes. In the gorilla it is positioned towards the back, pointing backwards, whereas in man it is set well forward and points downwards.

The earliest hominids

There is still much doubt about when the first hominid (or creature in man's direct lineage) emerged from ape stock. Some authorities claim the honour for *Ramapithecus* which lived 12 to 14 million years ago; others dispute this. The australopithecines, however, which lived as long as four or five million years ago, as shown by specimens found at Kanapoi and Lothagam Hill, Kenya, are members of the same family as modern man.

The first specimen, found in 1924 by Raymond Dart, was named *Australopithecus africanus* (southern-ape of Africa). Individually, australopithecine fossils have been given many names and assigned to separate genera but they may be different species of the same genus. As a group they show differences in size, teeth, jaw and brain capacity and these have given rise to the popular division into gracile or delicate (*africanus*) and robust forms [1].

When the first robust specimen was discovered at Kromdraai, South Africa, in 1938 it was given the name *Paranthropus robustus* and this generic name is still used by some anthropologists and palaeontologists, although others regard it as *Australopithecus robustus*.

The name *robustus* was given because, although it attained the same height as *africanus* (1.5m [4.9ft]), it weighed 60–70kg (132–154lb) compared with the 30–40kg (66–88lb) of *africanus*. An even larger form was found by Louis and Mary Leakey at Olduvai Gorge, Tanzania, in 1959 [5]. Originally

**1 *Australopithecus robustus* [A] and *Australopithecus africanus* [B] are hominids. *Africanus*, which is on the same evolutionary line as modern man [C], was the smaller and is known as a gracile australopithecine. Its brain was relatively small.

2 Piltdown man, discovered at Piltdown in Sussex, is one of the most famous of all hoaxes. In 1912–13 finds were made by Charles Dawson (1864–1916), a local lawyer, Arthur Smith Woodward (1864–1944) (British Museum) and Pierre Teilhard de Chardin (1881–1955). These were accepted as genuine because they supported a contemporary evolutionary theory; this suggested that the enlargement of the braincase occurred before modification of the jaw [A]. Piltdown was said to represent an intermediary stage. It is now known that jaw modification preceded braincase enlargement [B]. Tests undertaken on Piltdown man between 1949–53 revealed a man's skull of about AD 1330 together with an orang-utan's jaw of AD 1450.

3 The Pleistocene period lasted from about two million years ago to 10,000 years ago. It is during the earlier part of this period that man's evolution passed from *Australopithecus* to *Homo*.

- Homo sapiens sapiens
- Homo sapiens neanderthalensis
- Homo erectus
- Homo habilis
- Australopithecus

- Mousterian Tayacian
- Blade tool industries
- Flake industries
- Hand-axe industries
- Pebble tool industries

Geological sub-epochs		Absolute age BP	Glacial periods	Hominid fossils	Cultures		Stone industries
Holocene		10,000	Post Glacial		Middle & Upper	Mesolithic / Neolithic / Bronze / Iron	
Upper Pleistocene		50,000	Würm Glacial				
		150,000					
	late	250,000	3rd Interglacial / Riss Glacial				
		350,000	Great Interglacial				
Middle Pleistocene	early	450,000	Mindel Glacial				
		550,000	1st Interglacial				
		650,000	Gunz Glacial				
		750,000					
		850,000					
		950,000					
		1,050,000					
Lower Pleistocene		1,150,000			Lower Palaeolithic		
		1,250,000					
		1,350,000					
		1,450,000					
		1,550,000					
		2,000,000					

called *Zinjanthropus boisei*, it is now called either *Australopithecus boisei* or *Paranthropus boisei*.

Careful analysis of the plant and animal remains now suggests that Sterkfontein and possibly Taung are the oldest of the various South African sites, with an age of about 2.5 million years, while Makapansgat, Swartkrans and Kromdraai are successively younger, the last being about one million years old. In East Africa, where the sites are much older, the difficulties of dating have been less marked because of the various volcanic rocks that are embedded with the hominid-containing layers. These volcanic rocks provide a number of good specimens that enable absolute dating to be carried out.

Australopithecine's life-style

The australopithecines walked erect although some authorities have cast doubt on their ability to walk with the typical striding gait that characterizes modern man. *Africanus* has many similarities to later men, especially in the size of the teeth and the general shape of the dental arcade. The

canines had been reduced to a size similar to those of present-day man.

The life-style of the australopithecines which has been pieced together from their remains reveals that they had reached several important landmarks in the development of *Homo sapiens*.

The three most important of these are the switch to life in open country and savanna, the beginnings of co-operative hunting techniques and the addition of meat to the diet. All these opened up new resources – physical, intellectual and social – hitherto largely unexplored by primates.

Africanus lived in small groups. The females protected the young and searched for food, while the males hunted when a suitable opportunity presented itself – or more likely scavenged from another animal's kill. Bones were split open to obtain the marrow. By contrast the highly developed teeth and jaw of *robustus* show that he fed almost exclusively on vegetable matter – seeds, roots, nuts, and so on. Evidence suggests that *africanus*, not *robustus*, continued the evolutionary line to modern man [5].

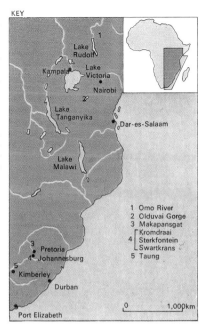

KEY

1 Omo River
2 Olduvai Gorge
3 Makapansgat
4 ⌈ Kromdraai
  ⌊ Sterkfontein
    Swartkrans
5 Taung

0   1,000km

The first confirmed australopithecine remains were found in 1924 in the Cape Province. Since then four other sites in South Africa have been unearthed. *Africanus* remains were found at Taung (Cape), Sterkfontein and Makapansgat (in the Transvaal), while *robustus* is known from Swartkrans and Kromdraai (Transvaal). In recent years other important australopithecine discoveries have been made in Tanzania, Kenya and Ethiopia. The first site to be developed was Olduvai Gorge which has since yielded a wealth of material. Recent work in the Lake Rudolf area, Kenya, and the Omo River region, Ethiopia, added to our knowledge of these first hominids.

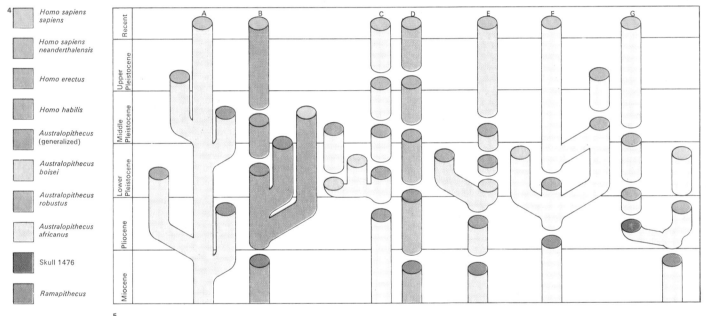

4

Homo sapiens sapiens

Homo sapiens neanderthalensis

Homo erectus

Homo habilis

Australopithecus (generalized)

Australopithecus boisei

Australopithecus robustus

Australopithecus africanus

Skull 1476

Ramapithecus

Recent | Upper Pleistocene | Middle Pleistocene | Lower Pleistocene | Pliocene | Miocene

A | B | C | D | E | F | G

4 Man's evolution has been variously interpreted. [A] Le Gros Clark suggests that no known hominid is a direct descendant of any other; [B] John Napier, that *Australopithecus* is two separate offshoots; [C] Philip Tobias that *Australopithecus* gives rise to the main hominid line; [D] Loring Brace, that each hominid is directly descended from the one before; [E] Desmond Clark, that there are two separate lines of *Australopithecus*, one leading to more advanced hominids; [F] Louis Leakey that there are interweaving branches, and that *Homo erectus* is not the forerunner of modern man; [G] Richard Leakey, that the main stem separated early from *Australopithecus*.

5

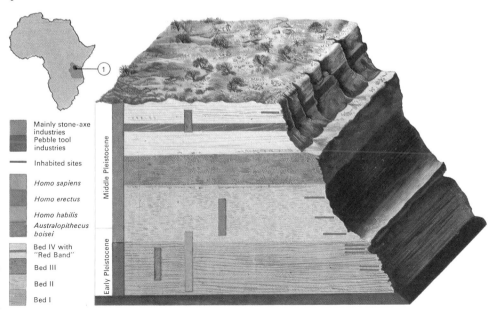

Mainly stone-axe industries
Pebble tool industries
Inhabited sites

Homo sapiens
Homo erectus
Homo habilis
Australopithecus boisei

Bed IV with "Red Band"
Bed III
Bed II
Bed I

Middle Pleistocene | Early Pleistocene

5 Olduvai Gorge on the Serengeti Plain of Tanzania [1] is 40km (25 miles) long and 100m (320ft) deep and is the most famous site in the world for the discovery of the remains of prehistoric man. The gorge is cut through lake and volcanic sediments. The lake deposits provide an ideal medium for the preservation of fossils and the volcanic rocks yield excellent material for accurate dating of remains. Louis Leakey (1903–72), a British anthropologist and his wife were largely responsible for the sensational finds of hominids made since 1959 in the rocks of Olduvai Gorge, including the 1964 unearthing of *Homo habilis*.

The first men

The immediate predecessors of *Homo* or man are known to have been the australopithecines – tool-using, upright-walking, ground-dwelling creatures found mostly in Africa. But to reconstruct the evolutionary line between them, and to fix the point at which *Homo* emerges, involves as many difficulties as there are in tracing the ancestry of the australopithecines themselves.

Homo habilis and Homo erectus
In 1964 Louis Leakey (1903–72) announced a new find which he called *Homo habilis* (handy man) [1], thereby placing this new hominid in the same genus as present-day man. The specimen, which has an estimated brain size of 600cc, came from Bed I of Olduvai Gorge, Tanzania, and was dated as being between 1.8 and 1.2 million years old. Others of a younger age have been added to this original find, some of which show characteristics in between those of *Homo habilis* and *Homo erectus* (Chellean man), which was discovered at the top of Bed II and is therefore a little older than half a million years. Many authorities believe that these

finds, with the exception of Chellean man, do not warrant a separate generic or specific division from the gracile (or delicate) australopithecines and hence call them *Australopithecus africanus*; others regard them as a separate species which they call *Australopithecus habilis*.

There is evidence to suggest that *Homo erectus* (erect man) did have an australopithecine ancestor and the African finds from Olduvai Gorge seem to confirm this. But there are those who suggest that the two are not directly linked. This view has been strengthened by recent discoveries particularly that of a hominid from Lake Rudolf, Kenya, with a brain size of just under 800cc and an age of three million years. This could mean a very early split in the australopithecines and not a gradual transition between *africanus*, *habilis* and *erectus*.

Even in eastern Asia, where most of the *Homo erectus* material has been unearthed, clues about its origins are almost non-existent [5]. The oldest specimens have been found on Java where a find that could be an australopithecine was also made. But lack of

material makes it tantalizingly difficult to link the two. Yet it is possible to show advancement between the populations of Java and the later ones of China. Javanese forms are given the name *Homo erectus erectus* and the Chinese forms *Homo erectus pekinensis*.

Java man was a wanderer like the other men of that period. Moving from place to place gathering food he undoubtedly hunted and this may have stimulated co-operation within the group. Unfortunately, no implements were discovered with the finds at Trinil or Modjokerto in Java. But chopping tools [3], resembling those found with Peking man, have been discovered in younger deposits.

The life of Peking man
Remains of nearly 45 hominid individuals were discovered between 1927 and 1937 by palaeontologists at Choukoutien, near Peking, in China [2]. These specimens were originally called *Sinanthropus pekinensis* but are now called *Homo erectus pekinensis*. They have average brain volumes of 1,000cc compared with the 880cc of the Java specimens. The tools and animal remains discovered

1 **Homo habilis** [A] is the earliest of the prehistoric men to be placed in the same genus as modern man, that is *Homo*. *Homo erectus* [B] shows a considerably larger brain size than the australopith-ecines, averaging about 1,000cc. In general the size of *erectus* is not un-like that of modern man [C] and he cer-tainly walked with a striding gait. There are significant changes in his skull although it may still be termed robust, par-ticularly with regard to the face and jaw. It has been sugges-ted that the initial increase in brain size might well be linked to increasing size of the body.

2 **Dragon Hill, Chou-koutien,** 40km (25 miles) southwest of Peking, in China, is the site where David-son Black (1884–1934) first discovered the remains of what is now known as *Homo erectus pekinensis*. The caves, the en-trance to one of which is shown here, were discovered accidentally while quarrying was being carried out. The site is a most sig-nificant one and has yielded a large amount of human material as well as the remains of deer, rhinoceros and other mammals. There is also evi-dence of tool-making, using both cores and flakes. The most important find with-in the site is the evidence of man's use of fire.

3 **Peking man's tools** were made of quartz, greenstone and quartzite. Quartz was the most common material used by the Choukoutienian cul-ture as shown by an-alysis of 100,000 tools found at the main site. The tools, which belong to the so-called chopper culture, were made by simple means. One boulder was hit with another to chip off a few flakes and so form a rough cutting edge. Sometimes it was this remaining core that would be most useful; at other times the finer flakes, which could be re-touched, were used. Bones and antlers were also shaped into tools. The hand-axe traditions of Europe, Africa and western Asia are not found in eastern Asia.

4 **The original finds** of *Homo erectus* at Choukoutien, near Peking, were unfor-tunately mislaid dur-ing World War II. In order to keep them safe their owners, the Geological Survey of China, de-cided to ship them from China to Ameri-ca. They disappeared somewhere between Peking and the ship.

Many attempts have been made to solve the mystery. Some be-lieve that they were lost when a barge taking goods to the main ship capsized. Thanks to the diligent work of Franz Weidenreich (1873–1948) excellent casts, such as this skull of an adult female, are still available to the scientists of today.

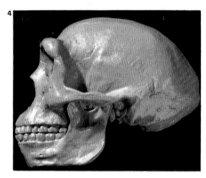

5 **Homo erectus** was first found in 1891 at Trinil on the River Solo, Java, by Eug-ène Dubois (1858–1940). When it was discovered, the speci-men was called *Pithe-canthropus erectus*, which means erect ape man. The site lies at the foot of an active volcano, Mt Lawu, and fossils of elephants, rhinoceroses and deer were also found. Other examples of *erectus* have been excavated at Sangi-ran and Modjokerto, with younger speci-mens at Ngan-dong and Wadjak.

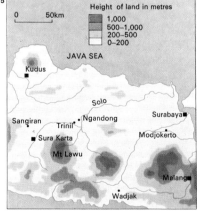

with the human fossils show that Peking man was a successful hunter. They certainly suggest one great advance over all the other groups – the first use of fire. It would seem likely that Peking man had not learnt how to make fire but "captured" it from some naturally occurring blaze such as a brush or forest fire. The presence of fire in the cave would not only have provided light and heat but would have kept wild animals at bay.

The remains of *Homo erectus* elsewhere in the world are somewhat meagre. One of the most important recent finds is the Tetralona skull from a cave in Greece. This was discovered in 1959 and for some time was considered to be of a Neanderthal man. Further work on dating the deposits has revealed it to have a much earlier age. Re-examination of the skull seems to indicate an *erectus* that may have been a contemporary of Heidelberg [Key]. Another discovery, at Vertesszöllös in Hungary in 1965, is considered by some authorities to represent an intermediate form between *Homo erectus* and *Homo sapiens*. This is reflected in the name *Homo sapiens* (syn *sapiens*) *palaeo-hungaricus*. Evidence of

the use of fire was also found at Vertesszöllös. Other finds at Nice, France, have only recently been fully investigated [6].

Dating the finds

Homo erectus lived during the Middle Pleistocene, which stretched from a little more than one million years ago to about 200,000 years ago. As is the case with the australopithecine sites there has been considerable difficulty in the dating of the various finds and hence comparison on age grounds is difficult. Java man is usually regarded as between one million and 500,000 years old and Peking man between 400,000 and 200,000 years old.

Much research still has to be undertaken to provide the links between the ancestors of *Homo erectus* and his successor, *Homo sapiens*. However, because of his brain size and type of culture, *Homo erectus* was undoubtedly the first man, although whether *erectus* developed gradually from *Australopithecus africanus* and *habilis* or from an earlier splitting off as suggested by the find at East Rudolf, is still an open question.

KEY

Homo erectus is known from Africa, Asia and Europe. Much of present knowledge is from the first specimens, those found in China [1] and Java [2]. Originally regarded as completely separate

types, they are now considered to be subspecies of *Homo erectus*, being called respectively *pekinensis* and *erectus*. In Africa finds have been made in Olduvai Gorge, Tanzania [3], Ternifine, Algeria [4]

and at Swartkrans in South Africa [5]. In Europe the most famous find is that of a jaw-bone from Heidelberg [6]. Other finds have come from Tetralona, Greece [7], and from Vertesszöllös, Hungary [8].

6 Palaeolithic camps 300,000 years old have been found at Terra Amata (beloved land) in Nice, France. The finds were made in 1966 during building excavations, and a rescue operation involving 300 workers over a six-month period was

mounted. The site covered 221 square metres (264 sq yds) and 35,000 objects were recovered. Although no human remains have been found it would seem that the occupants of the site must have been *Homo erectus* from a footprint that has been

found. An examination of the pollen content of the fossilized human faeces has revealed that the hunters came to the camp in late spring or early summer. Evidence suggests that part of the site was occupied during 11 consecutive years.

The huts, which were oval shaped, ranged in size from 7.8–14.7m (26–48ft) in length and from 3.9–6m (13–20ft) in width. The walls were made of branches pushed down into the sand and propped up by stones around the outside. There also

appear to have been central supporting posts [1]. Each hut has a hearth [2] protected on the windward side by a screen of stones. Although seafood [3] formed part of the diet, there seems to have been a preference for the young

of big game: boar (*Sus scrofa*) [4]; Merk's rhinoceros (*Dicerorhinus merki*) [5]; deer (*Cervus elaphus*) [6]; wild ox (*Bos primigenius*) [7]; ibex (*Capra ibex*) [8]; and the extinct elephant (*Elephas meridionalis*) [9]. The tools were made

on the spot and include scrapers [10], cleavers, projectile points and biface tools [11]. Wooden staves may have been hardened in the fire [12]. Owing to the contemporary level of the sea the camps were closer to the coast than at present.

From ancient to modern man

There is a considerable gap in time between specimens of *Homo erectus*, the earliest man to walk erect, and unmistakable forms of *Homo sapiens* (intelligent man). In physical terms the dividing line between the two is rather arbitrary and is related largely to brain size. The various sub-species of early *Homo sapiens* have a brain case about a third larger than that of *Homo erectus*.

Incomplete picture of ancient man

In tracing the early evolution of *Homo sapiens* a major difficulty has been that the older finds are limited mainly to Europe [Key], although significant evolution certainly took place elsewhere. Also, most of these finds were made when archaeological and palaeontological techniques were less advanced than they are now. As a result, information that would have been valuable has been lost forever. Increasing interest, on a worldwide scale, in palaeontology and the history of man has led to a more balanced picture. As more and more finds are made pointers to significant trends in the evolution of *Homo sapiens* are beginning to emerge.

The two specimens usually regarded as having claims to being the earliest sapiens were found at Swanscombe, Kent, England, and at Steinheim, near Stuttgart, Germany. Both of these came from the Mindel-Riss interglacial period, about 250,000 to 200,000 years ago. The brain volume of the Steinheim specimen is calculated at 1,150cc, and that of the Swanscombe specimen at about 1,300cc, but these estimates are based only on fragments of the skull. Although the brow ridges are more like those of *Homo erectus*, the rounded shape of the back of the skull is more modern than that of Neanderthal man who appears later. Another find at Fontéchevade, in France, of even more modern appearance, is from the Riss-Würm interglacial of about 90,000 to 80,000 years ago. It cannot be placed with certainty on any evolutionary line.

From the early upper Pleistocene there appear to have been a number of different populations of sapiens. The best known are the Neanderthals who lived during part of the last and most intense period of glaciation, the Würm, from about 70,000 to 35,000 years

ago. There are a few finds that indicate that the Neanderthal type was established earlier than this. Finds at Saccopastore, near Rome, are undoubtedly Neanderthal and have been dated to the Riss-Würm interglacial. Other finds of the early Würm glaciation have come from Morocco, Libya and Israel. The majority of finds have been made in western Europe, however, and the picture of Neanderthal man so gained is limited. The face and skull features of western and northern European Neanderthals may have been no more than local adaptations to extreme cold; the broad nose, for instance, may have developed to moisten and warm more air before it reached the lungs.

Beginnings of culture

The first specimen of Neanderthal man was found in 1856 [2]. Because anti-evolutionists did not want to recognize it as a relative of man, it was the subject of a long and heated scientific debate before being accepted as *Homo sapiens neanderthalensis*. Neanderthal man became the cave man of the cartoonists – a shuffling, stupid creature. This view was

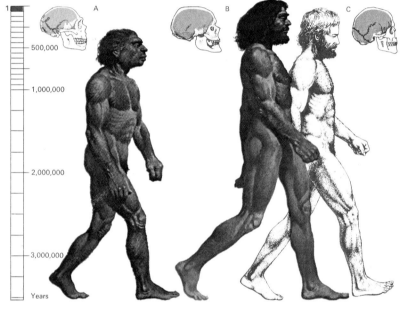

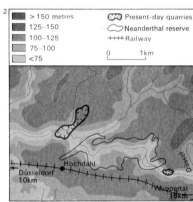

1 **Neanderthal man** [A], with his large brain but squat build and heavy brows, gave way about 30,000 years ago to Cro-Magnon man [B] whose physical make-up was closer to that of modern man [C].

2 **The site** of the first Neanderthal find was the valley of the River Düssel, called Neandertal, between Düsseldorf and Wuppertal, where farm workers discovered a skullcap in 1856.

3 **The hunting skills** of Neanderthal man must have been well developed. Among his opponents were the woolly rhinoceros and the cave bear which was 1m (39in) taller than Neanderthal adults. These animals provided not only food but also materials for clothing and for use in the cave. Hunting techniques required a great deal of co-operation within the group. Among different methods used were the stampeding of animals towards narrow gorges or cliff edges and the digging of pits or other forms of ambush. Fire-hardened spears and stone clubs were employed to kill the prey and stone tools to skin it.

4 **Cannibalism** can be traced back at least as far as Peking man. Skulls like this one have been found with holes drilled in their bases. These holes probably allowed extraction of the brain.

5 **Ceremonial burial** of the dead is suggested by the discovery of a Neanderthal skeleton clasping the jawbone of a boar. The Neanderthals were the first to practise such rites. Ritual burials have been found at Monte Circeo, Italy, and at several sites in France.

distinctly false. Although, with his sturdy build and massive brow ridges he was certainly different from modern man [1], his brain size was, on the average, slightly larger. Not only was he a good hunter [3] but, more important, there was an awareness of life and death and perhaps the beginnings of magic or religious ritual within his group [5, 6]. It was in 1908, at Le Moustier, near Payrac in the French Dordogne, that the grave was discovered of a young man positioned as if for sleep. At Drachenlock, in the Swiss Alps, the skulls and bones of cave bears were carefully arranged on small walls; similar sanctuaries have been found in Austria and Germany.

Although the number of finds of this "classical" Neanderthal have made them justly famous, the various specimens do not provide the vital information necessary to trace the evolution of modern man. Tantalizing clues exist. Some of the most interesting finds, in which a mixture of Neanderthal and modern features are displayed, have come from the caves of Tabun and Skhul at Mount Carmel, Israel. Other finds of Neanderthaloids have been made in China

and Java, while in Africa, specimens have been discovered at Broken Hill, Zambia (Rhodesia man); Hopefield, South Africa (Saldanha man); Kanjera in Kenya; and Omo in Ethiopia. Claims and counterclaims have been made for these specimens. But dating is difficult because the normal aids, such as layers, are often missing, the finds having been made in collapsed caves or on bare ground that has registered little change.

Crop cultivation and animal husbandry

In Europe the Mousterian culture disappeared abruptly about 35,000 years ago. Neanderthal man was replaced by men of the upper Palaeolithic, probably from the Near East, such as Cro-Magnon, little different from modern man. Cro-Magnon man, like his predecessors in hominid evolution, was a hunter and a gatherer. Nomadic groups moved from place to place in the continual search for food. All of this changed about 7000–8000 BC when, in the Middle East, Neolithic man began to cultivate crops and domesticate animals. The relatively short road to modern societies had begun.

1 Swanscombe 10 Spy
2 Channel Islands 11 Neandertal
3 La Chaise 12 Steinheim
4 Le Moustier 13 Gánovce
5 Lacave 14 Sipka
6 La Ferrassie 15 Krapina
7 La Chapelle- 16 Gibraltar
aux-Saints 17 Monte Circeo
8 Montmaurin 18 Kiik Koba
9 Malarnaud 19 Mount Carmel

0   1,000km

Finds of specimens that can be placed in the *Homo sapiens* group are widespread, particularly in Europe. The earliest in geological age, almost 200,000 years old, are those of Swanscombe in England and Steinheim in Germany. Famous sites of Neanderthal and related species are the Neandertal valley, Germany, Spy in Belgium, La Chapelle-aux-Saints, La Ferrassie and Le Moustier, all located in France, Monte Circeo in Italy, and Gibraltar. Finds of Neanderthaloids are also widespread in Asia and Africa. Upper Palaeolithic men are represented by a number of races such as Cro-Magnons and Predmostians. Man as an agriculturist first appeared in the Middle East about 7500 BC.

6 Caves used for shelter by early man are rarely found unchanged by later natural events. But it is still possible, with experience, to read the story they contain. In this hypothetical cave it is assumed that the river was close to its present level [1] 25,000 years ago when cave painting became widespread. A river terrace [2] shows that the cave was once submerged, leaving an insoluble limestone residue on

At the cave mouth, rock debris conceals layers of ash built up by the fires of the early men who used the shelter of the entrance both to keep a lookout and to cook their food. The ashes have accumulated in three main layers, each denoting a long period of use.

Just inside the mouth of the cave a bird of prey, possibly a buzzard, built its nest. The evidence lies directly below it on the slope of the rock debris where small rodent bones are scattered.

Cave art is found well within the cave and appears to have been part of the hunters' semi-religious efforts to secure both success and safety in hunting the animals depicted.

Stalactites hanging from the roof of the cave have developed over thousands of years as a result of water containing calcium carbonate seeping through the limestone above.

Animal remains are littered through the cave but are heaped at this point because animals have fallen through a hole left by a rock fall.

Rock falls can provide a new entrance to a cave after the mouth has been blocked. Here rock debris and earth from above have totally filled the shaft.

the floor [3]. As the river cut into its valley, it gradually fell [4], depositing a silt bed [5]. Continued deepening of the valley left the floor of the cave dry. Rock falls blocked the entrance after man had abandoned the cave.

A burrow in the cave floor has been left by a small animal. It was deflected sideways by harder rock, until it could continue downwards, throwing up bones from older deposits. It died at the end of its burrow. This kind of mixing of bones of different ages makes dating difficult.

Evidence of a fairly widespread cave bear cult has been found in carefully prepared arrangements of bear skulls (here in a stone compartment), leg bones and other fragments. Men could hardly have picked a more dangerous opponent, and they could find meat much easier.

Early men buried their dead in various ways. Some buried skulls only, arrayed with possessions or ornaments; others buried men but not women. This skeleton shows evidence of careful arrangement in a sleeping position similar to that of the Grimaldi remains in the Grotte des Enfants, Monaco.

Even the interior of a structurally stable cave changes over a long period of time. Upright stalagmites, produced by the slow seepage of calcium carbonate-rich water through the limestone above, have built up on a large pile of rock that has fallen from the roof.

7 The origins of art go back to at least 30,000 BC. Carvings on antlers [A] are among the oldest forms, dating from the Aurignacian period (30,000–22,000 BC). Cave paintings like that of the wounded aurochs [B] were produced by the Gravettian people (22,000–18,000 BC). These paintings may have been magical (helping to trap game) or educational, but they certainly show man's knowledge and sympathy with animals. The Gravettians dec- orated ivory and produced some of the famous Venus figurines [C] and body ornaments [D]. Solutrean people (18,000–15,000 BC) produced fine friezes and the Magdalenians (15,000–8000 BC) were painters and modellers.

8 Mesolithic man lived about 10,000 years ago. His main hunting weapon was the bow and arrow. Other tools that have been discovered reflect changes in the food-gathering pattern which included fishing, fowling and sealing. Canoes were used and, in the subarctic regions, skis and sledges were made. Farming of crops and animals was yet to come, but the domestic dog was bred from the wolf by the people of Mesolithic times. They also formed large social groups.

27

Spread of man: 1

The evidence so far collected by archaeologists and palaeontologists suggests that the cradle of mankind was in East Africa about five million years ago, when the australopithecines evolved. From his beginnings there man has spread to populate and dominate the world. The speed and extent of this dispersal have been largely controlled by two factors: climate, and man's own ability to exploit the advantages and overcome the disadvantages presented by it. For millions of years the wanderings of the family Hominidae as nomadic hunter-gatherers were severely limited by environmental factors. For without the ability to fashion tools, build dwellings or make clothes early man was completely at the mercy of the elements.

The tools of survival

The restraint of ignorance was eventually broken – long before the evolution of *Homo sapiens* some 50,000 years ago. Remains of *Homo erectus* (who emerged more than a million years ago) found in Europe and Asia show that he was eventually able to make the journey away from Africa. More flexible communication within the group (which immensely expanded the range and effectiveness of co-operation) together with an improvement in tools were among the factors that enabled him to become a more efficient hunter, and thus to follow new game, feed more mouths and rear larger families.

The discovery of fire by *Homo erectus* was a crucial weapon in man's fight for survival in a harsh environment. The survival of Neanderthal man, a sub-species of *Homo sapiens*, during a glacial phase (an ice age) is a testament not only to the steadily growing skill of early man but also to the importance of fire in his life.

Neanderthal man, as well as fashioning crude clothes from animal skins, used fire as a means of warmth. He also devised several ingenious ways of trapping animals and began co-ordinating groups for hunting. These people proved themselves well able to cope with their harsh environment, and so could live a kind of life which earlier hominids like the australopithecines would not have been able to cope with. The key advance here is that man had begun to create his own environment, although no matter how able he became he could not overcome the fact that climate dictated the type and abundance of the flora and fauna he needed to eat in order to stay alive.

Migrations during the ice ages

There is evidence of migration of man into Europe from Asia during the early part of the upper Palaeolithic. But it was during a warmer period of the last glaciation that the Eastern Gravettian culture of Europe flourished [1]. The warmer conditions caused a retreat of the ice sheets and permitted the movement of game along the steppe lands of eastern Europe, between the northern ice sheets and the glaciated mountains of the south. The mammoth was the staple beast of this period (in the same way that the reindeer is for the Lapps today), providing meat, skin for clothes, ivory for carving-tools and mighty bones for the framework of shelters.

This phase of migration was brought to an abrupt end by the final major advance of the ice. Man was now able to combat the cold, however, and could adapt his activities to sur-

1 The Eastern Gravettian culture, which flourished during the upper Palaeolithic (35,000–10,000 BC), is remarkable for its sophisticated solution to the problem of shelter. These peoples, living on the edges of the glaciers of Czechoslovakia and southern Russia, an area without natural caves or rock shelters, constructed huts from movable poles covered with animal skins sewn together. The bottom edges of these skins were weighted down with the bones of the mammoths, the reindeers and the occasional rhinoceros that men hunted. When they were forced to move on in the search for game the huts could be dismantled and transported. The hut floors were scooped out of the soil as a protection against draughts and there may have been more than one fire in each hut as well as the communal hearths outside. There is evidence that surface coal and mammoth tusks provided fuel for these fires. The clothes of these peoples were made from skin, much as those of the Lapps and Eskimos are today. Statuettes were made from clay and ochre. The dead, often painted with ochre, were buried not far from the camp in a shallow ditch protected by mammoth bones and tusks. Included in the graves were everyday goods such as food, arms and ornaments. Hunting techniques developed considerably during this period, with foliage-covered pits being used to trap big game against which their weapons (primarily blades, spears and clubs made from bone, antlers and flints) would have been ineffective. Needles, spoons, borers and end-scrapers were the basic tools at that time.

vival under the deteriorating climatic conditions. The culture of the Magdalenian people, for example, resembles that of the Eskimos under similar conditions. In the manufacture of their many tools and implements, they made extensive use of bone.

Once the last retreat of the ice began, man was once again able to resume his migration northwards. It was once assumed that when the reindeer hunters followed the herds northwards they left western Europe unpopulated. It is now known, however, that some peoples stayed behind and that they developed the so-called Mesolithic cultures.

After the ice retreated

The Mesolithic period is characterized not by the rigours of an ice-bound existence but by the problems of survival as forest cover increased. Man was forced to the sides of the rivers and lakes, to the sea shore and other open locations [2]. The upper Palaeolithic tools had to be adapted to new uses – the hafted stone axe was developed to fell trees and for use in working wood. The Mesolithic peoples hunted with bows and arrows,

catching birds and fish as well as the smaller animals such as red deer and roe deer that lived in the woods. The dog was domesticated during Mesolithic times and was undoubtedly used in hunting. The rich vegetation of this period meant that there were plenty of fruits and berries as well as roots to be found.

Pollen counts from sediments of the time show a decrease in tree pollen and an increase in herbaceous pollen at a number of sites and provide evidence that Mesolithic man made clearings, possibly by use of fire, to set up temporary camps. With the benefit of ice-free seas he also made skin-covered boats and dug-out canoes and so began to exploit the natural resources of the seas.

It was in the Near East, however, that man took the next great step in controlling his environment by domesticating plants and animals. The introduction of agriculture took place in the Neolithic period. The first cities also date from this period and from this point on the world's population began to increase more rapidly. Trade played its part by influencing the movement of peoples and the diffusion of ideas between cultures.

Nomadic tribesmen cross an arid up- land plain in Iran on the arduous trek to find summer pasture in Afghanistan.

2 The forests that grew up as the glaciers retreated to where they are today presented almost as many difficulties as the ice. The new environment forced man to occupy sites on the sides of rivers and lakes until his technology had developed sufficiently to enable him to clear the trees. Lepenski Vir, a site southeast of Belgrade in Yugoslavia, is typical of the Mesolithic settlements in the Danube basin. Discovered in 1960, the site appears to have been built in about 5000–4600 BC. To achieve level foundations the houses had to be constructed on terraces cut into the sloping bank of the river. They were aligned in rows with their entrances facing the river and varied in size from 5.5 square metres (6.6 sq yds) to 30 square metres (35.9 sq yds). The floors were of hard limestone plaster covered by a thin red or white burnished surface. The hearths within the houses were elongated pits lined with limestone blocks and were often surrounded by a pattern of thin sandstone blocks. These fireplaces may have been constructed primarily to smoke and dry the fish that were abundant in that part of the river. Lepenski Vir is the name of the large whirlpool that existed at that point in the Danube; it churned up the small organisms on which fish feed. The inhabitants also hunted the red deer, aurochs, roe deer and wild pigs that were plentiful on the thickly wooded hills. Their hunting methods are not known but blocks of stone may have been used to club fish that had been landed alive after being caught in nets and fish traps.

Spread of man: 2

Man is now scattered over the whole earth, but he was, in the earliest known stages of his development, confined to eastern and southern Africa. During the last glacial phase Neanderthal man, a sub-species of *Homo sapiens*, began to penetrate the then inhospitable regions of Europe. The real expansion began, however, with the evolution of modern man (*Homo sapiens sapiens*) and continued in explosive fashion during the latter part of the Palaeolithic or Old Stone Age. Man migrated and colonized whole new areas of Africa, Asia and Europe and spread to Australia and the Americas [Key, 2].

Following the herds

Those early migrations were not purposeful feats of exploration but probably food-searching expeditions undertaken by hunter-gatherers who had to follow the game if they were not to starve.

They took early man surprisingly far afield. For many years it was thought that the wastes of Siberia had been inhabited only recently. However, archaeological investigations reveal that man has lived in the area for at least 20,000 years. Although barren and cold, particularly in winter, the area supports the type of plant growth that is ideal for grazing animals. It is likely, therefore, that man followed the roaming herds to those regions. In order to carry out such journeys, man needed at least a rudimentary technology: he had to be able to clothe himself, build a shelter and fashion suitable tools.

The effects of a changing climate

The fluctuations of climate during the last glacial phase had a profound effect, not only on the distribution of animals and plants but also on the level of the sea relative to the land. During the periods of intense glaciation, when vast quantities of water were frozen, the sea-level was much lower than at present. Numerous land bridges were thus created and in this way the land called Beringia, linking eastern Siberia with Alaska, came into existence.

The coastal area was not covered with ice at the periods of maximum glaciation. The area would certainly have been like today's tundra, with many lakes. Both plants and animals spread from Asia to the Americas, but man the hunter would primarily have been concerned with following caribou, musk oxen and mammoths.

The earliest diffusion of man into North America cannot, as yet, be dated with certainty because much of the evidence, if it exists, is now below the waters of the Bering Strait. But it seems certain that *Homo sapiens sapiens* is the only member of the family of man to set foot in the Americas. There is no evidence of the presence of *Homo erectus* or even of Neanderthal man.

Penetration from Beringia into the American mainland was impossible at certain times because of intervening glaciers, but at other times (for example between 20,000 and 28,000 years ago and 32,000 and 36,000 years ago) an ice-free corridor would have existed between the glaciers of the Rockies and the main ice sheet to the east. The way through may well have been via the Yukon and Mackenzie Rivers.

The colonization of the Americas occurred in a number of waves. These intermittent arrivals may be one reason for

1 Sittard, in the Netherlands, is a site dating from the end of the fifth millennium BC when the first mixed farming communities settled in the Meuse valley in Dutch Limburg. These people were part of the Danubian culture complex which spread from Moravia, Bohemia and central Germany throughout most of central Europe. They practiced a very primitive agriculture without plough or fertiliser, which was dependent on the soft, fertile loess found across the region. Ten Danubian villages have been discovered in the area, each usually includes about 20 large houses. Plan and construction of these varied, but timber posts and daub walls were basic. Danubian-made artifacts include a distinctively incised pottery-ware [A]; simple bone ornaments and jewellery [B]; grinding stones; stone axes; flint sickle blades [C]. They kept cattle, sheep, goats and pigs, and grew a variety of crops, including barley.

the great diversity of American Indians now found on the continent. If the earliest migrations of "Palaeo-Indians" began as long as 35,000 years ago, then those early migrants would not have evolved very marked Mongoloid traits. The fact that Indians farther south in the continent generally have features that are less Mongoloid than those in the north, whereas the most recently arrived inhabitants, the Eskimos, are the most Mongoloid, tends to support the "wave theory".

Certainly, once the northern ice barriers were penetrated there would have been a population explosion in reaction to the plentiful game and the lack of competitors.

Migration in the Southern Hemisphere
Already well established in East Asia, man was helped in his migration southwards to Australia by a series of land bridges that linked many of the islands of Indonesia and also attached New Guinea to Australia. The first people to reach Australia did so more than 20,000 years ago and perhaps even as long as 30,000 years ago. They arrived at a number of different points and gradually

spread out over the continent, possibly reaching Tasmania before it became isolated from the mainland.

The inhabitants of Melanesia also have ancestors from the Asian mainland as do the peoples of the more remote areas of Micronesia and Polynesia, although there is some debate about the Asiatic origins of the peoples of Polynesia.

Africa, the continent that has been inhabited by the Hominidae for five million years, nevertheless presents no clear picture of the distribution of *Homo sapiens.* Recent finds in the Omo Valley, Ethiopia, have provided evidence of a form that may be ancestral to both the neanderthaloid Rhodesian man and the Negro-Bushman types. It is not easy to relate these finds to the present-day types found in Africa, the Negroids, Hottentots and Bushmen and the Caucasoids. The last are found only in the northern parts and spread there from Asia. There are affinities between the Negroids and Bushmen. The latter may represent a group, once widespread, which has been pushed southwards by Negro expansion from the western forests.

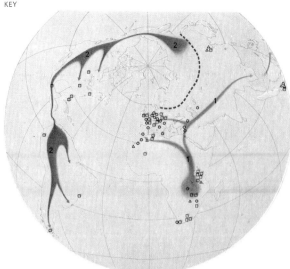

KEY

Early migrations
▲ Fossils of *Homo erectus*
● Fossils of Neanderthal man
□ Fossils of early *Homo sapiens*
– – – Expansion of early *Homo sapiens*
1 Limits of man's expansion 100,000 years ago
2 Mongoloid migrations to America 15,000 years ago

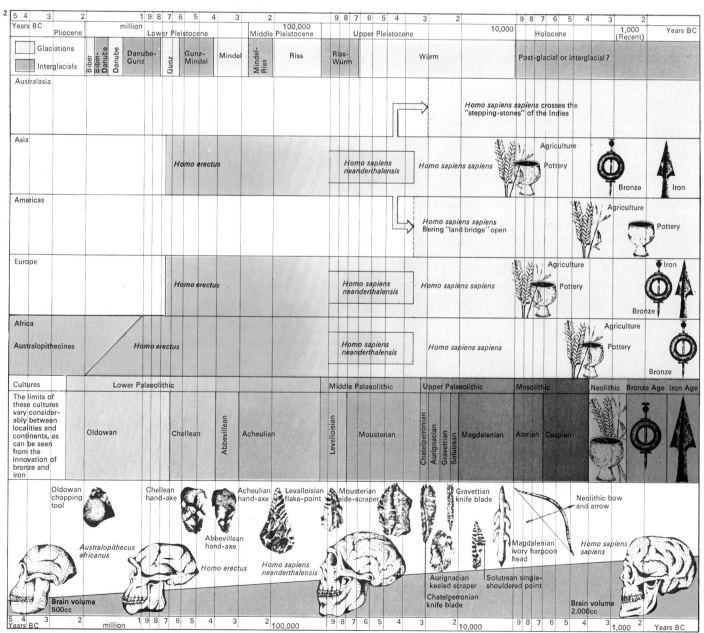

Family of man: today's peoples

The great dispersal that has made man the most widely distributed of all species began between 40,000 and 50,000 years ago. As he moved into different regions he came into contact with a wide range of conditions from the intense cold of the Arctic to the arid heat of the Australian and Kalahari deserts, and settled in habitats as different as high mountain ranges and flat salt marshes.

To live in these varied environments man's body shape, his skin pigmentation and probably his physiology altered, giving rise to the wide range of human types that exists today. Despite this great diversification, however, every human belongs to a single sub-species, namely *Homo sapiens sapiens.*

Three basic human types
The enormous numbers of anomalies and exceptions to every generalization about physical types makes classification difficult. There do seem, however, to be three major forms: Caucasoid, Mongoloid and Negroid [Key]. Some of the differences between these are easily visible. Mongoloid peoples have yellowish skin, straight black hair and dark brown eyes which have an epicanthic fold of the eyelids. Caucasians have a greater pigmentation range; from the very pale-skinned individuals of northern latitudes to the darker peoples of the Mediterranean, Middle East and India. The Negroids have dark skin, woolly hair, a protruding lower jaw (known as prognathism) and thicker lips. Within these main groups there are an infinite number of subdivisions and some anthropologists have given many of these small groups separate racial status [2].

The Caucasian form is widespread in Europe, North Africa, the Middle East and India. With the exceptions of the Finns, Turks, Basques and Hungarians, they all speak Indo-European languages. The range of physical form is wide and this has led some authorities to divide the group into several sub-groups: for example, Nordic, Alpine, Mediterranean or Dinaric. The accuracy of these divisions is erratic. In one study only 11 per cent of army conscripts in Sweden, supposed home of the Nordic type, fitted the typical description of fair-haired longheads with blue eyes. The term Caucasian, however, still serves to describe the light-skinned inhabitants of an area stretching from India to northern Scandinavia. This group, the members of which probably share some common ancestry, also spread into Asia and probably extended across the Far East to the shores of the Pacific Ocean.

Adapation to a cold climate
The term Mongoloid includes a large number of sub-groups from the Chinese to the American Indians. The typical Mongoloid face, many anthropologists feel, is an adaptation to an extreme climate and it has been called a "mask against cold". The bones and soft tissues of the face have become remodelled to give the greatest protection against cold air. The nose is small and low, the eyes are protected by a layer of fat in the lids (epicanthic fold) and the features are generally flat; all of these adaptations would tend to lessen the risk of frostbite.

In the southern Chinese and Japanese these changes are less marked than they are in the more northern peoples. This may be because the northern peoples in Siberia were

CONNECTIONS

See also
34 Family of man: how peoples differ
268 Predudice and group intolerance

In other volumes
28 The Natural World
30 The Natural World
34 The Natural World

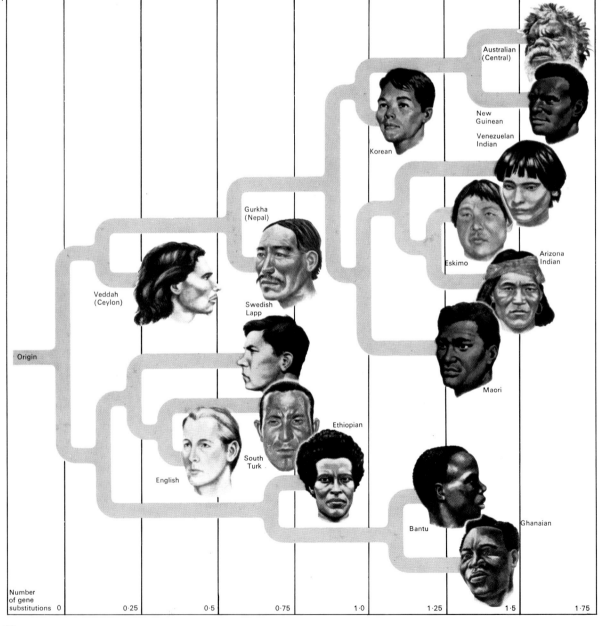

1 **The relationship** between various populations can now be assessed accurately by the computer analysis of a wide range of inherited traits. These characteristics may be under the control of several genes, as are skull shape and fingerprints, or controlled by a single pair of genes, as are the genetic factors of the blood (such as the various blood groups and enzyme and protein variation). Previously it was possible to compare the variation of only one or two inherited characteristics between a few populations. This computer reconstruction took 15 populations, three from each major area: English, Lapps and Turks from Europe; Veddahs, Gurkhas and Koreans from Asia; Eskimos and Venezuelan and Arizona Indians from the Americas; Bantu, Ethiopians and Ghanaians from Africa; and Australian Aborigines, Maoris and New Guineans from Oceania. These populations were examined for five blood group systems and the results used to calculate the relationship between them. This is expressed as the number of gene substitutions (variations in the genetic material controlling those particular factors). The final results fitted in well with the geographical distribution and expected relationship of the groups. The computer readings made clear the relatedness of the African, European, Asian and American population groups.

Number of gene substitutions 0   0.25   0.5   0.75   1.0   1.25   1.5   1.75

Australian (Central)
New Guinean
Venezuelan Indian
Korean
Gurkha (Nepal)
Eskimo
Arizona Indian
Veddah (Ceylon)
Swedish Lapp
Origin
Maori
Ethiopian
South Turk
English
Bantu
Ghanaian

stranded by the last glaciation and selection favoured those individuals with the characteristics best suited to the rigorous climate; those who lacked them would have died from frostbite or pneumonia. This theory has also been used to explain the physical differences between American Indians and Eskimos. The ancestors of the Northern and Southern American Indians crossed into the Americas before the changes provoked by the last glaciation, and the Eskimos, a much later migration, came after the Mongoloid facial form had altered.

The peoples of Africa

The Negroid peoples occupy much of Africa south of the Sahara and include the tallest and shortest people in the world. The Nilotes living in the hot deserts of Ethiopia and Somalia are tall and thin with a slender physique. This, it has been suggested, is the ideal body shape for hot, dry climates as it permits rapid heat loss. Farther south in central, east and southern Africa, the typical Negro is of medium height and heavy build. In the Congo are found the Pygmy tribes who

exhibit very few differences from the typical Negro, except in their stature; the maximum height of a Pygmy is only 1.38m (4ft 8in).

The most puzzling of African groups are the Bushmen and Hottentots who have been classified as a separate racial group. They still live a Mesolithic life, using stone tools and weapons in a hunter-gatherer economy. They have many distinctive features including the "click" language and the ability of the women to store fat in their buttocks (steatopygia). It appears that the Bushmen were once more widespread than they are today and that they have been driven into their present areas by the Bantu expansion which started in the second century BC.

The people of Oceania present their own problems as to origin. Many of these, such as the Melanesians, are Negroid in appearance with dark skins whereas others, such as the Polynesians, may have migrated from mainland Asia and could contain a Caucasian element. The Australian Aborigines seem to show variation from north to south and it has been suggested that these differences could in fact represent two separate racial groups.

2

KEY

Caucasoid   Negroid   Mongoloid

2 Nine distinct races of man are commonly distinguished within the Caucasoid, Mongoloid and Negroid peoples. This classification is based on the geographical distribution of population prior to modern migrations. For convenience, many of these races have been further divided and the subgroups given separate racial status. Shown is a typical classification of the races of man.

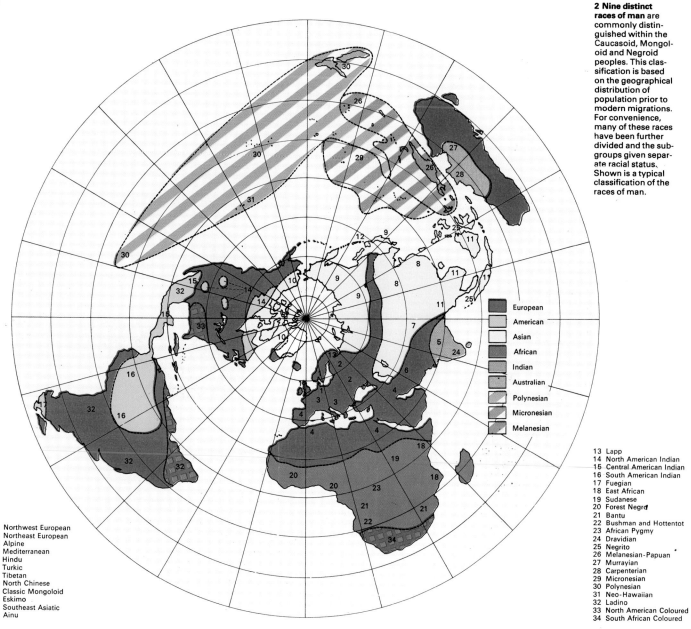

European
American
Asian
African
Indian
Australian
Polynesian
Micronesian
Melanesian

1 Northwest European
2 Northeast European
3 Alpine
4 Mediterranean
5 Hindu
6 Turkic
7 Tibetan
8 North Chinese
9 Classic Mongoloid
10 Eskimo
11 Southeast Asiatic
12 Ainu

13 Lapp
14 North American Indian
15 Central American Indian
16 South American Indian
17 Fuegian
18 East African
19 Sudanese
20 Forest Negro
21 Bantu
22 Bushman and Hottentot
23 African Pygmy
24 Dravidian
25 Negrito
26 Melanesian-Papuan
27 Murrayian
28 Carpenterian
29 Micronesian
30 Polynesian
31 Neo-Hawaiian
32 Ladino
33 North American Coloured
34 South African Coloured

Family of man: how peoples differ

Man has studied the physical differences between the races of his own sub-species *Homo sapiens sapiens* for many centuries. In the past this study did not extend much beyond the comparison of measurements of various parts of the body and assessments of skin, hair and eye colour. Measurement of skulls and other bones allowed comparison not only between contemporary peoples [1], but also between contemporary peoples and those of much earlier ages.

Measuring the human skull
In modern studies the length and breadth of human skulls are measured between specific points and these are used to calculate the cephalic index (breadth/length × 100). This index allows skulls to be divided into three groups: long heads (dolichocephalic) have an index of less than 75; median heads (mesocephalic) are 75–80; and broad heads (brachycephalic) more than 80.

Although it is possible to find all types in one population, various races tend to have a characteristic cephalic index. Most values in western Europe lie between 75 and 80,

whereas in central Europe they are generally more than 80 and may even reach 85. Most African values are less than 75. Asiatic populations usually have indices of more than 80, whereas Australian Aborigines have values of less than 75.

The colour of human skin has also been used to characterize populations [Key]. Skin colour results from the presence of melanin, a dark brown pigment found in the epidermis. All peoples have the same number of melanin-forming cells, but Negroes have more granules of melanin in their skin. It is thought that the melanin content acts as a protection against ultra-violet rays in sunlight. The light colour of northern European peoples may facilitate the formation of vitamin D by the action of sunlight, since Negroes living in northern latitudes have been found to suffer from rickets, a disease resulting from a deficiency of vitamin D.

The human blood groups
Human blood can be classified into groups by various systems that analyse different constituents. The first such classification – the

ABO grouping – was discovered in 1900. More recently the technique of electrophoresis (the separation of charged particles in an electric field) has revealed many characteristics that vary from person to person. These include variations in haemoglobin, red cell enzymes and serum proteins.

On the ABO system [2], the *O* antigen has its highest frequency in the indigenous populations of the Americas, whereas the *A* antigen occurs more often in Europe, and particularly in Anatolia (the region of European Turkey). The *B* antigen increases in frequency eastwards throughout Europe and is also high in India. Other classifications include the Rh (Rhesus) blood group system, which is determined by the R_o gene complex. Although present in most populations, the R_o complex is more than 70 per cent in most African groups. Rh negatives have an incidence of about 12 per cent in Europe but are virtually absent from Oriental, Oceanic and American Indian peoples.

The Duffy blood group system is also of interest. The Fy^a and Fy^b genes that govern it have a similar frequency in Europe, but in

CONNECTIONS

See also
32 Family of man: today's peoples
62 Heart and blood circulation
60 Skin and hair
268 Prejudice and group intolerance

1 Until recently anthropological race analysis concentrated on physical features and races were classified as Caucasian, Negroid, Mongoloid and Oceanic. But such external differences are often vague, owing to the wide variety of types that occur in each "race". More recently races have been classified geographically and this classification is supported by genetic evidence from blood groups. Shown here are some common physical features, although they are inevitably only broad generalizations. Stature and body build [A] also seem to depend on geographic area, narrow build being associated with warm climates. Facially [B], two clear-cut differences are that Asians and American Indians have inner eye folds and that extreme lip eversion occurs in Africans. In all races there is a wide range of skin colour [C]; a "typical" example of each is shown here. There are three basic skull shapes (seen from above, the wavy line representing the joins between skull bones) [D], while head hair [E] has many different colours and degrees of curliness. Genetic differences between races are revealed in the frequencies with which blood groups [F] such as the ABO group and the Rhesus negative (R_o), Diego (Di^a) and Duffy (Fy^a) factors occur. The patterns of such frequencies provide a distinctive profile for each group.

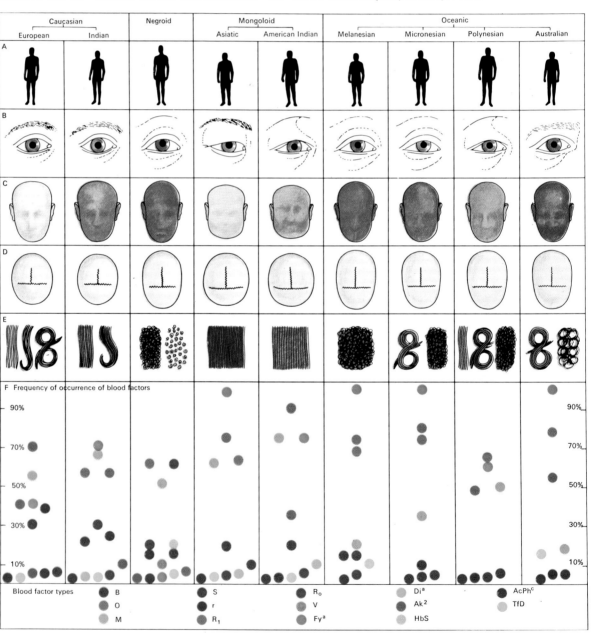

New Guinea Fy^a is by far the most common. Both are, however, almost completely absent from Negroes and are replaced by the Fy gene. The Diego (Di^a) gene is found almost exclusively in Mongoloids and American Indians. The protein and enzyme systems of human blood show similar variations. One variant, for example, has been found only in the peoples of Europe and India.

Variation and survival

Interest is now moving from using these inherited characters in the tracing of racial origins and affinities to determining how and why such variations may have arisen. One fruitful line of enquiry is to investigate whether differences of genetic make-up endow any positive advantage, such as a greater resistance to disease, in a particular environment. The only completely proven association is that between malaria and the abnormal haemoglobin S.

Haemoglobin S is a variant of normal adult haemoglobin; if two recessive genes for haemoglobin S (HbS) are present, it causes a serious condition called sickle cell anaemia, a disease that is invariably fatal in early childhood. Individuals with only one HbS gene do not suffer from the disease and can lead a normal life. The HbS gene seems, on this evidence, to be disadvantageous because it is fatal in those who possess it in a double dose.

In some populations, however, particularly in Africa, the HbS gene appears frequently. On investigation it was discovered that the distribution of this haemoglobin variant coincided with that of the malaria parasite *Plasmodium falciparum*. Careful collection of data showed that children with a single HbS gene were protected from this type of malaria whereas those without the abnormal haemoglobin were more susceptible to infection. The child carrying a single HbS gene is therefore at an advantage over those carrying either two or none at all. This particular type of gene distribution is called balanced polymorphism.

Geneticists and anthropologists are continually searching for similar situations, because it is through genetic variations that they detect the reasons for the diversity in the family of man [3].

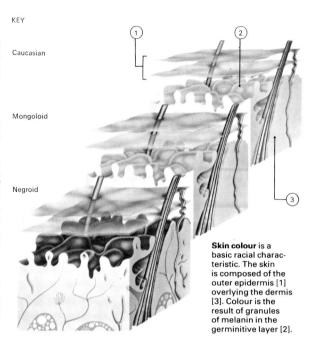

KEY
Caucasian
Mongoloid
Negroid

Skin colour is a basic racial characteristic. The skin is composed of the outer epidermis [1] overlying the dermis [3]. Colour is the result of granules of melanin in the germinitive layer [2].

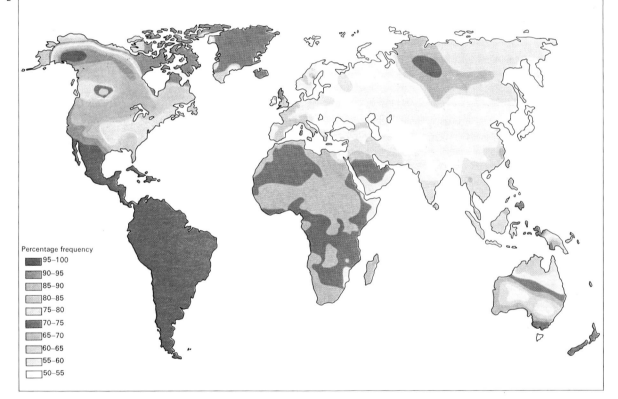

2 There is more information available for the ABO system than for all the rest of the blood grouping systems added together. This is largely due to its early discovery but also because it is of major importance in blood transfusion, where it is the main criterion for compatability of blood types. Shown is the distribution of the O antigen in the world. Central and South American Indians have few or no A or B antigens and are therefore very high in O. This is probably due to the fact that the relatively few members of the original colonizing population lacked the antigens. The A antigen is found more often among North American Indians. In Europe the O antigen reaches its highest frequency on the Atlantic margins. This may be due to a remnant population that has been forced westwards by population pressures from the east.

Percentage frequency
- 95–100
- 90–95
- 85–90
- 80–85
- 75–80
- 70–75
- 65–70
- 60–65
- 55–60
- 50–55

Bantu   European   Chinese   American Indian   Australian Aborigine   Eskimo   Bushman

3 When seen together the outer similarities and differences between the subdivisions of man become apparent. Shown here are the cold-adapted faces of the Chinese and Eskimos contrasted with the American Indian, the Bantu Negro, the Australian Aborigine and the fair-skinned European. Some Bushmen have highly developed buttocks (steatopygia) as a fat reserve. All of these are adaptive changes to the environment, which has moulded the human body into its modern variety.

An introduction to body and mind

The human body – even when a mere seven weeks after conception [Key] – is a highly complex machine, an apparatus of intricate, interlocking systems [1] in which millions of vital processes take place. But despite this complexity its mechanisms function with inimitable efficiency and precision, and its basic structure is relatively uncomplicated.

The interaction of body and mind

While our knowledge of how the body and the mind work mechanically is reasonably advanced, the way in which they interact to produce "the person" remains an unsolved puzzle. That the two are not separate and are related is obvious from what we now know of the psychosomatic diseases. These are physical disorders such as hypertension and ulcers that are, in part, caused by psychological factors. We are also aware that attitudes of mind affect the body's susceptibility to attack by diseases. But we are far from unravelling the extent and complexity of the relationship of body and mind – indeed almost all current research points to the fact that it would seem impossible to draw a line clearly between the

two. Too little is known for the way in which they relate to be controlled or influenced with any degree of success.

What is now better understood is the way in which the brain controls the body and how changes in body function are achieved to maintain optimum working efficiency. Two main systems are involved. The first of these consists of nerve pathways carrying controlling impulses from the brain to the organs themselves. This type of control includes the regulation of breathing and heart rate. Centres in the brain constantly monitor the body's performance and alter breathing and heart rates accordingly.

The second great system, the endocrine system, is chemical and makes use of the bloodstream. The pituitary gland at the base of the brain is its co-ordinator and also provides the link between it and the brain. In response either to impulses from the brain or to changes in body chemistry, it secretes hormones into the bloodstream. The hormones flow round the body and cause changes in its organs. Many of these changes, such as the control of urine secretion, the regulation of

digestion and metabolism, the secretion by the adrenal gland of hormones to combat stress, are important in maintaining life.

The Hippocratic view of the brain

The problem of how body and brain interact exercised the mind of Hippocrates (c. 460– c. 377 BC), the most admired physician of the ancient world. His insight into brain function was remarkable. He wrote: "I hold that the brain is the most powerful organ of the human body. . . . Eyes, ears, tongue, hands and feet act in accordance with the discernment of the brain . . . I assert that the brain is the interpreter of consciousness." This amazing insight into the nature of consciousness was forgotten or ignored by later philosophers and doctors until the eighteenth and nineteenth centuries.

In the Middle Ages the precise observations of the ancients had been lost and speculation took their place. Costa de Luca (864–923), misinterpreting the work of Galen (c.130–c.200) who was, after Hippocrates, the most distinguished physician of antiquity, described the function of the brain

1 The body has six major systems. The skeletal system is made up of more than 200 bones joined by fibrous ligaments. At the joints cartilaginous plates on the bone ends glide over each other to allow movement. The muscles are the body's motors, making possible its concerted internal and external responses to the environment. The digestive system is concerned with the intake of food, its digestion and then the absorption of its energy-giving and body-building substances. The circulatory system's principal organ, the heart, pumps blood to every part of the body through arteries, arterioles and, finally, capillaries. These are connected to the venules and veins that return blood to the heart. The skin protects the underlying tissues, regulates body temperature and helps excrete wastes. The nervous system includes the brain, spinal cord and nerve-network and receives and responds to all internal and external stimuli needing either conscious or unconscious response. The body's healthy and proper functioning depends on the close and efficient interaction of the more than 50,000 million cells that contribute to the interaction of the six major systems of the human body.

1   Digestion

Skin and hair

Heart and blood

Nerves and brain

Skeleton

Muscles

as based on a valve action of the movements of "Pneuma" between the ventricles.

The idea that messages from the brain were conveyed by fluid flowing from the ventricles down the nerves to the muscles continued until the late eighteenth century when Luigi Galvani (1737–98), working in Italy, discovered that electricity applied to a frog's leg made the muscles twitch.

Recent theories about the brain

With the coming of the twentieth century scientific knowledge of how the body functions increased. The detailed physiology of respiration and nutrition and the finer details of anatomy and tissue structure came to be understood. Charles Scott Sherrington (1857–1952) in his Oxford laboratories investigated the nervous system and defined in great detail the way in which messages flowed to and from the brain. John Hughlings Jackson (1835–1911), clinician and philosopher, through observations of patients with brain disorders, constructed models of the organization of brain function and the formation of consciousness.

Since the 1940s the understanding of brain function has been changing rapidly. The discoveries of the valve amplifier and, more recently, the transistor have allowed detailed examination of cellular activity within the brains of both animals and man. The science of biochemistry has led to understanding of the processes that take place deep within the cell and has succeeded in defining the very nature and make-up of the cell itself. Present theories about the functioning of the nervous system rely heavily on ideas from computer technology.

The brain is now seen as an organ constructed to process the vast amount of information fed to it by the senses. It then constructs a model of the outside world. This model is continually updated by incoming information. There is little doubt that what the brain "sees" depends not only on what is fed to it by the senses, but also on how it wishes to process this information and to interpret its own private world. It would seem that while most people have a common perception of the world, each individual in addition has his or her own personal "reality".

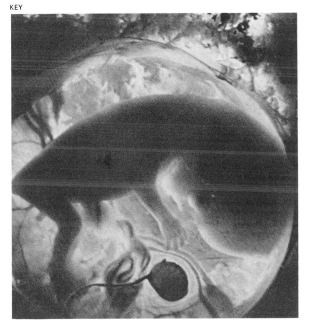

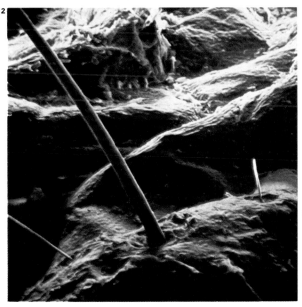

2 Human skin is the body's first line of defence and acts also as an organ of excretion and water regulation. It protects the body from many potentially dangerous chemical, physical and biological substances. It is waterproof, enabling the body to exist in dry air or be immersed in fresh or salt water. Within the skin, shown here magnified 60 times, lie touch- and pain-sensitive cells and hair follicles.

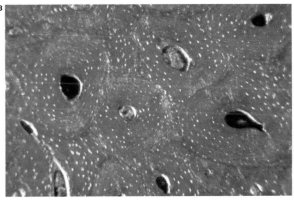

3 This cross-section through the compact bone tissue of a long bone shows the concentric layers of several Haversian canal systems. At the centre of each is a blood vessel surrounded by bone-forming cells.

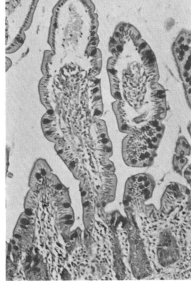

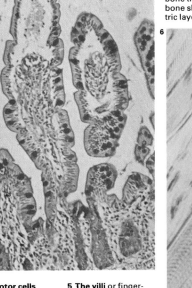

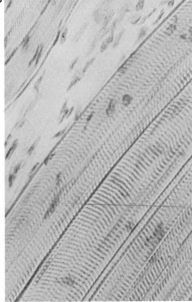

4 The motor cells controlling muscle movement lie deep within the spinal cord. These cells provide the final pathway for impulses arising in the voluntary motor area of the brain. It is these that are destroyed in polio.

5 The villi or finger-like projections and folds of the small intestines are seen here in cross-section. They present a large area for the absorption of nutriments from the partially digested food reaching the intestine from the stomach.

6 Voluntary muscle consists of bundles of fibres. These are divided into bands, which appear as dark lines under a microscope. They contain actin and myosin, two important contractile proteins.

37

The nervous system

The nervous system is the circuit that brings our bodies alive. It allows us to perceive the world outside and monitor the world within. Through its efficiency we can respond and react to changes in the environment around us by moving our muscles and co-ordinating the actions of our various organs. Without it survival would be impossible.

The structure of the nervous system

The nervous system can be thought of as a number of separate but integrated sections that monitor the environment and instigate and co-ordinate the body's multiple activities [Key]. The central nervous system (CNS) consists of the brain and spinal cord [1]. Connected to it and running throughout the body is the peripheral nervous system which has two main parts – the somatic or voluntary nervous system and the autonomic or involuntary system which deals with the unconscious control of the body's organs [2]. The somatic deals largely with awareness of sensation and the voluntary control of muscle activity. Information is fed into it through the sensory nerves from the eyes and ears, taste buds, balance organs and the millions of contact, temperature, pressure and pain receptors in the skin which together make up the sense of touch. There are also receptors for sensing the degree of tension in muscles and tendons and for monitoring blood pressure and the levels of oxygen, glucose and carbon dioxide in the blood. Motor nerves, originating in the CNS, carry information to the muscles and in this way initiate movement.

The cells or neurons that make up the nervous system consist of a cell body, which contains the nucleus, and a long tail or extension called the axon [4]. On the cell body and at the end of the axon are short branches called dendrites. Contact between nerve cells, from the axon of one to the cell body of the next, is established via these dendrites but takes places across a small gap called a synapse. The nerves of the peripheral system consist mainly of axons that run the whole length of the nerve (in the leg this distance may be as much as one metre [39in]), while those of the CNS consist mainly of cell bodies with short tracts or bundles of axons. Where cell bodies occur outside the CNS, as with sensory and autonomic nerves, they are collected together in groups called ganglia.

How nerves send their messages

Although nerves can be thought of as wires or telegraph cables carrying messages in the form of bursts of electricity, in fact a nerve impulse is more complicated than a surge of electrons travelling along a copper wire. The transmission of an impulse through a living cell involves the movement of electrically charged particles – ions – across, not along, a membrane, in this case the wall of the axon.

At rest (that is, when not conducting impulses) a nerve cell is polarized. In other words the outside of its membrane bears a different electrical charge from the inside. This is because of the different concentrations of sodium and potassium ions within the membrane and outside it. Inside is a high concentration of potassium and a low concentration of sodium while the reverse is true outside. When a nerve is stimulated, the arrangement of the molecules in the membrane is altered, allowing potassium ions to

CONNECTIONS

See also
40 How the brain works
96 Diseases of the nervous system

In other volumes
148 Science and The Universe

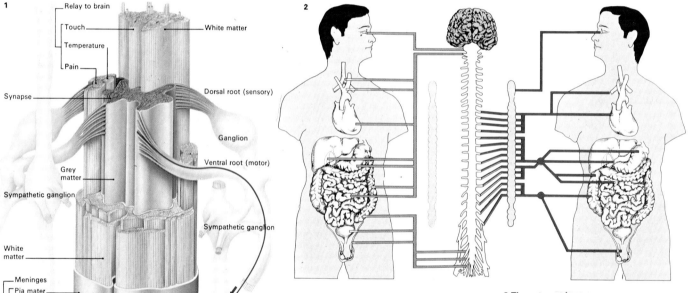

1 — Relay to brain — Touch — Temperature — Pain — Synapse — Grey matter — Sympathetic ganglion — White matter — Meninges — Pia mater — Arachnoid mater — Dura mater — Synapse — White matter — Dorsal root (sensory) — Ganglion — Ventral root (motor) — Sympathetic ganglion — Reflex pathway — Ganglion — Cell nucleus — Nerve fibres

3 — A — B — Cervical — Thoracic — Lumbar — Sacral

1 The spinal cord is made up of numerous bundles of nerve fibres carrying messages to and from the brain. The pathways cross so that sensations from one side of the body register in the cortex of the brain on the other side. The nervous system needs careful protection. The brain and spinal cord are wrapped in tough membranes called meninges and bathed in a liquid called cerebrospinal fluid. This adds an effective watery cushion against jarring, as well as augmenting the supply of nutrients and oxygen provided by the blood. Most tracts of axons, except some of the autonomic system, are protected by a white, fatty substance called myelin. Cell bodies form the grey matter of the brain.

2 The autonomic nervous system is made of two opposing parts; the sympathetic and the parasympathetic. The sympathetic system prepares the body for fight or flight. It can limit the blood supply to the digestive organs, increasing the amount of blood available for muscles and limbs. The parasympathetic system conserves resources and relaxes the body after an emergency or for sleep. Sympathetic nerves [red lines] lead from the spinal cord to nearby nerve chains, the sympathetic chains of ganglia. The parasympathetic nerves [blue lines] lead directly to the organs. This system also controls the heartbeat.

3 Spinal nerves carrying messages to and from the spinal cord serve specific areas of skin known as dermatomes. The 31 pairs, one for each of the bones of the spine (the vertebra), are subdivided into five groups. As the area served by the last pair of spinal nerves is ill-defined, only the four major groups are shown for the front [A] and back [B] of the

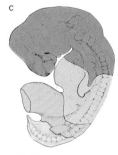

adult. The human embryo [C] shows the same divisions.

leak out and sodium to leak in. At this moment, the nerve membrane becomes depolarized and the electrical change causes an alteration in the molecular structure of the next section of membrane, which in turn becomes depolarized. In this way an impulse travels rapidly along the nerve fibre.

Nerve cells are remarkable because they can "communicate" with each other. A stimulated neuron sends messages, in the form of the tiny rapid pulses of the electrical activity described, along its axon, the synapses linking it to other neurons. The electrical impulses cannot themselves jump across synapses, but the signal is passed by causing the release of a chemical transmitter substance that makes the surface of the next neuron develop an impulse [5].

The role of the nervous system
Everything we do requires the mediation of the nervous system, from the simplest flick of a finger to highly co-ordinated, sophisticated activities. Some simple responses use only particular parts of the nervous system. If someone touches something hot, he pulls his hand away quickly. This simple but essential response is called a spinal reflex [6], since impulses from the sensory nerve endings in the skin need reach only the spinal cord to be acted upon. Impulses generated within motor cells in the spinal cord pass back down the motor neurons of the arm and activate the muscles that move the hand. This reflex is automatic [7] (it can happen during sleep) and is unlearned: a baby behaves identically.

The behaviour of very simple animals is made up entirely of reflex movements but higher animals and man have greater freedom of action and can respond in a variety of ways to most situations. Reflexes continue to be important in emergencies and for such vital activities as breathing and bowel movements. But most human behaviour falls into another category – it is voluntary, learned and non-reflex. This kind of behaviour is made possible because the nervous system can learn from experience and direct its own activities. Since no two people have the same range of experiences, and no two brains are exactly alike, every nervous system behaves in a unique way.

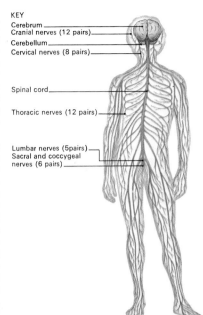

KEY
Cerebrum
Cranial nerves (12 pairs)
Cerebellum
Cervical nerves (8 pairs)
Spinal cord
Thoracic nerves (12 pairs)
Lumbar nerves (5pairs)
Sacral and coccygeal nerves (6 pairs)

The nervous system is divided into two parts: central and peripheral. The central nervous system (CNS) includes the brain and spinal cord. It receives information, makes decisions and transmits instructions. The peripheral nervous system consists chiefly of nerve fibres leading to and from the CNS. It cannot make "decisions" and acts only as a message transmitter. There are 12 pairs of such nerves radiating from the brain and 31 pairs from the spinal cord. The peripheral nervous system includes the somatic system, which deals with voluntary actions, and the autonomic nervous system, which controls automatically such organs as the heart, lungs and intestines.

4

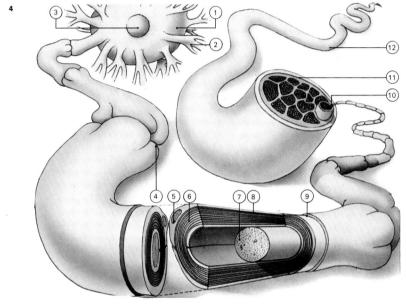

6

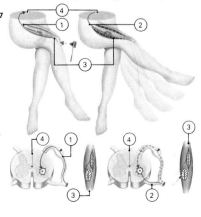

7

5 Nerve impulses are transmitted by bioelectricity. At the junction between nerve and muscle fibres (the motor end plate) a chemical transmitter is released from the nerve in response to the electrical change produced when the nerve is stimulated. As a result, the muscle fibre contracts. [A] shows a longitudinal section through a motor end plate. [1] shows a nerve axon, with an impulse travelling towards its tip. [2]

shows a section through a terminal button of a nerve cell, the part containing acetylcholine, a chemical transmitter. [3] is the synaptic gutter – a muscle fibre membrane greatly folded to form a "gutter" that accommodates the terminal button. [4] is the muscle fibre itself in a relaxed state. [B] is a detailed diagram of the terminal button region, showing the nerve cell about to release the chemical transmitter. [C] de-

monstrates the release of the chemical transmitter [5] as the nerve impulse reaches the tip of the terminal button. The transmitter causes changes in the permeability of the muscle fibre membrane. [D] shows the contraction of muscle fibre [6]. The permeability change results in an altered electrical state across the membrane stimulating muscle fibre contraction. The muscle fibre filaments pull the fibre units closer together.

4 The structure of a single nerve and myelinated cell fibre is shown. [1] Cell body. [2] Dendrites conducting messages to the cell body. [3] Cell nucleus. [4] Node of Ranvier, a constriction of the myelin sheath [5] Schwann cell nucleus – the cell secreting the myelin sheath [6] that insulates the axon [7]. The latter conducts impulses from the cell body. [8] Nerve fibre. [9] Endoneurium – inter-fibre tissue. [10] Perineurium, sheathing fibre bundles. [11] Epineurium sheathing the nerve [12].

5

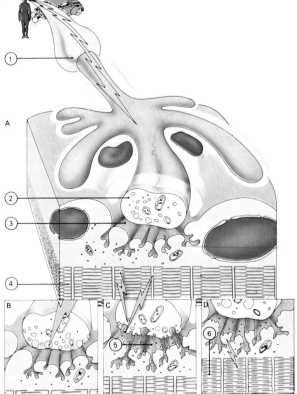

A

B   C   D

6 Nerve impulses from a pricked finger travel to the brain via the spinal cord, but because the sensory and motor nerves connect at synapses in the spinal cord itself reflex withdrawal occurs before the message reaches the level of consciousness.

7 The tendon jerk is the simplest reflex action, involving only a sensory receptor neuron [1] and a motor neuron [2]. Impulses running to and from the muscles [3] traverse only one segment of the spinal cord [4]. This reflex is independent of the brain.

How the brain works

Within the bony protective casing of the skull the brain has a shape and surface rather like that of a huge walnut kernel. The major part of the brain has two symmetrical and linked halves with fissures, folds and wrinkles, and is covered with thin layers of membrane.

Anatomy of the brain

The brain, weighing an average of 1,380gm (3lb) in a man and 1,250gm (2lb 12oz) in a woman, is made up of about 30,000 million cells called neurons. Intelligence is thought to be related in some way to the complexity of the microstructure, the connections between its units and its biochemistry. At birth the structure of the brain is almost complete, but it continues to grow until the age of 20, both by increasing the size of individual cells and by augmenting the amount of tissue connecting the neurons, which finally makes up some 40 per cent of the volume.

The brain is not a homogeneous mass, but has several distinct parts that have developed during man's evolution [1]. The oldest areas, responsible for such life-supporting functions as breathing, circulation of the blood and

sleeping, are found at the base of the brain joined to the spinal cord, while the more recently developed parts are wrapped round the older areas. They have many folds and thus they have a large surface area in relation to their volume.

The two hemispheres of the cerebrum, the major part of the brain, are mirror images of each other, each being chiefly concerned with movements and sensations of only one side of the body. The left hemisphere controls the right side and vice versa. Complex behaviour such as speech is controlled by only one of the two hemispheres – in most people the left hemisphere – which also controls the dominant hand, the right hand. Thus, if a right-handed person suffers a stroke affecting the left side of the brain, he usually loses the use of the right hand and arm – and also the power of speech.

Structure and function

The hemispheres make up 70 per cent of the whole brain and nervous system, including the nerves of the body. They consist of the cortex [5], an outer layer of grey matter sur-

rounding a thicker layer of white matter made up of nerve fibres, and are connected by a bundle of fibres, the corpus callosum.

Beneath the cortex there are the four lobes of each hemisphere. The occipital lobe at the back of the brain receives and analyzes visual information. The temporal lobes on the side of each hemisphere deal with sound. Certain vivid auditory and visual sensations have also been located here. The frontal lobes are mainly concerned with the regulation of voluntary movements and also have something to do with the use of language. The prefrontal areas of this lobe are thought to have something to do with intellect and personality, but their specific functions are still unknown. The parietal lobes are mainly associated with our sense of touch and balance. The senses of taste and smell, poorly developed in man compared to other animals, are represented by small areas buried in the frontal and temporal lobes.

At the base of the brain is the brain stem, in the medulla, which controls essential activities such as breathing, coughing and heartbeat. Behind and slightly above this is

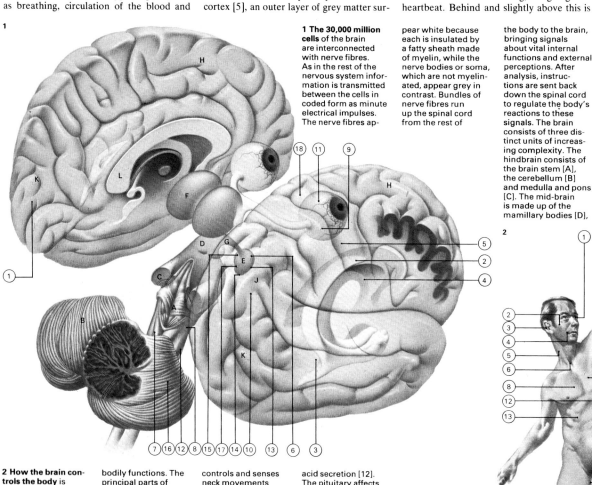

1 The 30,000 million cells of the brain are interconnected with nerve fibres. As in the rest of the nervous system information is transmitted between the cells in coded form as minute electrical impulses. The nerve fibres ap-pear white because each is insulated by a fatty sheath made of myelin, while the nerve bodies or soma, which are not myelinated, appear grey in contrast. Bundles of nerve fibres run up the spinal cord from the rest of the body to the brain, bringing signals about vital internal functions and external perceptions. After analysis, instructions are sent back down the spinal cord to regulate the body's reactions to these signals. The brain consists of three distinct units of increasing complexity. The hindbrain consists of the brain stem [A], the cerebellum [B] and medulla and pons [C]. The mid-brain is made up of the mamillary bodies [D], the pituitary [E], the thalamus [F] and the hypothalamus [G]. At the top and (covering the other two) is the forebrain which is made up of twin structurally similar, but functionally distinct units–cerebral hemispheres. Each of these two is divided into four lobes: the frontal [H], temporal [I], parietal [J] and occipital [K]. The two are connected by a massive bundle of nerve tracts called the corpus callosum [L].

2 How the brain controls the body is still largely a mystery despite modern surgical and laboratory techniques. Partly because its functions are interlinked in a most complex way and partly because of the difficulties of experimenting with humans, the brain remains very much *terra incognita*. It is possible to identify some connections between certain parts of the brain and some bodily functions. The principal parts of the brain have been moved apart to show their appearance and relationships, but in real life they are packed tightly in the skull. Optic tracts carry stimuli from the retina of the eye [1] to the occipital lobes. The cortex controls and senses face movements [2], and also hearing [3]. Broca's area of the cerebral cortex controls speech [4]. The cortex controls and senses neck movements [5]. The pituitary controls the thyroid gland [6], which in turn regulates the body's metabolism. The brain stem controls heart rate [7] and respiratory rate [8]. The cortex controls arm movements [9]. The parietal lobe controls judgment of weight, shape size and feel [10]. The cortex controls the trunk [11]. The brain stem controls stomach motion and acid secretion [12]. The pituitary affects the adrenal glands [13] which protect the body from stress and control body fluid chemistry. Hormones from the pituitary control the kidney's urine secretion [14]. The pituitary controls the testes [15] and ovaries. The cerebellum controls movement and balance [16]. The pituitary controls the growth of long bones [17]. The cortex controls legs and feet [18].

the cerebellum, which is important for co-ordinating bodily movement and for maintaining posture and balance. It does not initiate movements but is responsible for their smooth and balanced execution, for maintaining the tension of muscle tone and for integrating movements into a complex action such as walking. The two sides of the cerebellum are united by the pons.

Emotional controllers
Through the brain stem runs the recticular formation, a network of fibres carrying the sensory pathways from the spinal cord to the brain. This diffuse area monitors incoming information from the body's sense perceptors and regulates the level of response. It is also thought to have profound influences on emotional behaviour [3].

In the middle of the brain, grouped round the fluid-filled cavities called the ventricles, are regions controlling many of our basic drives. The hypothalamus controls hunger, thirst, temperature, aggression, sex drive and, by regulating the activity of the pituitary gland, is responsible for controlling the secre-

tion of many hormones from other glands.

Looped round the hypothalamus is a collection of structures forming the limbic system – the septum, fornix, amygdala and hippocampus. These are thought to be involved in emotional responses such as fear and aggression and to interact closely to produce mood changes. Also in the middle of the brain lies the thalamus, a tightly packed cluster of nerve cells that relays impulses from the sense organs to the cortex.

The brain normally consumes 20 per cent of all the oxygen extracted from the blood. Without oxygen, brain cells are damaged irreparably and die: the brain has no capacity for cell regeneration.

The brain has been likened to a computer, but it is an infinitely more complex structure and much of its functioning still remains a mystery. Although we know to some extent what each part of the brain does, it would be wrong to think that these parts act independently. The brain works as a highly efficient interactive unit, its many parts performing a multitude of complex functions that together comprise what we know as human life.

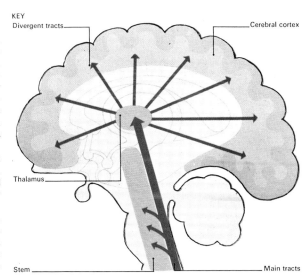

KEY
Divergent tracts
Cerebral cortex
Thalamus
Stem
Main tracts

When the brain is awakened from sleep, sensory information carried by the main tracts stimulates the reticular formation deep in the stem. This area, which regulates the brain's activity, then sends showers of impulses to the thalamus. From here divergent tracts arouse the cerebral cortex.

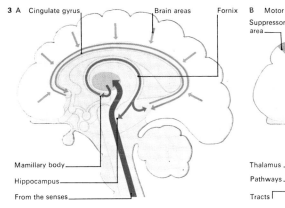

3 A Cingulate gyrus   Brain areas   Fornix   B Motor area   Suppressor area   C Limbic system
Mamillary body
Hippocampus
From the senses
Thalamus   Pathways   Tracts   Cerebellum   Spinal cord
Input   Pituitary gland   Hypothalamus   Heart and lung rates   Stem   To other organs

3 Identifying areas of the brain that are responsible for such functions as memory, thought, consciousness, judgment, and personality is extremely difficult. Over the past few years, however, various functions have, at least tentatively, been associated with specific areas of the brain [A]. Memory, for example, seems to be linked with the limbic system. One explanation is that impulses from the senses and brain areas enter the limbic system in the middle of the brain and are then passed through the mamillary body, around the fornix to the hippocampus and cingulate gyrus. It is these limbic structures that record impressions and recall them [B]. Movement starts as complex signal patterns in the motor area. These are modified by the suppressor area and then transmitted via nerve tracts to the muscles. Return information reaches the cerebellum through the spinal cord and then, via the pathways and thalamus, helps to coordinate body movement. [C] As far as the emotions are concerned, it has been established that the hypothalamus controls thrist, appetite, sex drive, aggression and emotion generally. It is thought that impulses from the frontal lobes are integrated in the limbic system and fed to the hypothalamus which regulates the pituitary gland. Through this gland many hormones are affected. Limbic impulses are also responsible for affecting heart and lung rates, and other organs.

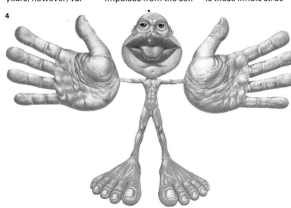

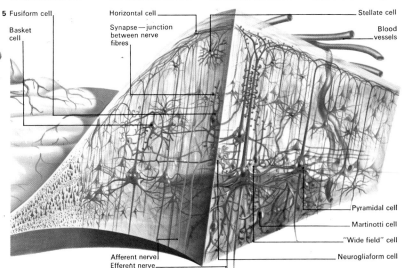

5 Fusiform cell   Horizontal cell   Stellate cell
Basket cell   Synapse—junction between nerve fibres   Blood vessels
Pyramidal cell
Martinotti cell
"Wide field" cell
Afferent nerve   Neurogliaform cell
Efferent nerve

4 This grotesque figure, a homunculus, might also be called "sensory man". It shows the parts of the body as they appear to the brain. The physical size of each part is related to the size of the area of the brain controlling that part's activity. The brain clearly assigns importance to particular areas like the hands and fingers which are very "sensitive" in order to assess an object's weight, size, texture and rigidity. This figure shows how the brain "sees", "feels" and "moves" our bodies. The body image on the brain is inverted, so that the feet are controlled by the top of the cortex and facial features at the bottom. Similarly sensations from the left side of the body are "felt" on the right side and vice versa.

5 The cerebral cortex, as if peeled from the surface of the brain, exposes the various different types of nerve cells with their distinctive organization and fibres running horizontally and vertically. The blood vessels on top of the cortex are also shown. The cortex varies in thickness from 1.5mm to 3mm (0.059in to 0.012in); it is thickest in the motor region of the frontal lobe.

41

Mind and brain

Although psychologists, educationists, philosophers and others cannot agree exactly what the mind is, they all acknowledge that the more they find out about it the more amazing it becomes. It has been said, for example, that if mankind wished to use all its current resources to build a computer that could do everything done by the normal human mind, such a computer might have to be the size of our planet – and even then nobody would know how to program it.

Mind and matter

It was once thought that man was made up of mind and matter, matter being something that could be seen and felt, occupying space and having weight, and mind being a substance present in a person but taking up no space and not able to be weighed, seen or touched. During the last two centuries the argument went to extremes, some saying that the mind did not exist at all and that everything could be explained in terms of the body, and others saying that everything physical was an illusion and that the only reality was the mind. These arguments have subsided a

little today, and the most common view suggests that both mind and body exist, each being dependent upon a harmonious interaction with the other. For convenience, the brain is often described as the purely physical matter inside the skull, and the mind is described as "what the brain does".

Until quite recently it was thought that the larger the brain the more able the mind. This view was discarded when the brains of deceased people were weighed and no significant correlation was formed between the intelligence and mental abilities of the people when alive and the size and weight of their brains. The next theory was that the number of individual brain cells determined the ability of the mind. Again it was found that variations from the average number of cells per brain seemed not to be closely related to a person's mental capacity.

Possibly the number of permutations and combinations of the connections between thread-like filaments called dentritic spines that branch off from each brain cell determine a person's mental ability [2, 3]. The average number of connections for a normal

human brain is now known to be so enormous [Key] that previous estimations of individual potential have had to be revised drastically. According to one estimate, by an Israeli microbiochemist, there are between 100,000 and one million different chemical reactions taking place every minute in the brain.

Sleep and dreams

It has been suggested recently by an American psychologist, Robert Ornstein, that the two halves of the brain, the left and right hemispheres, deal with different mental functions. The left side deals with the more academic processes, while the right deals with the more artistic and imaginative activities [1]. Current interest in the activities of the right side has led to investigations of the more imaginative, rhythmical and colourful aspects of the mind.

Much of the work on consciousness has involved studies of sleep. Investigation of sleep deprivation by John Lilly and his colleagues showed, to the surprise of the experimenters, that people seemed to be able to make do with very little sleep but suffered

1 The brain is composed of two halves, the left and right hemispheres, and it has been thought for some time that the left hemisphere controls the right side of the body and vice versa. By recording the origin of brain waves associated with specific types of thought it has been indicated more recently that each hemisphere is also responsible for a different range of mental activities. The left hemisphere is mainly responsible for language, logic, numbers, analysis, critical thinking and academic activities. The right side of the brain is mainly concerned with imagination, spatial relationships, form and more artistic and intuitive activities. Humour seems to be a product of both sides of the brain. Western societies have tended to emphasize left-hemisphere activity with an education system based mainly on reading, writing and arithmetic. As a consequence the more artistic side of human endeavour has suffered some neglect to the detriment of the potential capacity and functioning of the brain as a whole. There is some evidence that if both sides of the brain are encouraged to co-operate, creativity, productivity and the level of general intelligence can be raised. An interesting research side-line is that when both sides of the brain are working in harmony, the body works better too. A child with jerky handwriting concentrating too hard on academic activity can achieve a remarkable improvement by concentrating on colour, while people suffering from such stress-related ailments as ulcers and backache can gain relief apparently through co-operation between left and right hemispheres of the brain.

if they were not allowed to dream [4]. Those who had little sleep and were allowed to dream were slightly more irritable than usual but otherwise normal. Those who had little sleep and who were not allowed to dream became, within a few days, very disturbed. A number of them began hallucinating during the day. In other words, the mind needs to dream and if it cannot do so during sleep then it has to compensate in waking hours.

The studies concluded that although the normal requirements for sleep vary enormously, most people need to exercise their creative and imaginative faculties through the process of dreaming. Some people, however, require little sleep and seldom dream.

Further work on dreams and imagination has shown that rather than being things about which people should feel fear and guilt, dreams may well be the playground of the mind – stories, plays and fantastic shows and panoramas that amuse, educate and sometimes advise and give warning. The more people come to terms with their dreams and cease to regard them as irrelevant or shameful the more relaxed, creative and

"whole" they may become. Dreams have provided inspiration for many works of imagination, notably for the poem *Kubla Khan* by Samuel Taylor Coleridge (1772–1834). Some stories of Edgar Allan Poe (1809–49) were based on his nightmares and the artist Salvador Dali (1904–) also makes use of the landscape of his dreams. The German chemist Friedrich Kekulé (1829–96) envisaged the cyclical structure of the benzene ring in a dream.

Potential of the mind
Recent research indicates that people may be using only a small fraction of their mental potential. But with new understanding of the way in which the mind works and with the beginning of detailed investigation into recall processes, retention, real learning abilities, creative processes and general mind/body functioning, boundless prospects are opening up. It appears from all the evidence available that the human brain may be the most versatile and finely constructed object known. Only now are some of the basic elements in its design beginning to be recognized.

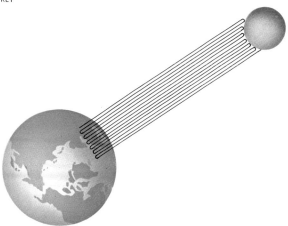

The capacity of the human brain can be expressed in the number of permutations and combinations of connections of which it is capable. A 1974 Soviet estimate put the possible figure at one, followed by 10.5 million kilometres of noughts. This is a number so large that it would stretch from the Earth to the Moon and back more than 13 times. Such an estimate may even be conservative as it is certain that the full potential of the mind has not yet been realized.

2 Information is transmitted in the brain by electrical pulses at the synapse [A] or tip of the dendritic spines in the conducting cells of the nervous system. Some regions of the cortex have large numbers of these spines [B]. An enlarged diagram of the synapse [C] depicts the transmission of information from one dendrite to another. The more of these connections there are, the greater the mental ability.

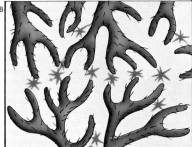
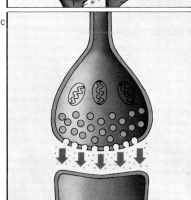

3 A brain cell is like a tiny, multi-tentacled octopus. From the centre, or nucleus, of the cell, the tentacles, or dendrites, stretch out in all directions and on each dendrite there are thousands of minute protuberances called dendritic spines These connect with each other to form linkage networks between the 10×10^{12} cells of an average brain. Mental activity may increase the number of such connections and facilitate learning and ability to relate facts and understand new ideas.

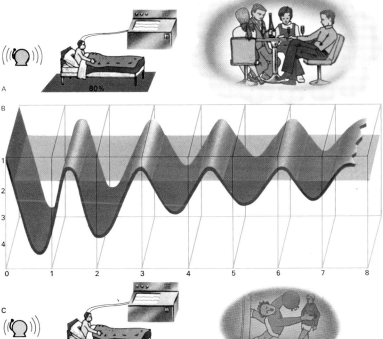

4 Depth of sleep can be measured by the fluctuations shown by brainwaves throughout the night [B]. When the waves rise from the deeper levels, rapid eye movements (REM) underneath the eyelids of the sleeper can be seen (pink areas on the graph). Sleep studies have shown that on 80% of the occasions when subjects were woken up during an REM period [A], they gave lengthy narratives of their dreams. Only seven per cent of subjects woken during other levels of sleep [C] could recall any dreaming and these were only fragmentary memories compared with the reports from the REM awakenings. It seems clear therefore that dreaming is associated with REM sleep. On the graph, the vertical numbers indicate stages in the depth of sleep and numbers on the horizontal axis indicate the hours of sleep.

43

Memory and recall

The human memory, much maligned and generally thought to be rather inefficient, is in fact more sophisticated than any computer memory ever devised. Memory can be divided into two main parts: retention – the ability to store information; and recall – the ability to retrieve it again. Studies of the human memory reveal that it is excellent at storing information but less reliable at recall – at least without special practice.

Experiment, theory and experience
Researchers approaching the problem from a number of different angles have all concluded that there is much more stored in our minds than is generally supposed. A Canadian neurosurgeon, Wilder Penfield (1891–), showed that by stimulating the brain electrically he could produce total recall of specific events from people's lives.

Dreams, in which characters and events "forgotten" for many years suddenly reappear in perfect clarity, similarly indicate that the information had been stored all the time. Another example is the surprise recall of events until then forgotten that is often trig-

gered by a particular smell, colour or sound, or by returning to a first school, childhood home or once familiar town.

There are probably different mechanisms for "short term" and "long term" memory: short term resulting from active brain processes and long term from chemical changes. But the physical basis of memory is not yet known. One theory is that the astonishing storage capacity of the brain is a product of the almost limitless combinations of the interconnections between brain cells. These connections are stimulated by patterns of activity; repeated reference to information therefore helps recall. It may be that this improvement is due to strengthening of the chemical bonds involved in memory.

Remembering and forgetting
Most people find their memory is faulty when it comes to recall. The problem is probably not an inherent one, however, but seems to arise from a misunderstanding of how the mind works. Recall can itself be divided into two major areas: recall during learning and recall after learning. During a learning period

the mind, like the body when it is exercised, needs periods of activity and rest. If the right period of activity is followed by an appropriate rest period, recall performance improves in detail and speed considerably.

Recall after learning rises for a brief time as the information "sinks in", and then it drops dramatically [2]. This loss of detail can be minimized if certain review techniques are combined with periods of activity and rest. When reading, recall can be improved, for instance, by breaking up learning periods into sessions of between 20 and 40 minutes, during which notes are made. A ten minute gap is followed by a ten-minute recall period allowing everything remembered to be noted down and compared with the original notes. Memory is reinforced by a two- to four-minute review of the same material next day and then a two-minute review period during the following week.

Faster reading also improves concentration and retention of information, as well as allowing more time for revision of important passages [4]. Aids to faster reading include practice in rhythmical reading with the help

CONNECTIONS

See also
42 Mind and brain
46 Potential of the mind
152 Thinking and understanding
40 How the brain works
178 An active old age

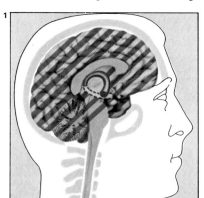

1 Memory processes may occur throughout almost the whole brain (striped area) but certain areas within and bordering on the limbic system – notably the hippocampus and the mamillary bodies – appear to be indispensable to memory. These areas, together with the fornix, form the loop indicated in red at the top of the brain stem. The dotted area recedes into the left hemisphere.

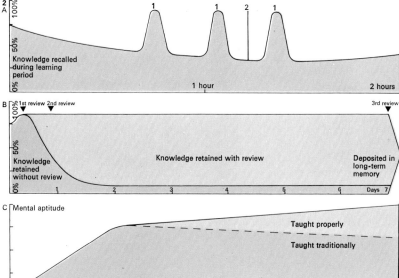

2 Recall of information being learned varies considerably as time passes [A]. Usually, we recall more from the beginning and end of a learning period and more items that are in some way linked to each other [1] or that stand out from the background material [2]. A period of 20 to 40 minutes seems to maximize understanding and recall. A properly organized review programme can prevent a rapid decline in recall of detail [B]. Without it, 80 per cent of the material is usually lost within a day. Although mental ability, including memory, generally declines with age [C], improvement can continue with a review and recall approach.

Knowledge recalled during learning period

1 hour   2 hours

1st review 2nd review   3rd review
Knowledge retained without review   Knowledge retained with review   Deposited in long-term memory
Days 7

Mental aptitude
Taught properly
Taught traditionally
25   50   75   Age in years 100

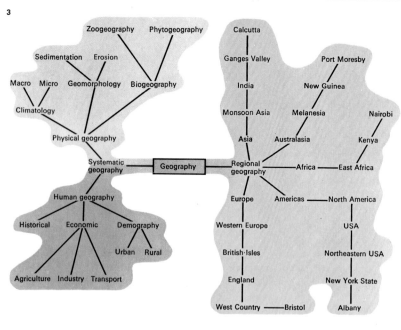

3 A new form of note-taking, called Mind Patterns, is based on the way the mind itself organizes information. Key words and key images are linked to form easily recalled networks of facts. Several pages of standard notes on the subject of geography, for instance, can be expressed in a single, compact pattern.

Zoogeography Phytogeography
Sedimentation Erosion
Macro Micro Geomorphology Biogeography
Climatology
Physical geography
Systematic geography — Geography — Regional geography
Human geography
Historical Economic Demography
Urban Rural
Agriculture Industry Transport

Calcutta
Ganges Valley Port Moresby
India New Guinea
Monsoon Asia Melanesia Nairobi
Asia Australasia Kenya
Africa — East Africa
Europe Americas — North America
Western Europe USA
British-Isles Northeastern USA
England New York State
West Country — Bristol Albany

4 A slow reader who follows only a word at a time is prone to mental wandering because information is entering the mind too slowly. A reader who takes in several words understands and recalls more because information enters the mind in "wholes" in the way it is stored. Advanced reading involves taking in areas of print that the mind "scans", selecting key words that it needs.

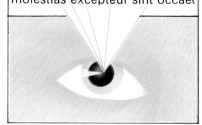

Loem ipsum dolor sit amet, cor eiusmod tempor incidunt ut lab Ut enim ad minim veniam, quis laboris nisi ut aliquip ex ea com irure dolor in reprehenderit in vo illum dolore eu fugiat nulla paria dignissum qui blandit praesent molestias excepteur sint occaec

of a metronome, time tests to encourage higher speeds and (contrary to popular belief) the use of a finger to follow words and eventually blocks or even pages of words.

When making notes, it helps to use a key word or phrase to sum up assemblages of diverse information, since this is how the brain stores it [3]. Special memory techniques called mnemonics often use similar methods to accentuate the normal recall processes of associating or linking ideas [5]. Recent studies have shown that a person who can associate ideas in this way will, apart from improving his memory, also improve his creative ability.

Much of what we "forget" we have in fact never fully taken in because we have not been concentrating. Many people have considerable difficulty remembering names of people they meet, for example. As a result, they experience mild fear when meeting new people and almost "rehearse" an introduction situation in which they will forget the names. The embarrassment of the imagined situation may also cause them to avoid eye contact with those they are meeting and to give their faces only the most fleeting glance. All this behaviour is the opposite of what is needed for recall, for we tend to forget anything that is associated with fear.

Paying attention

In order to remember difficult things such as people's names it is necessary to pay attention, repeat the names and develop or discover any associations that they may have. This means that during introductions, complete attention should be paid to the face of the other person and any links between outstanding characteristics of the face and name should be made. Sensible repetition of the person's name is also helpful. Repetition tends to build up a more solid "imprint" of a name and will result in a recall of considerably greater duration. These techniques of mental rehearsal, concentration and repetition can be generally applied to anything an individual wishes to commit to memory. We are not born with patterns of memory, recall and retention that are fixed. It is possible to educate the memory in the same way as other skills can be developed by practice.

The complexity of the human memory has been underestimated just as the scale of the universe was once underestimated. The more each is explored, the vaster it becomes. Like the astronomer, the psychologist is increasingly using terms such as "limitless" to describe memory.

5 Mnemonic (memory) systems of various kinds assist recall by organizing information into units that are as easy to store as possible. A simple mnemonic technique is the use of rhyme to learn numbers, dates or difficult names. Another method is to associate unfamiliar names or facts with familiar ones by making key words into readily remembered images. One form of this technique which has a long history is the placing of objects in regular order within a well-known scene. To recall the objects, the mnemonist moves through the scene "seeing" the objects as he comes to them. Often where abstract things have to be remembered, a visual interpretation will be used. A simple example of the technique is a scene that might be used to remember characters of the nursery rhyme "The House That Jack Built". The characters are placed in order for the mnemonist to come across as he weaves his set path through the mental scene he has constructed. The final verse of the nursery tale (which is already arranged in a mnemonic rhyming system) is: "This is the farmer sowing his corn that kept the cock that crowed in the morn that waked the priest all shaven and shorn that married the man all tattered and torn that kissed the maiden all forlorn that milked the cow with the crumpled horn that tossed the dog that worried the cat that killed the rat that ate the malt that lay in the house that Jack built."

1 Farmer
2 Corn
3 Cock
4 Priest
5 Man
6 Maiden
7 Cow
8 Dog
9 Cat
10 Rat
11 Malt
12 House
13 Jack

Potential of the mind

The most intriguing fields of research into the potential of the mind [Key] are concerned with perception, vision, illusions, hallucinations and controversial paranormal phenomena such as telepathy and extrasensory perception.

The sense organs are stimulated by various types of physical energy: the eyes by electromagnetic wavelengths; the ears by mechanical vibrations; the nose and tongue by chemicals; and the skin by combinations of pressure and temperature. Perception is the translation of such sensory stimuli, signalled by nerve impulses, into an organized picture of the world.

The problems of perception
Interest in the problems of perception is shared between philosophers, physiologists and psychologists. Philosophers ask questions about whether things exist independently of our experience of them and how we can test the truth or validity of our observations. Physiologists are concerned with the mechanisms of the nervous system, while psychologists are more interested in the pro-

cesses relating to learning and knowledge of perception. Illusions and errors are important to them. When we judge something to be a certain size, for instance, is it actually that size? And if there is a discrepancy between the actual and the perceived, how does that discrepancy arise? Illusions occur if the physiological mechanisms are upset or if there is a misapplication of knowledge.

Some psychologists, such as the so-called "atomists", believe that we perceive in units of simple sensations – such as dots, lines, colours and so on – and from these simple sensations we construct a face or a tree or a house. Opposed to this view are the "constructionalists" who think of even basic sensations, such as coloured patches, as being derived from neurological processes of which we have no consciousness.

Recent theories of perception do not regard experience as primary. Sensations of colour, or of shape of objects, are derived from physical stimuli coded by the nervous system and "read" – rather as a message is read from the dots and dashes of the Morse code. Physiologists have found out a great

deal about the form of the neural code – trains of electrical impulses – by which the world of objects is signalled to the brain. In 1958 two American physiologists, David Hubel and Torsten Wiesel, reported that neural cells in the brain's striate cortex of a particular orientation respond to lines of a different orientation. The detailed mechanisms of feature recognition are thus beginning to be understood in physiological terms – but how these neural signals are "read" to give perception remains largely mysterious.

Gestalt theories of perception
The Gestalt psychologists of the 1920s and 1930s, such as Kurt Koffka (1886–1941) and Wolfgang Köhler (1887–1967), held that we see organized wholes or configurations (*Gestalten*). These are made up of simple sensations but the whole is more important than the sum of its parts. Patterns are more important than the elements of those patterns. We "see" dots but perceive a dotted line. They also postulated a variety of principles of organization that govern our perceptions. The most general of these, called *Prägnanz*

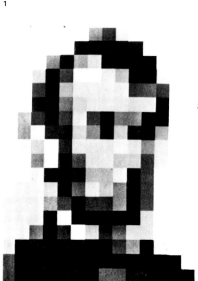

1 Computer-processed pictures are part of a research project to discover the least amount of visual information a picture may contain and still be recognizable. Shown is a portrait of Abraham Lincoln.

2 Proximity [A] and similarity [B] are two examples of the principles put forward by *Gestalt* psychologists to explain the organization of our perceptions. [A] The horizontal distance between the dots is greater than the vertical distance. As a result, the dots are perceived as vertical columns. [B] The dots are equally spaced, but because of similarity of colour they are still "seen" as vertical columns.

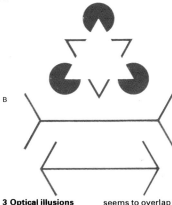

3 Optical illusions arise from tricks of perception which occur either in the retina of the eye itself or at the higher cognitive levels of the brain. [A] The bright white triangle which seems to overlap the dark circles and V-shaped lines is an illusion. [B] The central lines in both arrows are the same length, though the bottom one appears to be distinctly the shorter.

4 The mind adjusts size to apparent distance. The moon thus looks smaller on a near horizon [A] than on a far one [B] as the observer makes a bigger enlargement of the same retinal image [R] in his mind.

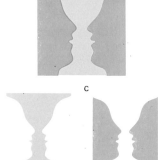

5 A reversal effect can occur when information about foreground and background is ambiguous. The picture [A] may first appear to be a vase [B] but may then begin to look like two faces [C].

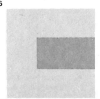

6 Colour is a product of the mind's response to varying wavelengths of light reaching the eye, rather than a property of objects themselves. Perception can be changed in the process of adapting to sudden changes of colour or to colour contrasts. The central colour in each pair of pictures is of the same shade and intensity on the left as it is on the right, for instance, but the surrounding colour fools perception and produces the illusion of slight differences in the central hues.

(exactness), states that the particular perceived whole will be as good as circumstances allow. "Good", in this context, does not have one all-embracing definition but is rather a series of qualities such as unity, simplicity, symmetry, regularity, proximity and similarity [2]. These principles govern what we see from the mozaic of stimuli at the eye.

The *Gestalt* psychologists believed that perceptions are organized into the (usually simple) shapes of objects by electrical "brain fields" that tend towards simple shapes, rather as bubbles are spherical because this is their most physically stable form. These *Gestalt* "brain fields", as little pictures or models in the brain, have now, however, been abandoned. Their laws of organization may just as easily be the result of computerlike rules, by which sensory signals are accepted, rejected and modified to serve as data for perception. For example, similar dots close together are likely to be the edge of an object and are accepted as belonging together because they probably are part of a single object. We see these groupings in arrays of dots, because such rule-following generally leads to

appropriate perception of objects from the patterns of stimulation they cause at the receptors.

It is now believed that a complex mixture of innate knowledge and data-processing rules underlies perception, though there is disagreement about the relative importance of these two elements. Research on human infants certainly indicates that they are born with an innate perceptual framework since they can discriminate even in the first few days of life between complex stimuli.

Differences between individuals
Finally, it must be stressed that there are many differences between individuals, according to their age, culture, expectations and motives. These include the fact that susceptibility to some types of illusion increases with age, while that to others decreases; that an Eskimo will have many more words for "white" than someone who is not always surrounded by snow; and that literal mistakes in type are often difficult to see because we gloss over what is actually printed and perceive what we expect to see.

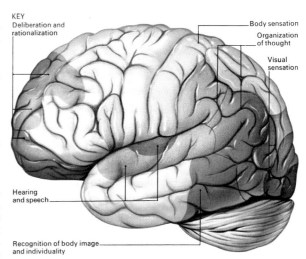

KEY
Deliberation and rationalization
Body sensation
Organization of thought
Visual sensation
Hearing and speech
Recognition of body image and individuality

Five senses were thought to be man's equipment for experiencing the world: sight, hearing and balance, taste, smell, and touch, which includes the perception of heat and cold. It is now thought that we have more than five, including senses of time, direction and motion. Other senses, such as the ability to respond to magnetic and electrical waves, may lie undeveloped. Shown here are skills tentatively associated with various brain areas.

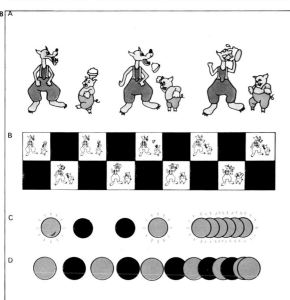

7 After-images can produce strong visual effects in the mind. Stare intently at the centre of the flag for 40 seconds. Then look at a blank white area where the flag will be seen in its true colours.

9 Changes in perception seem to occur when stimulation to the brain is reduced to a minimum. In a watery, stimulation-free environment designed for research in sensory deprivation, subjects reported strong hallucinations. Some also experienced changes in their mental image of their own bodies; arms felt disconnected from the body and the body actually seemed to change shape, becoming larger, smaller or distorted. Some subjects also experienced auditory imagery, hearing music, buzz-saws and the chirping of birds. It has been suggested that these changes in perception show that minds need some form of meaningful contact with the environment.

Soundproofing
Microphone
Experimental control room
Air supply
Subject in water tank

8 Cinematic technique relies on basic visual processes to create the illusion of movement. Pictures of objects in different spatial relationships [A] can be seen as moving because of persistence of vision. In the laboratory this is simulated [C] by switching two lights on and off in close succession. They appear as a single moving light because the image persists on the retina for several seconds. The shutter of a cinema projector [B], by cutting out light between frames, causes visual fusion of images with no flicker. This can also be shown in the laboratory by switching a light on and off [D]. At about 70 flashes per second it ceases to appear to be flashing and seems instead to be a steady light.

Sight and perception

Sight is one of man's principal means of interpreting the world about him. Its instruments are the eyes, which form in the embryo as two "buds" from the brain. It is in the brain that nerve impules from images taken in by the eyes are decoded to give vision.

The anatomy of the eye

Some parts of the eye are directly involved with receiving light, whereas others protect delicate structures [Key]. The protective elements include the eyelids, which are fringed with eyelashes – coarse dust-collecting hairs. The eyebrows, above the lids, may help prevent sweat from running into the eyes. Lining each lid is the conjunctiva, a transparent membrane that loops back to cover the front of the eye. Its low-friction surfaces help the lids to open and close easily and, more important, protect the front of the eye. Under the eyelids are the lacrimal, or tear-producing, glands. Tears, which wash away foreign particles and kill bacteria, are continually secreted on to the conjunctiva through 12 ducts and are normally removed through two canals in the nasal corner of each eye.

The outer coating of the eyeball is the sclera, the tough, fibrous white of the eye. The sclera provides attachment for the six extrinsic muscles [6] that move the eyeball, maintains eyeball shape and guards delicate inner layers. At the front it is transparent, to allow the entry of light. This is called the cornea and is the most important part of the eye for giving the focused image. The central canal joins the lens to the blindspot; it is essential in pre-natal development, when it carries blood to the lens. Lining the inside of the eyeball is the choroid. Towards the front of the eye the choroid becomes the ciliary body. The muscles of the ciliary body also suspend and alter the shape of the lens, a tough biconvex structure made of an elastic capsule filled with fribrous tissue that adjusts the focusing, or "accommodates" [1]. By muscular contraction the lens is made thicker for near vision to give maximum focusing power and is thinnest when a distant object is being viewed.

The lens lies behind the iris, a coloured, muscular continuation of the choroid which, according to inherited characteristics, gives the eye its colour [9]. The pupil, a circular opening in the iris, controls the entry of light, for it can change in diameter, by reflex action, from 1 to 8mm (0.04 to 0.32in) [3]. The iris is smallest in bright light, largest in dim light. It closes down for near vision, increasing the depth of field and in some conditions increasing sharpness of vision (acuity).

Light-sensitive cells

After penetrating the lens light passes through the vitreous humour, the clear jelly filling the eyeball behind the lens. It then falls on the retina which contains light-sensitive photoreceptor cells, the rods and cones [4]. The 125 million rods are responsible for recognizing light and dark, whereas the seven million cones are responsible for colour vision. The rods are most numerous towards the edges of the retina, and the cones are concentrated at the centre, where they are clustered in the fovea, the small area in which vision is most acute.

The rods and cones contain visual pigments whose structure is altered by light. As a result nerve impulses are generated which

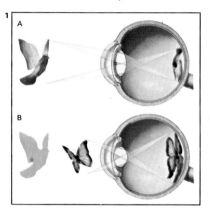

1 Focusing of light rays from distant objects [A] is mainly done by the cornea with little help from the lens. Ciliary muscles encircling the lens relax and stretch ligaments, which pull the lens flat. Rays from a near object [B] are bent by a thick lens produced when the ligaments slacken as the ciliary muscles contract. This process, which is called accommodation, is essential for sharp focusing.

2 Light rays from a point on an object are received by the temporal zone of one eye[1] and the nasal zone of the other [2]. The impulses from each, some of which "cross over" at the optic chiasma in the brain [3], then arrive at the opposite side of the brain and can be interpreted by the visual cortex. It is in this way that we perceive objects in terms of their height, width, depth and colour.

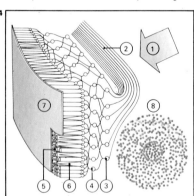

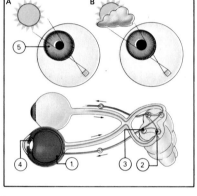
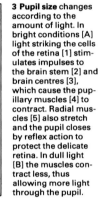

3 Pupil size changes according to the amount of light. In bright conditions [A] light striking the cells of the retina [1] stimulates impulses to the brain stem [2] and brain centres [3], which cause the pupillary muscles [4] to contract. Radial muscles [5] also stretch and the pupil closes by reflex action to protect the delicate retina. In dull light [B] the muscles contract less, thus allowing more light through the pupil.

4 Light rays [1] falling on the eye pass through optic nerve fibres [2], ganglia [3] and bipolar nerve cells [4] before reaching the rods [5] and cones [6], the eye's photoreceptor cells found in front of the pigment cell layer [7]. Cones [orange], concerned with colour vision, are concentrated in the middle, forming the fovea [8]. Rods [green], effective in black and white, are mostly on the periphery. The ratio of rods to cones is 18:1.

5 The visual field [1] is here divided into four colours. That of each eye [dotted lines] is dissimilar but overlaps [2, 3]. The image from the side is indistinct [4] as it falls on the edges of the retina. The clearest image is formed at the fovea [5]. Information from the retina is taken by the optic nerve fibres [6] to the chiasma [7] where "crossing over" takes place. The fibres leave the chiasma and form the optic tract [8], which divides into the lateral geniculate body [9]. The visual cortex in the right hemisphere [10] receives information from the right side of each eye (the left side of the visual field), and the left hemisphere [11] recieves information from the left side of each eye (the right side of the visual field).

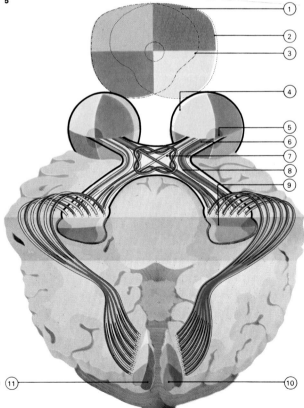

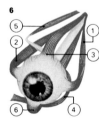

6 The eyeball is rotated by six extrinsic muscles. The lateral rectus [1] moves it away from the midline of the body and is thus in opposition to the medial rectus [2], which moves it towards the midline. The superior rectus [3] and inferior oblique [4] move it upwards, while the superior oblique [5] and the inferior rectus [6] move it downwards. The movements of the eye are important in perception, for they ensure that a "fresh" image is constantly presented to the retina.

pass to the brain for interpretation as vision. The pigment in the rods is called visual purple or rhodopsin, and light splits it into retinene and opsin. The cones are of three kinds – "red", "green" and "blue" – and respond to light of those colours. The broad range of tints that we perceive results from "mixing" of the three primary colours [5]. The cones provide precise vision in daytime, but are of little use at night or in dim light.

Both the rods and the cones are connected with ganglion cells and these give rise to one million nerve fibres which leave the eye at the optic nerve. At this junction, the blind spot, the eye detects nothing. The nerves from each eye lead, after "relay" stations, to the occipital lobe of the cerebral cortex at the back of the brain. Optic nerve fibres are so arranged that impulses from the left side of the visual field of each eye travel to the right side of the brain and vice versa. Crossing over takes place at the optic chiasma situated behind the eyes. Information from each eye is combined by the brain to create stereoscopic vision.

Eye movements are vital for perception.

The eyes follow moving objects smoothly; but move in saccadic jerks to select regions of interest and to prevent adaptation of the receptors by continuous stimulation.

The process of perception
Vision results from the stimulation of nerve cells in the retina, signalling patterns of light intensity and colour, which are "decoded" by the brain to give perception of separate objects. The most remarkable feature of perception is its ability to convert continuous patterns of energy at the receptors into individual objects and events, in space and time, all from the same kinds of pulses of electricity running along nerve fibres.

Edges – contours – are very important to perception. Orientations of contours stimulate "feature detectors" in the striate cortex. Orientations, angles and movement are represented separately, by feature detectors "tuned" to these characteristics. Combinations of these signalled characteristics are then put together to form our perceptions of objects. This also requires a great deal of experience, stored in the memory.

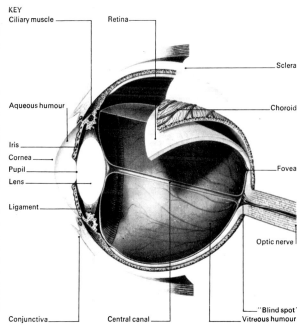

KEY
Ciliary muscle
Retina
Aqueous humour
Sclera
Choroid
Iris
Cornea
Pupil
Lens
Fovea
Ligament
Optic nerve
"Blind spot"
Conjunctiva
Central canal
Vitreous humour

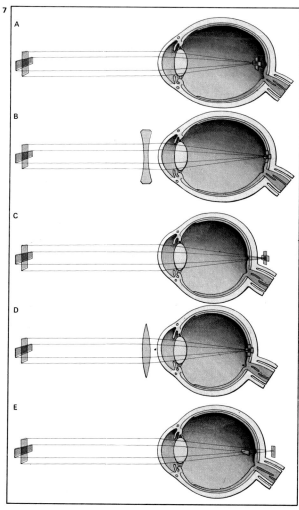

7

A

B

C

D

E

8

A

B

C

D

9 Iris pigments, and hence eye colour, are inherited. In humans the brown gene is dominant over the blue. A man [1] with two genes for brown and a woman [2] carrying two genes for blue will bear brown-

eyed children [3]. Browned-eyed parents [4] carrying the recessive genes for blue eyes produce on average one blue-eyed child to three brown-eyed. Two blue-eyed parents will produce blue-eyed children.

9

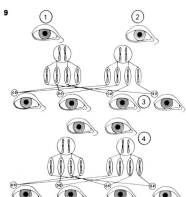

10

A

B

8 Colour blindness, of which there are two basic types, means that the individual is unable to tell all colours apart. Normal vision [A] requires red, blue and green to reproduce all the hues of the spectrum. Most colour blind people can see only two basic colours and tend to confuse others. This is called dichromatic vision. The most common version of this is the confusion of reds and greens [B] but a few people confuse blue and yellow [C]. Total colour blindness or achromatic vision [D] is the rarest defect. In this the world becomes nothing but shades of black and white. Approximately 4 per cent of men are colour blind compared to 0.5 per cent of women.

10 Contact lenses are made of hard or soft plastic, glass or a glass-plastic laminate and float on the surface of the eye. Two types are the small corneal [A] and the scleral [B] covering the front of the eyeball.

11 Cataracts, cloudy changes in the lens, may hinder vision badly, depending on their size and shape. Shown is the normal lens [A] compared to two different cataracts, the cortical [B] and dense nuclear [C].

11

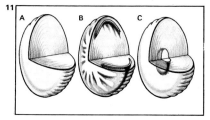

A

B

C

7 Normal vision depends on the refractive, or light-bending, power of the eye and eyeball. Where no defect exists, light rays entering the eye focus clearly on the retina. Myopia [A], or nearsightedness, occurs when the refractive power of the eye is too

strong or when the eyeball is elongated. The image then forms in front of the retina. A concave lens is needed [B]. Long-sightedness [C] is due to weak refractive power or a flattened eyeball. The image "forms" behind the retina. A convex lens [D]

is needed. In astigmatism [E], which occurs in several forms, refraction is unequal for different orientations due to unequal curvature of the eye in different planes, resulting in distorted vision. This can be corrected with oppositely astigmatic spectacle lenses.

Hearing and balance

The ear is concerned not only with hearing but also with posture and balance. The visible ear flap or pinna, consisting of skin and cartilage, and the auditory canal that it guards are known as the outer ear [Key]. The canal is protected by hairs and modified sweat glands that secrete wax or cerumen. These trap potentially harmful particles and bacteria.

For protection the delicate mechanisms of the ear are set deep inside the bone of the skull. At the end of the auditory canal lies the eardrum or tympanic membrane. On the other side is the cavity of the middle ear which communicates with the back of the throat through a canal known as the Eustachian tube. This opens into the nose and throat and provides a complementary flow of air from the inside to match the pressure of air from the outside of the eardrum. This is most dramatically demonstrated if a sudden change in external air pressure – experienced, for instance, in an aircraft – causes "popping" of the ears and momentary deafness. Swallowing opens the tube and restores the balance. Another important function of the Eustachian tube is the replacement of oxygen used by the cells lining the middle ear.

Across the cavity of the middle ear, linking the eardrums with the inner ear, there are small bones, the malleus or hammer, the incus or anvil, and the stapes or stirrup, collectively known as the auditory ossicles [1]. These, the smallest bones in the body, transmit the vibrations of the eardrum to a membrane, the oval window, that closes off the inner ear. Inside the inner ear is a fluid-filled coiled tube, resembling a snail's shell, called the cochlea [2]. Here the vibrations are converted to nerve impulses that the brain perceives as sound.

How sound reaches the brain

Sound is wave-like vibrations of air and it has the qualities of amplitude (the size of the waves) and of frequency (the number of vibrations that occur every second). If a sound is loud then it has a large amplitude, and the higher the pitch of a note the greater the frequency. Most people can detect sounds that have a frequency of between 20 and 20,000 hertz (cycles per second) [9].

The pinna not only protects the ear but also serves to direct sound into the auditory canal. Sound waves pass along the canal and vibrate the eardrum. The ossicles respond to the movements of the eardrum and, acting as a series of levers, magnify the intensity of the vibration more than 20 times, although leaving its frequency unchanged. The stirrup passes vibrations to the fluid in the cochlea through the oral window.

The cochlea is divided into two cavities by the basilar membrane, running the length of which is a complex structure called the organ of Corti [3, 8]. Sound waves travelling through the fluid along the membrane cause the sensitive cells of the organ of Corti to vibrate. High notes cause the basilar membrane to vibrate only at its lower end, low notes set the whole of the membrane into activity. A complex of more than 30,000 separate nerve fibres carry the information to the brain.

The intensity of sound, determined by the amplitude of the wave, is measured in decibles (dB). A barely audible whisper is about 20dB, ordinary conversation 60dB and a jet aircraft at take-off 140dB. Prolonged exposure to noise above 90dB can lead to tem-

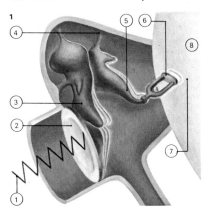

1 Sound waves [1] first strike the eardrum [2] and are transferred to the inner ear [7] by vibrations of the hammer [3], anvil [5] and stirrup [6]. These tiny bones or ossicles are held in place by ligaments [4].

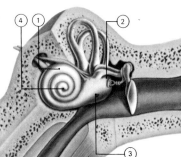

2 Waves pass through the cochlea [1] from the oval [2] to the round window [3]. High frequencies activate only the base [4].

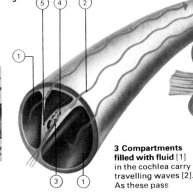

3 Compartments filled with fluid [1] in the cochlea carry travelling waves [2]. As these pass along the basilar membrane [3] they activate the sensory cells [4] in the organ of Corti [5].

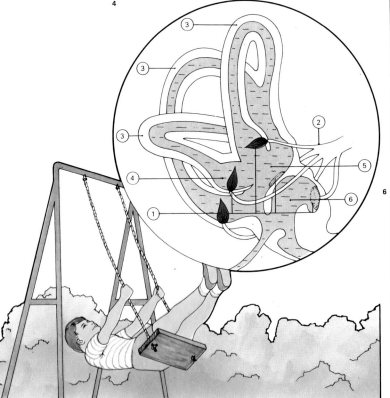

4 The inner ear contains five chambers filled with fluid – the three semicircular canals, the sacculus and the utriculus. These all contain receptor cells [1] which monitor head movement and relay this information along nerves [2] to the brain. The semicircular canals [3] are set at right angles to each other and can therefore monitor angular movement of the head in any direction. At the base of each canal is a chamber, the ampulla [4], which contains sensory receptors. The utriculus [5] and the sacculus [6] are interconnected and also contain sensitive receptors. These register the position of the head with respect to gravity and thus allow us to know, even with our eyes closed, if we are standing on our head or our feet.

5 The position of the semicircular canals, the utriculus and the sacculus are shown here [in red] in relation to the other parts of the ear. Balance also relies on vision, stimuli from the position sensors of limbs and on cells in the soles of the feet. In some cases, even when the inner ear is defective and not sending adequate signals to the brain, the other senses can maintain equilibrium.

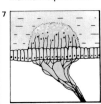

6 Body movement is detected by the ampulla when fluid in the semicircular canals bends the hairs of its receptors. The canals are positioned in three planes, so that any movement will affect the fluid in at least one of them.

7 The utriculus and sacculus contain hair cells which include small "stones" of calcium carbonate, otoliths. When the head is upright, otoliths press on particular nerve receptors. When the head tilts, they press on others.

porary deafness. Intensive noise – above 100 dB – can cause permanent damage.

Deafness and other hearing defects

Of the two main types of deafness, the conductive form occurs where there is interference with the passage of sound *en route* to the inner ear. This may be caused by a blockage of the auditory canal (often by wax), perforation of the eardrum, fluid, inflammation or abscess in the middle ear, or defective growth of one or other of the three bones in the middle ear. Probably more common is perceptive or sensorineural deafness, which is caused by damage to or incomplete development of the inner ear. Apart from congenital malfunction, this is brought about by disease, head injuries, virus infections, prolonged exposure to loud noise, or simply ageing [10].

The sense of balance

In addition to the cochlea, the inner ear also contains three fluid-filled canals, the semicircular canals and two other chambers, the utriculus and sacculus [4, 5]. These struc-

tures, which function separately from the hearing mechanism, are concerned with balance and posture.

The three semi-circular canals are set at right angles to each other and are thus able to detect movements of the head in any particular plane. At the ends of the semicircular canals are chambers that contain tufts of hair supplied with nerve fibres. Any movement of the head sets the fluid swirling through one or more canals. This moves the hairs whose nerves tell the brain the body's position [6].

The utriculus and sacculus are the means by which we are able to detect changes in the posture of the body relative to gravity. Like the semicircular canals, these two chambers contain tufts of hair connected to nerves. The hairs are surrounded by a jelly-like substance containing minute crystals of calcium carbonate, called otoliths [7].

As the crystals move under the influence of gravity, they stimulate the hairs which trigger nerve impulses to the brain. Here the information from all these balance organs is integrated with nerve impulses from the eyes to monitor the body's posture.

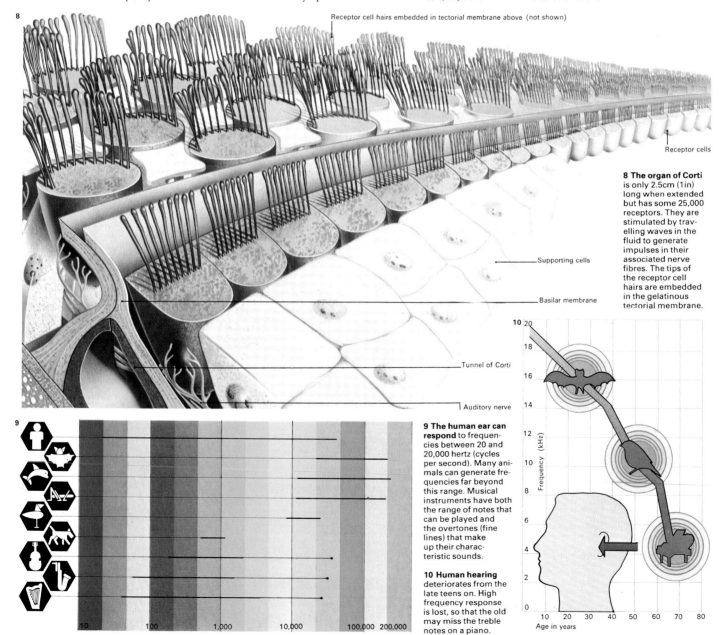

The outer ear [1] collects sound waves and directs them to the eardrum [2]. Its vibrations are transferred by three bones [3] to the cochlea [4]. From here, impulses run along the auditory nerve [5] to the brain. Movement of fluid in the semicircular canals, the sacculus and utriculus [6] is responsible for the sense of balance.

8 — Receptor cell hairs embedded in tectorial membrane above (not shown)

Receptor cells

Supporting cells

Basilar membrane

Tunnel of Corti

Auditory nerve

8 The organ of Corti is only 2.5cm (1in) long when extended but has some 25,000 receptors. They are stimulated by travelling waves in the fluid to generate impulses in their associated nerve fibres. The tips of the receptor cell hairs are embedded in the gelatinous tectorial membrane.

9

9 The human ear can respond to frequencies between 20 and 20,000 hertz (cycles per second). Many animals can generate frequencies far beyond this range. Musical instruments have both the range of notes that can be played and the overtones (fine lines) that make up their characteristic sounds.

10 Human hearing deteriorates from the late teens on. High frequency response is lost, so that the old may miss the treble notes on a piano.

10

Frequency (kHz)

Age in years

51

Touch, pain and temperature

Amid the wealth of information that is constantly fed to the brain is a stream of messages about the sensations of touch, pressure and temperature. These stimuli, which often demand minor adjustments in our everyday lives, originate in the skin, the body's main contact with the outside world. It is in the skin, therefore (and in the mucous membrane, such as that lining the mouth and vagina, that is a modification of surface skin), that some of the body's most sophisticated sensory equipment is located.

Types of receptors

Physical sensations are detected by sensory receptors – specialized nerve extremities that respond best (but not exclusively) to a particular type of stimulus. In the skin these include mechanoreceptors that are involved in detecting touch [2] and pressure [3], and thermoreceptors for sensations of heat and cold [5]. Itch, tickle and vibration have no specific nerve endings of their own; itching is caused by mild stimulation of the pain receptor, tickle is the result of light agitation of touch receptors and vibration is felt via

pressure receptors. Receptors are also concentrated in the ear, eye, nose and tongue.

Sensitivity to a particular stimulus is not necessarily uniform all over the body. Receptive fields may overlap, thus boosting the level of sensation, and in many cases the density of nerve endings varies. There are more touch receptors on the tongue and fingertips, for example, than in body areas such as the back (which explains why it is so difficult to judge how many fingers someone is using to touch your back). Most receptors can adapt to unimportant stimuli and it is for this reason that we do not continuously feel the presence of our clothes.

Significant stimuli are converted into nerve impulses to travel, by way of the spinal cord, to a centre in the brain known as the thalamus. The thalamus provides vague awareness (as opposed to fine discrimination) of sensory stimulation, determines its quality, pleasant or otherwise, and may also be responsible for interpreting pain. But the essential function of the thalamus is that of integrating sensory material – grouping

together impulses of the same nature from different parts of the body – for onward transmission to the sensory cortex, the region of the brain that analyses sensations [1]. Underlying the thalamus is a smaller structure called the hypothalamus, which, among other activities, regulates body temperature.

Pain: variety and interpretation

Pain is the most dramatic and in some ways the most puzzling of the senses that play a part in keeping us alive. Pain is felt in different ways: "bright" or pricking pain such as that associated with a cut finger is intense, short-lived and easily localized to the part of the body affected; "burning" pain is slow to develop, longer-lasting and less easily localized; and "aching" or visceral pain is persistent, often nauseating and may be referred away from the pain source [7]. The anomaly of referred pain is due to the entry of impulses from the affected area into the spinal cord and their stimulation of neighbouring nerve fibres from the skin.

Both the sensory end organs and the neural pathways that conduct pain sensations

CONNECTIONS

See also
60 Skin and hair
194 Meditation and consciousness
46 Potential of the mind

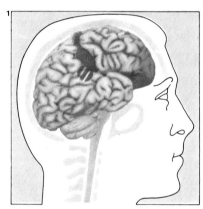

1 The region of the brain that analyses sensations such as touch and pressure and heat and cold is the sensory cortex. The primary area is coloured solid red; the secondary area is shown striped.

3 Some of the largest sensory receptors in the body are pressure receptors known as Pacinian corpuscles [1], mainly found in the dermis [2] of the skin; they also respond to vibration and stretch.

Pacinian corpuscles are distributed [3] both on the body surface and in some internal organs and are particularly abundant in hairless zones such as the palms of the hands and soles of the feet.

2 The sensation of touch usually results from mechanical distortion of various types of sensory receptors in the skin. Touching a hair [A] affects free nerve endings that form a web [1] around

hair follicles beneath the epidermis [2]. These hair end organs are distributed over the body [3]. Mostly found in hairless skin [B], Merkel's discs [4] and Meissner's corpuscles [5] are ter-

minal bulb receptors plentiful in such areas as the fingertips that are especially sensitive to touch. Merkel's discs detect continuous touch while Meissner's corpuscles signal the point and texture

of contact. Some free nerve endings can detect both touch and pressure. In the eye [C] receptors of this type [6] serving the epithelium of the cornea [7] above the lens [8] respond to the lightest contact.

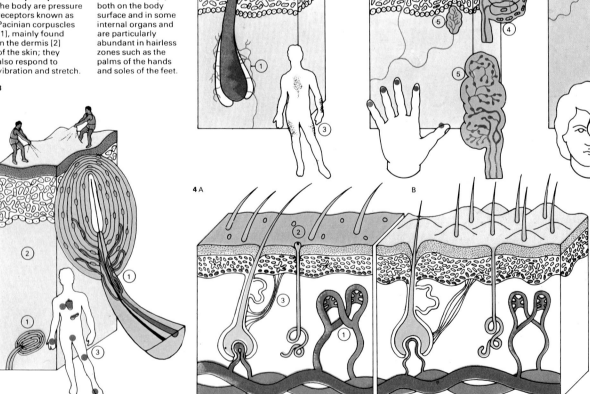

4 Body temperature is adjusted by the hypothalamus of the brain which signals changes in the skin in response to sensations of heat and cold. A rising temperature [A] causes blushing of the skin as blood vessels [1] dilate to allow heat to be radiated from the surface. Evaporation of sweat [2] and relaxation of hair erector muscles [3] cool the skin. When cold [B], blood vessels constrict making the skin look blue and hair stand on end, causing "goose-flesh".

to the central nervous system – the spinal cord and the brain – are so varied that there is dispute about the existence of specialized "pain nerves". Pain may be the result of excessive stimulation of any nerve fibre. Sometimes damage to the neural pathways gives rise to spontaneous pain, referred to the area supplied by the pathway and coupled with sensory loss. Mostly burning in character, spontaneous pain arising from an old injury to peripheral nerves is known as causalgia and is nearly always referred. Similarly, spontaneous central pain occurs when more central structures, such as the spinal cord or the thalamus, have been damaged. The bizarre sensations that are a common legacy of amputation are known as phantom-limb pains. Causalgia, central pain and phantom-limb pain are known collectively as neurogenic pains.

Research has shown that both human beings and animals quickly learn to evaluate the experience of pain as a result of events in infancy. There is evidence that undue protection from damage reduces the ability to feel pain or to evaluate painful stimuli. Sensory

deprivation of this kind is potentially dangerous in that pain is an important protective sensation – securing withdrawal of the tissues from harmful agents – as well as a useful indicator of disease. Some people suffer from a congenital nervous abnormality that drastically reduces or totally excludes the ability to sense pain – a rare condition known as hypoasthesia. They face a far higher risk of injury and without pain to signal the need for treatment their life expectancy is shorter than average.

The psychology of pain
Pain, like all other experience, is entirely subjective and often considerably affected by psychological factors [6]. Soldiers, for example, can suffer fearful wounds in battle without noticing them until later. All the techniques so far available for measuring so-called pain thresholds are open to criticism and all that is known for certain is that men are more sensitive to pain than women, that office workers are more sensitive than manual workers and anxious people more than calm types.

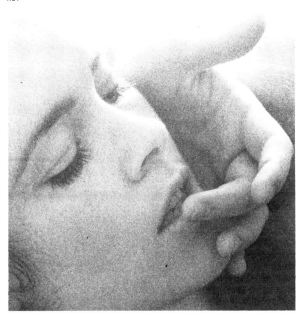

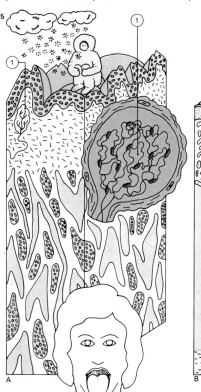

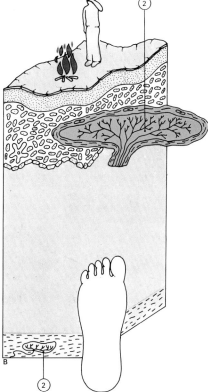

5 Two kinds of skin receptors specialized to detect thermal changes are Kraus end bulbs [1], which respond to cold, and Ruffini corpuscles [2], which react to heat. Kraus end bulbs are most prevalent in the mucosa of the tongue, the conjunctiva of the eye and the external genitalia. Ruffini corpuscles, found deep in the dermis of the skin, or even in the subcutaneous layer, are flatter in shape and particularly abundant in the soles of the feet.

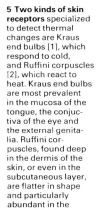

6 Pain is both physical and psychological. Physiologically, it originates from sensory receptors all over the body, but psychological and cultural attitudes seem to override the most powerful pain stimuli to the extent that subjugation of physical sensation is almost a commonplace occurrence at religious ceremonies in the East. This Hindu, taking part in a festival of penitence of Kuala Lumpur, shows no evidence of pain from needles embedded in his skin.

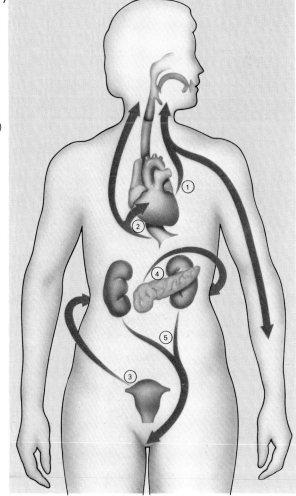

7 Referred pain is the term used for the sort of visceral pain that makes itself felt somewhere other than the site of the trouble. For example, the condition known as ischaemia [1] (a shortage of blood to the heart) causes pain in the base of the neck as well as in the shoulders and arms; acid irritation of the oesophagus [2] is felt in the throat and in addition over the heart and in the arms; womb trouble [3] and one or two disorders of the pancreas [4] give rise to backache; and stones in the kidney may set up pain in the groin [5]. As a sign of some internal disturbance referred pain is an important factor in diagnosis, to be considered together with other symptoms in the accurate location of a disorder.

53

Smell and taste

The senses of smell and taste together form part of a system that samples the world about us and reports to the brain. They are both chemical senses, separate but closely linked. In humans neither can normally operate at a great distance and taste, the weaker sense, depends to some extent on the sense of smell to augment it.

The sense of smell (olfaction), centred on two small patches of olfactory epithelium [1] in the nose, detects any airborne odours that reach it. The sense of taste, centred in the tongue, detects chemicals only in the mouth, but the two senses are so interdependent that a foul smell can leave a nasty taste in the mouth and enjoyment of food is a complex response to sensations from the tongue, palate and nose. The person suffering from a cold, for example, is unable to "taste" food because he has a "stuffed nose".

The sense of taste

The sense of taste enables us to appreciate food and drink and warns us if food has gone stale or is bad. The principal organs of taste are microscopic nerve endings called taste buds situated on the tongue and also, to a lesser extent, on other surfaces of the mouth. They are housed in small projections called papillae [2].

Only liquids can be tasted. A solid in a dry mouth creates no sensation of taste for taste buds are stimulated only when chemicals in food are dissolved by saliva and washed out over the taste buds.

When the receptor cells are stimulated they cause signals to be transmitted along two of the 12 pairs of cranial nerves – the glosso-pharyngeal and lingual nerves – to the taste centres in the brain [4].

The taste buds register only four basic sensations – sweet, salt, sour and bitter – although some experts believe that metallic and alkaline tastes can be detected too. Most tastes, however, can be considered as combinations of the four basic sensations. Different parts of the tongue detect different tastes [6] and some tastes do not combine. The centre of the tongue has no sense of taste. Taste buds are concentrated in large numbers toward the back of the tongue.

Touch receptors in the mouth convey information about temperature and texture and the various signals are combined in the brain to give the full flavour sensation. The tongue is also sensitive to irritants such as pepper or the ingredients of curry powder.

How smell works

Smell is less important to man than the other senses and it is the one about which least is known. It is not nearly as sensitive in man as it is in many species of animals. Male moths, for example, can smell female moths several miles away. During his evolution man's sense of smell has decreased in importance as he has come to rely more on sight and hearing.

The olfactory epithelia are housed in the roof of the nasal cavity [Key] – the narrow, lofty chamber of the nose. The floor of the cavity is the roof of the mouth and its ceiling is the brain case. The whole cavity is lined with mucous membrane, a soft, warm adaptation of the skin that is well supplied with blood vessels.

Glands in the mucous membrane secrete a film of mucus, the viscous fluid that keeps internal surfaces of the cavity moist. Cilia –

CONNECTIONS

See also
40 How the brain works
52 Touch, pain and temperature

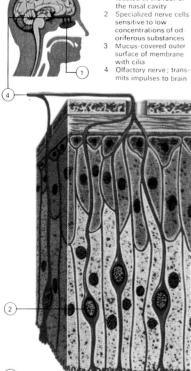

1
1 Olfactory membrane; located at the roof of the nasal cavity
2 Specialized nerve cells sensitive to low concentrations of od-oriferous substances
3 Mucus-covered outer surface of membrane with cilia
4 Olfactory nerve; trans-mits impulses to brain

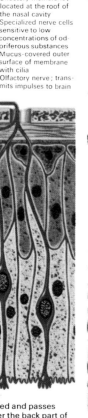

1 Ability to detect smells comes from small organs called olfactory receptors, occupying in the adult about 12sq cm (2sq in) in the top of the nasal passages. (In compar-ison, a young rabbit's olfactory surface has the same area as the whole of its skin.) These receptors con-sist of thousands of hair-bearing cells em-bedded in a layer of mucus-secreting cells. Odorifer-ous substances such as hot foods give off molecules that float through the air. When this air is in-

haled and passes over the back part of the nasal passages the molecules dissolve in the mucus. It is believed that a chemi-cal reaction then stimulates the hair to transmit nerve im-pulses to the ol-factory bulbs which are the two centres dealing with the sense of smell. These pass signals to the brain. A substance must reach the nose as a gas or a vapour for a person to be aware of its odour. Our sense of smell is closely linked with that of taste.

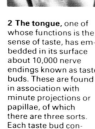

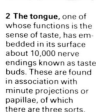

2 The tongue, one of whose functions is the sense of taste, has em-bedded in its surface about 10,000 nerve endings known as taste buds. These are found in association with minute projections or papillae, of which there are three sorts. Each taste bud con-

sists of a cluster of taste receptor cells exposed to the surface via a small pore. As food enters the mouth it is moistened by fluid produced both by serous glands in the tongue and by salivary glands in the cheeks (parotid), under the tongue (sublingual)

and in the lower jaw (submandibular). Sub-stances in solution then enter the taste buds and stimulate taste receptor cells. Nerve impulses travel via neighbour-ing nerve fibres to the brain. Taste buds can detect up to one part of a chemical in two

million. The taste buds in the front and middle of the tongue send messages along the lingual nerve and those at the back of the tongue use the glosso-pharyngeal nerve. The brain sorts out these signals and identifies them as different tastes.

1 Epiglottis, a flap that covers the entrance to the windpipe when food is being swallowed
2 Filiform papilla
3 Fungiform papilla
4 Vallate papilla
5 Entrance to a taste bud
6 Taste receptors
7 Nerve taking taste sensations to the brain
8 Serous gland cell
9 Serous secretion; helps to moisten food

microscopic hairs growing from the epithelium – waft the mucus continually back towards the throat; sniffing and swallowing help it down towards the stomach.

Air drawn through the nostrils is first filtered by a network of guard hairs, then warmed and partly cleansed by its passage over the sticky surfaces of the nasal cavity. Flowing back towards the lungs, the air passes over the two patches of sensory epithelium in the roof of the nasal cavity. These are made up of many thousands of cells bearing hairs that are connected to the deeply embedded sensory cells.

The detection of smells

How the sensory cells detect odours is not known, but it is believed that molecules of chemical vapour carried in the airstream are deposited on the mucus and in some way sensitize the hairs; these communicate with the cell body beneath them and this in turn initiates an impulse in the associated nerve fibres. The fibres pass from the olfactory membrane to the olfactory bulbs, which are linked directly to the brain [3].

The receptors of the sense of touch in the nose also contribute to the sense of smell. The sharp odour of ammonia, for example, is partly identified by stimulation of the pain receptors in the nose and the smell of menthol is partly a sensation of coldness.

Although the sense of smell is no longer important in terms of our basic survival, large perfume and deodorant industries thrive on it. The smell receptors in man can detect an amount of artificial musk equivalent to one drop being diffused throughout a room the size of a concert hall.

The sense of smell adapts rapidly to new odours. If two smells are encountered simultaneously first one is recognized and then the other. But when the receptors become saturated with an odour it quickly fades and this explains how one can tolerate an unpleasant odour after the initial aversion to it. The sense of smell is backed by an excellent memory and well-trained sense receptors can recognize more than 10,000 different odours. Sensitivity to smells differs between the sexes: most women are less aware of the odours of coffee and tar than are men.

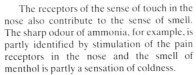

KEY

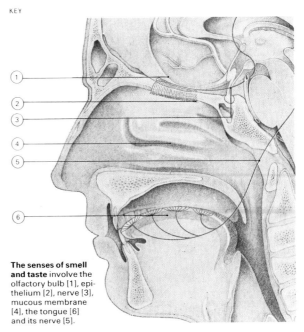

The senses of smell and taste involve the olfactory bulb [1], epithelium [2], nerve [3], mucous membrane [4], the tongue [6] and its nerve [5].

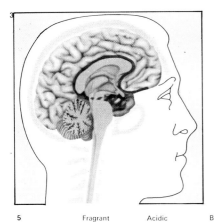

3 The smell pathway (the red line) can be traced from the olfactory bulb to the rhinencephalon (the area shown striped here), located deep within the brain.

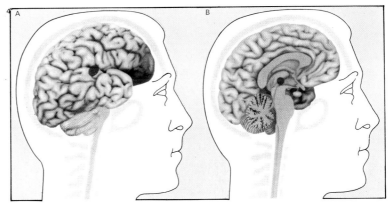

4 Taste centres in the brain (those areas most concerned with sensing taste) are part of the somatosensory cortex [A] and arcuate nucleus of the thalamus [B]. The former controls sensations from the mouth but the latter is thought to be more important.

Fragrant Acidic Burnt Caprilic

5 One classification of odours based on subjective judgments has suggested four basic qualities: fragrant, acidic, burnt, and caprilic (pungent or animal-like). The intensity of each quality has been graded from numbers 0 to 8. In this way the vinegar illustrated here may be designated as 4813 showing that it is "half" fragrant (4), very acidic (8), slightly burnt (1) and moderately caprilic (3).

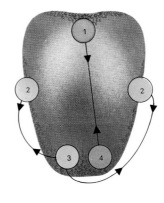

6 Different areas of the tongue are sensitive to different specific tastes. Bitterness is detected at the back of the tongue [1]; sourness is detected at the sides [2]; saltiness is tasted all over the tongue, but particularly at the front [3]; and sweetness is detected at the tip [4]. Not all taste qualities can fuse with one another but bitter and sweet blend together to produce a fused sensation, as do sour and salty.

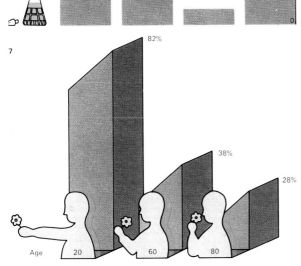

82%
38%
28%
Age 20 60 80

7 Olfactory receptors atrophy as man ages, beginning immediately after birth, so that the sense of smell deteriorates. Shown here are the percentages remaining at the ages of 20, 60 and 80.

8 Taste buds decline in number as man ages especially after 60 and it is largely for this reason that his sense of taste becomes less acute. Here are the number of taste buds in the trench wall of one papilla at the ages of 20, 60 and 80.

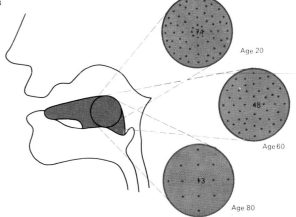

Age 20
Age 60
Age 80

55

Skull, skeleton and joints

The bones of the skeleton form a strong scaffolding of protection and support for the human body. Because bone has a high content of calcium, phosphorus and other minerals, it appears hard, dry and rigid. But one-third of its composition is a fibrous protein called collagen and this gives bone a meshwork of pliable threads.

Bone structure and design
If a long bone is soaked in acid, so that the minerals are dissolved away, the remaining protein framework ("gelatin") is so pliable that it can be tied in a knot. If, on the other hand, a bone is heated strongly, destroying the protein, the remaining mineral structure is hard and brittle. The combination of these qualities gives living bone its unique nature. It has strength and rigidity, with resilience and the ability to withstand bending and crushing forces.

Integrated with the bony skeleton, and equally important, are connective tissues. Strong and supple ligaments link bone to bone; springy cartilage lines and cushions the joints where bones meet; and fibrous, inelastic tendons (sinews) are cables securely attaching muscles to the various levers that they move.

There are four main types of bones. The flat bones such as the shoulder blades, skull vault [1] and hips, are both strong and light and safely house and support delicate internal organs, as well as providing anchorage for muscles. The irregular bones that form the flexible vertebral column or spine [2] are separate from each other but are jointed together by fibrous ligaments and muscles. The short bones, found in the wrist (carpals) and foot (tarsals) are not only strong and light, but provide resilience. Finally, the long bones in the limbs and chest are strong, hollow and light – an essential requirement for weight transmission, leverage and movement [4].

Bone formation and the skeleton
Bone develops from a foundation of cartilage [5], which usually forms a temporary skeleton for the fetus while it grows and expands. Cartilage has an enveloping sheath (the perichondrium) and contains bone-building cells called osteoblasts. When these cells are activated, they absorb soluble calcium salts from the blood and convert them into insoluble salts that are then deposited in the cartilage causing it to harden. This bone-hardening process is called ossification.

Bone tissue develops from centres of ossification [8]. A long bone has three such bone-producing centres. The shaft, or diaphysis, contains one centre that becomes active at about the sixth week of fetal development; and each of the cap-like epiphyses, separated from the shaft by a plate of cartilage, also contains one that becomes active, usually after birth. The shaft gradually lengthens until all the cartilage becomes ossified, usually between the ages of 18 and 25 years, after which the bone-length can no longer increase.

Adult bone, being hard and rigid, is more liable to fracture than that of a young child in whom the proportion of elastic organic matter is still high, and overstressing causes only a partial or greenstick fracture. In the elderly a gradual loss of elasticity makes bone more brittle, and so less resistant to fracture.

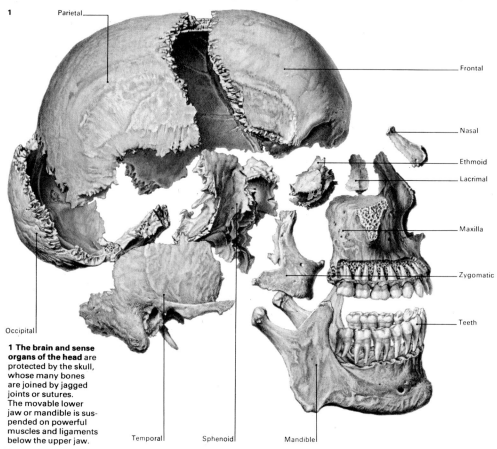

1 The brain and sense organs of the head are protected by the skull, whose many bones are joined by jagged joints or sutures. The movable lower jaw or mandible is suspended on powerful muscles and ligaments below the upper jaw.

Labels: Parietal, Frontal, Nasal, Ethmoid, Lacrimal, Maxilla, Zygomatic, Teeth, Occipital, Temporal, Sphenoid, Mandible

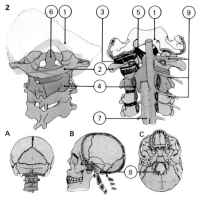

2 The vertebral column supports the head and rib cage and protects the spinal cord. The bowl-shaped occipital bone [1] at the base of the skull is attached to the first vertebra or "atlas" [2] by membranes and by a fibrous capsule [3] that is sandwiched between projections or "processes" on each bone. The second vertebra or "axis" [4] is similarly attached to the atlas by a combination of such joints and by ligaments [5]. Turning the head involves movement of the skull and atlas round the central process [6] of the axis. Nodding involves movement of the occipital bone on the atlas. The spinal cord [7] passes through the opening [8] in the occipital and down through the vertebral canal. Spinal nerves [9] leave through spaces between the vertebrae. The back view [A], side view [B] and bottom view [C] of the skull are shown. The spine is made up of 32-34 vertebrae: 7 in the neck, 12 in the thorax, 5 forming the small of the back, 5 fused in the sacrum and 3-5 fused in the tail or coccyx.

3 Simplest of the freely movable joints are gliding joints [A], such as those between the carpal bones at the wrist. Movement is restricted by ligaments. Synovial joints [B] contain fluid to lubricate bearing surfaces. Ball-and-socket joints [C] are found in the hip and shoulder. The socket grips the ball firmly, assisted by ligaments. The hip joints support the trunk's full weight. Internal ligaments [D] help to stabilize the knee joint. They cross within the joint, enclosed in synovial membrane where the femur and tibia articulate. Pivot joints [E] are found where the atlas and axis vertebrae articulate and between the radius and ulna. The former allows the head to turn, the latter the forearm to twist. Hinge joints [F] allow movement in only one plane, as at the knee, between humerus and ulna at the elbow and between the bones of each finger and toe. The saddle joint [G] at the base of the thumb is so shaped that combinations of movements in different planes are possible allowing the pad of the thumb to meet the finger pads.

1 Bone
2 Cartilage
3 Synovial fluid
4 Synovial membrane
5 Tendon
6 Ligament

The arms, legs and skull (appendicular skeleton) are supported by the spine (the central axial skeleton) [Key]. The vertebrae support the mobile shoulder framework at the pectoral girdle: this includes a pair of collarbones (clavicles) at the front, and a pair of shoulder blades (scapulae) at the back. The clavicles articulate with the breastbone (sternum) at one end, and the scapulae at the other, and act as struts to prop back the shoulders. Attached to the vertebrae are the seven pairs of "true" ribs, three pairs of "false" ribs and two pairs of "floating" ribs.

At the base of the spine, the pelvic girdle is made up of a number of fused bones, which together form the massive framework of the pelvis. The weight of the body is transmitted to the legs through the cup-shaped acetabulum on each side.

Joints and their actions

Joints are formed wherever one bone meets another; some have no movement at all and others have varying degrees of mobility. In this, cartilage plays a vital role: for less freely moving joints, a pad of fibro-cartilage unites bone directly with bone. Displacement of one of these protective pads between the vertebrae can result in the well-known "slipped disc". Where joints are virtually immovable, such as those between the bones of the skull, the bones are tightly dovetailed together [1].

In freely movable joints two bones with friction-free hyaline cartilage at the ends are surrounded by a capsule and ligaments [3]. This capsule is lined with a membrane that secretes a fluid in the joint cavity, acting as a lubricating buffer against friction. In large joints such as the knee, external sacs called bursae give additional protective cushioning [7]. (The complaint "housemaid's knee" is due to inflammation of the bursa in front of the knee-cap.)

Most bones have hollow cavities that are filled with a spongy bone marrow. Some of this (not in the long bones) forms new red and white blood cells. Little more than 250 grammes (0.5lb) of bone marrow provides the five billion red cells a day needed to replace old cells that have worn out after their 120-day lifetime in the body. Some white cells are formed in the lymph nodes.

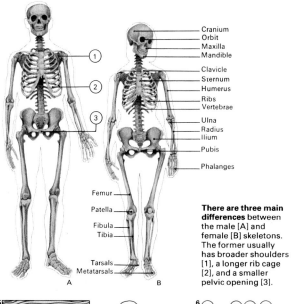

Cranium
Orbit
Maxilla
Mandible
Clavicle
Sternum
Humerus
Ribs
Vertebrae
Ulna
Radius
Ilium
Pubis
Phalanges

Femur
Patella
Fibula
Tibia
Tarsals
Metatarsals

A   B

There are three main differences between the male [A] and female [B] skeletons. The former usually has broader shoulders [1], a longer rib cage [2], and a smaller pelvic opening [3].

4 A magnified cross-section of bone [A] shows that it is composed of rod-like units [1] which when further magnified [B] are seen to have a central channel [2] containing blood vessels [3]. These are surrounded by concentric layers or lamellae of collagen fibres, each arranged in a different direction from those in adjacent layers [4]. Calcium salt crystals and bone cells [5] are embedded between the fibres.

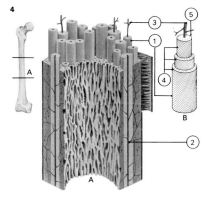

5 The three types of cartilage are composed of cells or chondrocytes that are embedded, either singly or in groups within cavities or lacunae, in different matrices. The enlarged cross-sections show [A] hyaline cartilage; [B] fibro-cartilage, which contains many collagen fibres; and [C] elastic cartilage with its cells surrounded by dense elastic fibres. [D] shows the locations of the different types in the body.

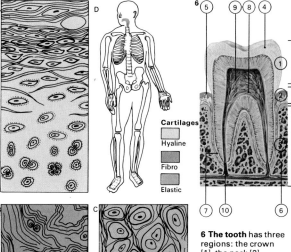

Cartilages
Hyaline
Fibro
Elastic

6 The tooth has three regions: the crown [1], the neck [2] and the root [3]. The crown, capped by enamel [4], projects above the gum [5], the root being fixed in the jaw [6] by a layer of cementum [7]. The connective tissue pulp [8] contains capillaries, nerves and lymphatics [9]. These enter the tooth at the base through the root canals [10].

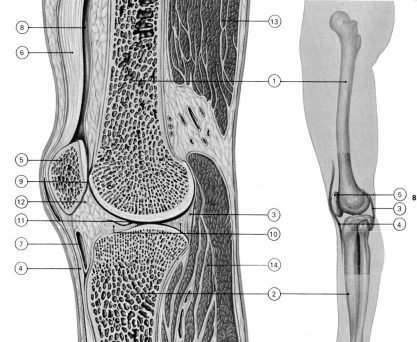

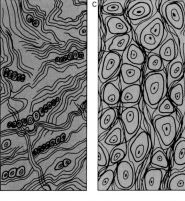

7 At the knee the femur [1] and tibia [2] are joined by ligament [3]. Further ligaments [4], attached to the patella [5], form the tendon for the quadriceps muscle [6]. The synovial membrane forms the infrapatellar [7] and suprapatellar [8] bursae. Cartilage [9] covers the articular surfaces and there are two crescents of cartilage [10] between the femur and the tibia. Synovial membrane and fluid [11] lubricate; fatty pads [12] pack the joint. Biceps [13] and gastrocnemius [14] muscles are shown.

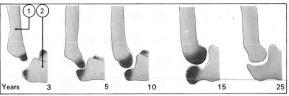

Years   3   5   10   15   25

8 Bones lengthen by growth at the epiphyses [red], which are originally cartilage but develop into bone. Shown is the growth of the arm at the elbow with the humerus [1] and radius and ulna [2] giving the stages of growth at different ages.

Muscles and action

Muscle, the raw material of movement, is a contractile tissue that makes up from 35 to 45 per cent of the body's weight and powers all its actions – from the merest flicker of an eyelid to the sustained effort needed to run a marathon race. Even during sleep some muscles remain active in order to power vital support systems.

There are three types of muscle, all different in structure and function [1]. Skeletal muscle – the meat on our bones – is strongest and most abundant [3]. Smooth muscle, found in the arteries, intestines and other internal organs, performs slow, sustained contractions. Cardiac muscle is specific to the heart, generating the powerful contractions that pump blood throughout the body.

Voluntary muscles

Only skeletal muscle is under conscious control – governed directly by the central nervous system – and is therefore described as voluntary muscle. It is networked with blood vessels and nerve endings and consists of fibres up to a maximum of about 30cm (12in) long [Key]. These fibres have the ability to contract along their whole length in response to nerve stimuli, sometimes shortening by as much as a third [4]. Under a microscope the fibres appear striped, which accounts for the name "striped" or "striated" muscle. Each voluntary muscle is bound within its own tough, elastic sheath and is further protected (and separated from its neighbours) by layers of connective tissue.

Skeletal muscles, mostly attached to bone either directly or through tendons, include flexors to bend joints and extensors to straighten them. Abductor muscles are needed to pull the arm, for example, away from its natural position at the side of the body, and adductors return it to the side. Limbs are twisted by rotators. Each fibre in a muscle has a separate nerve branch; these branches meet at what is called the motor end plate. A nerve impulse from the brain or spinal cord releases a chemical transmitter substance that makes the fibre contract.

No single muscle, however, is of much use in isolation. Because contracted fibres require an opposing force to expand, skeletal muscle is mostly arranged in antagonistic blocks with flexors working against extensors and abductors against adductors.

When muscles contract, using energy, heat builds up and carbon dioxide, water and lactic acid are formed. After strenuous exertion muscles are affected by an accumulation of lactic acid because there is not enough oxygen available for metabolism. This results in an aching sensation and in "heaviness" of the limbs. A build-up of lactic acid, combined with deprivation of salt and water or oxygen, may be the cause of cramp.

Involuntary muscles

Involuntary muscle, the basis of internal support systems, has an equally vital role to play. Both types of involuntary muscle – smooth and cardiac – are in continuous use, maintaining such functions as respiration, digestion, circulation and heart contraction.

Smooth muscle fibres are arranged in two different patterns, depending on the delicacy of their task. There are multi-unit smooth muscles, for example, in the arteries and in the iris of the eye. There the tiny individual fibres are separated, each contracting only

CONNECTIONS

See also
56 Skull, skeleton and joints
94 Diseases of the skeleton and muscles
62 Heart and blood circulation

In other volumes
154 Science and The Universe

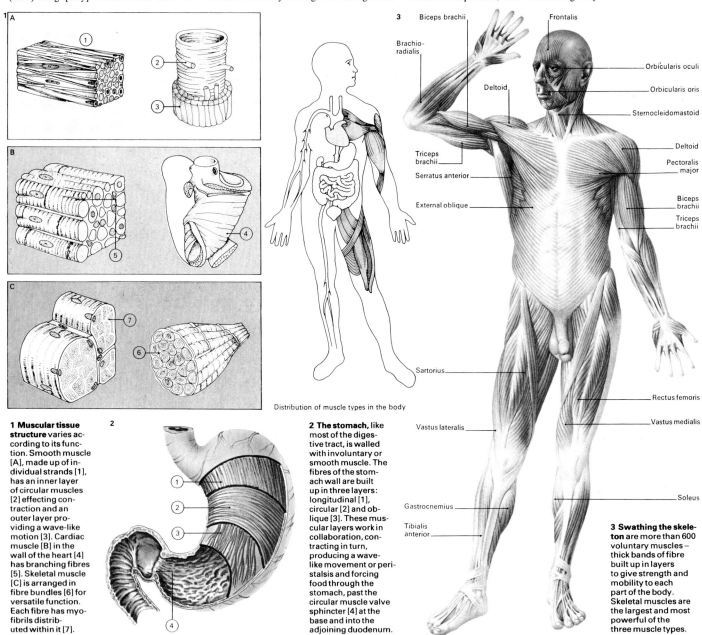

1 **Muscular tissue structure** varies according to its function. Smooth muscle [A], made up of individual strands [1], has an inner layer of circular muscles [2] effecting contraction and an outer layer providing a wave-like motion [3]. Cardiac muscle [B] in the wall of the heart [4] has branching fibres [5]. Skeletal muscle [C] is arranged in fibre bundles [6] for versatile function. Each fibre has myofibrils distributed within it [7].

2 **The stomach,** like most of the digestive tract, is walled with involuntary or smooth muscle. The fibres of the stomach wall are built up in three layers: longitudinal [1], circular [2] and oblique [3]. These muscular layers work in collaboration, contracting in turn, producing a wavelike movement or peristalsis and forcing food through the stomach, past the circular muscle valve sphincter [4] at the base and into the adjoining duodenum.

Distribution of muscle types in the body

3 Biceps brachii
Brachio-radialis
Frontalis
Orbicularis oculi
Deltoid
Orbicularis oris
Sternocleidomastoid
Deltoid
Triceps brachii
Pectoralis major
Serratus anterior
Biceps brachii
Triceps brachii
External oblique
Sartorius
Rectus femoris
Vastus medialis
Vastus lateralis
Gastrocnemius
Soleus
Tibialis anterior

3 **Swathing the skeleton** are more than 600 voluntary muscles – thick bands of fibre built up in layers to give strength and mobility to each part of the body. Skeletal muscles are the largest and most powerful of the three muscle types.

when stimulated by a nerve impulse.

In the kind of smooth muscle found in the viscera – the intestines, bile ducts, ureters and other internal organs – smooth muscle fibres are packed together so densely that they are almost indistinguishable from each other. This is the kind of muscle that powers the wave-like movement known as peristalsis. The antagonistic mechanism vital to muscle function is achieved by the arrangement of two sets of muscle fibres – longitudinal and circular – that compose the muscular wall, and by two sets of autonomic nerves, the sympathetic and parasympathetic.

Early on in the digestive process there is an ingenious mixture of muscle types working in collaboration. This is in the oesophagus or gullet, the canal running from the mouth to the stomach. Voluntary muscle rings the upper third of this tube, but sensation of movement ceases as food reaches the middle section, which consists of mixed voluntary and involuntary muscle. The remaining third is made up of involuntary muscle.

Voluntary and involuntary features are also combined in the bladder. The smooth layers must first relax, allowing the bladder to stretch as it fills with urine, and then contract to expel the contents. The sphincter, a ring of striated muscle that is under conscious control, is the governing mechanism. Kept contracted while the bladder is filling, the sphincter seals off the outlet until, under voluntary direction, it is relaxed and urine is passed.

The muscle of the heart

Cardiac muscle – the fabric of the heart – is striated like voluntary fibres but it is not under conscious control. Highly adaptable, cardiac muscle is stretched as blood enters the heart chambers. The degree of expansion is signalled by pacemaker tissue that sends impulses by conducting tissue to ordinary cardiac muscle cells, ensuring adequate and co-ordinated constriction to empty the heart completely. Once again the antagonistic principle comes into play and changes in the rate and strength of contraction are governed by the sympathetic and parasympathetic nerves. In the heart, as in the intestine, stimulation of one set of nerves increases contraction while the other set decreases it.

An exploded view of skeletal muscle shows how the bundles are built up from elongated fibres. In turn each individual fibre consists of many myofibrils. Each of these myofibrils contains filaments of protein that overlap to give the characteristic striations or striped pattern seen under the microscope. The two proteins involved, actin and myosin, interact in response to nervous stimuli, causing the muscle to contract. Skeletal muscle is of a voluntary type.

4 Contraction of voluntary muscle [A] is achieved through many elongated cells or fibres [B], made up of fibrils bound together by a membrane. Each fibril [C] contains two different proteins – actin and myosin – laid down in filaments that are arranged in parallel, creating dark bands. Actin filaments show a spiral configuration and the strands of myosin are studded with tiny, knob-like projections [D]. These projections on the myosin filaments [pink] attach themselves to active sites along the actin [brown] in the ratchet-like movement that causes shortening of the muscle fibres [E, F]. Nerve impulses trigger the two protein partners.

Levator scapulae

Trapezius

Rhomboid major

Latissimus dorsi

Gluteus maximus

Biceps femoris

Iliotibial tract

Gastrocnemius lateral head

Gastrocnemius medial head

Soleus

Tendo calcaneus (Achilles)

4 A

B

C

D

E

F

5 Chewing and biting, though routine movements, nevertheless require intricate collaboration of skeletal muscles. On each side of the head are five powerful muscles that are chiefly involved in the business of mastication. The buccinator [1] forms the side walls of the mouth. By compressing the cheek, it modifies the position of the food in the mouth. The medial pterygoid [2], which lies horizontally across the side of the face, contracts to lift the jaw. It also produces sideways grinding movements by pulling the lower jaw to one side. The external pterygoid [3], almost at right angles to the medial pterygoid, contracts to protrude the lower jaw and open the mouth. The masseter [4] lifts or closes the jaw by contraction. The temporalis [5], at the side of the head, performs the same function as it contracts.

5

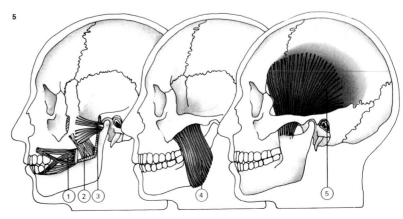

Skin and hair

Skin is one of the largest of the human organs. In an adult it covers a surface area averaging about 1.75 square metres (2,750 square inches), makes up about seven per cent of the total body weight and receives roughly one-third of the fresh blood pumped from the heart. Versatile in its range of functions, human skin is a waterproof fabric that serves as a first line of defence against injury or invasions by hostile organisms, and also has important roles to play as a sensory organ, as an agent of secretion and excretion and as a modifier of body temperature.

The skin's layered structure
Microscopic examination of the epidermis, or outer skin, reveals five layers through which cells migrate to replace the dead cells shed from the surface of the skin [1]. In the deepest level of the epidermis, the stratum basale, cells constantly divide to provide the steady supply of fresh cells forming the next layer, the stratum spinosum. As the cells move farther away from the source of blood and nutrients in the underlying dermis they gradually degenerate – by filling with granules of protein waste – and die. Next they form the stratum granulosum, and then the shiny, almost transparent stratum lucidum.

Flattened and laden with keratin, the durable protein of hair and nails (and animal horn), these cells finally emerge to form the visible skin – the thick, impervious stratum corneum or "horny layer". Dead skin flakes away from the surface all the time: it is estimated that the epidermis is totally replaced once every three weeks and that an average lifetime sees the shedding of 18 to 22kg (40 to 44lb) of dead skin cells.

The underlying dermis, built around a network of protein fibres, gives the skin its elasticity and strength – loss of elasticity of the dermis is a feature of ageing, characterized by folding or wrinkling of the skin. About three millimetres (0.1in) thick, the dermis is interlaced with blood and lymphatic vessels and contains numerous nerve endings, glands and hair follicles.

Deeper still is the subdermal layer between skin and muscle. Composed of loose connective tissue, this region is heavy with fluid and fat cells that provide insulation, cushion muscles and nerves and serve as an energy store. It is this subdermal layer of fat, which is thicker in women, that gives shape and substance to the body as a whole.

Hair, nails and pigmentation
Hair and the nails [2] on the fingers and toes are derivatives of skin: both are made up of modified epidermal cells filled with hard keratin. But whereas hair and nails are nerveless and dead and can be cut without pain, the skin as a whole is acutely responsive to external sensations such as touch, pressure, heat and pain.

Hair develops as a downgrowth of epidermal cells into the dermis to form a hair follicle set at an angle in the skin [3]. A tube of keratin is pushed up from the follicle to form the hair shaft. Each hair is supplied with its own minute erector muscle and is capable of independent movement – standing on end in response to cold or alarm. What we call "gooseflesh" is equivalent to a bird fluffing out its feathers or a cat's fur standing on end. At maturity the only parts of the skin free of hair are the undersides of the fingers and

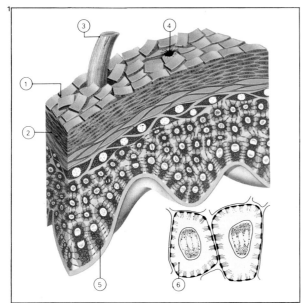

1 The epidermis or outer skin is covered with flattened (squamous) or tile-like dead cells [1] filled with the protein keratin. It is keratin that toughens and waterproofs the skin. Millions of dead cells are shed (desquamated) from the surface all the time, to be replaced by new cells making their way up from below [2]. Resistant to bacteria, the skin is penetrated only by hairs [3] and sweat pores [4]. The epidermis and the deeper skin, or dermis [5], fit tightly together in a series of corrugations. Epidermal cells [6] cling to each other by means of special pronged attachments.

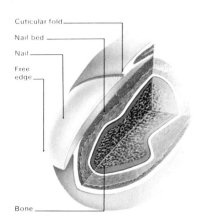

Cuticular fold
Nail bed
Nail
Free edge
Bone

2 Nails, like hair, are derivatives of skin. A useful protective covering for the ends of fingers and toes, nail is made of hard keratin. It rests on the nail bed, which is formed from the two deepest layers of the epidermis. The cuticle is above the root of the nail and the sensitive quick is located beneath the free edge. The pale half-moon called the lunula lies under the root and represents the active, growing region of the nail.

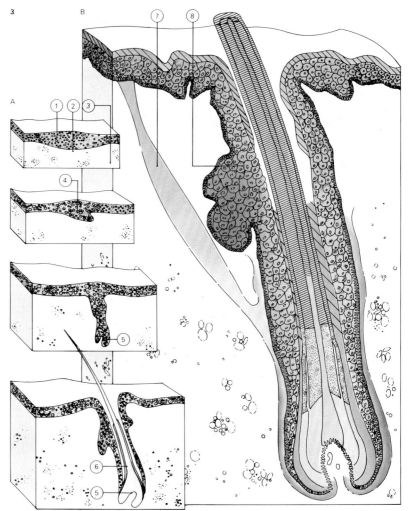

3 The presence of hair on the skin is a distinctive mammalian characteristic. The development of hair [A], which begins in the third month of fetal life, is prefaced by the downgrowth of thickened cells of the epidermis [1] into the underlying dermis [2] and connective tissue [3]. The hair [4] is the result of a multiplication of cells that clump together to produce a papilla [5] at the base of the follicle. Constantly dividing, these cells push upwards towards the surface, becoming impregnated with the protein keratin, to form the hair shaft [6]. Hair grows because of the continued division of these "matrix" cells. A cross-sectional view [B] of the hair shaft shows how each hair is equipped with its own erector muscle [7], endowing independent movement. An associated sebaceous gland [8] secretes oily sebum to lubricate the hair and the surrounding skin. The scalp alone is covered with up to 2 million hairs growing about 0.3mm a day.

toes, the palms of the hands, soles of the feet and parts of the external genitalia.

Closely correlated, hair and skin colour are genetically determined and depend on the amount of pigment – mainly melanin [4] – present in the individual (in the case of red hair there is an additional pigment). Production of melanin in the hair follicles diminishes with age and minute bubbles of air in the shaft make the hair look grey or white.

It is a failure in pigment synthesis that accounts for the albino state. In albinos melanin cannot be synthesized because an enzyme is missing due to gene mutation. Albinos, therefore, have white skin and hair. Also, with no pigment in the iris, the eyes are extra sensitive to light and they appear pink.

Glands of the skin
Closely associated with all types of body hair are the sebaceous glands, which discharge an oily secretion known as sebum into the hair follicles. Besides lubricating the hair and surrounding skin, sebum also inhibits excess evaporation from the three million sweat glands lying coiled in the dermis.

More widespread of the two types of sweat gland are eccrine glands [7], producing the salty sweat which has an important cooling function in hot weather and which also carries away small amounts of waste products such as urea and lactic acid. Apocrine sweat glands, larger and secreting much thicker fluid than the eccrine glands, are concentrated in the armpits and the anal and genital regions. Secretions from the various glands in the dermis help to keep the skin supple and, by absorbing a certain amount of radiation, to prevent sunburn. They play host to many different kinds of bacteria, most of which are quite harmless.

Skin does not end at the lips and anus but continues into the body orifices in the modified form of the mucous membrane that lines the nose and mouth, the alimentary canal, urogenital organs and anal canal. Lacking the heavily keratinized outer layer of skin proper, this membrane is rich in glands secreting mucus which moistens and protects delicate surfaces and is well endowed with nerve receptors that give added sensitivity, particularly to touch and temperature.

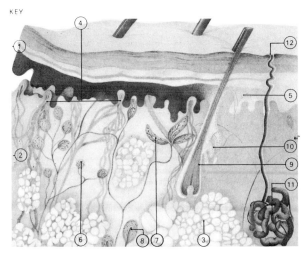

The skin consists of the epidermis [1], a tough, waterproof and germ-proof outer fabric, and its underlying dermis [2]. A deeper layer of fat cells [3] insulates the body and shapes its contours. Networked in the skin are blood vessels [4] and four types of sensory receptors [5, 6, 7, 8]. Hair pushes up from a follicle [9] and a sebaceous gland [10] secretes lubricant sebum. Sweat glands [11] empty by way of the pores [12].

4

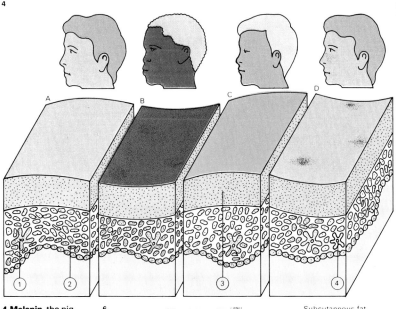

4 Melanin, the pigment mostly responsible for skin and hair colour, is common to all racial types. It is produced in the epidermis by cells known as melanocytes [1] and injected into the surrounding epidermal cells [2]. When the skin is exposed to sunlight melanin production speeds up to give the Caucasian [A] a protective tan. The skin of the Negroid type [B] radiates heat more efficiently than the lighter skin of his Caucasian and Mongoloid [C] neighbours. Carotene, a yellow pigment found in the horny layer [3] of skin, is more abundant in certain Asian races. Freckled skin [D] is due to pockets of pigment surrounded by inactive melanocytes [4].

6

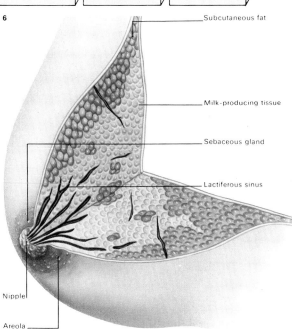

Subcutaneous fat

Milk-producing tissue

Sebaceous gland

Lactiferous sinus

Nipple

Areola

5
A

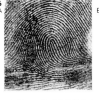
B

C

D

E

F

5 Ridges in the dermis or deeper skin of the fingertips show through on the surface as fingerprints and these are specific to the individual. Fingerprints, first used to solve a crime almost a century ago, made their debut in police work in Argentina in the 1890s. This method of identification was introduced to Britain in 1901. Distinctive patterns include the loop [A], which is most common, the double loop [B], the whorl [C], the central pocket loop [D], the arch [E] and tented arch [F]. Fingerprints remain the same throughout life unless the dermis tissue is accidentally destroyed

6 The breast – a feature that is rudimentary in the male – contains a collection of special secretory glands. These lie deep within the skin where they are protected by fatty tissues. Each gland is made of lobes and becomes active in the secretion of milk in the period following childbirth. Milk is collected by a network of lactiferous sinuses and taken to the nipple, which is kept lubricated by sebaceous glands. These glands show through as tiny swellings in the areola, which is the dark area surrounding the nipple. The areola darkens still further during pregnancy and remains more deeply pigmented in women who have given birth to children.

7

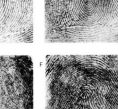

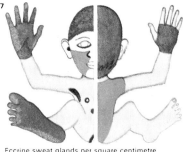

Eccrine sweat glands per square centimetre

☐ Less than 100   ☐ More than 200
☐ More than 100   ☐ More than 300
■ Specialized glands of the skin

7 Glands of the skin include the abundant sebaceous glands and – less common – ceruminous glands which produce ear wax to protect and lubricate the skin of the auditory canal. Most widespread of the sweat glands is the coiled eccrine type, producing salty sweat. Larger, and yielding a much thicker secretion, the apocrine sweat glands develop in association with the coarse hair of the armpits and the anal and genital regions. The apocrine glands begin to function at puberty.

61

Heart and blood circulation

It was the English physician William Harvey (1578–1657) who, in 1628, first deduced that the movement of blood is "constantly in a circle". At the time this was a remarkable observation, since Harvey could not have known of the presence of the vast network of capillaries that make up the 96,500km (60,000 miles) of human micro-circulation [Key], nor of the oxygen that is exchanged between capillaries and tissues.

The function of blood

Every living organism, however primitive, needs a circulatory system to fuel its parts and to remove waste. In the human body, this vital exchange is transacted by the versatile fluid known as blood – about 5 litres (9 pints) of it in the adult. It is blood, circulating deep in the tissues, that carries oxygen and nutrients to keep the cells alive and removes carbon dioxide and other waste materials for elimination from the body.

Made up of specialized muscle, the heart provides the motive power for the circulatory system. It consists of two upper chambers or atria and two lower chambers known as

ventricles [5]. Blood-flow between chambers and into the aorta and the pulmonary artery is controlled by the four heart valves [7]. The rhythmic action of the heart is regulated by the pacemaker – modified muscular tissue responsible for the impulses that trigger first the atria and then the ventricles to contract.

How the heart works

The heart is, in effect, a single pump powering a double circuit. In the space of one heartbeat freshly oxygenated blood arriving from the lungs enters the left side of the heart, for dispersal to organs and tissues, and "stale" or deoxygenated blood is pumped via the right side of the heart direct to the lungs where it is reoxygenated [3].

About the size of its owner's fist, the heart begins work in the fourth week of fetal development and continues throughout life. In the adult the heart rate averages 70 to 80 beats a minute although this resting rate can increase dramatically up to two and a half times – in response to stress or exertion. The force of the heartbeat is such that if the body's largest artery – the aorta, arching

away from the heart – is severed, a 2m (6ft) jet of blood is released.

Freshly oxygenated blood is conducted away from the heart by arteries – thick-walled, muscular vessels most of which run deep in the tissues to minimize the possibility of damage [6]. Where arteries are present near the surface – as at the wrist – there is a pulse echoing the heartbeat. Arteries branch into arterioles, which in turn break up into a complex network of capillaries [1]. These are the minute structures that thread their way to every part of the body to carry out the vital function of the blood, trading oxygen and other materials for carbon dioxide and waste.

On the venous return side the deoxygenated blood, now carried at low pressure by the capillaries, passes through the larger venules and into the veins. Ultimately it converges in two major vessels, the inferior and superior venae cavae, that join above the heart and drain spent blood into the right atrium to be pumped to the lungs.

Drawn fresh from the body, blood is a dense, sticky fluid that congeals quickly when exposed to the air. Separated out in the

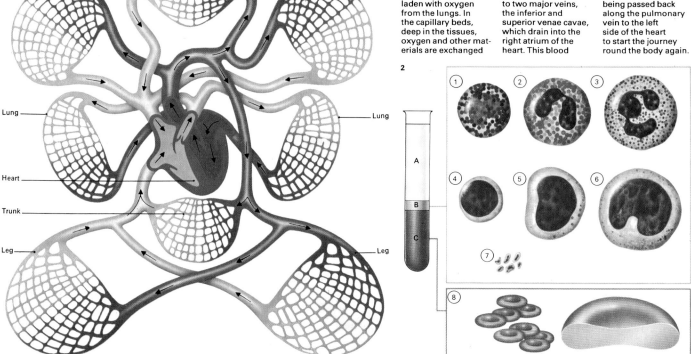

1 Fresh blood [red] starts on its long journey through the circulatory system laden with oxygen from the lungs. In the capillary beds, deep in the tissues, oxygen and other materials are exchanged for carbon dioxide and waste. Stale or deoxygenated blood [blue] is carried back to two major veins, the inferior and superior venae cavae, which drain into the right atrium of the heart. This blood then enters the pulmonary circulation, to be reoxygenated in the lungs, before being passed back along the pulmonary vein to the left side of the heart to start the journey round the body again.

2 Spun in a high-speed centrifuge, blood separates out into plasma [A], layers of white cells and platelets [B] and red cells [C]. Fluid plasma, almost 90% water, contains salts and proteins. Three main types of white cells, shown in magnification, are granulocytes [1–3], small and large lymphocytes [4, 5], which are involved in the body's defences, and monocytes [6]. Platelets [7] are vital clotting agents. Red cells [8] are the most numerous.

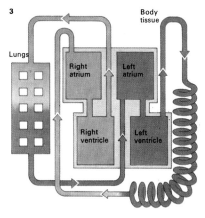

3 The heart is a dual-purpose pump. In the space of a single beat, blood is pumped from the left to the systemic circulation and from the right to the pulmonary circulation for recycling.

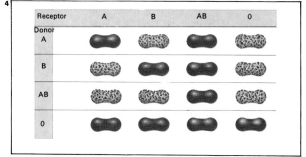

4 The foremost system of blood group classification in use today is the ABO system. The letters A and B relate to types of proteins present in the membrane of red cells. Blood group, therefore, is determined by the presence of A or B proteins or both; people lacking both A and B proteins belong to group O. Accurate cross-matching is vital in transfusions, where incompatible combinations can cause destruction of red cells (represented here by mottled cells), with serious consequences to the recipient patient.

laboratory, it settles into the yellowish fluid known as plasma, which constitutes 55 per cent of the volume, and a reddish-brown sediment of fine particles [2].

Plasma, corpuscles and lymph
Blood plasma is a solution of salts and proteins, similar to the fluid matter of living cells. Serum, the colourless fluid that oozes from a scrape on the skin, is plasma minus fibrin, one of the proteins that helps blood to clot.

The dark sediment consists of vast quantities of red blood corpuscles (erythrocytes), rather less of the defensive white blood corpuscles (leucocytes), and platelets (thrombocytes), which play a vital part in blood clotting. Formed in the bone marrow and measuring less than one ten-thousandth of a millimetre in diameter, red blood corpuscles are among the smallest cells in the body. It is the red cells, composed of one-third haemoglobin, which act as vehicles for oxygen. Carried in the bloodstream as oxyhaemoglobin, oxygen quickly detaches on reaching target tissues to diffuse through the fine capillary walls.

White blood corpuscles are two to three times the diameter of red cells but almost a thousand times less numerous. Colourless and transparent, they have a quite distinct role to play, gliding in and out of blood vessels to accumulate at sites of injury or infection, where they are part of the body's defences. Also present at the site of a wound are platelets, smallest of the blood particles, which are concerned in the clotting process.

Closely integrated with the circulation of the blood is the lymphatic system. Lymph is the watery medium in which life-giving substances pass through the capillary walls to nourish the surrounding tissues. Basically plasma but with a low proportion of proteins, lymph bathes the cells, where it is known as tissue fluid, before draining away into the quite separate network of vessels that make up the lymphatic system. Also laden with waste materials – bacteria, dead cells and other debris – lymph is carried away to be returned to the bloodstream by way of the two jugular veins at the base of the neck. Other functions of the lymphatic system include its role in combating illness.

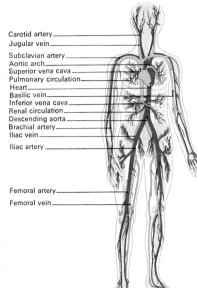

Carotid artery
Jugular vein
Subclavian artery
Aortic arch
Superior vena cava
Pulmonary circulation
Heart
Basilic vein
Inferior vena cava
Renal circulation
Descending aorta
Brachial artery
Iliac vein
Iliac artery

Femoral artery
Femoral vein

The circulatory system, a 96,500km (60,000 mile) network of vessels, takes life-giving blood into every part of the body to replenish tissues and remove waste. From the heart, driving force of the circulatory system, freshly oxygenated blood enters the 3cm-wide (1in) aorta. The aorta branches into major arteries serving the organs and limbs. Arteries divide into arterioles and finally into minute capillaries which, weaving throughout the tissues, serve the body's microscopic cells. Spent blood returns by way of venules and veins, to be pumped by the heart back to the lungs where it is reoxygenated.

5

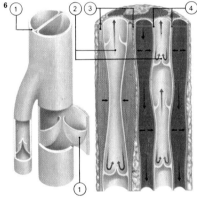

5 The heart contains four chambers – two atria and two ventricles – and four sets of valves. Blood from the body passes into the right atrium [1], via the venae cavae [2, 3]. Flow of blood into the right ventricle [4] is controlled by the tricuspid valve [5]. Pulmonary arteries [6] conduct blood from the right ventricle to the lungs, while the pulmonary veins [7, 8] carry oxygenated blood back from the lungs to the left atrium [9]. In a parallel manner, the mitral valve [10] controls the flow of blood between the left atrium and the left ventricle [11]. The aorta [12] conducts blood from the left ventricle to all parts of the body.

6 Major vessels of the blood circulatory system are arteries [red] and veins [blue]. Fresh blood flows under high pressure, most of which is lost by the time blood reaches the veins. Veins, therefore, are

6

fitted with cup-shaped valves [1] to prevent back-flow. Movement of blood in veins [2] is promoted by the massaging effect of surrounding skeletal muscle [3] or by the surge of adjacent arteries [4].

7 The sequence of a single heartbeat takes about 0.9 seconds and begins [A] as the atria [1, 2] fill with blood. Contraction [B] of the atria forces blood past the retaining tricuspid

[3] and mitral [4] valves into the ventricles [5, 6]. The thick muscular walls of the ventricles contract [C], snapping shut the atrio-ventricular valves and tending to cause the semi-lunar

valves [7, 8] to open. With both ventricles fully contracted [D], two streams of blood are forced along the separate routes; fresh blood into the aorta [9], spent blood into the pulmonary artery [10].

7 A

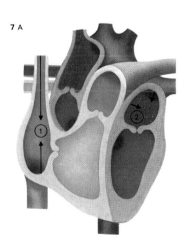

B

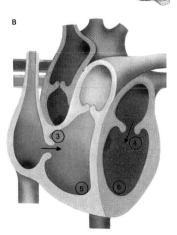

C

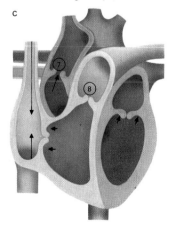

D

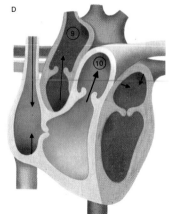

Glands and their hormones

The hormone-producing endocrine glands, working in collaboration with the nervous system, provide a means of controlling various body functions. But whereas the nervous system presides over fast-moving events, the endocrine system governs more widespread processes measured in hours, months or years. Some hormones are, it is true, involved in rapid events – the "fright, flight or fight" response, for example, is due to a sudden increase of adrenaline in the blood – but in general the endocrines govern such processes as metabolism, growth and development.

Unlike glands of external secretion, which liberate their products to the surface or to the body cavities by way of ducts, the endocrine glands are ductless. These glands of internal secretion take in raw materials from the blood to produce hormones which are released direct into the bloodstream.

The body's master gland
Master gland of the endocrine system is the pituitary [1], measuring only 1cm (0.4in) in diameter and located beneath the brain. The pituitary, sometimes described as the "gland of destiny" because of its importance in our overall development, exercises control over most hormone systems.

Most important of the pituitary's two main lobes is the anterior, which produces a group of hormones governing the thyroid, adrenal and sex glands. It is also the source of somatotropin, which controls the growth of tissues and influences fat and sugar metabolism, and of prolactin, which stimulates and sustains milk production in the nursing mother. The posterior pituitary releases two hormones : vasopressin, which acts on the kidneys to control the body's water content, and oxytocin, which is mainly active during labour to help contraction of the womb or uterus. It also assists in making milk available to the suckling infant.

The two-lobed thyroid [2], operating under direction from the hypothalamus of the brain and the anterior pituitary, embraces the larynx and upper windpipe. Its function is to absorb iodine and other materials from the bloodstream to produce a hormone called thyroxine, and it is thyroxine that regulates rates of metabolism all over the body and maintains levels of heat production. The thyroid also produces calcitonin which plays a part in calcium regulation.

The parathyroids are four tiny glands. They cling limpet-like to the rear surface of the thyroid, but have little in common with the larger gland [3]. Parathyroid hormone (PTH) is vital to the metabolism of calcium and phosphate in the body.

Protecting the body
More versatile are the adrenal glands, located one on each side of the spinal column just above the kidneys. Shaped like a tricorn hat, each gland consists of an outer layer, or cortex, and a central medulla – two distinct areas that differ in their endocrine functions [5]. The cortex produces steroid hormones known as corticoids, including glucocorticoids, important in the metabolism of carbohydrates and proteins; aldosterone, another of the hormones involved in maintaining water-balance through its effect on the kidneys; and small quantities of sex hormones which supplement the larger amounts produced by the gonads.

1 Hormones from the anterior lobe of the pituitary [A] control the activity of other endocrine glands. They include [B] thyroid stimulating hormone (TSH) [1]; growth hormone (GH) [2]; prolactin, which influences lactation [3]; luteinizing hormone (LH) and follicle stimulating hormone (FSH), which stimulate secretion by the ovaries [4] and testes [5]; adrenocorticotrophic hormone (ACTH) controlling the adrenal cortex [6].

TSH
GH
Prolactin
LH
FSH
ACTH

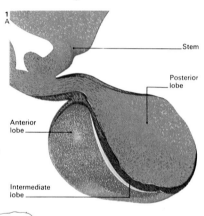

2 Principal hormone of the thyroid [1] is thyroxine, released under control of the pituitary gland [2]. Rich in iodine, thyroxine stimulates energy production in the body cells [3]. An overactive thyroid increases metabolic rate, producing rapid heartbeat, high blood pressure and weight loss. Thyroid deficiency lowers resistance to cold. Thyroxine also affects absorption of sugars by the intestine [4] and the blood-cholesterol level [5]. In addition it has psychological effects – due perhaps to its influence on the adrenal medulla [6]. In children thyroxine is vital to growth – partly due to its effect with the growth hormone [7].

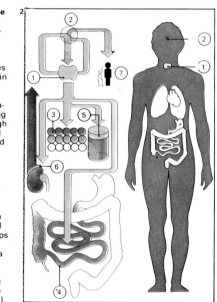

3 The thyroid [1], curving around the larynx and trachea, has twin lobes joined by an isthmus. Unrelated to their host are the four parathyroid glands [2] lodged on the rear of the thyroid.

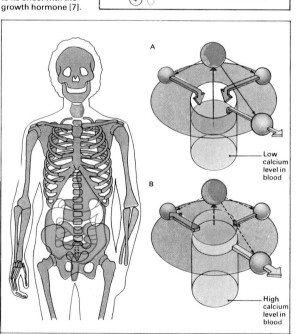

4 The hormone produced by the parathyroid glands [orange], abbreviated to PTH, has the important function of regulating the level of calcium in the blood. The mineral itself controls secretion of PTH by means of a feedback mechanism. A low blood-calcium level [A] triggers secretion of PTH. This in turn mobilizes calcium from the bones [blue] and aids its absorption from the intestines [green] and reabsorption from the kidneys [pink] to boost the amount that is circulating in the blood. With a high blood-calcium level [B] the sequence is reversed. Production of the parathyroid hormone is inhibited and consequently more calcium is laid down in the bones of the body.

Low calcium level in blood

High calcium level in blood

Unlike the adrenal cortex, the medulla is not essential to life. It yields two hormones whose effects resemble those produced by stimulation of the sympathetic nervous system. Adrenaline reinforces the role of the sympathetic nervous system in preparing the body for action in an emergency or stress situation [6]. The other product of the adrenal medulla, noradrenaline, is closely allied to adrenaline chemically and exerts similar effects.

The pancreas, sometimes called the "salivary gland of the abdomen", secretes pancreatic juice (made up of various enzymes) to aid the digestive process, and this is released to the duodenum by way of the pancreatic duct. But the pancreas also has an important endocrine function, secreting two hormones – glucagon and insulin – which are concerned in the maintenance of blood-glucose level. Sites of production are the tiny islets of Langerhans – small masses of special endocrine cells interspersed throughout the ordinary glandular tissue [7]. Glucagon encourages the breakdown of glycogen, a carbohydrate storage material manufactured in the liver, and this process raises the blood-glucose level. Insulin has the effect of lowering the blood-glucose level by facilitating the uptake of glucose in the muscles. It is failure to produce sufficient insulin that is the cause of diabetes mellitus, or sugar diabetes.

Maleness and femaleness

Stimulated by secretions of the anterior pituitary, the gonads – ovaries in the female, testes in the male – secrete sex hormones. Together with the sex chromosones, these are responsible for maleness and femaleness – for the bodily characteristics distinguishing the sexes. The testes produce the hormone testosterone. The two ovarian hormones, oestrogen and progesterone, are essential to the reproductive cycle.

Although the endocrine glands may appear to be independent of each other they do, in fact, achieve a close measure of collaboration, with hormones from one often influencing output from another [8]. With other constituents of the blood the secretions of these ductless glands form a supportive fluid medium for millions of body cells.

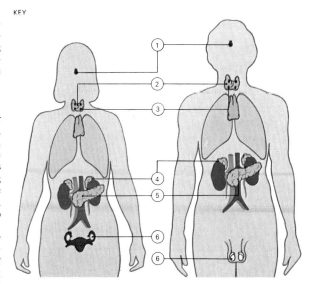

The endocrine glands secrete hormones directly into the blood, which carries them to specific "target" organs. The glands are: [1] pituitary; [2] thyroid; [3] para-thyroid; [4] adrenal; [5] pancreas; and [6] sex glands (gonads) – ovaries and testes.

5 Situated just above the kidneys, the two adrenal glands [1] are well supplied with blood entering directly from the aorta [2] and from tributaries of the renal arteries [3]. Each gland consists of two separate functional parts: an outer layer or cortex [4] and an inner medulla [5]. Only the cortex – secreting steroid hormones to control body chemistry – is essential to life. Taking up raw materials from the arteries, the different cell types of the cortex release their products into the veins [6]. The medulla secretes two hormones which prepare the body for action in emergency.

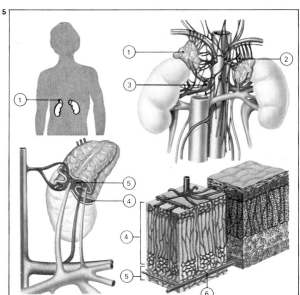

6 A vivid example of a hormone at work is the way in which adrenaline acts in response to fear, a powerful emotion that produces dramatic effects on the body as a whole. It is the sudden release of adrenaline from the adrenal medulla that meets the situation by mobilizing energy output and spurring the body to action. Specifically, fear triggers the following chain of events: [1] the frontal lobe is altered; [2] the hypothalamus of the brain orders the adrenal medulla into production; [3] there is a rapid release of adrenaline; [4] the pupils dilate; [5] hair stands on end; [6] blood readily coagulates; [7] the chest expands; [8] bronchioles dilate; [9] the heart dilates; [10] blood-pressure rises; [11] muscles contract; [12] skin capillaries constrict, causing pallor; [13] muscle blood vessels dilate; [14] the bladder evacuates.

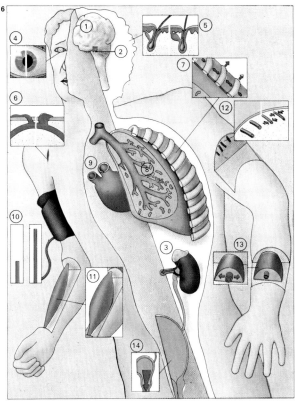

7 The pancreas, lying next to the duodenum [1], is a versatile gland producing both digestive enzymes and hormones. Pancreatic tissue consists of thousands of lobules [2], each containing glandular alveoli [3], where enzymes are produced and released to the duodenum by way of the main pancreatic duct [4]. Glucagon and insulin, the two pancreatic hormones, are produced in the islets of Langerhans [5] grouped around capillaries [6] within the lobules.

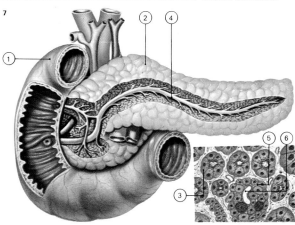

8 A schematic plan of the endocrine system shows the reciprocal action – or feedback mechanism – governing the release of hormones. Most influential is the anterior pituitary [1], secreting "trophic" hormones to trigger production in the thyroid [2], adrenal cortex [3], testes in the male [4] and ovaries in the female [5]. These then affect the hypothalamus [6] and the posterior pituitary [7] which in turn influence anterior pituitary activity.

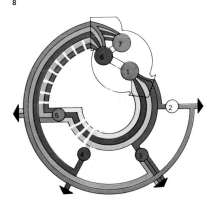

Breathing and the lungs

Respiration is not simply a matter of breathing in and out; it is something altogether more complex – the process by which energy is released from food. Respiration is, of course, to do with the chest, lungs and bloodstream; with breathing (ventilation) and the intake of oxygen and removal of carbon dioxide (gaseous exchange); but it also involves every other part of the body, because it is at the cellular level that the release of energy occurs. In its widest sense, respiration is one of the most important aspects of the chemistry of living matter.

The structure of the lungs

A continuous supply of oxygen is essential: deprived of oxygen, we rapidly lose consciousness and die. As the body surface is too small to absorb all the oxygen necessary to fuel millions of cells, humans and other complex animals have developed a special internal surface for breathing. This is the lining of the lungs, a vast expanse of up to 70sq m (753sq ft) in the adult – over 40 times the surface area of the body.

The additional value of this internal surface is that it remains permanently moist and is protected to a great extent from damage and from the intrusion of bacteria, fungi and other threatening organisms. Thin as tissue paper and interlaced with a network of fine-walled capillaries, the internal lining of the lung absorbs oxygen for rapid transfer to the haemoglobin of the blood. At the same time carbon dioxide, a waste product of cell metabolism, is released from the blood.

The lungs, which weigh about 1kg (2.2lb), are spongy and delicate rose-pink at birth, darkening to slate-grey or even black in later years. Covered with moist pleural membrane, they fill the deep thoracic (chest) cavity, corresponding to it in shape. The right lung consists of three lobes, the left lung – slightly smaller and indented to make room for the heart – has only two.

The lungs receive air, already warmed and filtered in the mouth or nose, by way of a series of pipes: the pharynx (throat), larynx ("voice-box") [3], trachea (windpipe) and, where the trachea forks, the right and left bronchi which enter the lungs. Inside the lungs the two primary bronchi branch first into smaller bronchi, then into bronchioles – "twigs" of the respiratory tree. In turn the bronchioles subdivide into alveolar ducts leading into the air sacs or alveoli.

How the lungs work

It is in these hollow air sacs, lying bunched like microscopic grapes deep inside the lungs, that the actual gas exchange takes place. Capillaries of the pulmonary artery enmesh the alveoli, bringing stale blood close to the fine surface membrane. The alveoli themselves secrete a thin film of liquid, with low surface tension, as a medium for diffusion. The reoxygenated blood then returns via the pulmonary vein to the heart so that it can be pumped once more throughout the body, refuelling distant cells with the essential oxygen that air contains [4].

When we breathe in, the thorax is enlarged due to a tightening of the muscles of the diaphragm (causing it to contract from a dome- to a saucer-shape) and of the rib-cage, which swings upwards and outwards. As the chest expands, the lungs must also expand, and the air inside them becomes rarefied,

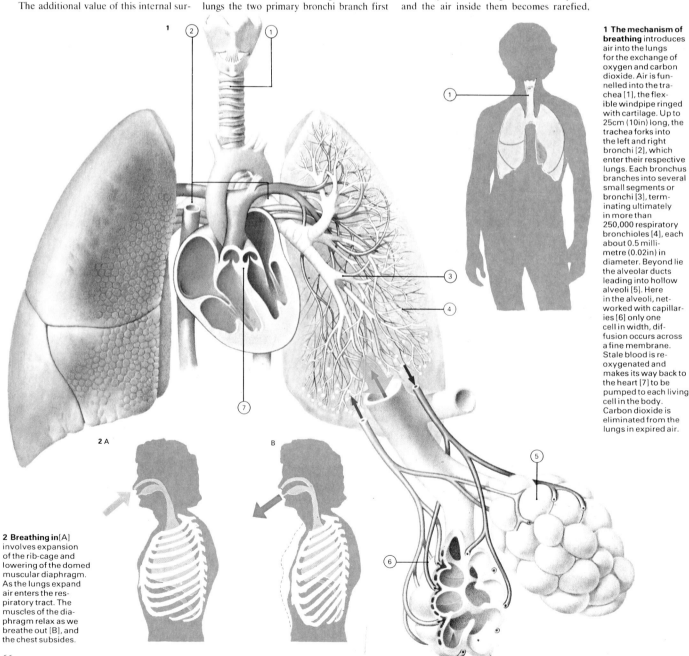

1 **The mechanism of breathing** introduces air into the lungs for the exchange of oxygen and carbon dioxide. Air is funnelled into the trachea [1], the flexible windpipe ringed with cartilage. Up to 25cm (10in) long, the trachea forks into the left and right bronchi [2], which enter their respective lungs. Each bronchus branches into several small segments or bronchi [3], terminating ultimately in more than 250,000 respiratory bronchioles [4], each about 0.5 millimetre (0.02in) in diameter. Beyond lie the alveolar ducts leading into hollow alveoli [5]. Here in the alveoli, networked with capillaries [6] only one cell in width, diffusion occurs across a fine membrane. Stale blood is reoxygenated and makes its way back to the heart [7] to be pumped to each living cell in the body. Carbon dioxide is eliminated from the lungs in expired air.

2 **Breathing in** [A] involves expansion of the rib-cage and lowering of the domed muscular diaphragm. As the lungs expand air enters the respiratory tract. The muscles of the diaphragm relax as we breathe out [B], and the chest subsides.

forming a low-pressure area [2]. Air comes rushing down the trachea to restore pressure. Tension in the diaphragm and rib-cage muscles slackens and the lungs revert to their resting shape, pulling the thorax back to shape with them and, as they recoil the air inside them is compressed and some of it is forced out through the trachea.

Quiet breathing – at intervals of from four to six seconds – involves intake of about 0.5 litre (0.75 pint) of air [1]. Perhaps a third of this lingers in the bronchi and bronchioles, while the remaining two-thirds finds its way to the alveoli. Similarly, we breathe out roughly 0.5 litre at a single expiration. But, with a residual volume of 1–1.5 litres (1.75–2.5 pints), the lungs are never completely drained of air.

Once air reaches the alveoli, oxygen is diffused into the bloodstream and carbon dioxide passes out from the blood to be expelled from the body. In addition to its main purpose – trading fuel for waste products – this exchange maintains the correct level of acidity in the blood. It is this acidity level that determines the rate of breathing. If breathing is too slow, a build-up of carbon dioxide causes slight acidification in the blood. This change in acidity level is monitored by special cells in the medulla of the brain and elsewhere, and these in turn signal deeper and faster breathing. In this way the acidity level is restored to normal.

Cleaning the air
To cope with the impurities in the air we breathe, the respiratory tract incorporates its own cleansing mechanisms. Supplementing filtration in the upper airways, the cilia – tiny hairs lining the bronchi and bronchioles – are at work all the time wafting mucus, laden with cell debris and foreign particles, upwards to the throat (this phlegm is swallowed and disposed of by way of the stomach). Larger particles, including germs, give an excess of mucus, and this has to be coughed up. Both the coughing and sneezing reflexes help blow detritus from the air passages. At the end of the network, the delicate alveoli are kept clean by patrolling phagocytes – scavenger white blood cells that engulf dust particles and bacteria.

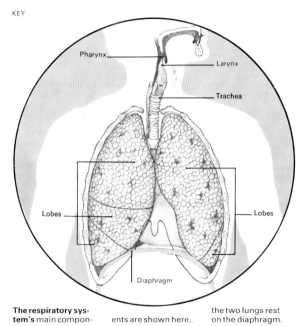

Pharynx — Larynx — Trachea — Lobes — Lobes — Diaphragm

The respiratory system's main components are shown here, the two lungs rest on the diaphragm.

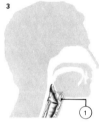

3 Air forced through the larynx or voice-box causes the vocal cords [1] to vibrate, producing sound. The vocal cords are fibrous strands that stretch between the thyroid cartilage [2] and the two moving cartilages [3]. The epiglottis [4] drops over the larynx during swallowing to keep out food. The trachea [5] is supported by rings of cartilage [6].

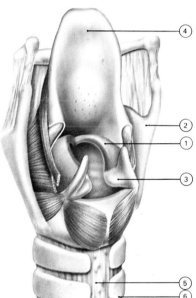

4 The lungs represent the hub of the respiratory cycle. Ventilation [A], a vital part of this cycle, involves: [1] exhaling stale air; [2] breathing in fresh air rich in oxygen; [3] the exchange of oxygen and carbon dioxide; [4] the elimination of carbon dioxide. A diagrammatic view of the alveolar surface [B], shows the rapid transaction of oxygen for carbon dioxide. Fresh blood is returned to the main circulation to fuel the body cells. Carbon dioxide is expelled, along with water vapour collected in the moist respiratory passages when we breathe out. For respiration to take place, chemicals and oxygen must react together; carbon dioxide and water, the end products of this reaction, must be removed [C]. The pulmonary artery conducts stale blood, dark with reduced haemoglobin, out of the heart. Reoxygenated in the lungs, and now bright red in colour, it returns to the heart by way of the pulmonary vein to be pumped afresh around the entire body.

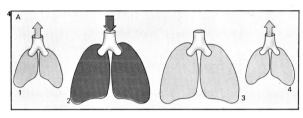

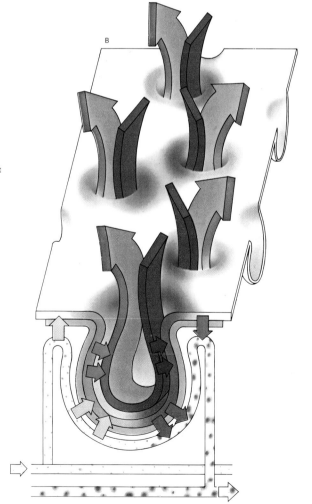

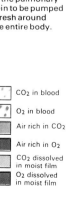

CO₂ in blood
O₂ in blood
Air rich in CO₂
Air rich in O₂
CO₂ dissolved in moist film
O₂ dissolved in moist film

The digestive system

Every one of the millions of living cells within the body needs energy to function – energy that is provided by the food we eat. Placed side by side ten million cells would measure only about 1 millimetre (0.04in), yet each cell is a complex biochemical factory which must be supplied with its food energy in a precise form. The vast variety of solids and liquids that are taken into the body therefore require a great deal of processing – both physical and chemical – before they can be assimilated by the cells and used in the vital processes of growth and metabolism.

The process of digestion
Digestion is the process whereby foods are gradually broken down into their basic components: proteins into amino acids (to provide the building blocks for new proteins); carbohydrates (starches) into simple glucose; and fats into fatty acids and glycerol to provide the body with energy. This involves groups of enzymes (the body's chemical catalysts), hormones (chemical messengers) as well as nervous impulses and controlled muscular action.

The intricacies of digestion are perhaps best understood by tracing the fate of a ham and lettuce sandwich during the 24 hours or so that it stays in the digestive tract [3]. This is a good example because it contains the three essential types of food: the ham is largely protein, the bread is rich in carbohydrate, the butter is fat and the lettuce provides a largely indigestible residue of cellulose.

As the sandwich is eaten it is chewed up and mixed with alkaline saliva, which is secreted by three pairs of salivary glands situated round the jaws. The saliva pours into the mouth via small ducts under the tip of the tongue and in the cheeks. [1].

At this stage digestion has already begun because saliva contains a digestive enzyme, known as amylase, which acts on starches (in this case the bread) and begins converting them to maltose, a soluble form of sugar.

Each mouthful is formed into a ball-like mass (bolus) which is pushed by the tongue to the back of the mouth and propelled down the gullet (oesophagus) into the stomach. It is conveyed not merely by gravity but by peristalsis – strong, rhythmic, wave-like contractions of the wall of the oesophagus.

When it enters the stomach, the food is thoroughly mixed with gastric juice by the churning action of the stomach's strong muscular walls. Between one and two litres (two to four pints) of juice are secreted daily by glands in the stomach lining [2]. This juice consists of strong hydrochloric acid, protective mucus and pepsin, the first digestive enzyme to act on protein in the food (the ham in the sandwich).

After an hour or so the sandwich has been reduced to a pulp (chyme) and is ready to go into the duodenum, the first short curved section of the small intestine.

In the small intestine
The small intestine is so called because of its diameter rather than its length, which totals some 7m (about 23ft). The bulk of digestion and absorption into the surrounding blood supply takes place within its length.

Chyme passing from the stomach into the small intestine encounters another dramatic change of environment, this time from acid to alkaline, as it mixes with digestive juices

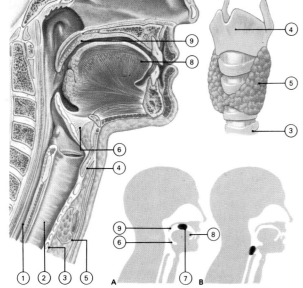

1

1 The mouth is a gateway not only to the digestive system but also to the respiratory system. The oesophagus [1], lies behind the trachea [2] which is supported by cartilage rings [3]. The larynx is at the top of the trachea and its front is formed by the thyroid cartilage [4], so called because of its proximity to the thyroid gland [5]. The flap-like epiglottis [6] is attached to this cartilage. Food, mixed with saliva, is formed into a bolus [7] and pushed by the tongue [8] into the pharynx. During swallowing [A, B] the soft palate [9] blocks entry to the nose and the epiglottis closes.

2 The digestion and absorption of food takes place within the digestive tract, a coiled tube some 10m (33ft) long which links mouth to anus. Food is passed down the oesophagus [1] to the stomach [2] where it is partially digested. Chyme is released into the duodenum [3] – the first part of 7m (23ft) of small intestine. The duodenum receives bile secreted by the gall bladder [4] in the liver [5] and enzymes secreted by the pancreas [6]. Most absorption occurs in the jejunum and ileum, the remaining parts of the small intestine [7]. Any residue passes into the caecum [8], the pouch at the start of the large intestine. At one end of the caecum is the 10cm (4in) long vermiform appendix [9] which serves no useful purpose in man. Water is reabsorbed in the colon [10]. Faeces form and collect in the rectum [11] before being expelled as waste through the anus [12]. Many of the wastes that result from cell metabolism are filtered from the blood by the kidneys [13] into urine. This passes down the ureters [14] to the bladder [15] where it is stored until voided via the urethra [16].

2

Muscular contraction of the stomach wall churns the food, mixing it with gastric juice which includes hydrochloric acid. This enables the enzymes to act and at the same time kills many of the microbes taken in with the food

Acid chyme squirts into the duodenum through the pyloric sphincter and is neutralized by alkaline secretions of the pancreas which enter the intestine through the same duct as bile from the gall bladder

The pancreas secretes important enzymes: trypsin and chymotrypsin (which act on protein); amylase (which acts on starch); and lipase (which acts on fats). Fat digestion is also aided by the action of bile

Digestion is completed in the small intestine. Further enzymes mix with the food, which is ultimately reduced to amino acids (from proteins), sugars (from starches) and glycerol and fatty acids (from fats)

The passage of residue through the colon produces putty-like faeces. Their odour is due to bacterial action and their colour to the presence of bile pigment. They are expelled via the rectum and the anus

The residue that passes into the colon contains much water that is reabsorbed into the bloodstream. The colon also contains bacteria that break down cellulose

3

24 hours

3 A typical time scale for the passage of food through the digestive tract relates each part of the system to the time on the clock. Correct timing is essential for two reasons: food must stay in the stomach and small intestine long enough to allow complete breakdown of protein, fat and carbohydrate; and the residue must pass through the large intestine slowly enough to allow water to be reabsorbed into the body.

Mouth
Oesophagus
Small intestine
Stomach
Large intestine
Caecum
4
8·5
9·5

All blood passes through the kidneys where waste materials are filtered out. This is achieved by more than two million kidney units or nephrons. It has been estimated that in the process between 170 and 200 litres (37·4 and 43·9 gallons) of fluid are filtered out every 24 hours of which 99 per cent is reabsorbed

pouring into the duodenum from the pancreas and gall bladder.

The pancreas is a gland about 18cm (7in) long. It plays a vital role in digestion by secreting digestive enzymes that act on carbohydrate, protein and fat.

The digestive process is further aided by the emulsifying effect of bile, a thick, green, bitter liquid formed by the liver and stored in the gall bladder.

The lining of the small intestine also secretes a whole range of digestive enzymes, collectively known as the succus entericus, and protein, carbohydrate and fat all finally become amino acids, glucose and glycerol and fatty acids.

Absorption of these digested substances into the bloodstream is the next essential step in the provision of nutrients to the cells. The lining of the small intestine is heavily folded – to increase its surface area – with thousands of minute finger-like projections known as villi [4, 6]. These waft to and fro in the intestinal contents, which are thus brought into intimate contact with the rich supply of blood capillaries and lymphatic vessels that lies within each villus.

Amino acids and glucose diffuse across the membrane of the villus into the blood capillaries and the fatty acids and glycerol enter the lymphatic cells to be transported to the liver which acts as the body's chief storehouse and also as a breakdown plant and as its chemical factory.

Eliminating waste

By now the sandwich has been digested, absorbed and processed by the liver. But most of the lettuce remains as indigestible cellulose at the end of the ileum. It passes together with great quantities of gastric secretions and other debris, through the ileocaecal valve into the caecum, the pouch at the start of the colon, the main part of the large intestine. The principal function of the lining of the large intestine is the retrieval of water and important chemicals by reabsorption into the bloodstream. The colon disposes of the waste material – all that remains of the ham sandwich eaten many hours before – through the anus in the form of faeces, a large percentage of which is dead bacteria.

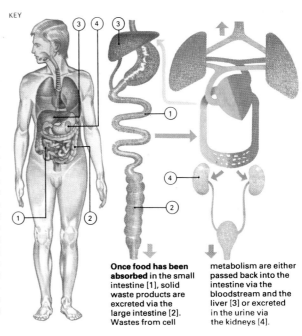

KEY

Once food has been absorbed in the small intestine [1], solid waste products are excreted via the large intestine [2]. Wastes from cell metabolism are either passed back into the intestine via the bloodstream and the liver [3] or excreted in the urine via the kidneys [4].

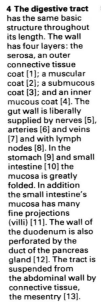

4 The digestive tract has the same basic structure throughout its length. The wall has four layers: the serosa, an outer connective tissue coat [1]; a muscular coat [2]; a submucous coat [3]; and an inner mucous coat [4]. The gut wall is liberally supplied by nerves [5], arteries [6] and veins [7] and with lymph nodes [8]. In the stomach [9] and small intestine [10] the mucosa is greatly folded. In addition the small intestine's mucosa has many fine projections (villi) [11]. The wall of the duodenum is also perforated by the duct of the pancreas gland [12]. The tract is suspended from the abdominal wall by connective tissue, the mesentry [13].

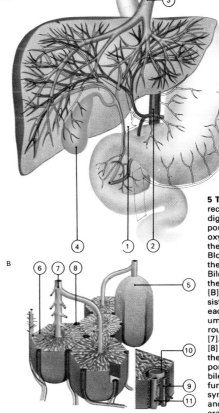

5 A

B

5 The liver [A] receives products of digestion via the portal vein [1] and oxygenated blood via the hepatic artery [2]. Blood is drained by the hepatic vein [3]. Bile is stored in the gall bladder [4]. [B] The liver consists of lobules [5], each formed by columns of cells [6] round a central vein [7]. Portal tracts [8] carry branches of the hepatic artery [9], portal vein [10] and bile duct [11]. Liver functions include synthesis of fats and storage of minerals and vitamins.

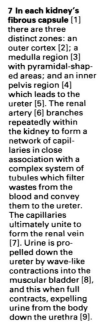

6 The small intestine's internal surface area is greatly increased by the presence of microvilli, microscopic structures covering the surface of each villus. Amino acids and sugars pass into the capillary network [1]; fatty acids and glycerol enter the lymphatic system via a central lacteal [2].

7 In each kidney's fibrous capsule [1] there are three distinct zones: an outer cortex [2]; a medulla region [3] with pyramidal-shaped areas; and an inner pelvis region [4] which leads to the ureter [5]. The renal artery [6] branches repeatedly within the kidney to form a network of capillaries in close association with a complex system of tubules which filter wastes from the blood and convey them to the ureter. The capillaries ultimately unite to form the renal vein [7]. Urine is propelled down the ureter by wave-like contractions into the muscular bladder [8], and this when full contracts, expelling urine from the body down the urethra [9].

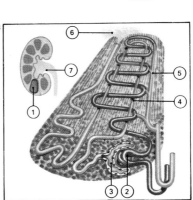
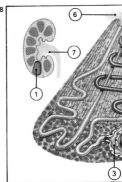
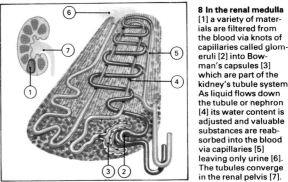

8

8 In the renal medulla [1] a variety of materials are filtered from the blood via knots of capillaries called glomeruli [2] into Bowman's capsules [3] which are part of the kidney's tubule system. As liquid flows down the tubule or nephron [4] its water content is adjusted and valuable substances are reabsorbed into the blood via capillaries [5] leaving only urine [6]. The tubules converge in the renal pelvis [7].

A healthy diet

All living things need food to grow and to maintain life. Man must eat to obtain the energy that fuels his muscular activity and his basic bodily processes – tissue growth, breathing, heartbeat and so on. The aim of good nutrition is a balanced diet [Key] supplying him with just as much energy as he needs and no more.

Energy sources

Foods vary widely in their energy value [1]. The most commonly known unit for measuring these differences is the kilogramme calorie or kilocalorie (Cal) – the amount of heat needed to raise the temperature of a kilogramme of water one centigrade degree. In most diets, the principal energy foods (though not necessarily the highest in calorific value) are carbohydrates, the simplest of which are sugars like glucose. A molecule of sucrose (cane or beet sugar) consists of two simple sugar units chemically linked together. Starch molecules are chains of up to a few hundred such units. Most of the staple foods which provide the bulk of people's diets are plant organs containing starchy food reserves; for example, cereal seeds like wheat, rice, millet and maize, or tubers like potatoes and yams.

Energy is not available from all carbohydrates. Cellulose is a carbohydrate occurring in all green plants, making up cell walls, the tough fibres of leaves and stems and the outer coats of cereal grains (bran); these materials are almost unchanged in the human gut. Although without food value, this roughage is useful in helping the passage of food through the digestive system.

Fats are energy-rich foods and certain constituents of fats are probably essential in a balanced diet. They are important foods in wealthy countries, but people living in poverty may eat very little fat and sometimes suffer from a lack of fat-soluble vitamins, especially vitamins A and D.

How much fat is desirable in a healthy diet has still to be established. Many believe that there would be nutritional advantages in a reduction in the high intake of fats common among prosperous communities and in the replacement of some animal fats with those of vegetable origin.

Protein molecules are large assemblages of simple units called amino acids. Different proteins contain varying proportions of the 20 or so common amino acids. During digestion, food proteins are broken down into their constituent units, which are absorbed into the blood and then reconstructed into the various proteins needed by the body, including enzymes, contractile muscle protein and blood-plasma proteins.

Essentials of health

Some amino acids are interconvertible but about ten are synthesized in the body only very slowly. Sufficient of these "essential amino acids" must therefore be present in any balanced diet. Animal proteins like those in meat, fish and eggs are rich in essential amino acids. Cereals and vegetables contain comparatively little protein; most plant proteins, moreover, are relatively low in essential lysine and methionine. Many poor people rely largely on plant foods, with meat and fish as rare luxuries, and their protein intake is thus both restricted and of poor quality [7]. Protein-deficiency symptoms [6] usually

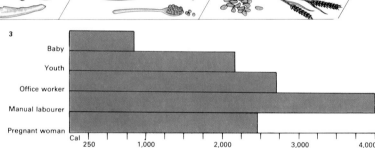

2 Certain foods are important sources of carbohydrate [A], fat [B], protein [C] or roughage [D]. Except for some refined products like cane sugar, most foods contain several different nutrients.

Milk provides carbohydrate, fat, protein, calcium and some vitamins. Even potatoes contain protein and vitamin C as well as energy-rich carbohydrate.

1 Weight for weight pure fat supplies 2.25 times more energy than carbohydrate [A]. Any protein surplus to the body's need for tissue growth and repair can also supply energy. Simply to keep their metabolism going, children use far more energy per kilogramme of their body weight than older people do [B]; they also need much more protein in proportion to their size.

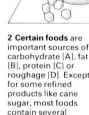

1 A

Carbohydrates — Proteins — Fats — Cal

9
8
7
6
5
4
3
2
1

Production of energy per gram

1 B

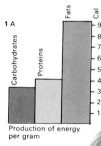

3

	Cal			
Baby				
Youth				
Office worker				
Manual labourer				
Pregnant woman				

Cal 250   1,000   2,000   3,000   4,000

3 People's daily energy needs vary widely. At work, a labourer uses more energy than a clerk does. A big man will consume more energy walking a kilometre than will a child or a small woman. A restless, excitable person may also need more energy than his placid brother.

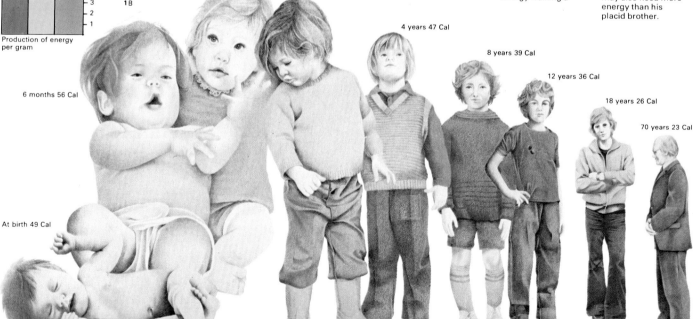

6 months 56 Cal
1 year 54 Cal
2 years 51 Cal
4 years 47 Cal
8 years 39 Cal
12 years 36 Cal
18 years 26 Cal
70 years 23 Cal
At birth 49 Cal

appear in such communities, especially among young children, who need plenty of protein during the process of growing and building up new cells and tissues.

Dietary influences

A diet of carbohydrates, fats and proteins alone will not keep people healthy. A wide range of organic compounds, collectively called vitamins [4], and a number of inorganic substances – minerals [5] – are essential in a balanced diet, though many of them are needed in minute amounts only. Mineral and vitamin deficiencies occur even among otherwise well-fed people, especially if they choose a restricted range of foods rather than a varied diet. Iron-deficiency anaemia, for instance, is widespread in the developed countries. Many governments therefore require by law the fortification of certain staple foods. Calcium and iron are added to bread-making flour and vitamins A and D, normally present in butter, are added to the margarine that may replace it. Fluoride is sometimes put into water supplies that do not naturally contain the element and iodine is

added to common table salt in some districts.

People's customs and taboos can influence their diet in ways quite unconnected with the needs of their bodies. Many people like the sweet taste of sugar, for instance, and eat large amounts, even while being aware of the probable consequences – obesity and dental decay. Most Europeans refuse to eat caterpillars and a devout Hindu will not eat beef, although caterpillars and beef are both nourishing high-protein foods. Only in extreme conditions, such as famine, are such behaviour patterns likely to break down. Physical idiosyncrasies can also affect diet by making certain foods unacceptable: wheat and wheat products are poisonous to people with coeliac disease, while in parts of Asia and Africa, there are many adults who, though quite healthy, cannot digest milk.

When a meal is eaten, the proportion of nutrients actually absorbed by the body depends partly on the efficiency of digestion. This is reduced by anxiety, grief, weariness or ill-humour and enhanced by comfort and tranquillity. Digestion also improves when the meal is eaten with enjoyment.

A balanced diet for a fairly active man would supply about 3,000 calories daily with adequate protein, minerals and vitamins. The caloric value of foods common in Western diets makes this intake comparatively easy to achieve.

495 Cal
55 Cal
440 Cal
150 Cal
360 Cal
396 Cal
12 Cal
255 Cal
207 Cal
24 Cal
220 Cal
218 Cal
25 Cal
80 Cal
136 Cal
42 Cal
195 Cal

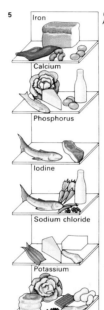

4 Vitamin K, Vitamin D, Vitamin A, Vitamin E, Vitamin B₂, Vitamin B₁, Vitamin C, Nicotinic acid, Vitamin B₁₂, Folic acid

4 Vitamins essential to health are found in great quantity in liver, milk, green vegetables, legumes and corn. Fatty foods provide the chief source of vitamins K, D, A and E, which are soluble in fats and oils. Vitamin K is produced by micro-organisms in the gut, while any ordinary diet contains adequate vitamin E. Symptoms of deficiency of these two vitamins cannot therefore be positively identified. Some vitamin D is produced in the skin on exposure to sunlight, but this is insufficient except in sunny climates. Children receiving too little vitamin D cannot absorb calcium properly and their bones become soft and deformed, with swollen joints. This condition (rickets) is also produced by lack of calcium. Vitamin A deficiency leads to eye damage with eventual ulceration and blindness (keratomalacia). Though they are essential nutrients, vitamins A and D are poisonous in large doses. The vitamin B group of water-soluble compounds, including niacin and folic acid, is involved in cell enzyme systems. As vitamin B is generally found in meat and the bran of cereals, deficiencies of it are more likely in poor communities where staple cereals are milled and little meat is eaten. Lack of niacin produces skin eruptions and mental disorder (pellagra). Vitamin C, found in fresh vegetables or fruit, is easily lost in cooking and storage. Lack of it caused sailors to suffer from scurvy, with swollen gums and internal bleeding.

5 Iron, Calcium, Phosphorus, Iodine, Sodium chloride, Potassium, Fluorine

5 Vital minerals are found in several foods. Those supplying iron help oxygen exchange. Calcium and phosphorus are major components of bone, together with magnesium. Iodine is needed for the hormone controlling metabolic rate. Although sodium chloride (salt) is found in most diets, extra quantities may be needed in hot countries where people lose salt in sweat. Potassium is important in nerve and muscle function and fluoride, found chiefly in drinking water, protects children's teeth.

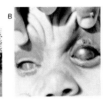

6 Deficiency diseases caused by lack of vitamins include rickets [A], keratomalacia [B], pellagra [C] and scurvy [D]. Kwashiorkor [E] appears in small children who lack protein, often on abrupt weaning. The child's stomach is swollen and its muscles wasted. Growth may be stunted and the brain damaged. Iodine deficiency causes goitre, a chronic neck swelling [F].

7 The average daily diet of a child in India (shown here) is poor, consisting of rice with some vegetables and dried fish. Since this lacks essential proteins and vitamins, health and life expectancy is low.

8 An over-rich daily diet produces a surplus that the body stores as fat. Obesity is the most common kind of malnutrition in wealthy societies fond of fatty and sugary foods.

9 New sources of protein for the world's expanding population are being investigated. Farming of fish and big game like antelope is increasing. Geneticists have developed more efficient livestock breeds and new cereals like triticale, which is a heavy-yield wheat-rye hybrid. Yeasts and bacteria from petrochemicals or industrial wastes can be made into animal feeds but the products are not necessarily suitable for human consumption.

Keeping fit

The purpose of "keeping fit" is to avoid ill health, resist the mental and physical fatigue that makes us vulnerable to infections and feel more pleasure in being alive. Research the world over affirms the value of a balanced diet, fresh air, sunshine, adequate rest and some form of regular exercise for everyone, male or female, young or old, fat or thin. There is no doubt that if such a regime were universally followed the numbers of people seeking medical assistance would plunge dramatically, hospitals would be less crowded and many debilitating and fatal illnesses could be avoided.

Unfortunately, although the advantages of physical fitness are generally acknowledged, modern man tends to endorse them from his armchair; sedentary occupations, spectator sports, labour-saving devices, mechanical transport – these all tend to encourage more preaching than practice.

The heart of the matter
One of the reasons for the longevity and good health of the physically fit is that they do not put the heart under excessive strain [Key]

and so the possibility of heart and blood-vessel diseases is reduced. They also avoid the consequences of weakened muscles which are the cause of so much back and abdominal trouble.

Some 640 assorted muscles account for about 45 per cent of our body weight, and for top efficiency and health these must have strength, endurance (or the ability to store energy) and elasticity, and be kept continually supplied with fuel by the blood. Sensible exercise, carefully suited to each individual (preferably on the advice of a doctor), approached slowly at first and only gradually building up to a sustained level of steady benefit, is the best insurance that all these requirements will be properly met.

Working out the best method
Most people have learned basic exercises in school or gymnasium – stretching, running on the spot, lifting weights and so on – and these are still fundamentally useful, but in recent years several new keep-fit techniques have proved especially popular and effective. Isometrics, developed by the US Marine

Corps, is a form of exercise without movement in which one muscle group is fully exerted against another tensed or contracted muscle group for approximately five to eight seconds [2]. The exercise strengthens or "tones" a particular muscle or set of muscles. This form of exercise is suitable for both men and women, is safe to carry out without supervision and can be performed comfortably during the most crowded schedule and almost anywhere – even sitting at a desk.

Exercises involving vigorous movement and extended physical effort – such as running, jumping, fast walking, chopping wood and most competitive sports – are called isotonics. They can be combined with isometrics for the development of both strength and endurance. This combined method is used widely by athletes in training programmes.

The Royal Canadian Air Force has also developed a famous keep-fit plan of exercises: 5BX, an 11-minute-a-day plan for men, and XBX, a 12-minute-a-day plan for women [4]. These require no special equipment or environment and are well suited to urban living. The selection of exercises is

CONNECTIONS

See also
70 A healthy diet
66 Breathing and the lungs
90 Diseases of the circulation

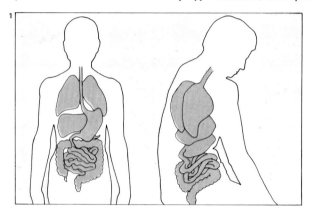

1 Despite his upright posture for hundreds of thousands of years, man still shows the strain of this "new" position by slumping forwards, curving his back, rounding his shoulders and shrinking his chest into his abdomen. The results are neck and back pains and hindered breathing, in turn causing oxygen starvation and impeded blood circulation. Poor posture can lead to sagging or prolapse of stomach muscles; and long hours of sitting, to constipation and haemorrhoids. Conversely, over-rigid posture can strain the spine and cause lumbago, fibrositis and liver troubles.

2 Isometric exercises need only take a few moments and can be done almost anywhere, either sitting, standing or lying down. Props may be just as casual: a towel, desk, doorway or table can all be useful aids for creating the dynamic tension that strengthens muscles. A doorjamb assists the contraction that helps a woman [A] raise her bust-line and also the contractions that firm up leg and stomach muscles [B]. Palms pressed down with great force on a desk or table [C] strengthen the upper arms and, for a woman, raise the bust-line. Regular isometric exercise should quickly tone up any muscle.

3 The aim of yoga exercises is precision and slow harmonized motion, rather than vigour and force. Each one is performed once only and all are combined with deep inhalation and exhalation. The cobra firms breasts, increases spinal flexibility and tones abdominal muscles. The cat tones throat and chin-line and stretches the back and stomach muscles. The half locust firms thighs and buttocks, while the half plough flattens stomach muscles. The pose of the camel makes the spine flexible, strengthens the neck and firms the jaw line. The half shoulderstand invigorates and improves the circulation. Yoga exercises should always be learned with the help of a trained teacher.

The cobra

The cat

The half locust

The half plough

The pose of the camel

The half shoulderstand

carefully formulated to effect an overall tuning of muscles and the gradual building up of stamina.

For men, the 5BX plan is composed of six charts of five exercises each, variously modified for differing ages and gradually increasing in difficulty and effort demanded. XBX, for women, is composed of four charts of ten exercises similarly arranged for gradual progress so that there is no undue strain, stiffness or soreness. Properly followed, these programmes maintain weight balance, improve the appearance and keep the body trim, strong, and energetic.

Yoga and the sauna

The exercises or "postures" that are one part of the ancient Hindu philosophical system of yoga have been growing more and more popular in the West. Originally designed to relax the mind and body for extended meditation, they emphasize harmonious co-ordination of movement, stretching and correct breathing. Even a few of the exercises, practised for ten or 15 minutes a day, will improve the whole physical system and induce a state of relaxation and well-being [3]. They should, however, be learned under the supervision of a teacher with a full understanding of their overall aim, and should never be attempted too quickly, without adequate preparation by the gradual loosening of unused muscles.

Other methods of keeping fit can be combined with the foregoing – massage [5], for instance, can provide a daily tonic for slack muscles and sluggish circulation, as can the Scandinavian sauna [6], which, if properly used, is invaluable for cleaning the skin of impurities, calming the nerves, promoting good circulation, invigorating muscles and joints and relaxing the entire system.

Ordinary everyday activities pursued with vigour, and any recreation that exercises muscles and stimulates circulation, such as dancing, swimming, riding, cycling, golf or tennis, are excellent means of keeping fit. The main objective should be the balancing of all ingredients – rest, calorie intake in proportion to energy output, maintained good posture and breathing, regular exercise and, above all, an optimistic state of mind.

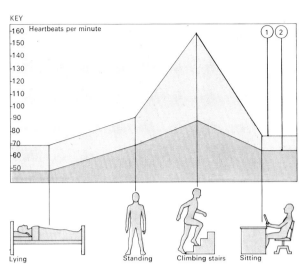

When a person is physically unfit, his heart is given an extra work load as shown in the number of heartbeats per minute required for various everyday activities. The comparison between someone in poor condition who takes little or no exercise [1] and a person who follows a programme of regular exercise [2] shows the value of keeping fit.

4 The exercises for both men and women developed by the Royal Canadian Air Force are carefully graded to increase fitness gradually and without fatigue in 11 to 12 minutes a day. A few are shown here.

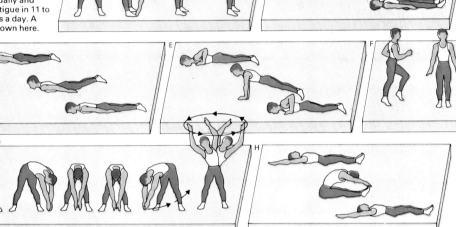

A   Feet astride, arms upward; then touch the floor and straighten.
B   Lie flat on floor, feet apart, then raise head 15cm (6 ins). off the ground.
C   Lie on front, palms under thighs, then raise head and each leg alternately.
D   Lie on front, palms under thighs, then raise head and both legs together.
E   Press up slowly and evenly from floor until arms are fully extended.
F   Running on the spot, interspersed with astride jumps.
G   Feet astride, arms upward, then make full circle with hands touching the ground outside each foot.
H   Lie on back, arms outstretched, raise trunk.

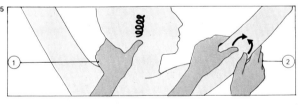

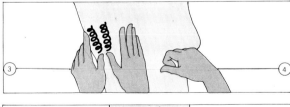

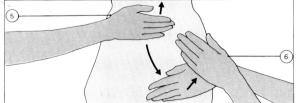

5 Massage, the manipulation of body tissues, is an ancient keep-fit method which can be physically and psychologically therapeutic. Among the methods that are most widely used, often in combination, are: friction (circular movements) [1, 3]; petrissage (kneading) [2]; tapotement (tapping) [4, 6]; and effleurage (stroking) [5].

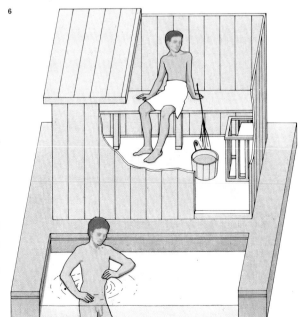

6 In a sauna bath stones are heated on top of a stove to produce a dry heat of 80-110°C (176-230°F) Cold water thrown on stones results in steam making the bath feel hotter, although in fact it is not. There is a 10°C (50°F) difference between the floor and the ceiling. The Finns use bundles of twigs to tone up the skin. A cold plunge follows the sauna.

Reproduction

Sex has two attributes that are sometimes separate and sometimes indissolubly linked. One is strictly physiological – the making of a new human being. The other is emotional – the expression of the affection, love and passion of two people. Few cultures have sought to make children without, at least, affection; many cultures have sought to make love without making children as a necessary consequence.

The male and female reproductive systems
Reproduction is possible only if a female germ cell (the egg or ovum) is fertilized by a male germ cell (the spermatazoon). The reproductive system of the human female is designed for the monthly production of these eggs by the ovaries and for the accommodation and nurturing in the uterus of the growing fetus for the nine months until birth. The male system is required to produce sperm and transfer them to the female reproductive tract [Key] where they can come into contact with an ovum.

The female external genitals are known collectively as the vulva. At their front limit is the mons veneris or mons pubis, a fatty mound over the pubic bone. Running down from the mons veneris are two folds of skin, the labia majora, which surround two smaller folds, the labia minora. Behind these lies the clitoris, a small organ which is an important source of arousal and corresponds to the male penis.

The vaginal opening lies within the labia and is almost closed off in young girls by the hymen, a thin membrane normally torn the first time that a woman has intercourse, though it can be ruptured earlier by violent exercise or injury.

The vagina [2] is a muscular tube some 10cm (4in) long which grips the penis during intercourse. It is here that semen is deposited at ejaculation and the sperm must then pass through a narrow neck or cervix before entering the pear-shaped, 8cm (3in) long uterus. Two Fallopian tubes about 10cm (4in) long connect the uterus with the two walnut-sized ovaries that are situated within the protection of the woman's abdomen. Every 28 days the ovaries shed one ripe egg, which enters one of the Fallopian tubes and passes down into the uterus. The ovaries are also responsible for the production of the female sex hormones: progesterone and oestrogen essential to fertilization.

Most of the male reproductive system [1] lies on the outside of the body. The visible parts are the penis and the testes, which are suspended in the bag-like scrotum. Normally the penis is limp and flaccid but it becomes erect when the man is sexually excited. Erection [3] is brought about inside the penis by spongy tissue, called corpora cavernosa, becoming gorged with blood. The two testes are continually producing sperm within their numerous tiny coiled tubes and these sperm are then stored in a long tube, the epididymis, which winds over the surface of each testis. The bulk of ejaculated semen, however, is composed of a fluid which is produced by the seminal vesicles, the prostate gland and Cowper's gland which lie within the body.

Sperm, egg and fertilization
In a single emission of semen a man may release about 250 million sperm. However, only a few hundred of these minute tadpole-

1

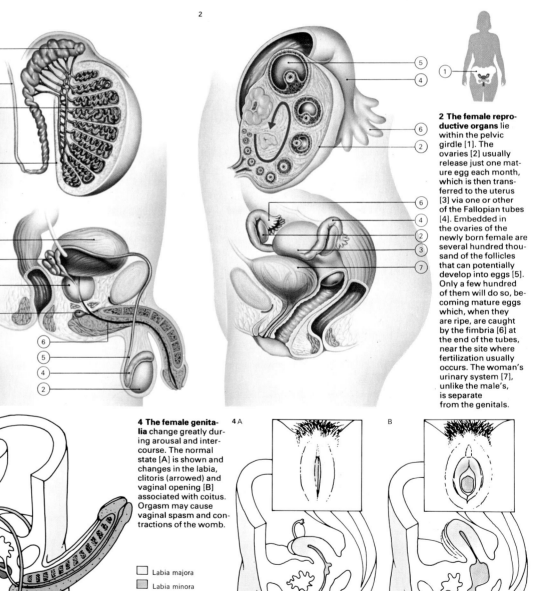

1 The male reproductive system is situated both inside and outside the pelvic region [1]. Outside the body are the testes [2], which produce every day millions of sperm in the convoluted seminiferous tubules [3]. The sperm are formed from the cells lining the outer wall. They mature gradually and are pushed towards the centre by newly formed cells. Once ready they are stored in the adjoining epididymis [4]. During coitus, these pass along the vas deferens [5] to the urethra [6], which also carries urine from the bladder [7]. The seminal vesicles [8], prostate [9] and Cowper's glands [10] all secrete fluids into the urethra, helping to make semen.

2 The female reproductive organs lie within the pelvic girdle [1]. The ovaries [2] usually release just one mature egg each month, which is then transferred to the uterus [3] via one or other of the Fallopian tubes [4]. Embedded in the ovaries of the newly born female are several hundred thousand of the follicles that can potentially develop into eggs [5]. Only a few hundred of them will do so, becoming mature eggs which, when they are ripe, are caught by the fimbria [6] at the end of the tubes, near the site where fertilization usually occurs. The woman's urinary system [7], unlike the male's, is separate from the genitals.

3 Just before orgasm, the penis, which is at its maximum length, exudes drops of seminal fluid. These may contain viable sperm which is why withdrawal is ineffective as a contraceptive method. On orgasm, muscular contractions occur at the base of the penis and bring about ejaculation.

3

- Erectile tissue
- Vas deferens
- Prostate
- Seminal vesicle
- Epididymis
- Testis

4 The female genitalia change greatly during arousal and intercourse. The normal state [A] is shown and changes in the labia, clitoris (arrowed) and vaginal opening [B] associated with coitus. Orgasm may cause vaginal spasm and contractions of the womb.

4 A   B

- Labia majora
- Labia minora
- Vagina
- Uterus
- Clitoris

like cells, with flattened heads and long tails, will actually reach the egg high up in the Fallopian tube and only a single sperm will penetrate the egg to produce a viable zygote [6]. After penetrating the membrane of the ovum the sperm loses its tail and middle section as it enters the protoplasm. The head swells in size and is now known as the male pronucleus. The nucleus of the ovum undergoes similar changes and then both pronuclei fuse together. Fertilization is now complete and the zygote begins to divide into a number of cells, at the same time moving down the Fallopian tube to the uterus. This journey takes about a week to complete, by which time the fertilized ovum has developed into a ball of 32 or 64 cells. The ball fills with fluid and the cells lie on the surface. It is at this stage of development that the young embryo, called the blastocyst, attaches itself to the lining of the uterus. If the ripe egg is not fertilized then it is this uterine lining that is shed during menstruation, which occurs, on average, every 28 days [5].

The monthly shedding of the egg by the female occurs between the ages of puberty, at about 12 years, and menopause in the middle forties – a lifetime total of maybe only 500 eggs are produced by the ovaries.

Physiological changes during coitus
Only in recent years have the physiological changes that occur during coitus been scientifically studied. The initial excitement phase is caused by imagination, stimulation of the senses and close body contact. Once aroused, the penis becomes erect and the vagina moist and distended [4]. During the following plateau phase tension and excitement mount; if stimulation continues and is not deliberately controlled, orgasm results and tension then subsides.

Infertility, the inability to conceive, is variously caused. In about 40 per cent of cases it is due to male sterility. In women it is due to hormonal deficiencies or physical obstruction. Often childless couples are subfertile and surgery or hormonal treatment (fertility drugs for the female) change the situation. Impotence, the inability to perform coitus, can be caused physiologically (through disease or drugs) or emotionally.

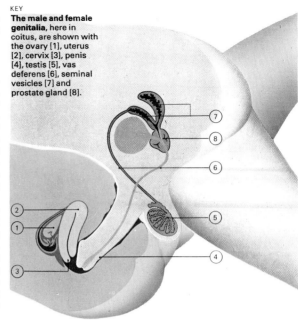

5 The menstrual cycle lasts on average 28 days and occurs regularly from puberty to the menopause. A ripened follicle in an ovary bursts open, releasing a single, minute egg. Empty, the follicle now turns into a corpus luteum (yellow body) and secretes two hormones, progesterone and oestrogen, under the influence of LH (luteinizing hormone) produced by the pituitary gland. These work together in the bloodstream to build up the endometrium or uterine lining. If the egg is fertilized, a rich bed is ready for it. If not, during the next 14 days, the corpus luteum withers, ceasing hormone production. The uterine lining then breaks down, causing the menstrual flow of blood to pass through the vagina. A new cycle of some 28 days has now begun. FSH (follicle stimulating hormone) promotes new follicle development.

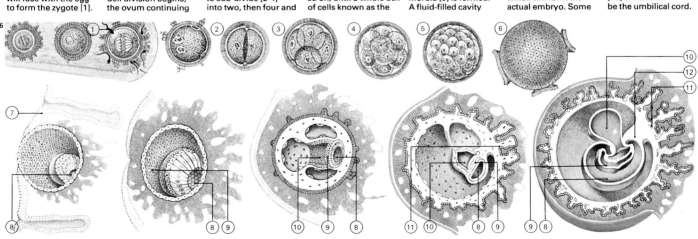

6 Fertilization usually occurs high in the Fallopian tube. Many sperm reach the ovum, but only one will fuse with the egg to form the zygote [1]. Cell division begins, the ovum continuing to sub-divide [2-4] into two, then four and so on, until a whole ball of cells known as the morula [5] is formed. A fluid-filled cavity develops within the structure, now termed a blastocyst [6]. One week after fertilization, it reaches the uterus and enters the lining [7]. An amniotic cavity [8] develops, and the ball of cells splits to become the two-layered disc [9] which will form the actual embryo. Some six days later, the bottom layer of this germ disc grows to form the yolk sac [10]. Outer cells form the placenta [11]. The amniotic cavity grows, surrounding the germ disc until it is joined to the outer layer by the body stalk only [12] which will be the umbilical cord.

Pregnancy

Pregnancy is the nine-month period during which a single cell is transformed into a human being. A missed menstrual period is usually considered the first sign of pregnancy, but there could be other reasons for a break in the cycle – emotional factors, ill health or the approach of the menopause. The most reliable tests are the biological and immunological tests carried out on a sample of urine. A pregnant woman's urine contains chorionic gonadotropin (CG), a hormone produced by the placenta. In the biological tests a little urine is injected into a young mouse, rat or toad which will ovulate in a few hours if CG is present. In the immunological tests CG inhibits the combination of rabbit blood serum containing CG antibodies with particles of red blood cells coated with CG.

Early stages

After confirmation of the pregnancy, it is then relatively straightforward to calculate the approximate date of birth, as long as the cycle is the normal 28 days you merely count forward nine months and seven days from the first day of the last menstrual period.

Even at this early stage the expectant mother is generally advised to go to her doctor or clinic for examination. Height and weight will be measured, blood pressure recorded and the breasts examined. An internal examination is also common practice in order to detect any infection or abnormality such as the uterus being tipped backwards or an ovarian cyst. Usually, no other internal examination will be given until the 36th week of pregnancy.

Blood tests will be taken at regular intervals. The mother's blood group must be noted in case, at any later stage, a transfusion becomes necessary. The tests will also tell whether the mother is anaemic and, further, whether she is Rhesus positive or Rhesus negative. Most people have an Rh (Rhesus) factor in their blood and are therefore termed Rhesus positive. But some do not. They are the Rhesus negative group. Special care is needed in the case of an Rh negative woman having a child by an Rh positive man. If the baby's grouping turns out to be Rh positive, then any of the baby's blood which chances to enter the mother's circulation through the

placenta will cause production of antibodies. In themselves the antibodies do no harm, but during a subsequent pregnancy these antibodies may get into an Rh positive baby's circulation, causing anaemia and jaundice. Potential danger can be avoided by giving an affected child transfusions at birth (or, in extreme cases, while still in the uterus) or by giving the mother a special injection to prevent these antibodies being formed.

Physical changes

During the 40 weeks of pregnancy there is an increase in the breast size due to hormone activity and increased blood supply in preparation for feeding [4]. Some women also notice an increase in pigmentation of the skin, and a darkening of the area around the nipples will usually begin in about the 14th week, although it may be earlier in dark-skinned women. Morning sickness is another fairly common symptom of pregnancy. Nausea of this kind may be due to hormonal changes, deficiency of vitamin B, a sudden drop in blood pressure on getting up in the morning or pressure of the enlarged uterus. It

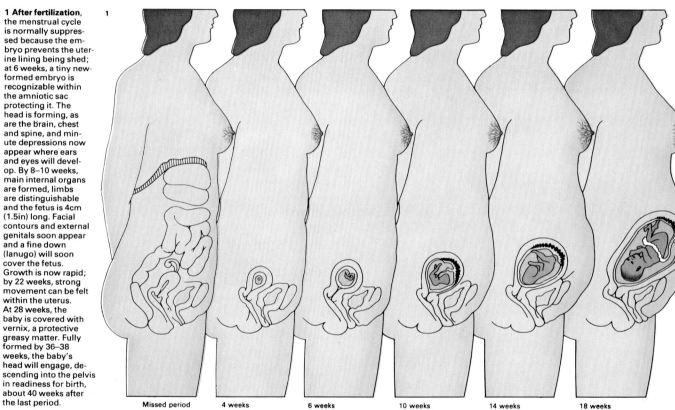

1 After fertilization, the menstrual cycle is normally suppressed because the embryo prevents the uterine lining being shed; at 6 weeks, a tiny new-formed embryo is recognizable within the amniotic sac protecting it. The head is forming, as are the brain, chest and spine, and minute depressions now appear where ears and eyes will develop. By 8–10 weeks, main internal organs are formed, limbs are distinguishable and the fetus is 4cm (1.5in) long. Facial contours and external genitals soon appear and a fine down (lanugo) will soon cover the fetus. Growth is now rapid; by 22 weeks, strong movement can be felt within the uterus. At 28 weeks, the baby is covered with vernix, a protective greasy matter. Fully formed by 36–38 weeks, the baby's head will engage, descending into the pelvis in readiness for birth, about 40 weeks after the last period.

Missed period | 4 weeks | 6 weeks | 10 weeks | 14 weeks | 18 weeks

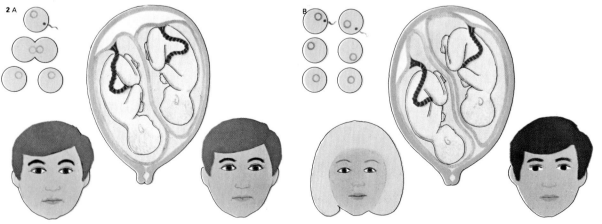

2 Twin births occur approximately once in every 90 deliveries. Identical twins [A] are always the same sex and develop from one egg which splits into two after fertilization, each fetus sharing the same placenta but with its own amniotic sac. Fraternal twins [B] need not be of the same sex; they are only as alike as any two children from one family. They develop as a result of fertilization of two separate eggs by different spermatozoa.

rarely continues for more than three months.

Clearly weight will increase, but the general aim should be to gain no more than 8-9kg (18-20lb) throughout pregnancy. Backache is a common complaint in the later stages, but can be prevented by correct posture and well-fitting shoes. The increased urge to urinate is caused by the enlarged uterus pressing on the bladder, and a tendency to constipation may be due to pressure on the bowels, although this can be avoided by eating plenty of fresh fruit and vegetables.

Toxaemia of pregnancy is an extremely serious condition with symptoms of speedy weight gain, swelling of the hands, feet and face and an increase in blood pressure. Treatment involves rest, reduction in salt intake and perhaps the taking of drugs to remove excess fluid and ease the swelling.

A healthy pregnancy

A wholesome diet is vital, since the fetus derives all nourishment from the mother; but there is certainly no need for the mother to "eat for two". She does need a great deal of iron, as the fetus will take a large amount from her system to form its own red blood cells. Providing excess fatigue is avoided, a normal amount of exercise is perfectly safe. Smoking affects the blood supply to the uterus, so reducing oxygen supply to the fetus. Thus women who smoke more than ten cigarettes a day may retard the normal growth of their fetuses. Drugs should never be taken except under medical supervision.

Rubella (German measles) is known to be potentially dangerous to the fetus if contracted by the mother in her first 12 weeks of pregnancy: it may affect the baby's heart, sight and hearing. If rubella is contracted after 12 weeks there is no real risk, as the heart, ears and eyes are already developed.

Earliest fetal movements (quickening) will generally be felt at about 18–20 weeks, reaching their most vigorous at 30 weeks. As early as the 7th week, fetal heartbeats can be distinguished by ultrasonics. Usually at about the 30th week the baby turns so that its head is then down. But even if it does not turn, a doctor can usually move the baby quite simply and painlessly to this position. Normally, after 40 weeks birth is imminent.

KEY

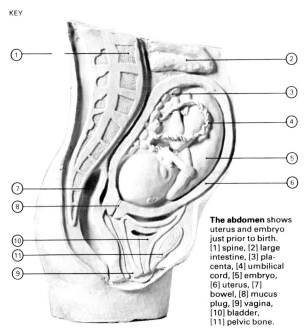

The abdomen shows uterus and embryo just prior to birth. [1] spine, [2] large intestine, [3] placenta, [4] umbilical cord, [5] embryo, [6] uterus, [7] bowel, [8] mucus plug, [9] vagina, [10] bladder, [11] pelvic bone.

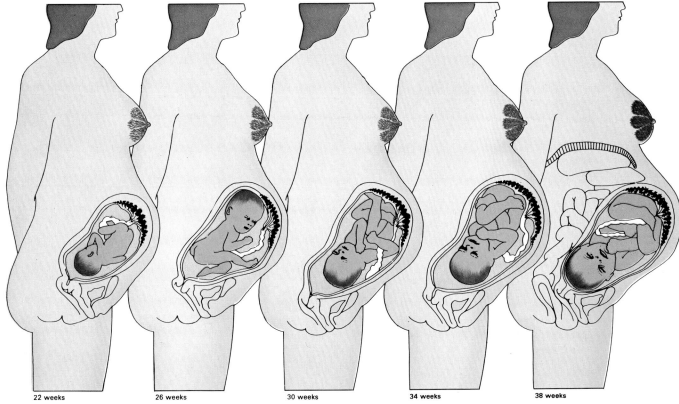

22 weeks   26 weeks   30 weeks   34 weeks   38 weeks

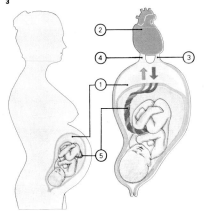

3

3 During pregnancy all nourishment is obtained from the mother via the placenta [1]. Nutrients are carried in blood pumped by the heart [2] into the placenta via the uterine arteries [3]. Food and oxygen [red] pass continuously from the maternal blood [pale yellow] to the fetal blood [green]. Waste products [blue] pass in the reverse direction entering the mother's bloodstream via the uterine veins [4]. The umbilical cord [5] carries blood.

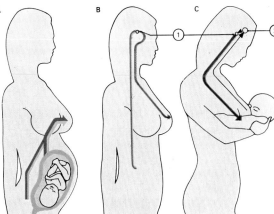

4 Breasts enlarge during pregnancy as a result of hormones (oestrogen and progesterone) from the placenta and ovaries [A]. After birth [B] loss of the placenta causes withdrawal of both hormones. Prolactin is then released by the pituitary [1] to stimulate milk production. Oxytocin, which promotes secretion of milk, is also released as the result of suckling [C]. Emotional factors, acting via the hypothalamus [2], are also important.

Giving birth

Birth, or parturition, marks the beginning of a new life in the outside world. It is achieved, at the end of the mother's pregnancy, through a sequence known as labour.

There are various signs that indicate that labour is imminent or already under way. Regular contractions of the womb, gradually increasing in frequency and strength, are the most common signal. These contractions are definitely rhythmic and cause discomfort, so a mother will usually have no difficulty in distinguishing between the contractions heralding labour and other uterine action commonly experienced during pregnancy. These actions occur spasmodically, cause no discomfort and are thought to be a normal and natural aid to fetal growth.

Three stages of labour

The birth process is a sequence of three stages. During the first stage [1, 2], when the uterus begins to contract vigorously, first the cervix (the neck of the womb), then the vagina dilate. This stage can continue for two to 20 hours, but towards the end of it there is a short transitional period during which the

amniotic sac that surrounds the fetus is ruptured. The fluid protecting the fetus then drains away, if it has not already done so – the so-called "breaking of the waters". A feeling of nausea, even slight vomiting, may occur.

The second stage of labour may last from just a few minutes to two hours. Contractions are commonly experienced every one or two minutes and last for about 60 seconds. During these the mother has to push as the baby's head descends through the cervix into the markedly enlarged vaginal passage [3]. This is the actual stage of delivery. In a straightforward birth the head emerges first and is moved so that the baby faces a direction in which the shoulders can emerge more easily [4, 5]. Generally, the rest of the body then slips out quite readily. The baby is held upside down for a short time by the obstetrician or midwife so that any mucus or liquid can drain from the mouth and upper respiratory tract. He then takes a first breath of air, no longer reliant on the mother for oxygen, and gives his first cry [Key]. Meanwhile, the umbilical cord will have been clamped near to the baby's body before being

cut quite painlessly [8]. Once the mouth and nasal passages are cleared the baby should breathe normally and his colour should change from bluish to a natural colour.

The third and final stage, which begins about 15 minutes after the baby has been born, involves the painless expulsion of the placenta (afterbirth) and the remains of the umbilical cord [6]. The placenta has to be carefully examined to make sure that it is complete. If parts are left inside the mother's body a haemorrhage may result. An intra-muscular injection is given to help the uterus contract and to prevent excess bleeding.

The period of six to eight weeks after birth is known as the puerperium, and during this time the mother's organs will return to normal. After delivery, a mother commonly feels elated; but a day or two later she may experience deep depression, due largely to a temporary chemical imbalance. This soon disappears. During the puerperium the uterus shrinks until it is only a little larger than before conception and the ovulation cycle begins again. Milk is secreted from the mothers breasts from about two days after

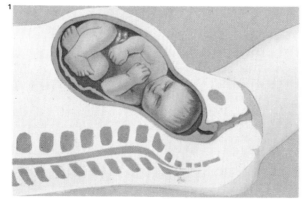

1 At full term the baby's head is engaged in the bowl of the pelvis ready for the journey through the birth canal. In this first stage of labour there may be 5 to 30 minutes between contractions.

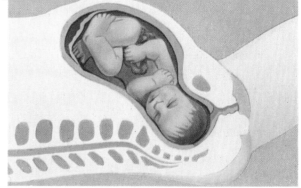

2 Uterine contractions continue to force the baby along and his head will normally turn sideways, enabling it to pass through the pelvis. Meanwhile contractions are more frequent.

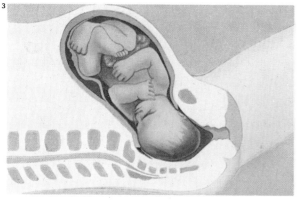

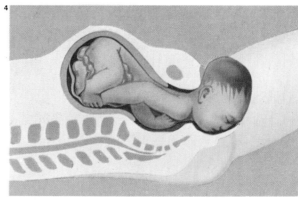

3 The cervix and vagina have now dilated so that they can receive the baby's head as it passes through the pelvis. The mother, "bearing down" with each contraction, helps the baby along.

4 The baby's head is now "crowned" and can be seen by the midwife. Once his head has emerged the shoulders are turned to facilitate birth and contractions force him into the world.

5 The midwife or obstetrician will normally assist with the birth of head and shoulders, after which the body slips out fairly easily. Forceps are sometimes used to help with delivery.

6 In the third stage of labour both the placenta and remains of the umbilical cord are expelled. An injection is normally given to the mother to stem blood loss and to contract the uterus, which returns to normal size after six weeks.

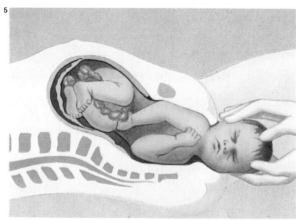

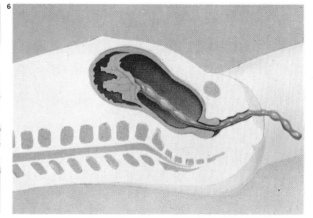

birth and continued stimulation of the breasts through feeding helps the distended uterus return to its normal position.

Pain and labour

Most women experience some degree of pain during childbirth but in a happy and healthy mother this is overshadowed by feelings of excitement and joy at the birth of her baby. During labour, however, most women need some form of pain relief. One recently developed technique is that of epidural anaesthesia [10]. Here a local anaesthetic freezes the nerves to the uterus while the mother remains conscious. Thus she is able to witness the birth and help in the process.

Inhalation anaesthesia is also sometimes provided for relief of pain both late in the first stage of labour and throughout the second stage, as required. Analgesic drugs – such as Pethidine – are commonly given by intramuscular injection. These usually reduce discomfort effectively, but care is always taken not to give the drug too late in labour as it might affect the baby's breathing after birth.

Hypnosis and natural childbirth dispense with drugs and anaesthesia, for it is claimed that if the mother is totally relaxed and free of tension she will feel little pain. Hypnosis achieves this through "suggestion"; natural childbirth by encouraging the mother during pregnancy not to be afraid of the birth process and to use special breathing techniques to relieve pain and relax muscles. Both of these methods have been used successfully.

Problems at birth

Difficult births are rare but if the baby is born feet or buttocks first (a breech position) [9], if the mother's pelvis is very small, or if the baby has to be delivered quickly to prevent suffocation, certain medical procedures are carried out. Forceps – large surgical tongs – may be used to help the baby out or an incision made in the vagina which eases the pressure on the baby's head. If a normal birth is impossible a Caesarean operation [11] is performed. This involves making a slit in the lower abdomen through which the baby is removed. Labour is often induced artificially if there are special reasons for doing so in the interest of the mother or her baby.

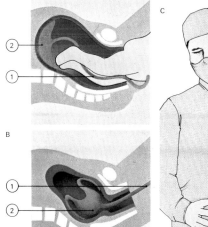

The baby's body [A] emerges after the head [B], leaving behind the umbilical cord [1] and placenta [2]. These are soon expelled [B]. The baby is then held upside down [C] and will usually start breathing.

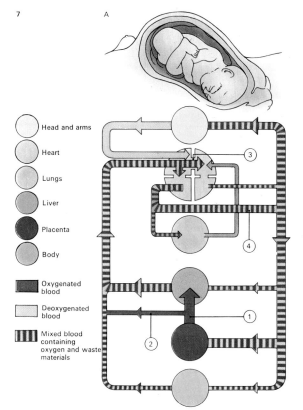
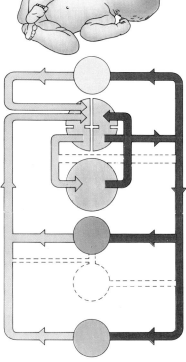

Head and arms
Heart
Lungs
Liver
Placenta
Body

Oxygenated blood
Deoxygenated blood
Mixed blood containing oxygen and waste materials

7 Supply of oxygen to the fetus takes place via the placenta, with blood circulation functioning as in [A]. Highly oxygenated blood reaches the fetus via the umbilical vein [1], most passing through the liver, head and extremities. A small amount of blood bypasses the liver through the ductus venosus [2] which closes after birth. In the heart most blood passes through the foramen ovale [3], a movable flap also closing at birth. Blood reaching the right ventricle is also diverted away from the non-functioning lung [4]. Once the umbilical cord is cut the baby's own lungs take on the task of oxygenation of blood with changes as shown in [B]. A rise in carbon dioxide concentration stimulates the brain's respiratory centre; the baby cries as he takes his first breath.

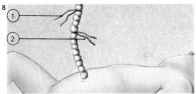

8 The umbilical cord must be cut following delivery, all the mucus having first been removed from the baby's mouth and nose. With the baby lying down, two clamps or knots [1, 2] will be put on the cord by the midwife at 15cm and 23cm (6in and 9in) from the baby so that it can be severed between these two points. The baby, now separated from the placenta, is totally on his own.

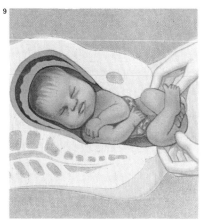

9 A breech baby is one born with feet or buttocks first. Between the 30th and 34th week of pregnancy the baby usually turns so there will be a head presentation. But roughly 1 in every 30 babies fails to do so and remains with the head still lying in the upper part of the uterus. Breech births are not usually difficult, but occasionally the obstetrician will turn the baby to avoid any damage to the arms or umbilical cord.

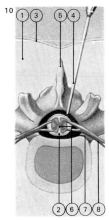

10 Epidural anaesthesia provides a pain-free labour. The diagram shows: back muscles [1], dura [2], skin [3], needle in the epidural space [4], vertebra [5], spinal cord [6], fluid [7], nerve [8].

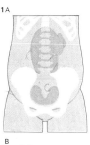

11 In the Caesarean section the baby is usually in a normal position [A]. A transverse cut is made in the abdomen [B]. The surgeon widens the cut and places his hand in the uterus [C, D], easing out the baby and pushing on the uterus [D].

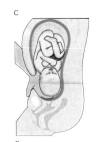

Birth control

All human societies have faced the problem of balancing the need for more hands with the difficulty of feeding more mouths. And most have tried to regulate fertility for personal or economic reasons. Before scientific knowledge of birth control a wide range of methods was used either to prevent conception or to limit intercourse by ritual taboos.

Development of birth control
In simple societies early methods were often based on sheer superstition. To ward off unwanted pregnancies women wore "magic" amulets made from the tooth of a child, for example, or the testicle of a weasel [3].

Other early methods of controlling conception reflected some knowledge of human biology, or at least close observation of cause and effect. Medical recipes for counteracting sperm date back to about 2000 BC. Vaginal douches of varying efficacy were used, as were simple versions of condoms to sheathe the penis or diaphragms to block off the uterus. The "safe period" of women was recognized by a number of primitive peoples and for hundreds of years men practised the

oldest method of all, coitus interruptus (removal of the penis before ejaculation).

Hostility towards "artificial" methods of birth control led to heated controversy during the nineteenth century when urban overcrowding and soaring birth-rates brought the issue of birth control to the attention of social reformers [Key]. The courageous work of women such as Aletta Jacobs (1849–1929), Margaret Sanger (1883–1966) and Marie Stopes (1880–1958) brought gradual acceptance of family planning by women.

The advent of the Pill
It was not until the middle of the twentieth century that the emphasis switched from attempts to stop fertilization to methods of intervening in the production of fertile eggs and sperm. In 1955, an American team of biologists led by Gregory Pincus (1903–67) discovered that the hormones oestrogen and progesterone, when taken orally, were highly effective in preventing ovulation [2]. By 1966 ten million women throughout the world were estimated to be taking the Pill; the

number today may be as much as 40 million.

The success rate of the Pill is such that women have been able to enjoy intercourse without the fear of unwanted pregnancy and without having to employ barrier methods. Described by some authorities as a safer drug than aspirin, the Pill is effective and generally easy to use. It does, however, have some side-effects and its use is not advisable for women with a history of blood-clotting, diabetes, cystic fibrosis, liver disease or cancer of the breast or reproductive organs. In addition, since it decreases normal vaginal secretions, venereal diseases are more readily transmitted. Although evidence about its long-term effects is still limited, fears about loss of fertility seem unfounded. One study has shown that 60 per cent of women formerly on the Pill became pregnant in their first cycle while only 40 per cent of other women conceived as quickly. Some women on the Pill also report increased sexual desire, although reactions are mixed. There are about a dozen "Pills" with varying amounts of oestrogen and progesterone and finding the most suitable one for a particular woman may require

1 The main methods of birth control are vasectomy [A], the Pill [B], the intrauterine device [C], the cap [D], the condom [E] and the rhythm method [F]. Coitus interruptus (withdrawal of penis before ejaculation) is practised but not reliable and often causes undue strain, both physical and emotional. Spermicidal creams, foaming tablets and aerosol sprays are equally unsatisfactory on their own, as is the douche – an attempt to wash out sperm from the vagina. The belief, once widely held, that a woman could not conceive while breast-feeding is not true. Prostaglandins, hormone-like substances now under study, can theoretically be used to induce menstruation in a pregnant woman but such a method of "birth control" could be seen as a form of abortion. Female sterilization (cutting and tying the Fallopian tubes) is effective but often irreversible.

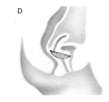

	A	**B**	**C**	**D**	**E**	**F**
How it works	Ducts taking sperm from testicles to penis are cut and tied. There is no effect on ability to enjoy intercourse.	Taken every day for most or all of the monthly cycle, the Pill inhibits ovulation and/or hinders penetration of the sperm into the cervix.	No one yet knows exactly how the IUD works. But it is thought either to inhibit implantation or to affect the sperm in some way.	All types of cap, inserted into the vagina and placed over the cervix, provide a barrier to sperm, preventing them reaching the womb.	The condom or sheath, rolled on to the penis just before intercourse, provides a barrier to sperm, preventing them from reaching the vagina.	Basically, a couple refrain from intercourse when the woman is most likely to conceive (at least 10 days each month and probably longer).
Possible effects	There need be no psychological problems. Physically, the operation is not risky but there is no guarantee it can be reversed.	Migraine, weight-gain, depression, mid-cycle "break through" bleeding, nausea plus possible extra risk of thrombosis. Change of Pill often helps.	Heavy periods do sometimes occur and removal of the intra-uterine device may be required. No long-term harmful effects are known.	Generally, caps are thought to be harmless – except in rare cases where there is an allergy to rubber or plastic, or sensitivity to a chemical used.	Condoms are harmless – unless the user or his partner is sensitive to rubber. Some couples also find condoms lessen sensation.	Some couples find that restricting intercourse in this way imposes severe strain and tension on their marital relationship.
How to use	A doctor or clinic will advise on surgeons, who carry out this operation. It does not usually involve a stay in hospital.	The right variety must be prescribed by a doctor or family planning clinic. Periodic check-ups advisable on grounds of general health.	The IUD must be fitted by a gynaecologist. Regular check-ups will be advised. Many IUDs have threads enabling their position to be checked.	Caps have to be fitted either by a doctor or the family planning clinic. To be effective they must be used together with a spermicidal cream.	Perhaps the most widely sold contraceptive available over the counter. It should only be put on when erection occurs.	Careful records of menstrual cycle and temperature needed. Difficult to use if periods are irregular or when sickness alters temperature.
Success rate	Tests are carried out after a few months to check seminal fluid is sperm-free. The chance of conception is about one in 30,000.	The risk of becoming pregnant is almost nil providing the Pill is taken exactly as directed. Failure rate 1 in 300 woman-years.	IUDs have been known to come out unnoticed with the result that in a few cases pregnancies have occurred. Failure rate is 1 in 20 woman-years.	Caps are thought to be about as effective as the condom and are dependent on the care with which they are used. Failure rate is 1 in 8 woman-years.	Needs care in use. Is safest with a spermicide. Some condoms have been known to leak. Failure is about 1 in 8 woman-years.	Timing cannot be guaranteed: so the calendar method is risky, but temperature is more reliable. Failure rate 1 in 4 woman-years.

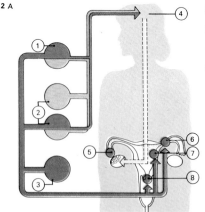

2 A

- Progesterone
- Oestrogen
- Gonadotrophins
- Effective sites of action

2 The Pill works by means of the contraceptive effect [A] of hormones similar to natural oestrogen and progesterone. Three main types are widely available using synthetic forms of these hormones with varying dosage schemes [B]. The combined pill [A1, B1] has both oestrogen and progesterone and one pill is taken each day for about three weeks in each monthly cycle. The hormones act together on the anterior pituitary gland [4] to inhibit normal production of gonadotrophic hormones and thus suppress ovulation [5]. Even if an egg is released the Pill acts further to prevent pregnancy by affecting the oviducts [6] and womb lining [7] and by altering cervical mucus [8]. During the fourth week withdrawal of hormones produces a lighter and often shorter menstrual period, beginning within three or four days. A month's supply of sequential pills [A2, B2] consists of oestrogen taken alone for 15 days and then a progesterone-oestrogen combination for the following five days. The mini-pill [A3, B3], in contrast, relies on direct progesterone effect only. Dosage is continual but there is a normal ovarian cycle and menstruation takes place as usual, (the first five pills are placebos), although the cycle may be more variable than normal.

a considerable amount of experimentation.

Various other methods of birth control remain popular and in most cases reasonably effective – these include vasectomy, intrauterine devices (IUDs), the cap, the condom and the rhythm method [1]. Choice of a birth control method is influenced by availability, religious attitudes and personal aesthetics as well as safety and effectiveness. The most suitable method is not only a matter of personal choice but also of medical advice.

Outlook for developing countries
Researchers are investigating the possibility of a pill that could be taken only once a fortnight and of long-lasting hormone injections or implants that could prevent ovulation (the last would be particularly useful in developing countries). As well as the possibility of a post-coital pill, scientists may soon develop a pill, to be taken by men, that would suppress the production of sperm. Psychologists are doubtful, however, whether women would readily trust men to take a pill as directed.

Family planning has been most effective in the developed countries where a strong force for the widespread use of contraception has been the Women's Liberation Movement, which stresses the woman's health, welfare and status. In the Third World, where the threat of starvation is so much greater, the struggle to persuade people to limit family size has unfortunately been less successful [4]. One problem is that the message of population control comes from former colonial powers and is sometimes seen as an attempt to retain dominance. In addition, children are seen variously as symbols of virility (especially in South America), a justification of a woman's life, an insurance policy against old age and a work force.

The problem of birth control exists not only on the biological front but also as a serious social dilemma in a world of limited space and resources. Education alone seems to offer a long-term solution [5]. The Indian government, which offers incentives for sterilization [6], considered a law demanding compulsory vasectomy for every male with three or more children. The ensuing outcry indicates the improbability of such a law.

Large families like this were common in the West at the end of the last century but are rare now. Family planning began in The Netherlands with the establishment in 1882 of a clinic founded by Aletta Jacobs.

Use of a diaphragm – the "Dutch cap" – was explained at the clinic but this kind of instruction in birth control made slow headway elsewhere. Margaret Sanger set up the first clinic in America in 1916

but it was closed soon afterwards and Mrs Sanger was arrested. Similar public hostility, including charges of "obscenity" was faced by Marie Stopes who opened the first British clinic in 1921.

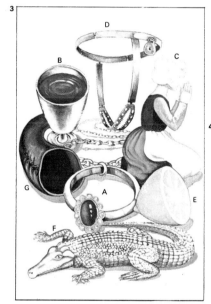

3 Early methods of contraception included the wearing of lucky charms [A], the drinking of supposedly "magic" potions [B] and even a resort to fervent prayer [C]. The chastity belt [D] sought to prevent intercourse. Casanova used a lemon [E] as a spermicide while the Egyptians used the faeces of a crocodile [F]. Early condoms [G] were made of sheep or pig intestines.

4 World population reached 1,000 million by 1850. But by 1970, due largely to a sharp decline in mortality rates, it had overshot the 3,500 million mark. Statistics show that in developing countries – in Africa, Asia and Latin America and parts of Oceania – the birthrate is far higher than in more industrialized countries of Europe, North America, Japan, Russia, temperate South America, New Zealand and Australia. This high birth-rate, combined with a fall in deaths due to better nutrition and medical facilities, has meant that the annual rate of increase in population is now far greater in those countries whose ability to provide food or adequate social services is already most strained. The aim of population control is to reduce this strain on resources and to promote better health and happiness.

The graphs show birth and death rates per 1,000 population in developed and developing countries since 1750. The silhouettes beside them represent the current growth in population every 60 seconds.

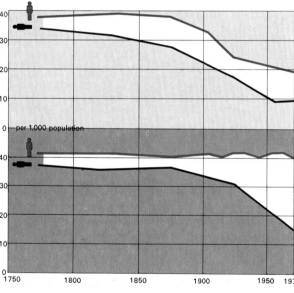

per 1,000 population

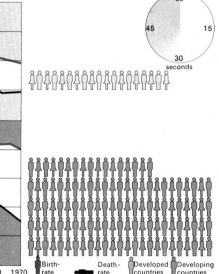

Birth-rate   Death-rate   Developed countries   Developing countries

5 Family planning posters are used in India to publicize the advantages of smaller and healthier family units. But some Indians said: "Look at the unlucky man with only three children."

6 After sterilization, a group of men from Kerala in southern India were given an array of gifts – rice, an airline bag, a plastic bucket, an umbrella, a lottery ticket and cash – all incentives in an intensive local campaign for birth control.

An introduction to illness and health

The men who inhabited the earth 20,000 years ago lived the life of the hunter, running after their prey and attacking it with stone axes. At the mercy of the weather, they took shelter in caves at night and protected themselves as best they could. For this life, men were as well, if not better, adapted than any other mammals, with bodies designed to endure long periods without food, to stand up to the heavy exertion of hunting and to respond quickly to the numerous emergencies with which their lives were fraught.

The dangers of comfort
The physical characteristics of modern man are little different from those of his forebears but are deployed in very different circumstances. Food is plentiful, exercise is rarely taken and tensions in an urban society are continually high. Early man's appreciation of sweet-tasting berries assured him of an adequate supply of water-soluble vitamins. The same craving for sweet things means that modern man's body is loaded with sugars and other carbohydrates, often in excess of his needs or his body's capacity to absorb them.

His sedentary, stressful urban life makes him overweight and tense. In order to feel better he may smoke cigarettes, drink alcohol or take tranquillizers: palliatives that must, in the long term or in excess, do him harm. The ingenuity of man's technology means that his body, which was designed to withstand the privation of a primitive existence, now has to function in a world that has been transformed by his intellect. The body is ill-equipped to deal with the results of that transformation.

The major health problems of developed societies stem from exactly this discrepancy between man's social and physical evolution. In industrially advanced countries most of the diseases with easily identifiable causes, such as tuberculosis, smallpox or cholera, have been virtually eliminated and the diseases that increasingly matter are those caused by what may simply be described as an unhealthy existence. High blood cholesterol levels resulting from an unbalanced diet, high blood pressure resulting from stress, obesity and smoking all add considerably to the chances of the occurrence of heart disease.

Health is difficult to define. It might be described as the subjective assessment of a person's ability to cope with his life: it is only when he feels ill that it is time to visit a doctor. This definition, however, ignores the fact that some diseases do not produce any noticeable symptoms at their onset and so the tendency in industrial societies is to screen people for diseases before they actually feel unwell so as to prevent their ever getting seriously ill.

Disease prevention and detection
It is obvious that health checks help to reduce the incidence of epidemic diseases, but such monitoring also helps to reduce mortality from more general causes. The first signs of the conditions that lead to many forms of heart disease or cancer – both scourges of developed societies – can be detected and steps taken to remedy the situation. Studies in the United States have shown that the mortality rate among those who have checkups is lower than normal and that it declines proportionately with the number of checkups that the patient undergoes. A campaign conducted in the United States to increase public awareness of health hazards now

CONNECTIONS

See also
142 Psychomatic diseases and retardation
122 Non-medical healing
126 Development of surgery
200 Astrology
116 Origins of curative medicine
118 Natural remedies
120 Man-made cures

In other volumes
82 History and Culture 1

1 Hippocrates, the "father of medicine", was born on the island of Kos c. 460 BC. He worked empirically, basing his judgments on observation rather than preconceived ideas. The Hippocratic Collection, including the Hippocratic Oath that binds doctors to keep their patients' confidences, are medical works by many authors and are not necessarily of Hippocratic origin.

2 Galen, who was born in Pergamum in AD 130, was both an anatomist and a physiologist. He proposed the theory that temperament was controlled by the balance of the four humours in the body (blood, phlegm and yellow and black bile).

3 Medicine and astrology were closely linked until the eighteenth century. An astrological chart would be drawn up for the patient and used in diagnosis and prescription. Each sign of the Zodiac is associated with a part of the body, so birth signs indicate which illnesses the subject is prey to. Astrological relationships are also now suggested between the Zodiac and the glandular and nervous systems. Aries rules the head and so Arians are said to be susceptible to headaches. Taurus rules the neck and throat, thus making Taureans vulnerable to colds. The arm, shoulder and lung region is ruled by Gemini, while Leo rules heart, spine and back. Cancer rules the stomach and chest, making Cancerians prone to indigestion and chills. Virgo rules the intestines and the nervous system, with the kidneys under the influence of Libra. Scorpio rules the sex organs and Scorpians are the most highly sexed of all the signs. Knees, bones and teeth are ruled by Capricorn, so those under this sign are likely to be troubled by dental and orthopaedic problems. Sagittarius rules liver, hips and thighs and Sagittarian women tend to have heavy thighs and hips. Pisceans are prone to trouble with their feet and Aquarians will probably suffer from varicose veins and hardening of the arteries because Aquarius is the sign ruling circulation.

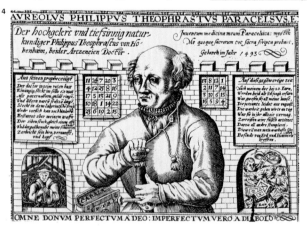

4 Paracelsus (c. 1493-1541), a Swiss physician and alchemist, broke with the traditions of Galen and revolutionized medical methods. Expelled from his post at Basel University in 1528, he continued with pharmaceutical experiments and encouraged research and development. He started the treatment of venereal disease by mercury and believed that the physician should be alchemist, astrologer and theologian in order to tend the body, soul and spirit.

seems to be contributing to the declining rate of fatal heart disease.

In contrast to those who believe in check-ups, some maintain that disease starts when the patient complains of symptoms and that over-anxiousness about health may actually lead to illness. A recent study, in which more than 25,000 English people were examined for diabetes, established that there were more undetected diabetics in the sample than there were known ones and that many of these people were symptom-free.

Another fact in favour of the argument that people are as healthy as they feel is that illness and mental strain are closely linked. Coronary thrombosis is clearly related to a stressful life-style and is often precipitated by a worrying event. The relationship between stress and illness has been so well defined that it is now possible to calibrate the chances of illness. Each event in a person's life is given a certain score: highest on the list are deaths within a family, followed by deaths of close relatives, changes in the home situation, court appearances, holidays, conferences and so on. A vulnerability score is obtained from

adding up all these points. Recent studies of sailors have shown that the assessment of their vulnerability to illness accorded well with the incidence of illness during voyages.

Fashions in illness

Medicine, as all human institutions, is subject to changing fashion, both in the type and symptoms of disease and in treatment. Before World War I it was fashionable for people suffering from mild psychological illness to display hysterical behaviour and no lady's handbag was complete without a bottle of smelling salts to cure a fit of the vapours. Today's disease is depression; the symptoms are a slowing down and withdrawal from the situation and the remedy is the anti-depressant tablet or tranquillizer. In the 1940s most children had their tonsils removed but tonsillectomy is now seldom performed unless medically necessary. Medical fashion varies not only with the times but from country to country. Circumcision, for example, is an operation widely performed in North America, whereas in Europe the trend has now moved against it.

Egyptian medicine, which was largely based on magic, relied on diet, hygiene and emetics (as prescribed by temple priests) for its remedies. Here Khemu fashions a man on a potter's wheel while ibis-headed Thoth fixes his life-span.

5 Leeches, collected from rivers by women who stood in the water and waited for the leeches to attach themselves, were used for blood-letting until recently. Leech salivary glands contain an anticoagulant, hirudin, which has been used to prevent blood clotting, but has now been superseded by new drugs.

6 Plethora, or excess of blood, was believed by Eristratus (an early teacher in the Alexandrine School of Medicine) to cause many diseases. He actually diminished blood by dietary methods, but his colleagues used blood-letting (phlebotomy) widely and so began a practice lasting for many centuries.

7 Mineral springs are to be found in almost every country and many cultures have exploited their curative properties. The Romans used hot springs as did the Crusaders, but spas were most popular in the 1700s and 1800s. Places such as Baden-Baden, Aix-les-Bains, Carlsbad and Bath became fashionable meeting places for the rich.

8 Franz Mesmer (1734–1815), Austrian physician and mystic, was the first man to use hypnosis and had remarkable successes with hysterical patients. He was convinced that cures were due to his "animal magnetism". His theories prompted charlatans to produce devices, such as this tub, which were claimed to cure all illnesses.

The causes of illness: 1

Few people today believe that "evil spirits" are the cause of epilepsy, that those who breathe marshland mists will develop malaria, or that smelly drains lead inevitably to an outbreak of typhoid fever. These ideas, quite firmly held no more than a century ago, have been almost completely dispelled by advances in modern medical science.

The scientific study of the causes of disease is known as aetiology. So successful has this science been that today doctors recognize thousands of disorders, and the discovery of new ones is still a fairly common event. Conveniently, however, nearly all the known causes can be grouped in a few general categories. For example, all the viruses, bacteria, protozoans and worms that cause disease are included in the category of infectious causes, and all the consequences of injury are grouped in a category called traumatic conditions [3].

Congenital disorders

Congential causes of disorders are traditionally considered first for they act within the womb and cause disorders which are usually obvious at birth. Hare-lip, cleft-palate and club-foot are examples of common congenital disorders. Mongolism [1], deformities of the heart, some of which are responsible for "blue babies", and abnormalities of the nervous system associated with spastic paraplegia are others. As a class they are caused either by some fault in the chromosomal structure of the fertilized egg or by damage inflicted on the developing embryo in the womb – it is not always easy to decide which.

A developing embryo in the womb can be damaged by diseases contracted by the mother. German measles contracted during the early months of pregnancy, for example, can cause abnormalities of heart and ears (and sometimes of mentality and sight). Other congenital diseases are caused by drugs taken during pregnancy, such as the deformed babies that were born to mothers who had taken thalidomide. The smoking of more than ten cigarettes a day by a pregnant woman may retard the normal growth of the fetus in the womb.

The disorders that are transmitted from the parents to the child from generation to generation are known as hereditary disorders. Perhaps the most well-known is haemophilia, a disorder in which the clotting mechanism of the blood is deranged so that even trivial injury is followed by prolonged bleeding. Another hereditary condition is phenylketonuria – in which there is an absence of enzymes that metabolize certain toxic phenyls. Untreated, it leads to severe mental retardation, but if detected soon after birth it can be dealt with quite successfully by a diet that is low in the animo acids that give rise to phenyls in the bloodstream.

Dietary deficiency

Starvation and malnutrition are obvious causes of disease. Even so, a seemingly ample diet may cause disease if it lacks certain vitamins, the nutrients essential for well-being. Vitamins occur only in certain foods and a diet that consistently lacks a sufficient quantity of a particular vitamin is certain, in time, to give rise to the corresponding vitamin deficiency disease [2]. The scurvy that afflicted sailors on long sea voyages and the beri-beri

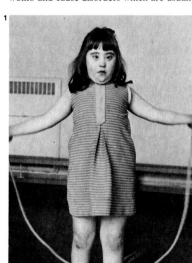

1 **Mongolism**, a form of mental deficiency associated with certain physical characteristics, is caused by a chromosome imbalance present at birth. Normal cells have 46 chromosomes, whereas mongoloids either have 47 or one of the 46 may be oversized. The excessive chromosome material interferes with the normal control of brain and body growth, producing the flattened face and folded upper eyelids that give the patient an Oriental look. Other features are short arms and legs, thick stubby hands and a large abdomen. The condition, also known as Down's syndrome, is more likely to occur in babies born to older women.

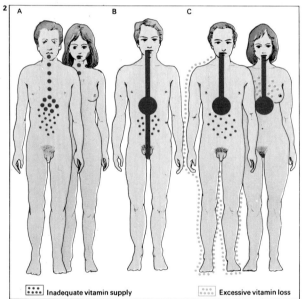

Inadequate vitamin supply   Excessive vitamin loss

2 **Vitamins are essential** to a healthy diet and lack of them causes a variety of deficiency diseases such as rickets (vitamin D deficiency), scurvy (vitamin C) and pellagra (niacin). Shortage of vitamins can be due to a primary food shortage or to diminished food intake [A] resulting from poverty or taboos against eating certain foods (beef for Hindus and pork for Jews and Muslims). Even when a diet is rich in vitamins, parasitic infections or impaired transport in blood or lymph vessels can reduce absorption [B]. Great physical activity, growth, or excessive loss through sweating or breast feeding [C] create the need for an increased vitamin intake.

3 **Accidents** are listed under the general heading of traumatic conditions. They range from simple cuts, burns and scalds to the drastic effects of a road accident. In the same category, doctors put injuries caused by lightning, electrocution, sunburn from ultra-violet light, frostbite and the air pressure problems of aircraft pilots or divers. Exposure to radiation from X-ray machines, radioisotopes or atomic explosions destroys various body cells. Some cells are more easily injured than others. Bone marrow, where red blood cells are produced, is most sensitive, then the lining of the stomach and intestines, the skin and sex glands and lastly cells of the brain and muscles.

from which thousands of prisoners in Asia died during World War II were caused in this way. The childhood disease of rickets is also caused by vitamin deficiency. In the industrial countries of the world, vitamin deficiency diseases are rare today but occasionally occur in people who, through preference or ignorance, choose to live on very restricted diets. Strict vegetarians, for example, are liable to suffer from lack of vitamins A, D and E which are contained in animal fats.

It is not only a lack of vitamins, however, that may cause deficiency diseases. There are several other constituents of the diet that are essential for health, and important among these is iron. A little of this mineral is needed to make the haemoglobin in the blood's red cells. A deficiency of iron causes anaemia. Another element needed by the body in small amounts is iodine, a shortage of which leads to an enlargement of the thyroid gland, a condition known as goitre.

Infections and their causes
An enormous number of living things gain entry to the body's tissues, grow there, and by so doing cause disease [4]. These infectious agents range in size from viruses so small that they can be seen only with the aid of an electron microscope, to bacteria and protozoa that can be seen with an ordinary light microscope and the various small creatures just visible to the naked eye, to the tapeworms which grow to several feet.

The way in which infectious agents cause disease is extremely varied. Viruses are intracellular parasites and the polio virus, for example, causes paralysis by growing in and destroying a particular type of nerve cell in the spinal cord. Bacteria, on the other hand, are largely extracellular and cause their ill effects either by secreting the powerful poisons known as bacterial toxins or by invading tissue. The malaria parasite destroys the blood's red cells and the amoebae of amoebic dysentry irritate and poison the bowel. The effects of infection or infestation with larger parasites are less well defined but there is little doubt that the debility associated with this type of infection is largely due to interference with nutrients and loss of blood from the intestinal wall.

KEY

Anatomy classes based on thorough human dissection, like this one at the Barber Surgeons' Hall, London, 1581, were the foundation of a proper understanding of how the healthy body worked, and consequently how diseases were caused. Previously such scientific knowledge was hampered by a strong taboo against human dissection and an over-reliance on such classical writers as Galen (c. AD 130– c. 200). The Belgian Andreas Vesalius (1514–64) was the first to break this tradition.

4 Infectious diseases are caused by invading organisms. Microscopic ones (germs) include viruses and bacteria. Larger are protozoan and fungal infections such as athlete's foot (tinea or ringworm). Larger still are flatworms and roundworms. Viruses cause many common diseases such as mumps, measles, chicken pox, influenza and the common cold. Some bacteria can live "dormant" in the body until fatigue or disease lowers resistance. Streptococci, for example, are often present but only sometimes cause a sore throat. Bacterial diseases include leprosy, tetanus, whooping cough and diphtheria. *Bacillus anthracis* [A] causes anthrax, an infection of animals which can be caught by those handling them. Symptoms are boils, fever and death in two or three days if untreated. Protozoa, minute one-celled parasites, cause many diseases. Among them are *Plasmodium*, responsible for malaria [B], which appears as minute "signet rings" in the red blood cells, and the trypanosome [C] that causes sleeping sickness. The mite *Sarcoptes scabiei* [D] burrows into the skin causing intense irritation. Undercooked, infected pork and beef are the cause of worm infection. *Trichinella spiralis* [E] lives harmlessly in the intestine but the larvae migrate to muscle where they form small painful cysts. The tapeworms *Taenia solium* and *Taenia saginata* [F] are man's largest parasites.

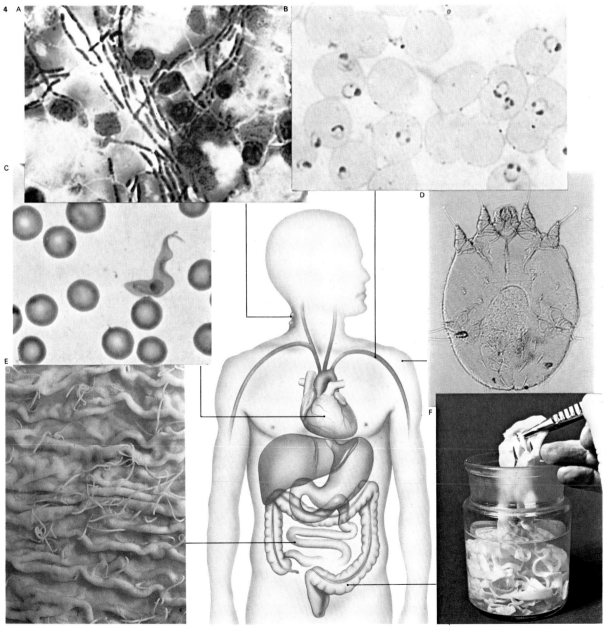

The causes of illness: 2

Many of the agents that infect human bodies produce poisonous substances but there are many poisons that originate outside the body and that are equally damaging [1]. Many of the chemicals used in industry are poisonous, most drugs are poisonous if taken in excess, many plants synthesize poisons in their leaves and fruits, and numerous animals, including some snakes and spiders, use venoms either for attack or defence.

Despite their diverse origins poisons have a marked similarity of action for nearly all of them interfere with one or more of the chemical reactions that take place in living tissues. As a consequence of this interference the poisoned tissue ceases to function properly and, in severe poisoning, may die.

Some poisons (the kind favoured by murderers and those attempting suicide) lead rapidly to death rather than to illness. Potassium cyanide is probably the best-known rapid poison and it works almost instantaneously by interfering with the intracellular oxidations responsible for the production of the energy used in all living tissues. Any cell it meets stops functioning immediately.

Carbon monoxide rapidly unites so strongly with haemoglobin that it prevents the blood from taking up any oxygen and swiftly results in the suffocation of the body's cells. Less immediately lethal poisons are those likely to be responsible for lasting illness. These slower poisons can affect many different chemical reactions and so produce different symptoms. Lead poisoning is cumulative, slowly disrupting the formation of red cells.

Tumours and cancers

Neoplastic (new growth) diseases are those that are associated with the growth of tumours. Generally speaking it is the extra growth that brings on the symptoms and so, as the underlying causes for the tumour are often not known [4], they are classified according to the size and site of origin.

Tumours vary from harmless warts commonly seen on hands and fingers to fatal cancers. A large proportion are slow growing with little tendency to spread to other parts of the body. This kind of tumour causes disease merely by its presence, and its

position is more important than its size [5]. A large tumour under the skin of the back may be harmless while a quite small one, in contrast, can cause disabling disease if it presses on an important or delicate structure such as a nerve and so deranges that structure's function. An acoustic nerve tumour, for example, is harmless in itself but causes deafness.

In contrast to the slow-growing, non-spreading tumours are those that grow rapidly and are capable of spreading to almost any part of the body. These are the cancers. A tumour of this kind causes disease not merely by its destructive effects on the tissue in which it originates and grows, but also by similar effects on any tissue to which it spreads. It is because cancers grow and spread so rapidly that they produce such serious symptoms and are so dangerous.

Degenerative and immunological diseases

Human tissues, like machines, suffer from wear and tear but, unlike machines, they are able in many cases to repair themselves. With increasing age, however, the reparative ability declines and as damage outstrips

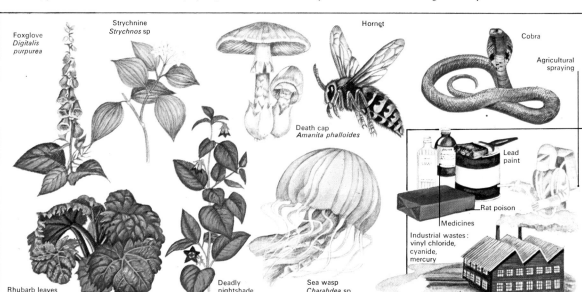

1 Poisons may be inhaled, injected, absorbed through the skin or swallowed; a few common ones are shown here. Some, such as hornet stings, are mild (although even these may kill), others so toxic that they can kill within seconds. Most deaths, however, are caused by weaker poisons in household goods. Children are particularly at risk since they are more sensitive to poisons than adults. The most difficult to avoid are those that occur undetected – for example, the bacterial toxins in contaminated stews and pâtés. In special circumstances harmless substances are poisonous: cheese can be lethal to anyone taking certain anti-depressant drugs.

Foxglove *Digitalis purpurea* · Strychnine *Strychnos* sp · Hornet · Cobra · Agricultural spraying · Death cap *Amanita phalloides* · Lead paint · Rat poison · Medicines · Industrial wastes: vinyl chloride, cyanide, mercury · Rhubarb leaves *Rheum rhabarbarum* · Deadly nightshade *Atropa belladonna* · Sea wasp *Charabdea* sp

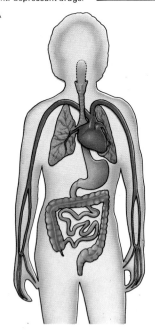
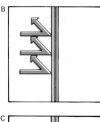
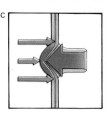
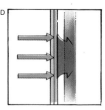

2 Allergies [A] can be caused by almost any substance. Pollen and dust affect the respiratory system [blue]. Some foods upset the digestive system [green]. Drugs, either injected into the bloodstream [red] or eaten cause general reactions. Many substances, on contact with the skin, cause local skin reactions [pink]. Potential allergens [arrows] have no effect on non-sensitive people [B]. Cells of "sensitized" people are coated with antibodies specific to the allergen [green layer]. Contact with the allergen causes these cells to release histamine which causes inflammation [C]. Antihistamine pills, sprays or injections suppress allergic reactions by blocking histamine action [orange layer] [D].

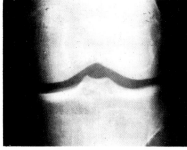
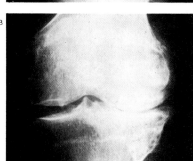

3 Osteoarthritis, a painful degenerative condition of the joints, is a common affliction of the elderly. Comparison of X-ray pictures of a normal knee joint [A] and an arthritic one [B] shows the characteristic loss of "joint space". The large joints at the knee, hip and shoulder are most commonly affected but small joints like those of the hand are often involved. The treatment consists mainly of less activity, heat therapy and pain killers. In severe cases cures include injections of cortisone drugs, replacing the joint with a new plastic one, or even ankylosis which involves removing any remaining cartilage and immobilizing the joint until the two bones fuse.

repair the symptoms of degenerative disease appear. Most common of all degenerative conditions are greying of the hair, baldness and the loss of elasticity of the lens of the eye which most middle-aged people suffer from. Far more serious, however, is the painful osteoarthritis that so commonly develops as the surfaces of the joints of elderly people begin to wear [3] and several degenerative conditions of the nervous system, such as Parkinson's disease.

Hardening of the arteries or arteriosclerosis, when the arteries lose their smoothness and elasticity, leading to angina pectoris, coronary thrombosis and stroke, is the most important of the degenerative diseases in industrialized countries, where it is responsible for more deaths than any other condition.

The immune defences of the body provide protection against infections and probably against the onset of many cancers. Sometimes, however, the defences become directed against quite innocuous "enemies" or even against the body's own tissues. Abnormal responses such as these are

responsible for hay fever, urticaria (nettle rash) and asthma [2] and also for the far more serious conditions of rheumatoid arthritis and haemolytic anaemia. The immune reactions that cause these diseases are similar to those responsible for the rejection of kidney grafts and heart transplants. And occasionally mothers respond to the red blood cells of their unborn infants by producing antibodies that cross the placenta, damage the infant's blood and result in the child becoming what is commonly known as a Rhesus baby.

Iatrogenic and idiopathic diseases
The final categories of diseases known to doctors are the iatrogenic diseases and idiopathic diseases. Iatrogenic diseases are those that are a consequence of medical care and range in severity from the acceptable side-effects of a drug necessarily given in the treatment of a serious illness to the disasters that sometimes follow accidents in operating theatres. Idiopathic diseases are those for which no cause has yet been found, such as the skin disease psoriasis. For this reason they pose great difficulties in treatment.

Cancer is a major killer, particularly in countries with a high life-expectancy. Research is currently centred on viruses, for more than 100 viruses have been identified as carcinogens, or external cancer-causing agents, in animals. Positive proof, however, of their ability to cause cancer in man has yet to be fully demonstrated although some researchers have associated herpes virus with cancer of the cervix.

4 The mechanisms by which normal cells mutate into cancer cells are not known but several carcinogens (external cancer-causing agents) and other factors that may have an effect have been identified [A]. These include various chemicals such as those released by cigarettes; hormones; radiation; chronic long-standing infection; air pollution; genetic factors (certain types of cancers run in families); chronic irritation or abrasion as caused in the mouth by a jagged tooth; viruses; age; sex (lung cancer is more common in men); race and geographical location. Normal cells have a specialized structure, appearance and function [B]. In benign tumours [C] this is maintained to an extent but in malignant tumours [D] cells are often grossly aberrant. They also lose their natural ability to stick together and so break away and spread the cancer.

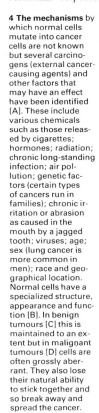

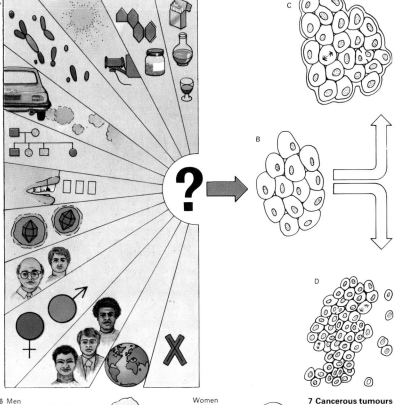

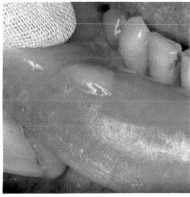

5 Lipoma is one of the most common of the slow-growing non-cancerous tumours. This tumour consists of large numbers of fat cells that grow in such a way that they form a mass, often just beneath the skin.

Such tumours can be unsightly but they are seldom dangerous. Surgical removal is usually feasible, especially when the tumour is small as here and this will probably prevent any further growth.

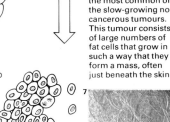

6 Cancer causes death by physical obstruction of an essential pipeline, by pressure on normal tissues (such as within the skull) and by crowding of normal tissue. This results in tissue death and consequent loss of the function performed by that tissue. Mortality for the ten most frequent sites of cancer, by sex, in 24 countries in 1969, is illustrated. More men die of cancer yearly than women and lung cancer is the most likely type.

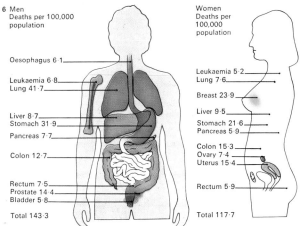

6 Men
Deaths per 100,000 population

Oesophagus 6·1
Leukaemia 6·8
Lung 41·7
Liver 8·7
Stomach 31·9
Pancreas 7·7
Colon 12·7
Rectum 7·5
Prostate 14·4
Bladder 5·8
Total 143·3

Women
Deaths per 100,000 population

Leukaemia 5·2
Lung 7·6
Breast 23·9
Liver 9·5
Stomach 21·6
Pancreas 5·9
Colon 15·3
Ovary 7·4
Uterus 15·4
Rectum 5·9
Total 117·7

7 Cancerous tumours of the skin are called epitheliomas because skin is an epithelial tissue. Such tumours occur with greatest frequency in particular sites such as the face, lips, ears and penis. The rapidly growing cells of the epithelioma spread readily to adjacent lymph nodes and for this reason it is important that these tumours are diagnosed and treated early. They often begin as small sores that refuse to heal.

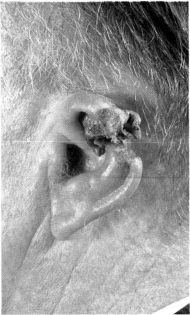

Diseases of breathing

The agents that infect the respiratory passages and the lungs [Key] are basically similar to those that cause disease in other systems. Ports of entry into the body for food, water or air – the mouth and nose – are also potential sites of weakness in the body's defences. The lining of the respiratory passages is especially vulnerable to various infections caused by invading micro-organisms.

The symptoms of infection
The tonsils and adenoids are prominent parts of the lymph system in the throat and help repel infection. Invasion by viruses or bacteria may result in sufficient inflammation to give tonsilitis [1] in which the throat and neck become sore and swallowing is painful. Many similar viruses of the rhinovirus group can affect the throat, nose and eyes to give "the common cold". Infections of the respiratory passages are named according to the part most affected – larynx (laryngitis), throat (pharyngitis), windpipe (tracheitis) and lung tubes (bronchitis).

When lung tissue itself becomes infected pneumonia develops [3]. Various bacteria can invade the lung air sacs causing inflammation, increasing blood flow to damaged tissue and preventing oxygen from entering the blood in sufficient amounts. Toxins, poisonous substances produced by the microbes, further affect the patient. Before the development of antibiotics recovery from pneumonia depended on the ability of the patient's own defences to fight the illness and mortality was high. Today the administration of appropriate antibiotics means quicker recovery, providing that pneumonia is not secondary to another serious debilitating disease.

Other bacterial infections of the lung tissues may not initially be as dramatic. Tuberculosis [6], once known as "consumption", may develop slowly before becoming obvious. Many children appear to come into contact with the disease and rapidly develop immunity. With careful treatment a cure can be expected with the drugs now available. Tuberculosis is more likely to occur in people not in the best of health – the elderly, those with chronic chest disease such as bronchitis and emphysema and in people with poor general nutrition such as chronic alcoholics.

The first symptom of a lung disorder is usually interference with breathing. This may take the form of pain, breathlessness, phlegm or catarrh. The breathlessness may appear suddenly or slowly, sometimes during rest or sometimes only on exertion.

Irritants of the air passages
Two lung diseases, bronchitis and emphysema [2], impede the passage of air to and from the lungs. In bronchitis the blockage is often due to the accumulation of excessive mucus; in emphysema to the compression and collapse of air tubes caused by unnaturally distended lung air sacs or alveoli in which air has become trapped. Chronic bronchitis is caused by continual irritation of the bronchial mucosa often due to tobacco smoke, dust, smoke and fumes.

It is not surprising that, as the years pass, diseases of the lungs develop, for the respiratory system suffers a considerable degree of abuse. This may take the form of large quantities of fine dust particles that are present in all air, but are often worse in

CONNECTIONS

See also
66 Breathing and the lungs
86 The causes of illness: 2
72 Keeping fit

In other volumes
146 The Physical Earth

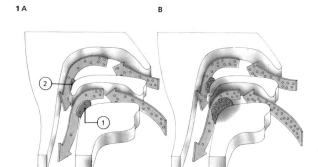

1 A   B

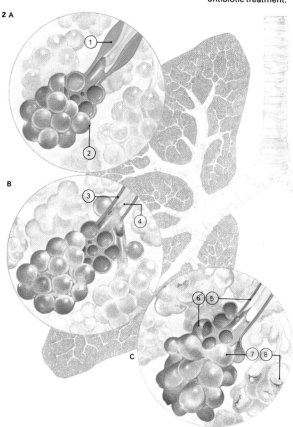

2 Viruses in the air 2 A
pass down the trachea (windpipe) into the lung tubes (bronchi) where they may cause asthma, acute tracheitis and bronchitis with symptoms of cough and fever. In asthma [A] the small air tubes become excessively narrow [1] causing difficulty in breathing. The alveoli overfill [2] but the condition is generally intermittent. Chronic inflammation of the tubes – called chronic bronchitis [B] – occurs mainly in older men who work in industrial areas and who smoke. A winter cough and wheezing occur, worsened by viral infections. The inflamed lining of the bronchial tubes [3] produces excess mucus [4] which is coughed up as phlegm. Inflammation and narrowing of the bronchi [5] lead to difficulty in moving air into and out of the alveoli [6]. The distended alveoli [7] are damaged and some may collapse [8]. The condition, called emphysema [C], stops oxygen getting into the blood normally.

1 Air breathed in through the mouth and nose [A] carries with it viruses and bacteria. The tonsils [1] at the back of the mouth and the adenoids [2] in a child are part of the body's defences against potentially infective microbes. These organisms may lead to inflammation of the tonsils with painful swelling [B], an increased temperature and difficulty in swallowing. The infection may require antibiotic treatment.

3 A 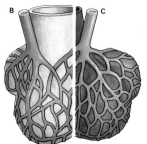 B   C

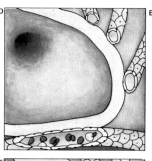 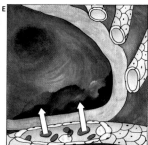
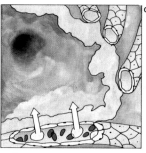 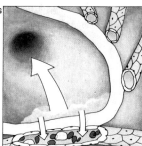

3 Infection may not affect only the upper respiratory system and the lung tubes but also the lung tissue and alveoli [1], resulting in pneumonia. This is normally an acute infection by invading bacteria but may also be caused by viruses. Normal lung alveoli [B] become engorged with blood, inflamed and thickened [C]. Sections through the alveoli seen under the microscope show the stages of the disease. Compared to the normal lung [D], tissue in a pneumonic lung [E] becomes inflamed and the alveoli are invaded by plasma and red and white corpuscles. These form a viscous mass that coagulates, causing acute congestion and airlessness. As the disease progresses [F] some tissue may die and excess of white corpuscles gives the lungs a grey appearance. During the patient's recovery [G] the dead bacteria, fluid and inflammatory cells are either coughed up or reabsorbed into the bloodstream. Treating the patient with antibiotics speeds recovery, while physiotherapy, by helping to ease breathing and by encouraging the coughing up of infected phlegm, minimizes permanent lung damage.

industrial areas [4]. The lining of the nose, which is equipped with mucus-secreting cells and long hairs, efficiently traps and filters most dust. Not everyone breathes through the nose, but the mouth and upper air passages are also quite efficient traps.

Lung disease can often be traced to a particular irritant; certain industrial processes produce silica particles that are a particular menace because they reach the alveoli in sufficient amounts to set up, over a period of time, chronic inflammation and scarring. This greatly reduces lung function and increases the risk of lung cancer. Other hazards are forms of asbestos which, when breathed in, increase the risk of cancer of the large air tubes (bronchi). These are also the major cause of a cancer of the pleura – the lining of the lungs and the chest wall. Restriction of the use of such substances, and removal and prevention of coal-dust and silica-containing dusts, are important medical preventive measures.

By far the greatest threat to the health of lung tissues is smoking [7]. Smoking increases the irritability of industrial dusts, is a major factor in lung cancer, greatly increases the risk of debilitating chronic bronchitis and emphysema and also has far-reaching effects outside the lungs – for example, it increases the risk of heart disease.

The incidence of different lung diseases varies from country to country and sometimes the reasons for this variation are obscure. Chronic bronchitis and emphysema, for example, have internationally been called the "English diseases", for they have a particularly high incidence in the United Kingdom. More males are affected than females and these diseases are more common in towns.

Allergies: asthma and hay fever
Other inhalants may produce allergic responses. In the summer months grass pollens, for example, produce hay fever whose symptoms are streaming noses and eyes, asthma [5] and general ill health in some people. Allergic responses can occur under conditions of prolonged exposure to a highly concentrated foreign material. Thus farmers may suffer asthmatic-type reactions when filling or emptying silos.

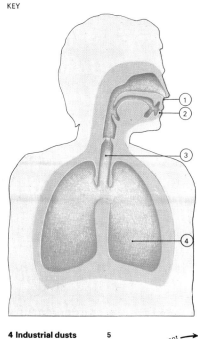

KEY

An oxygen supply is required for all the body's processes and is taken into the body through the lungs from which the waste gas carbon dioxide is breathed out. The nose [1] and mouth [2] can both be used for breathing in and out. The nose filters the air which passes to the windpipe [3], the larger and then the smaller bronchi, before reaching the many million air sacs (alveoli) of the lungs [4]. Many of the disease-causing agents enter the body through the same pathway. Various traps en route help to deal with these injurious agents and second-line defences are available to tackle any that get as far as causing disease.

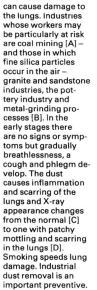

4 Industrial dusts can cause damage to the lungs. Industries whose workers may be particularly at risk are coal mining [A] – and those in which fine silica particles occur in the air – granite and sandstone industries, the pottery industry and metal-grinding processes [B]. In the early stages there are no signs or symptoms but gradually breathlessness, a cough and phlegm develop. The dust causes inflammation and scarring of the lungs and X-ray appearance changes from the normal [C] to one with patchy mottling and scarring in the lungs [D]. Smoking speeds lung damage. Industrial dust removal is an important preventive.

5 An attack of asthma can be precipitated by a number of factors. These may be psychological, infective or allergic, or occur in combination; varying in importance with individuals. While one factor, such as psycho-logical stress, may be most important, an attack becomes more likely if an infection or an allergy is also present. The size of the segments indicates the relative importance of each cause or combination of causes.

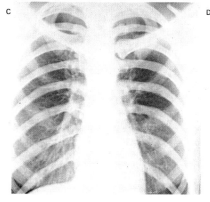

6 Tuberculosis normally affects the lung tissues but may spread throughout the body. A healthy person [1] can become infected by inhaling bacteria spread by a tuberculous patient [2] or by drinking cow's milk [3] containing bovine tubercle bacilli. A long, often chronic infection follows [red] with a 25 per cent mortality risk. Those who recover become immune [green]. Immunization [4] by modified bacilli [orange] will also give a person protection against tuberculosis.

7 Lung cancer most commonly affects smokers. Cancers develop from cells that alter their structure and multiply more rapidly. Changes may occur in cells of the inner or outer lining of the bronchial tubes [1]. Altered malignant cells increase in numbers [2] and extend locally [3] or invade adjacent tissue [4]. In addition the cancerous cells may spread outside the lung either through blood vessels [5] or through the lymphatic system [6].

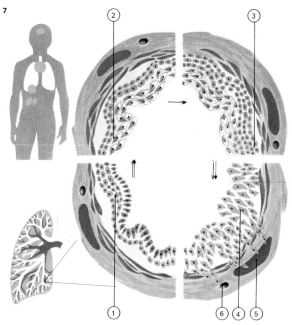

Diseases of the circulation

An obvious feature of the body's circulation system is the close interdependence of the heart, the blood vessels (arteries, veins and capillaries) and the blood. Abnormality or disease of any part [Key] often leads to malfunction of the whole system.

Congenital malformation that disrupts the flow of blood through the heart and/or its major blood vessels may lead to a baby appearing blue soon after birth because the blood is carrying too little oxygen. An operation may be needed soon after the disorder is diagnosed or perhaps later in childhood to correct the abnormality.

Types of heart disease

The heart and its valves are susceptible to damage by disease – a commonly known one being rheumatic fever which frequently leads to rheumatic heart disease. The damage it causes to the valves in childhood can lead to severe malfunctioning of the heart in later life but in many cases a valve-replacement operation can provide a remedy.

Occasionally the heart's pacemaker beats irregularly, or a defect in the conduction of the electrical impulse that causes the beat may lead to a chamber beating out of sequence. These so-called dysrhythmias may be helped by drugs (such as digitalis) but in some cases an artificial pacemaker is needed [2].

Any decrease in the supply of blood to the heart itself through the coronary arteries can lead to attacks of pain, called angina pectoris, a kind of cramp of the heart muscle that is usually worse after exercise or effort. Severe restriction in the blood supply can cause failure of the heart muscle. This is accompanied by sudden excruciating pain in the chest and is commonly called a heart attack.

The usual cause of a heart attack is a thrombosis or clot in a coronary artery [1]. Definite causes of coronary thrombosis are still largely speculative. A person who is overweight, takes little exercise, smokes, has high blood pressure, a high level of cholesterol in the blood and a family history of heart disease incurs a high risk of attack.

Defects in arteries and veins

Hardening of the arteries (arteriosclerosis) is a degeneration of the wall of the artery and is a normal part of growing old. It affects all the arteries of the body and can be accelerated and exaggerated by high blood pressure. Atherosclerosis (deposition of fats and cholesterol within an artery) is much more serious. It causes narrow, roughened vessels and may lead to thrombosis. If it occurs in the arteries of the legs, exercise may become painful and a graft may be needed to replace the damaged artery. Atherosclerosis of the aorta may reduce the elasticity of the vessel producing a bulging of the wall called an aneurysm, especially in the lower back. This may rupture spontaneously and cause severe haemorrhage needing emergency treatment.

In many cases, the cause of high blood pressure or hypertension is unknown but hardening of the small arteries of the kidney is certainly contributory. As high blood pressure makes this condition worse, treatment is needed to break the vicious circle. Several drugs can help. Some act directly to remove adrenaline (which causes the arteries to contract). Others tranquillize a patient and so reduce the amount of adrenaline he produces. Digitalis helps the heart to pump more

1 Thrombosis is the blocking of a blood vessel by a blood clot [1]. This commonly happens in an artery that is already narrowed and roughened by fatty deposits, a condition called atherosclerosis, which is quite common in old age but is not part of the natural process of ageing. It is aggravated by high blood pressure. If a thrombosis occurs in a coronary vessel [2] the part of the heart that it supplies will die (myocardial infarction) and the rhythms of the heart will be disturbed. This is what happens in a heart attack. Blood clots that form in one vessel, whether an artery or vein, may break off and then travel [3] to lodge in another vessel. A clot of this kind is called an embolus. If this occurs in cerebral vessels [4], it can deprive the brain cells of oxygen and this results in a "stroke". In a few cases, surgery to the damaged vessel may be helpful but physiotherapy and rehabilitation are the mainstays of treatment in most cases of this sort.

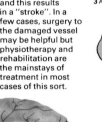
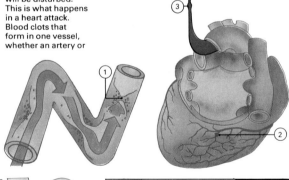
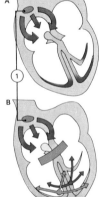
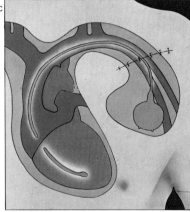

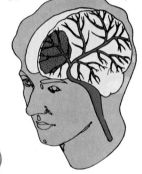
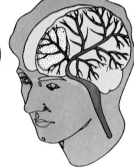
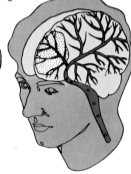
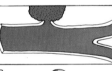
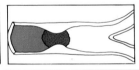

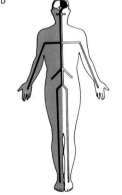
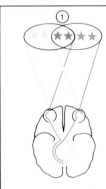
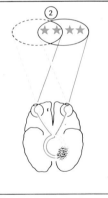

2 The heart has a natural pacemaker [A], the sinoatrial node [1], which controls heartbeat and rhythm. It transmits regular impulses through special conducting tissues, stimulating heart muscle contraction. Defective conducting tissue may result in atria and ventricles contracting in a dissociated way which may lead to complete heart block [B]. To avoid this possibility, a battery-operated artificial heart pacemaker can be used to stimulate ventricular contraction at a normal rate of 70 to 80 beats per minute. The pacemaker can be placed under the skin of the chest with the electrode leads passing through a great vein to the ventricle [C]. Batteries for pacemakers are charged every two years.

3 A "stroke" is a disruption of the blood supply to the brain and is due either to a haemorrhage or a cerebral blood vessel [A], a thrombus [B] or an embolus [C]. The damage [stippled area] may be permanent, but the outcome of a stroke depends on the extent and area of the brain affected by it. The symptoms can range from a temporary loss of speech or other brain function and paralysis of the limbs to sudden death. A stroke on one side of the brain affects the limbs of the opposite side of the body because of the crossing over of nerve tracts in the brain stem [D]. There is a similar effect on the visual cortex [E]. Normal vision [1] is impaired on the left side [2] by a stroke on the right side.

vigorously while drugs called diuretics reduce the fluid in the circulation.

Piles and varicose veins [4] are common conditions, both made worse by long periods of standing. Clots can also develop, especially in the legs, when the veins are inflamed (phlebitis). The leg will be swollen and painful but in plain thrombosis there may be no signs and diagnosis is difficult. Venous thrombosis is liable to occur in people who have prolonged periods in bed after an operation or a stroke. The danger of a clot (which can be 15-18cm [6-7in] long and 0.5cm [0.2in] wide) in the legs is that it may break off to become an embolus which travels through the heart and lodges in the lungs. Thrombosis is usually treated with anticoagulant drugs.

Blood defects
Blood is subject to a number of diseases. In anaemia there is usually either a reduction in the number of red cells or in the cell's haemoglobin content [5]. The reduction in the oxygen-carrying capacity of the blood causes tiredness and breathlessness. More

rarely, disorders of the bone marrow which makes the blood cells lead to anaemia; sometimes the red cells are destroyed more quickly than their usual four-month life-span and are not replaced fast enough.

Leukaemia [8] is the name given to a group of diseases characterized by proliferation of abnormal white cells. The patient is likely to suffer from anaemia, infection and bleeding. Drugs, including steroids, that interfere with the reproduction of white cells are used in the treatment of leukaemia.

Blood plasma contains various factors needed for the normal clotting mechanism. Deficiency of one, the anti-haemophilic factor, leads to haemophilia [6]. Deficiency in another causes the much rarer Christmas disease which has similar effects. Treatment with anti-haemophilic factor is needed as soon as possible after the start of a spontaneous bleed to prevent further damage [7]. Unfortunately the factor is in short supply and lasts in the body less than 24 hours. The hope for the future is that sufficient will be made available so that haemophiliacs can treat themselves at home.

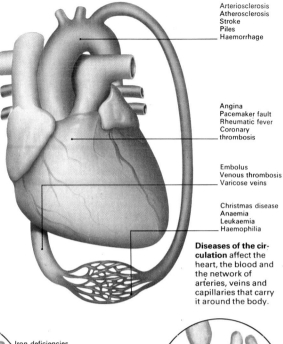

Arteriosclerosis
Atherosclerosis
Stroke
Piles
Haemorrhage

Angina
Pacemaker fault
Rheumatic fever
Coronary
thrombosis

Embolus
Venous thrombosis
Varicose veins

Christmas disease
Anaemia
Leukaemia
Haemophilia

Diseases of the circulation affect the heart, the blood and the network of arteries, veins and capillaries that carry it around the body.

4 Veins normally have a special system of valves that enables the blood to travel towards the heart but not back [A]. People whose occupation involve long periods of standing are prone to varicose veins. This condition occurs when the valves do not function properly, the veins becoming stretched and distorted [B]. Varicose veins of the lower end of the bowel are called piles (haemorrhoids).

If the affected veins fail to respond to simple treatment they will sometimes require injections or surgical removal. When varicose veins become infected and inflamed it is a form of phlebitis [C].

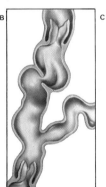

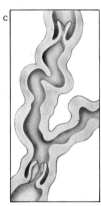

5

Iron deficiencies

Vitamin deficiencies

Normal

Indications of haemorrhage

Indications of bone marrow damage

6 The gene causing haemophilia [green] is carried on one of the female (or X) sex chromosomes controlling blood clotting. It does not show in a female carrier [1] because her other X chromosome [red] is normal, but if transmitted to the son [2] of a normal male, his Y chromosome [blue] cannot balance it.

7 In haemophiliacs, spontaneous bleeding occurs into the joints, which become hot, swollen and painful. The patient must receive anti-haemophilic factor quickly to stop the bleeding because it can lead to a crippling arthritis. The use of ice-packs and analgesics helps to relieve the pain.

7

5 Anaemia produces recognizable changes in the red cells of a blood film. Compared with normal blood, iron deficiency shows up in small, pale cells with reduced haemoglobin content. Iron deficiency may be due to hidden bleeding rather than dietary insufficiency and needs investigating. Lack of either of the vitamins B_{12}, or folic acid produces large, pale, abnormally shaped red cells and white cells with multiple nuclei. Diets that are lacking in liver and dairy products like eggs, milk and cheese (containing vitamin B_{12}) or fresh fruit and vegetables (folic acid), can cause these anaemias, but even with an adequate diet the vitamins that are needed may not be properly absorbed. Pernicious anaemia, for example, prevents the proper absorption of vitamin B_{12}. After severe haemorrhage or damage, bone marrow may release into the blood red cells that are immature or abnormal in some way.

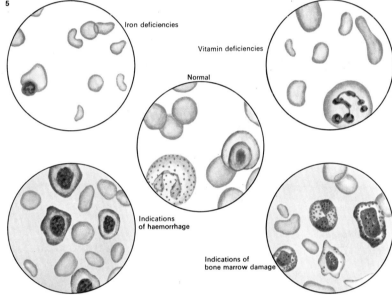

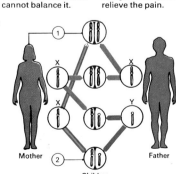

Mother
Children
Father

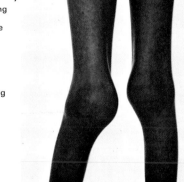

8 Treatment for leukaemia is aimed at preventing the abnormal white cells from reproducing themselves. Drug treatment interferes with the doubling of the genetic material (DNA) prior to division [A, B] or with the division process itself [C]. Steroid drugs also interfere with division [D]. The theory that leukaemia is caused by a virus is still only a tentative one, but if proved, could mean new and revolutionary treatment by the use of vaccines [E].

8 A   B   C

D

CH_2OH
H_3O $C=O$
H_3O
O

E

Diseases of the digestive system

Disorders and diseases can occur in any part of the gastrointestinal tract. "Indigestion" is something that nearly everyone suffers from occasionally but if it becomes frequent and persistent then the chances are that disease has affected the digestive system in the region of the oesophagus, stomach or duodenum or perhaps the pancreas, liver or gall bladder.

The formation of ulcers

The oesophagus passes from the throat to join the stomach through an opening in the diaphragm to which it is loosely attached. From middle age this attachment weakens and part of the stomach can become "pinched" in the opening; this hiatus hernia is quite common. After a meal the acid contents of the stomach can reflux back into the oesophagus causing pain and eventual ulceration. The pain, because of its position, is often called heartburn and is one of the symptoms behind the medical adage "if a patient complains of his heart, examine his stomach; if he complains of his stomach, examine his heart". The reflux may cause regurgitation of fluid into the mouth, prob-

ably after bending or straining. Occasionally surgery may be needed but the usual treatment consists of a weight-reducing diet plus advice on sleeping in an upright position.

The term "peptic ulcer" usually refers to an ulcer found at the lower end of the oesophagus, in the stomach or in the duodenum [1]. The cause of ulcers is thought to be increased secretion by the stomach of acid and the enzyme pepsin. This acid-pepsin mixture seems strong enough to overcome the normal protection of mucus in the stomach and "eats away" the digestive tract lining. Alcohol, smoking, stress and O group blood type are all associated with ulcers.

The pain that an ulcer causes is usually felt centrally in the upper part of the abdomen [5]. Its occurrence is related to meal times but often awakens a sufferer between 2 and 3 am. Taking food or antacids usually relieves the pain, as does vomiting.

The medical treatment of all ulcers involves relief stress with rest in bed, a bland diet, the stopping of smoking and the administration of drugs such as antacids and compounds derived from liquorice root.

There are many disorders in which the major defect is malabsorption from the intestine of one or more of the minerals, vitamins or other essential foods. Malabsorption tends to be associated with weight loss, anaemia, diarrhoea and vitamin deficiencies.

Malabsorption disorders

Two of the most important malabsorption defects are coeliac disease and Crohn's disease. Coeliac disease is due to an inability to cope with an allergy to gluten, a protein found in wheat and other cereals while Crohn's disease is an inflammation of the end part of the small intestine (the terminal ileum) and is also known as regional ileitis. Crohn's disease can sometimes affect other parts of the small bowel and, rarely, the colon. Typical symptoms include abdominal pain, diarrhoea, loss of weight, fever and anaemia. Medical treatment includes bed rest and a diet low in roughage.

Ulcerative colitis is an important disease affecting the large bowel. Its cause is a mystery but it involves inflammation that leads to ulceration. The rectum alone, or sometimes

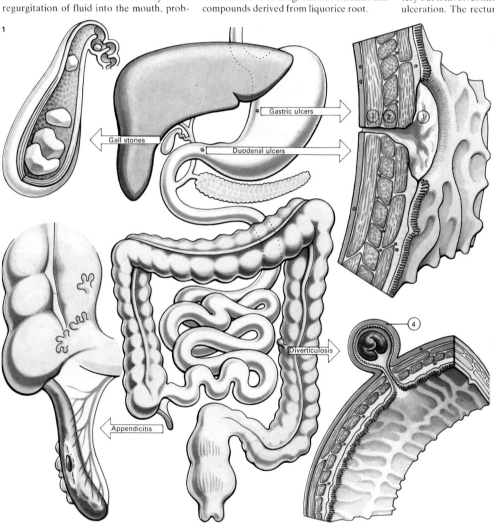

1 **Five common disorders** of the digestive tract are illustrated here. Ulcers of the stomach (gastric) or duodenum may penetrate the submucous coat [1], muscle layer [2] and finally perforate the wall [3]. Gall stones occur in 5–10% of the population and may be precipitated by fatty foods. They are a classic disorder of the "fs" – the fair, fat, forty, female and fecund. They are often associated with gall bladder infections. If stones move out of the gall bladder they cause obstruction to the bile duct and biliary colic. Treatment is surgical removal of the gall bladder. Appendicitis can also demand surgery. It is an inflammation and infection of the appendix. Diverticulosis, or the formation of narrow-necked, sac-like diverticulae, [4] commonly occurs in the colon. It is thought to be caused by weakening of the bowel wall and increased internal pressure due to lack of roughage. The diverticulae can become obstructed, infected and inflamed.

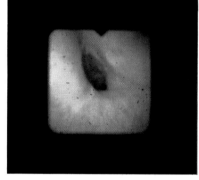

2 **A fibre-optic endoscope,** a lighted flexible tube, is used by physicians to look directly into the gastrointestinal tract. In the stomach the instrument can be rotated and pictures can be taken with a camera. In this way it is possible to compare a normal [A] with an ulcerated [B] lining. If an ulcer is present the instrument can remove a piece of tissue for examination in the laboratory.

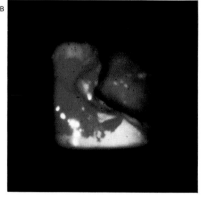

3 **Gall stones** are most commonly composed, like these, of a mixture of calcium, cholesterol and the bile pigment bilirubin. Stones can be made entirely of cholesterol, a fatty substance present in the blood or, more rarely, of bilirubin.

the whole colon, may be affected. Painful attacks of blood-stained diarrhoea are typical and in severe cases high fever, anaemia, exhaustion and collapse can occur.

Cancer of the intestine, especially of the stomach and colon, is second only in importance to cancer of the lung. Symptoms include non-specific pain, a change of bowel habits, loss of weight and intestinal bleeding. In cancer of the stomach loss of appetite, nausea and vomiting are usual and in cancer of the bowel blockage by faeces – intestinal obstruction – may take place. After diagnosis surgery is the usual treatment.

Jaundice diseases
The liver and gall bladder, organs vital to the orderly working of the digestive system, produce jaundice when they malfunction. Jaundice is a yellow discoloration, most noticeable in the skin, that arises when the amount of pigment bilirubin in the blood is above normal [4]. Liver disease or the presence of gall stones [1, 3] in the gall bladder will cause this. The liver diseases hepatitis and cirrhosis are well known for causing jaundice. Cir-

rhosis, associated with excess alcohol intake, involves the replacement of normal liver cells with fibrous tissue.

Hepatitis is of two types and both are caused by different viruses. Infectious hepatitis has a relatively short incubation period (15–35 days) and like many infectious diseases is spread by droplets (as in a sneeze). Serum hepatitis has a longer incubation period of 40–180 days and is passed in blood transfusion. Both diseases cause inflammation of the liver cells and this prevents them from functioning normally and causes an excess of bilirubin which in turn causes the typical jaundiced coloration of the skin [4].

Apart from the hepatitis viruses, other micro-organisms can affect the digestive tract. Most virulent of these are the *Salmonella* or "food-poisoning" bacteria, which are also responsible for typhoid [6]. Bacteria also cause the inflammation of the appendix and thus appendicitis [1]. Other invaders of the digestive tract are amoebae and parasites such as roundworms and tapeworms. All these organisms can cause severe loss of weight and debilitation.

KEY

Stomach
Over 35 years
Both sexes : ulcers
Males : cancer

Duodenum
Males 15–25 years : ulcers
All ages
Both sexes : parasites

Small intestine
All ages
Both sexes : viral and
bacterial infections
Under 35
Both sexes : appendicitis

Large intestine
Over 50
Both sexes :
diverticulosis,
colonic hernia, cancer
All ages : amoebic
diarrhoea

**Age can be a common
factor** in some disorders of the digestive system. The diagram shows a few well-known examples.

4 ● Red blood cells
🩸 Haemoglobin
ꞁꞁꞁꞁ Bilirubin
◯ Albumen
▭ Glucuronic acid

4 Excessive breakdown of red blood cells causes jaundice. Made in the bone marrow [1], the cells normally have a life of about 120 days before they are removed by the spleen [2]. The red pigment haemoglobin is metabolized to bilirubin and carried to the liver [3] attached to the blood protein albumen. In the liver albumen is removed and replaced with glucuronic acid. This passes to the gall bladder and from there to the gut. It is excess of bilirubin that produces jaundice.

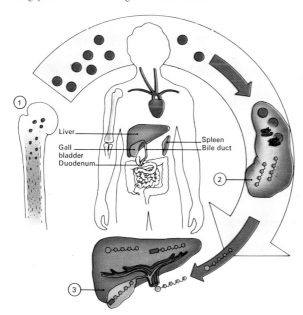

5 The acute abdomen is the name given to a group of disorders associated with abdominal pains. Some of them are shown here. The colics are pains that wax and wane in intensity. Intestinal colic can occur in food poisoning or intestinal obstruction. Biliary colic is not true colic but an intense, constant pain starting in the middle and moving to the liver. The pain from a perforated peptic ulcer or inflammation of the pancreas is persistent and "moves" through to the back. Appendicitis usually starts with a pain in the region of the umbilicus which moves to the area of the right groin. The pain of diverticulitis occurs in the left lower abdomen.

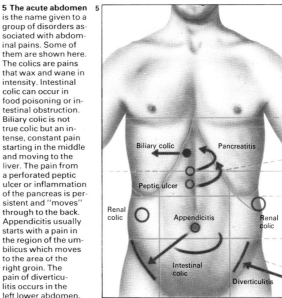

Biliary colic   Pancreatitis
Peptic ulcer
Renal colic   Appendicitis   Renal colic
Intestinal colic
Diverticulitis

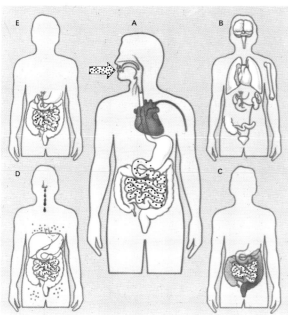

6 The bacterium *Salmonella* causes intestinal infections that may vary from mild gastroenteritis to severe typhoid. Bacteria enter the body [A] from infected food or water and incubate for up to 14 days. In typhoid [B] disease affects many body parts. After incubation [C] diarrhoea ensues. Other features of typhoid [D] are nose bleeds and rose spots. Bacteria remaining after an attack [E] make a person a carrier.

7 A barium meal X-ray is a common diagnostic procedure. The patient swallows a barium-containing mixture opaque to X-rays. Conditions that may be revealed include ulcers, various cancers and hiatus hernia of the stomach into the oesophagus.

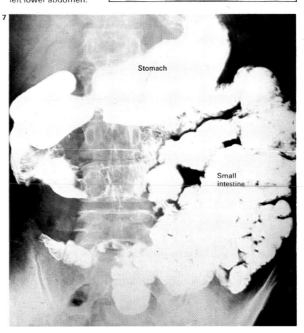

Stomach

Small intestine

Diseases of the skeleton and muscles

The skeleton is the framework on which the body is hung and supported, and the muscles provide movement. Considering the extent of movement in life, and the weight that is supported, it is not surprising that disorders of the musculoskeletal system occur. However, not all disorders are the result of wear and tear; inherited disorders may also occur, others may arise spontaneously during life.

Wear and tear in muscles

The wear and tear disorders are among the most common. Everyone suffers muscular aches and pains at intervals and severe and unaccustomed exercise usually leads to discomfort and stiffness in the muscles the next day. In due course such exercise of muscle will lead to the development of more powerful muscles. In the untrained, sudden movements will more commonly lead to muscle strains, tears of parts of the muscle mass, or tears in the tendons which attach muscles to bones. Such tears are often accompanied by bruising in the muscle, and healing can take two or three weeks.

Lumbago, or pain in the lower back, may follow muscle strain which results from unaccustomed activity such as digging. When back pain is accompanied by pain going down the back of one leg the condition is known as sciatica. Here the pain is due to pressure on one of the nerve roots that leave the spinal cord and supply the leg via the sciatic nerve. Treatment involves rest in bed on a firm mattress supported by wooden boards. The strain may be on one of the ligaments that run up the spinal column and support the vertebrae.

A weakness in the ligaments can allow the disc or pad which cushions each vertebra from the next to slip from its normal site and press upon the nerve root [3]. Treatment of a slipped disc requires rest while the tear heals and the disc returns to its normal place. If rest does not achieve this, surgical removal of the protruding portion of the disc may be necessary to alleviate symptoms.

Other sites at which injury can occur are in the joints between the bones. Bone ends are covered in cartilage and held together by strong fibrous capsules. The capsule of a joint may be torn, and the cartilage may be damaged. In the knee, specialized half-moons of cartilage act as cushions between the femur (thigh bone) and the tibia (shin bone). These two cartilages in each knee are attached to the edge of the wearing surface of the knee joint. Sudden twisting movements of the knee may be enough to dislodge a cartilage, tearing its attachment to the bone. This is painful, prevents full use of the knee, and if it does not settle the offending cartilage may have to be removed. Cartilage injuries are common in sports [6] in which the weight-bearing joints are frequently twisted.

Arthritis and its treatment

Over the years the degree of wear and tear becomes excessive and the consequence may be the premature development of osteoarthritis, also known as osteoarthrosis [1], in the joints. This type of arthritis develops in many elderly people in the ankle, knee and hip. The results are that the joint narrows, movement is restricted and pain and deformity occur. Osteoarthritis in the hip can be the most disabling, but it is now possible to replace the hip joint with metal and plastic substitutes and thus restore full mobility.

1 **Arthritis** means inflammation of the joints. In rheumatoid arthritis [A] the synovial membrane [1] becomes inflamed and thickened and produces increased synovial fluid within the joint [2]. The capsule and surrounding tissues [3] become inflamed, while joint cartilage is damaged [4]. Peripheral joints, as in the feet and hands, are involved. Blood tests reveal the presence of rheumatoid factor [RF] and the rate of red-cell sedimentation in a test-tube rises. Osteoarthritis [B], a degenerative disease, involves thinning of cartilage [5], loss of joint space [6] and bone damage [7]. Heavily used or weight-bearing joints are affected. Blood tests are normal.

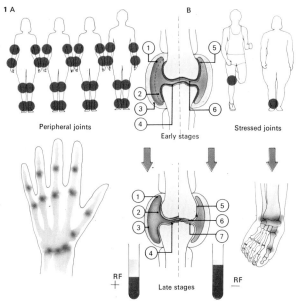

Peripheral joints   Early stages   Stressed joints

RF +   Late stages   RF −

2 **The hands** are common sites of rheumatoid arthritis. An X-ray shows the typical deformity of the joints and the deviation of the fingers towards the little finger side. The causes of rheumatoid arthritis (not clearly defined) may include infection, hereditary factors, emotional disturbances and physical overuse of the affected part. Treatment may involve the administration of anti-inflammatory drugs.

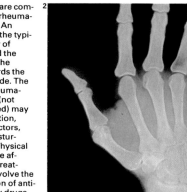

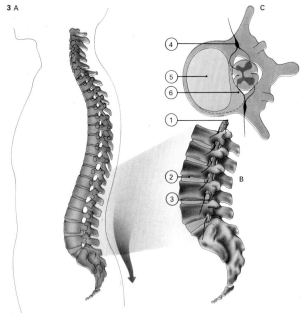

3 **The spinal column** [A] is a series of bones [B] surrounding the spinal cord [1] and separated by discs or cushion-like pads [2]. These may slip out of their normal place if the surrounding ligaments tear due to strain or degenerative changes. The disc between fourth and fifth lumbar vertebrae [3] here protrudes towards the back of the vertebrae. Pain, spasm of the muscles and restricted movement are common results. The fifth lumbar vertebra [C] (from above) shows the nerve roots [4] leaving the spinal cord. The position of the normal disc [5] is shown (pale), while the herniated disc [6] is seen pressing on the nerve root.

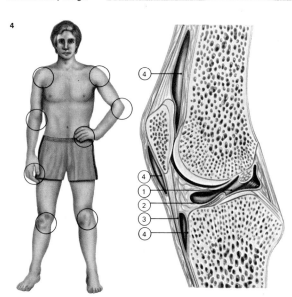

4 **Joints** are well lubricated by synovial fluid [1] produced by a synovial membrane [2]. Around many joints pass tendons [3] which are protected from rubbing and excess friction by strategically sited sacs or bursae [4], which are similarly lined and lubricated. Local damage, overuse or pressure can lead to inflammation and excess production of lubricating fluid (bursitis). Water on the knee, or housemaid's knee, is typical. Trigger finger and frozen shoulder are often due to synovitis, while tennis elbow results from damage at the attachment of a tendon. To avoid further strain the joint should not be overused but exercised gently; immobilizing it can lead to stiffness and may be harmful.

The other common arthritis, known as rheumatoid arthritis, is not a result of overuse. Women are more likely to be affected than men. The small joints of the hands and feet become damaged and deformed, and elbows, shoulders and ankles are also affected. The disorder is sometimes part of a more generalized body disturbance, and evidence of widespread inflammation can be detected in blood tests. In both types of arthritis, symptomatic treatment for the pain with analgesics is beneficial, but suppression of the inflammation may be required in rheumatoid disease. This is often achieved by administering other drugs.

Some musculoskeletal disorders are congenital, that is, present from birth; for example, congenital dislocation of the hip is sometimes found in infants [7]. The hip joint does not form fully for many years after birth, and initially the socket in which the ball of the femur is held is very shallow. It is quite easy to dislocate the head of the femur from the socket. In a few children this occurs spontaneously and, if not corrected, will cause maldevelopment of the hip. Treatment

is simply to splint the legs so that the joint is held correctly and cannot dislocate. Although ungainly, such treatment at an early stage is effective. The longer that diagnosis is delayed the more difficult is the cure and babies are examined shortly after birth for evidence of this uncommon condition.

Treatment by immobilization
Immobilization of an injured part of the musculoskeletal system can be required for many different troubles, such as fractures of bones [Key]. Plaster of paris casts may be adequate if the parts of the bones are well aligned. Otherwise internal fixation can be undertaken, in which the bones are held together with pins or plates screwed into the fragments. It is important at every stage after the operation, including the period during which the limb is splinted, that the patient is made to exercise as much as possible, because during the enforced rest much wasting of the unused muscles will have occurred. This can considerably slow a full recovery, for building up muscles always takes time, effort and dedication.

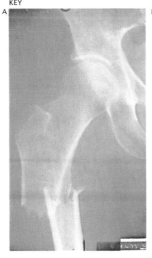

Fractures of the neck of the femur [A] are common in old people, who have quite brittle bones. The X-ray shows how pin and plate surgery [B] to the broken fragments can help to speed healing and recovery. The bones of young people break less easily than those of old people and when they do break they often cause less damage to surrounding tissue.

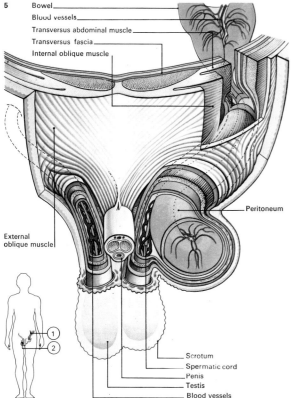

5
Bowel
Blood vessels
Transversus abdominal muscle
Transversus fascia
Internal oblique muscle

External oblique muscle

Peritoneum

Scrotum
Spermatic cord
Penis
Testis
Blood vessels

5 A hernia, often called a rupture, is the passage of any organ into an abnormal site. Common are those where the bowel [1] slides into the groin [2]. A weakness develops in the abdominal muscles. A loop of bowel insinuates itself along the path of the spermatic cord in the male and appears as a bulge in the groin. Such an inguinal hernia may be controlled by a truss or can be treated surgically.

6 In sports such as football, undue stress may be put on bone, tendon and muscle, and physical fitness is essential to minimize the possible risks. Shown here are the different kinds of damage that the various parts of the body suffer from. Sudden movements and collisions result in dislocated joints or broken bones in the shoulder, the arm and the leg. Cartilages in the joints may be displaced, while muscles and tendons are strained or torn.

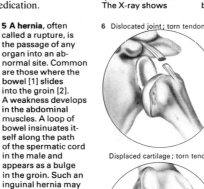

6 Dislocated joint; torn tendon

Dislocation; broken bone; bursitis
Broken bone; dislocation

Displaced cartilage; torn tendon

Dislocation; bursitis

Cartilage injury; bursitis

Torn tendon

Broken ankle; torn tendon

7 **Congenital dislocation of the hip** is a condition found at, or shortly after, birth. It is due to an increased ability of the head of the femur to slip out of the developing socket in the pelvic bone. In this X-ray, the joint on the left is normal, the one on the right displaced. Characteristic clicking of the hip is checked for in the weeks after birth.

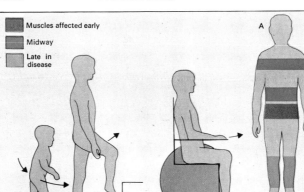

Muscles affected early
Midway
Late in disease

8 **Muscular dystrophies** are inherited degenerative diseases that cannot be cured. The Duchenne type [A] affects the pelvis, shoulders, trunk and later the limbs. The disease is first noticed as the child learns to walk with a characteristic waddling gait. He falls frequently and has difficulty in climbing stairs and rising from a sitting or lying position. A wheelchair becomes a necessity. The facio-scapulo-humeral type [B] affects the face, shoulders, trunk and finally the legs.

95

Diseases of the nervous system

The nervous system is made up of the central nervous system, comprising the brain and spinal cord, and the peripheral nervous system that carries messages to and from all parts of the body. These two parts of the nervous system vary greatly in their reaction to damage and disease, for while the cells of the peripheral nervous system are resistant to wear and tear, and have some ability to regenerate if partially destroyed, the cells of the central nervous system are susceptible to damage and, once lost, are irreplaceable.

Physical damage to the brain

The brain is often compared to a computer, but unlike a computer does not totally break down from minor damage. The skull provides fairly good protection but a knockout blow on the head may cause temporary amnesia (loss of memory) or concussion, which can be thought of as temporary brain damage. Even so, recovery is the rule.

Fracture of the skull [1] can cause physical damage to the brain. This, in turn, can result in cessation of activity in body muscles whose actions are controlled directly by the

brain. Because the skull in an infant is not fully closed the brain is more liable to damage, and physical damage, although rare, can occur during birth.

The more usual cause of damage is from within. In some people the arteries of the brain are fragile and can bulge out like weak tyres, a condition doctors call aneurysm. The great danger of this is that the artery may burst, causing bleeding into the brain. This is a cerebral haemorrhage in which the patient may become paralysed, and which can be fatal. A stroke, which involves a sudden loss of cerebral function, may be caused by the haemorrhage of a central vessel or by a blood clot or thrombus within a vessel.

Tumours in the brain cause physical damage by pressure they exert and other damage by the loss of brain tissue to the tumour. With modern techniques such tumours can in some cases be easily located and removed by surgery [2].

Effects of toxic chemicals

The brain is also sensitive to chemical damage. This can arise in various ways. Some

of the chemicals that find their way into the diet can affect the brain. Chronic lead poisoning, for example, causing tumours and paralysis, was once a problem in soft-water areas when lead piping was used for domestic water supplies, because soft water dissolves lead. In some disorders of metabolism normal body chemicals build up to such a level that they cause damage to the brain. Phenylalanine, a constituent of body proteins, is one such substance. In the inherited disease called phenylketonuria it is created to excess and causes mental retardation.

The chemicals harmful to the nervous system (neurotoxins) can be produced by bacteria that invade any part of the body and reach the brain via the bloodstream or via nerve trunks. The neuritis of diphtheria is caused solely by the toxin the diphtheria bacilli produce. Some bacteria release toxins that affect only nervous tissue [3]. The food-poisoning organism *Clostridium botulinum* produces a neurotoxin that causes fits, double vision, paralysis and death.

Some other micro-organisms seem to infect the central nervous system preferen-

CONNECTIONS

See also
38 The nervous system
90 Diseases of the circulation
124 Radiology and radiotherapy

In other volumes
36 The Natural World
102 Man and Machines

1 **Road accidents** are a significant cause of brain damage and much research is being conducted to lessen their severity. In a series of experiments a conventional car was shown to knock down a child dummy [A] but scoop up an adult dummy [B] at an impact velocity of 17km/h (10.6mph). The prototype car [C] with a low, rounded bonnet picks up the child dummy and reduces head impact. It also has an automatic restraining device.

2 **Brain surgery** is most often performed after injury to the brain associated with a skull fracture. Such fractures are most commonly caused by car accidents. Brain surgery is usually directed towards stopping bleeding or the removal of blood clots (haematomas). Other types of brain injuries (lesions) that necessitate surgery include brain tumours and the rupture of weak spots or aneurysms in the walls of blood vessels in the brain. Diagnosis of brain damage and disease has been revolutionized in the 1970s by the EMI scanner. With this machine the brain can be visualized in serial "cross-sections".

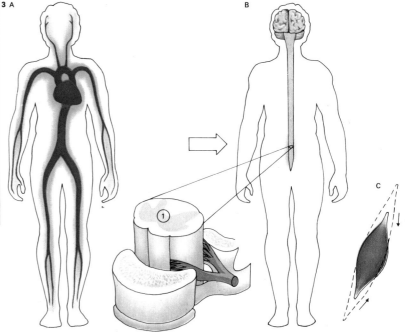

3 **Tetanus**, popularly called lockjaw, is caused by wounds infected with *Clostridium tetani* bacteria. These produce a poison that does not affect the wound but is carried [A], often by the bloodstream, to the central nervous system [B] where it becomes concentrated in the nerve cells of the spinal cord [1]. The normal control of muscle action that takes place in the spinal cord is completed by messages from the brain that modify signals passing out to the muscles via the motor nerves. The tetanus toxin acts by blocking the biochemical reactions essential to the completion of the nerve pathway. Thus the muscles cannot relax and go into spasm [C].

tially. The rabies virus, for example, specifically attacks the nervous system [7], as does the virus causing poliomyelitis [5] which, by attacking groups of motor nerves, paralyses the muscles whose movements they control.

Some micro-organisms that normally cause infections elsewhere find their way into the brain tissue (causing encephalitis) or its surrounding membranes or meninges (causing meningitis). Untreated syphilis can eventually cause an encephalitis, as can measles in about one case in 3,000. Pneumonia and tuberculosis can both cause meningitis, but as with all microbial infections the modern advances in vaccination [4] have reduced the incidence of disease, and antibiotics have in some cases controlled the spread of infection.

In degenerative diseases of the nervous system the nerve tissue undergoes chemical changes that stop it working properly. In multiple sclerosis, the most important of these, there is patchy hardening of the myelin that insulates the nerves, causing an intermittent loss of function whose locality depends on the exact bundles of nerves affected.

Muscle paralysis and speech disorders are two examples of degenerative diseases.

Symptoms of disease

Because some parts of the brain perform distinct jobs the symptoms of disease are often dictated by the area affected. Aphasia, for example, a disturbance in the ability to speak, write and comprehend words is due to damage in the area that controls speech. Parkinson's disease, in which the body trembles when at rest but becomes ridid at other times, is due to destruction of nerve cells at the base of the brain that are involved in the control of movement. The new drug L-DOPA now helps 40 per cent of all patients with Parkinson's disease. Yet one of the most remarkable things about the brain is that damage to some areas produces no detectable defect at all, as if the other cells take over the activities of the damaged ones.

Physiologists specializing in the study of the nervous system have made great advances in discovering the causes of disease, but epilepsy [8] remains a mystery, although scar tissue, a tumour or fever can be causes.

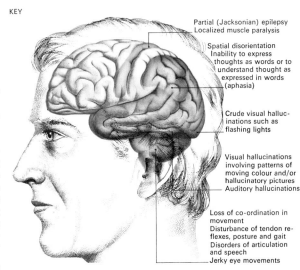

KEY

Partial (Jacksonian) epilepsy
Localized muscle paralysis

Spatial disorientation
Inability to express thoughts as words or to understand thought as expressed in words (aphasia)

Crude visual hallucinations such as flashing lights

Visual hallucinations involving patterns of moving colour and/or hallucinatory pictures
Auditory hallucinations

Loss of co-ordination in movement
Disturbance of tendon reflexes, posture and gait
Disorders of articulation and speech
Jerky eye movements

The symptoms that result from brain damage may reflect the particular function of the injured area of the brain. Some examples of such symptoms are indicated here. The specific effects of certain brain injuries have been of great use in producing physiological "maps" of the brain.

4 A vaccination campaign against polio, started in Europe and the USA in the 1950s, has almost obliterated the disease from those regions. Today babies receive polio vaccine in their first few months of life.

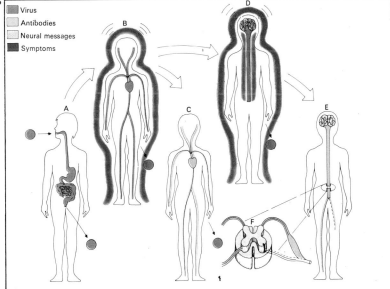

Virus
Antibodies
Neural messages
Symptoms

5 The polio virus enters the body by the nose or mouth [A], incubates in the intestine and then enters the bloodstream [B] causing headaches, fever and vomiting. The body now "fights back" by releasing antibodies [C], which may effect a complete recovery and also confer immunity against another attack. Or the virus may continue to multiply, reach the brain and spinal cord, and destroy nerve tissue [D]. The result is that the nerves can no longer send out messages to the muscles [E] to effect contraction and the muscles (usually of the limbs) become paralysed. The enlarged section of the spinal cord [F] shows the damaged area [black] and its "dead" nerve.

6 At a rehabilitation centre in Rwanda victims of polio – a disease also known as infantile paralysis because of its preferential attack on children – receive treatment to encourage paralysed muscles back to action.

7 Rabies is a fatal disease caused by a virus that selectively attacks the brain. It is caught from the bites of infected animals, particularly dogs. Louis Pasteur (1822–95) developed the first rabies vaccine from the saliva of "mad" dogs.

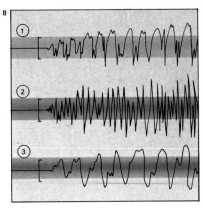

8 Epilepsy is not so much a disease as a symptom of a sudden disturbance in the brain's normal function. During an epileptic seizure the brain shows characteristic electrical rhythms that can be recorded on an electroencephalograph. During a "grand mal" [1] the patient usually loses consciousness and will fall unless supported. Most seizures last a few minutes. In a "petit mal" [2] the unconsciousness may be momentary. Focal epilepsy [3] is characterized by hallucinations.

97

Diseases of the skin

The skin, the body's covering organ, has a triple susceptibility to disease [Key]. First, because of its location; secondly because of its dependence on the rest of the body; and thirdly because the skin itself is prone to faulty functioning.

Attack from fungi

Infestations of the outer skin layer (the epidermis) are usually caused by microscopic fungi that use dead cells and skin secretions as food. These fungi are particularly prone to grow in humid protected regions, the armpits, groin, hair and feet in particular. Different types of fungi [5] grow in the different regions, and although the generic term ringworm [6] may be applied to them, they have nothing to do with worms. The mouth and vagina are also liable to colonization by the fungus *Candida albicans*, or thrush [7]. These fungal infestations are not usually harmful although they cause intense irritation. They must be treated, however, as severe complications can result.

The skin is also colonized by billions of bacteria. In general these do no harm; indeed, because their prescence makes it difficult for some dangerous bacteria to get a hold, they in fact do good. These beneficial bacteria use organic substances in sweat as food and the major disadvantage of their presence is that the chemical transformation of sweat they cause gives rise to the unpleasant odours of stale sweat. Other less wholesome bacteria can be present on the skin and if these manage to penetrate it through a cut or crack, they can infect a wound or may form an abscess. Similar bacteria result in the formation of spots or, more seriously acne [1], in which a hair follicle that has been blocked by excess secretions becomes infected. Impetigo is another bacterial skin infection. It causes itching and is contagious.

Itching and burns

In the UK insect infestations of the skin are now rare, but less than 50 years ago infestation by mites, lice and fleas were common. Itching caused by insects was so common that infestation by the scabies mite, burrowing into the skin, was called "the itch".

Itching is a non-specific response to any irritation of the skin. It can be psychological in origin, or be caused by a sudden fall in temperature or by excessive sweating. Where redness and swelling are present, they are likely to represent a local toxic reaction to plant or animal stings, for example, or an allergy [4] to a chemical. Eczema and dermatitis are both skin diseases often caused by allergic body responses. They may also be due to the irritant effect of chemicals that cause direct damage. Many chemicals, ranging from epoxy resins to washing powders, can be responsible.

Burns usually result from exposure of the skin to high energy (heat or radiation), although they may be caused by acids. They are divided into three degrees depending on their severity and the depths of the burn. First-degree burns involving redness of the skin occur in sunburn [11] and mild overexposure to other sorts of radiation such as X-rays or gamma-rays. Second-degree burns are more severe but still heal without a scar, while third-degree burns involve the peeling of all the epidermis and consequent painful exposure of the dermis and its nerve

CONNECTIONS

See also
60 Skin and hair
86 The causes of illness: 2
132 First aid
164 Adolescence

In other volumes
36 The Natural World
40 The Natural World

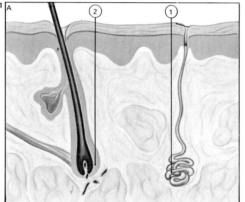
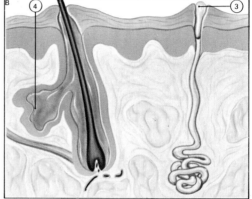

1 The sweat glands [1] and hair follicles [2] of normal skin [A] are liable to disorders and infections [B]. In hot, humid climates, the openings [3] of the sweat glands may become blocked, resulting in rupture of the gland and the inflammation known as "prickly heat". Bacterial infection of the oil-secreting gland [4] that is attached to each hair follicle is one of the causes of acne, in which pustular bodies develop.

4 Nettle rash or urticaria is a reaction shown on the skin in response to the absorption into the bloodstream of a substance to which the body is allergic. The substance is said to act as an antigen. In nettle rash, also called hives, the itching is caused by the release of histamine, while body enzymes produce the red rash. Strawberries, eggs, nuts and shellfish are other common allergy-producing agents.

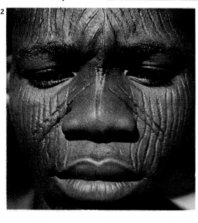
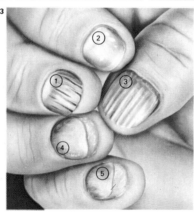
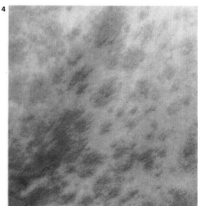

3 The nails, which develop from the skin beneath them called the matrix, may reflect deficiencies in the diet, for lack of calcium can make them brittle [1], or flecked [2]. A congenital abnormality of the nail-producing tissue can result in ridged nails [3]. Bitten nails often "ingrow" [4] into the surrounding skin, taking with them infective bacteria. The same bacteria gain entry if the cuticle is destroyed [5].

2 Scars form over wounds and burns. Corpuscles invading the wound produce fibrous tissue that joins the broken dermis, and this becomes covered with cuticle. At first the scar is delicate and the presence of many capillaries gives it a red appearance. As the wound heals, the fibrous tissue thickens and the capillaries are lost, so an old scar is hard and white. The faster the edges are brought together the less the scar shows.

5 Athlete's foot (misnamed *Tinea pedis* or ringworm of the foot) is the most common fungal infection of the skin. The affected skin becomes sodden and white, peeling off to reveal a red raw area.

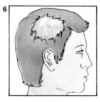

6 Ringworm of the scalp, also a fungal infection, causes lesions of the scalp and loss of hair from the affected areas. Common types of ringworm are easily cured and hair regrows.

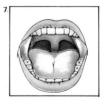

7 Thrush is the name given to the white patches of infection on the throat, tongue, lips and palate in young children. The same fungus causes an uncomfortable vaginal infestation in women.

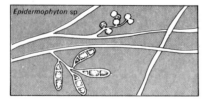

Epidermophyton sp

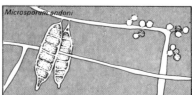

Microsporum andoni

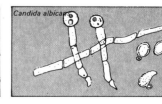

Candida albicans

endings, with formation of scars on healing.

Scar formation is the skin's method of repair to any severe damage on its surface. Scars are formed after third-degree burns and as a result of wounds, or due to rupture of the skin as when an abscess bursts. People in some cultures deliberately wound the skin to form scars: Prussian duelling scars, for example, or African tribal scars [2].

Infective agents
Skin problems by attack from within can be caused by infective agents. The viral diseases of childhood, as well as smallpox [9], are examples of diseases in which the infective agent reaches the skin through the blood-stream. The skin problems – rash [8], itching and the formation of pustules ("pox") – are only one part of the infection and its effects on other organs are the more severe, if less obvious, risks to health; arthritis or osteomyelitis (an infection of bone) can follow.

Direct viral attack of the skin is thought to be the cause of *verrucae* (and possibly warts), which seem to be contagious. There are also psychological aspects about them, for many

people never get them, despite exposure.

Wrong diet and upset metabolism are probable causes of malfunction in the skin's sebaceous glands, which can become blocked through over-production of oily sebum, producing a local build-up of sebum that can lead to dandruff or acne. Many of the less well-known skin rashes are caused from within the body and affect its metabolism. Red and greasy skin, for instance coarsened in texture (*rosacea*), is a disease apparently resulting from overindulgence and indigestion but it does affect tea drinkers and drunkards alike.

Several skin diseases are associated with a breakdown in normal function of the tissues of the skin itself. The most common of these are baldness, or alopecia, in which the hair follicles apparently wear out, failing to replace hair as it falls, and the whitening of the hair due to loss of pigment. These are both usually senile changes, with a strong hereditary component. Birthmarks, in contrast, are localized abnormalities that involve overpigmentation of the skin (moles) or both overpigmentation and disorders in the blood supply (port-wine stains).

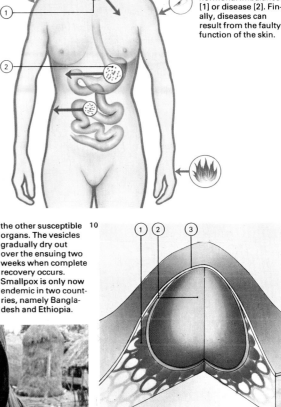

The skin is liable to damage from outside: from parasites, chemicals, burning or wounds. It is also liable to attack by germs that enter the body and is sensitive to upsets in normal physiology caused by hormonal imbalance [1] or disease [2]. Finally, diseases can result from the faulty function of the skin.

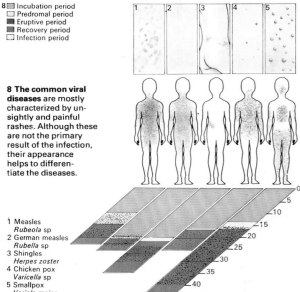

8 □ Incubation period
□ Predromal period
■ Eruptive period
■ Recovery period
□ Infection period

8 The common viral diseases are mostly characterized by unsightly and painful rashes. Although these are not the primary result of the infection, their appearance helps to differentiate the diseases.

1 Measles
  Rubeola sp
2 German measles
  Rubella sp
3 Shingles
  Herpes zoster
4 Chicken pox
  Varicella sp
5 Smallpox
  Variola major

9 Although the skin rash is superficially similar to chicken-pox, smallpox is a far more dangerous disease. By the time the characteristic pus-filled vesicles appear, the disease has already affected the other susceptible organs. The vesicles gradually dry out over the ensuing two weeks when complete recovery occurs. Smallpox is only now endemic in two countries, namely Bangladesh and Ethiopia.

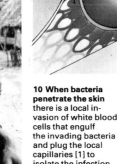

10 When bacteria penetrate the skin there is a local invasion of white blood cells that engulf the invading bacteria and plug the local capillaries [1] to isolate the infection. The mixture of white cells, dead bacteria and tissue fluids [2] build up inside the fibrous capsule [3] that develops round the infected area. If the abscess is not lanced, more pus is produced until the skin ruptures.

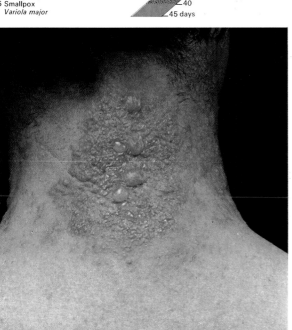

11 The most common form of burning is sunburn, caused by over-exposure to the ultraviolet rays of sunlight. Cautious exposure to the sun causes the development of brown pigment which is the body's natural defence against sunlight. Additional benefits include ultraviolet light and also the formation of vitamin D. Sunlight has been implicated, however, as one cause of cancer of the skin.

12 Albinism can occur in any population. It can be total, which is a congenital condition, or patchy. The lack of pigment causes sensitivity to strong sunlight. Pink coloration is due to reflection of light from blood vessels.

Diseases of the glands

The glands may be divided into two types according to where their secretions go. Those that release their secretion into a duct or tube are called the exocrine glands [8], while those without such tubes release their product, known as hormones, directly into the blood and are called endocrine glands. The exocrine glands normally release their products to the body surface – for example, sebum to lubricate hair and skin or tears to moisten the eyes – or into the alimentary tract, providing the digestive juices. It is alterations in the activities of the endocrine glands that are normally termed glandular disorders, although glandular fever is a disorder of yet another series of glands – the lymph glands – which are part of the body's defences.

Diabetes: causes and symptoms
The most common endocrine disorder is sugar diabetes (diabetes mellitus), in which insulin lack occurs [5]. The major part of the pancreas is exocrine, producing digestive juices, but about one per cent of the bulk is due to a million clumps of cells called the islets of Langerhans. Lack of insulin secre-

tion from these islets leads to a failure to control the use of the body fuels and building blocks (sugars, fats and amino acids). In diabetes these are overproduced at the expense of the body tissues and are under-used. Sugars, such as glucose, accumulate in the blood and are poured away as waste in the urine along with excess fluid and salts.

This great loss of water and urine led the Greek Aretaeus, in the first century AD, to call the disease diabetes, after the Greek word for "siphon". Thirst and weight loss occur and high sugar levels encourage infections. The situation can be corrected by insulin injections. When diabetes appears later in life some insulin is produced by the pancreas, but insufficient to keep the metabolism entirely normal. Such patients are often overweight, but with weight loss and controlled intake of carbohydrate foods, extra insulin is often not necessary to treat their diabetes.

While long-standing diabetes may lead to circulatory disorders, careful control helps to minimize such problems. Diabetes of varying severity may affect two to three per cent of

the population in Europe and the United States, but the opposite situation of overproduction of insulin is very rare. It is caused by an insulin-producing tumour of the pancreas, an insulinoma. Blood sugar falls excessively, resulting in hunger, sweating and altered mental ability.

The thyroid and pituitary glands
The next most common group of glandular disorders are those affecting the thyroid gland. Overproduction of thyroid hormone leads to increased demands for food and energy by the body tissues, increased heat production, weight loss, nervousness and irritability [2]. It may be treated with drugs or thyroid surgery. A lack of thyroid hormone, from failure of the gland, leads to slowness, apathy, weight increase and susceptibility to cold. It is more common in older people and in those with thyroids that have been overactive in the past. Treatment is by daily thyroid hormone tablets. An enlarged thyroid or goitre is caused by either overactivity or underactivity of the gland. Diets deficient in iodine are one reason for overactivity.

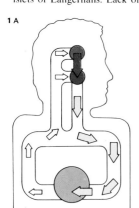
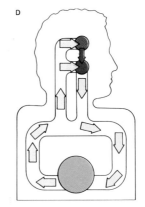
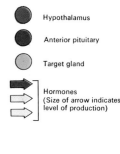

1A ... B ... C ... D

Hypothalamus

Anterior pituitary

Target gland

Hormones
(Size of arrow indicates level of production)

1 If one of the target glands on which an anterior pituitary hormone acts is underactive, the feedback of the hormone on the hypothalamus and the anterior pituitary is reduced. Both the hypothalamus (part of the brain) and the anterior pituitary (at the base of the brain) allow increased production of the pituitary hormone which attempts to stimulate the underactive glands [A]. Alternatively, the pituitary may be underactive despite signals from the hypothalamus, and the target gland is also underactive [B]. The opposite may occur, with glandular overactivity. The target gland is overactive despite feedback inhibition of the pituitary [C]. When this is overproducing, despite lack of signal from the hypothalamus, the target gland is stimulated to do likewise [D].

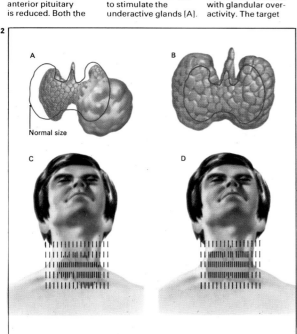

2 The thyroid gland in the neck may overproduce thyroid hormone because of a tumour (usually benign) [A] or a generalized enlargement of the organ [B]. One method of assessing this activity is to measure the uptake of iodine into the gland – iodine is necessary to produce thyroid hormone. Radioactive iodine is used in a small dose and the amount in the gland is indicated by the dashes in the thyroid "scans" [C, D].

Normal size

3 Body size may be affected by glandular disorders in childhood. Anterior pituitary overactivity can cause gigantism (as with this 2.82m [9.25ft] Dutchman); underproduction of growth hormone is one cause of dwarfism (as with his 0.9m. [36in] companion).

Normally the thyroid gland is directed by a hormone from the anterior pituitary gland at the base of the brain [1]. If the thyroid gland fails to produce sufficient thyroid hormone this is registered by the hypothalamus (part of the brain) and the pituitary which produces thyroid-stimulating hormone in greater amounts in an attempt to restore the normal balance.

Increased pituitary hormone production may, however, be a disorder in its own right, rather than a response to failure of a target gland. If a tumour is present, for instance, hormone production may be excessive. Often only one of the several hormones that the pituitary produces is increased. The effects of such pituitary gland abnormality vary according to which hormone is in excess and other body tissues and glands that respond.

One such hormone of the pituitary is that which affects the adrenal cortex, part of the adrenal gland that lies just above the kidney. This hormone is adrenocorticotrophic hormone (ACTH). Excess ACTH leads to excess steroid hormone production by the adrenal glands. Disturbance of many systems results, with retention of excess salt and water in the body and poor handling of sugar in the body. This disorder, called Cushing's syndrome, may result if the adrenal glands are overactive due to excess ACTH, but exactly the same pattern of symptoms arises if the adrenals are overactive in their own right.

Other pituitary hormones

Disturbance of one hormone can have marked effects on normal feedback mechanisms, as is seen with growth hormone from the anterior pituitary. Its lack during childhood leads to dwarfism, while overproduction may lead to gigantism [3]. Too much growth hormone in later life leads to acromegaly. This disease is usually first noticed because the sufferer's hats, gloves and shoes need replacing with larger sizes. The nose, lips, tongue, hands and feet broaden and enlarge. Treatment is by X-ray irradiation

The posterior part of the pituitary normally produces antidiuretic hormone which maintains the correct amount of water in the body. If it is deficient, excess water is lost in the urine; this is diabetes insipidus.

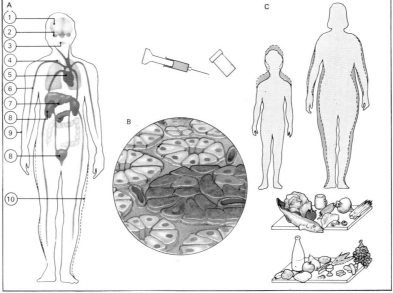

KEY

Daniel Lambert (1770–1809) weighed 335kg (737lb). He had a girth of 234cm (92in). The problem of obesity is often justified by sufferers as being due to a glandular disorder. This is rarely true: over-eating is the usual reason.

Disorders of the body glands give rise to groups of symptoms that are fairly characteristic for each condition.

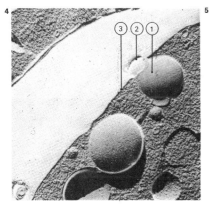

4 Hormones are stored in the cells of a gland in small sacs or vacuoles [1] until required. Shown is a specially prepared slice through a pancreatic islet cell that makes insulin. When blood glucose levels are high, after a meal, the cell discharges the insulin [2] from within these vacuoles through the cell wall [3]. In diabetes insufficient insulin is produced so blood sugar levels rise.

5 Diabetes mellitus [A] may result in drowsiness and coma [1], impaired vision [2], dry mouth [3], over-breathing [4], cardiac failure [5], high blood sugar levels [6], fatty liver [7], kidney and bladder infections [8], itchy skin and delayed wound healing [9] and loss of weight [10]. It is due to a lack of the hormone insulin produced in the beta cells of the pancreatic islets of Langerhans [green] [B]. Treatment [C] for juveniles includes a varied, low-carbohydrate diet, usually in conjunction with regular insulin injections. Adults also need a low-carbohydrate diet and weight loss, but drugs may help to control the disease. If not, then insulin may be administered.

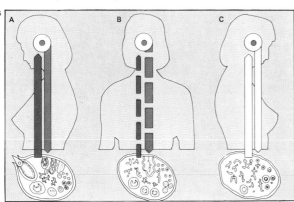

6 Before menopause [A] there is a balanced feedback mechanism between the ovaries, which produce the hormone oestrogen [red], and the anterior pituitary gland which produces ovary-stimulating hormones [blue]. At menopause (or the "change of life") [B], which usually occurs at 45 to 50, the ovaries cease producing an ovum and thus normal oestrogen amounts each month. As a result the pituitary produces an excess of hormones and the consequent high levels and hormone imbalance are thought to be responsible for the common symptoms of sudden heat sensations, headaches and weight gain. Oestrogens can be given as treatment. After menopause [C] hormone production ceases and the symptoms soon disappear.

7 Overactivity of the four parathyroid glands in the neck leads to raised levels of calcium in the blood [A]. Calcium is also lost in the urine where kidney stones and kidney damage result. The bones are weakened as calcium and phosphate are removed from the bone structure [B] in the body's attempt to maintain the high blood levels of calcium. Low levels [C] result if glands are damaged.

7 Parathyroid hormone
Blood calcium
Parathyroid gland

8 Apart from the endocrine or ductless glands, there are other glands that do not secrete their products directly into the bloodstream but via ducts to the appropriate site. These exocrine glands are affected by their own disorders. A sufferer from mumps has a viral inflammation of the parotid gland. This is one of the salivary glands which produce saliva to moisten the food and begin food digestion. Mumps may affect other glands such as the ovaries and testes.

Diseases of the urogenital system

While the urinary and genital systems serve very different functions, they are often considered together for they share anatomical parts [Key]. The openings of these systems, the exits for urine, the penis in men and the vagina in women, are potential points where infections from outside may enter [1]. Symptoms may include difficult urination as in cystitis (inflamation of the bladder), blood in the urine [2], back pain and disturbances such as fever, sweating and general sickness.

How infections start
Urinary infections are much more common in females, perhaps because the shorter urethra (the tube from the bladder to the outside) provides an easier path for invading bacteria than in the male. Of the 300ml (0.6 pint) or so of urine in the full bladder all but 1-2ml are normally passed. If bladder emptying is only partial, as may occur in the male with an enlarged prostate [4], any bacteria that reach the bladder are more likely to cause an infection as they are not adequately washed away. Congenital anatomical abnormalities present in the urinary tract, or the

development of stones in the kidney, favour bacterial infections.

Stones are formed from an accumulation of mineral substances filtered by the kidneys during the formation of urine. Stones may cause intense pain (ureteral colic) if they block or move down the ureter. If the kidney tubules, which are part of the kidney's two million or so filtering units or nephrons, are defective – either congenitally or as a consequence of pyelonephritis (severe kidney infection) – crystals of calcium salts may be deposited. Stones may form also in patients with gout.

The urogenital system can pick up infections during sexual intercourse. These venereal disorders are of various types [5, 6, 7]. They can produce both local symptoms at the time of infection and also permanent problems such as sterility. Other effects of syphilis may become apparent only after a period of years when widespread circulatory and nervous system diseases occur. However, the advent of antibiotics allows these diseases to be cured when they are promptly treated by specialists. Less serious but much more wide-

spread of vaginal infections causing irritation and discharge. These may be caused by fungi, especially *Candida albicans* (thrush) or microscopic amoeba-like organisms of the genus *Trichomonas*.

Sterility may be caused by venereal diseases and by other disorders of the urogenital system. These disorders include the retention of one or both testicles inside the male body (the testicles normally descend to the exterior before birth) and malformation, damage or disease of the testes or ovaries that impairs the production of sperm and eggs.

Kidney diseases
Kidneys may fail to function adequately either suddenly or slowly, and for many reasons. Less than 25 per cent of kidney tissue is necessary for normal removal of waste matter from the blood, and life is possible with even less than this. If kidneys fail completely, it is now possible to use a machine [3] to do their job of removing waste material. Alternatively, a diseased kidney can be replaced by a transplanted kidney. It is important that the transplanted kidney is

CONNECTIONS

See also
68 The digestive system
74 Reproduction
124 Radiology and radiotherapy

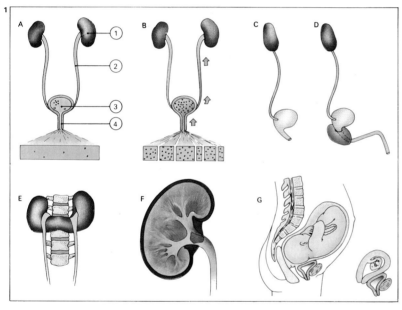

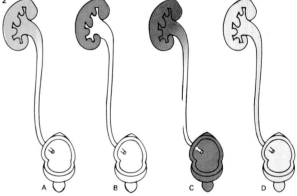

1 The urinary tract [A] consists of the kidneys [1], ureters [2], bladder [3] and urethra [4]. Infection of the tract [B] first affects the urethra and bladder giving symptoms of cystitis with painful and frequent

urination. If the infection spreads to the kidneys it causes inflammation and back pain. A number of factors may lead to infection. The short urethra in the female [C] provides easy access for bacteria, while enlarge-

ment of the prostate in the male [D] may prevent the bladder emptying completely. Inherited abnormalities of the kidney such as "horseshoe" kidney [E] and disorders such as kidney stones [F] predispose to urin-

ary infection. More frequent urination during pregnancy [G], especially in the first two to three months and in the last month, is partly due to distortion of and pressure on the bladder; it encourages infection.

3 A kidney dialysis machine is used if nearly all kidney units or nephrons fail, bringing the danger of death from an accumulation of waste products in the patient's blood. The patient can be connected in hospital, or later at home, to a filtering system that removes the impurities. Two or three overnight sessions a week are usually required to pass the blood with its impurities through a solution that has the correct concentration of salts.

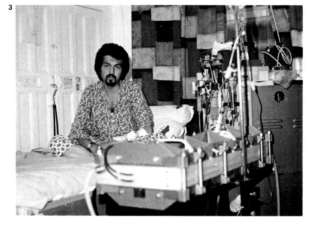

2 Indications of diseases or infections of the urinary tract can often be found by routine examination of urine. Abnormal contents such as blood, pus and proteins can give some informa-

tion on the type of disease and the part of the urinary system that is affected. Kidney diseases such as glomerulonephritis [A] are indicated by protein in the blood. In serious stages [B] the protein

coagulates into tiny cylindrical urinary casts. Damage to the kidney and bladder [C] is indicated by blood in the urine, and infection of the whole system [D] is indicated by the presence of bacteria in the urine.

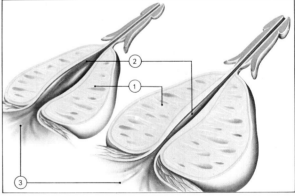

4 The prostate gland [1] is a partly glandular, firm structure that circles the male urethra [2] at the base of the bladder [3]. It produces some of the seminal fluid in which sperm are ejaculated. With increasing years, the gland

enlarges, normally a harmless process. However, it may constrict the urethra and prevent adequate urination. Urine becomes increasingly slow and difficult to pass, with dribble at the end of the stream. Infection can occur, the kid-

neys may be damaged and complete and painful stoppage of the flow may result. Treatment is usually required when this happens and also if urination is too frequent. This involves surgical removal of at least part of the gland.

carefully matched (like the matching of blood before transfusion), but drug treatment is necessary to help prevent the body's rejection of the graft.

Sudden kidney failure may occur if the blood pressure and flow through the kidneys drop too low – as after haemorrhage or severe burns. It is also a feature of acute glomerular disease where extensive swelling and inflammation is found in the glomeruli, which are parts of the filtering units. This can follow bacterial infection of, say, the throat by streptococci.

Many disorders can result from chronic renal (kidney) disease in which the ability of the kidney to remove the waste materials from the blood gradually deteriorates. The breakdown products of proteins are dealt with particularly badly by the failing kidney, and limitation of protein foods may be an essential part of the medical treatment.

Types of tumours

Like other parts of the body the urogenital system is at risk from tumour formation. In children, several rare types of kidney cancer are usually malignant, while in later life bladder tumours may be benign or malignant. They may be single or multiple polyps (wart-like growths) and can cause blood to appear in the urine. Often polyps can be removed or burned through a cystoscope – a tube inserted through the urethra by which the bladder can be inspected and treated.

The genitalia and reproductive organs may also show tumour growth. Tumours of the testes are rare, but the female reproductive organs are common sites for both benign tumours, often called fibroids, and malignant ones known as cancers. It has been found that cancer of the body of the uterus (womb) is more likely in women who have had no children. Similarly breast cancer is a little more common in those women who have no children, or in whom the first child was born when the mother was older than 30.

In contrast, cancer of the cervix, or neck of the womb, seems to occur in women with several children, and its incidence is influenced by factors in the male. Circumcision of the penis, or careful hygiene in the uncircumcised, appears to reduce the risk.

KEY

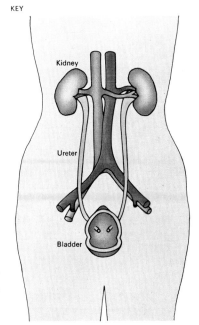

Kidney

Ureter

Bladder

The urinary system consists of two kidneys and ureters draining into a bladder and urethra (urine duct). Its function is mainly to produce and excrete urine, in this way maintaining the water and chemical balance of the body. Disorders of, or damage to, the kidneys may prevent normal urine production, while an enlarged prostate gland in the male may prevent the free passage of urine down the urethra. Stones, often of calcium salts, may form in the kidneys, while tumours may grow in the urinary tract. As in other tissues, infection can occur. The system is rather more vulnerable than some as the exit for urine may be an entry for bacteria.

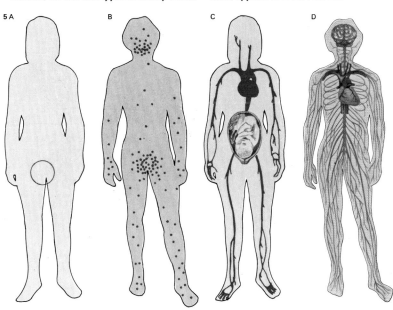

5 A B C D

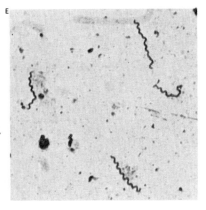

E

5 Syphilis is an infectious disease transmitted during sexual intercourse. Generally a sore or chancre appears painlessly at the site of the infection [A], usually on the genitalia, within ten weeks. The next phase occurs with a spread of bacteria by means of the blood, causing mild illness with skin and mucous membrane rash [B]. There may be no symptoms after this but a fetus in the uterus may be infected and born with congenital syphilis [C]. The disease may be dormant for up to 40 years but late effects can extensively damage heart and blood vessels, nerves and brain [D]. The bacterium is a spirochaete [E].

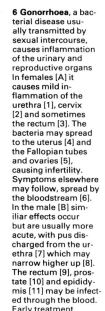

6 Gonorrhoea, a bacterial disease usually transmitted by sexual intercourse, causes inflammation of the urinary and reproductive organs In females [A] it causes mild inflammation of the urethra [1], cervix [2] and sometimes the rectum [3]. The bacteria may spread to the uterus [4] and the Fallopian tubes and ovaries [5], causing infertility. Symptoms elsewhere may follow, spread by the bloodstream [6]. In the male [B] similiar effects occur but are usually more acute, with pus discharged from the urethra [7] which may narrow higher up [8]. The rectum [9], prostate [10] and epididymis [11] may be infected through the blood. Early treatment for this is essential.

6 A B

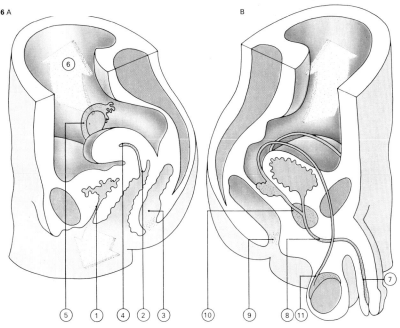

7 Number of cases at clinics in England and Wales, 1950-71

Gonorrhoea
NSU
Syphilis

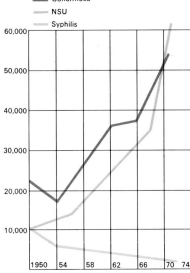

7 Cases of venereal disease have shown recent rapid increases in many countries. This cannot be accounted for simply by more open discussion of the condition and greater willingness to come forward and be treated. The rise may be related to increased promiscuity and the fact that oral contraceptives may decrease the naturally protective acidity of the vagina. Incidents of gonorrhoea and non-specific urethritis (NSU) have increased to a greater extent than those of syphilis.

Alcoholism and drug abuse

In every known society some people have formed the habit of taking substances that cause alterations of consciousness. There are two main groups of such substances, alcoholic drinks such as beers, wines and spirits, and drugs such as cannabis, mescaline, cocaine, heroin and LSD. Some of these are relatively mild and harmless, except when taken in repeatedly large doses; others, even in small doses, set up a dependence that can totally dominate the individual.

Susceptibility to alcohol
Alcoholics are drinkers who depend on alcohol to such an extent that eventually they show noticeable mental or physical disturbance. Most alcoholics are persistent heavy drinkers [1] but various patterns of alcohol abuse have been identified. Some drink steadily over many years and suffer physical damage in late middle age as a consequence. Others can function quite well without alcohol for quite lengthy periods but when exposed to alcohol cannot control the amount they drink. Bouts of heavy drinking lasting several weeks are not uncommon.

Most alcoholics are middle aged and male although in recent years there are signs that more young people and more women are abusing alcohol. People in certain occupations, such as waiters, salesmen, seamen, company directors, barmen and people working in the liquor industry, are particularly liable to alcoholism.

Alcoholism usually develops slowly. The alcoholic often begins by relying on alcohol to ease anxiety or depression, which it does briefly. However, tolerance to alcohol develops rapidly and increasing amounts are drunk to obtain relief. Gradually, the alcoholic develops physiological dependence on alcohol so that when deprived of it he exhibits withdrawal symptoms [3]. These include nausea and vomiting, tremors ("the shakes"), memory lapses, epileptic fits and delirium tremens [4]. Pathological drinking is invariably accompanied by difficulties at work due to absenteeism or drunkenness, family and marital disharmony, financial difficulties and mental and physical ill health.

Physical complications of alcohol abuse include gastritis, peptic ulceration, cirrhosis of the liver, inflammation of the pancreas and damage to the heart muscles. Brain damage with severe memory impairment is not uncommon. Psychiatric complications include severe depression with marked guilt feelings which sometimes ends in suicide.

Treatment of alcoholism is aimed at developing an awareness on the part of the alcoholic that he has a problem with drink. Some experts believe there is a form of metabolic fault in alcoholics and complete abstention seems to be necessary in most cases. Psychotherapy, aversion treatment and drugs are used to help achieve this.

The dangers of drug dependence
Drug dependence means the repeated non-medicinal use of a drug causing harm to the user or to others. All drugs can be dangerous and should be treated with extreme caution, irrespective of some current social attitudes that tend to glamorize the use of certain drugs. Drugs have been produced from a bewildering variety of plants but a large group, including morphine and heroin, are derived from the poppy and are called

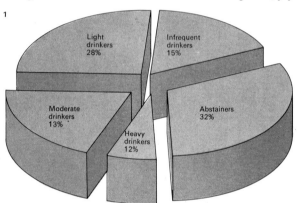

1 Drinking categories in a 1969 American survey assumed that abstainers drank less than once yearly; infrequent drinkers less than once monthly; light drinkers at least once monthly but not more than a drink (pint of beer or single measure of spirits) per session; moderate drinkers at least once monthly, three-four drinks per session; and heavy drinkers almost daily, with up to five drinks per session.

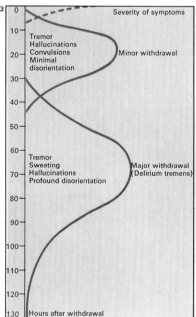

2 Short-term effects of alcohol on the body include dilation of skin blood vessels [A] and a drop in temerature. Gastric secretion [B] is increased at a low concentration (20mg alcohol/100ml blood) [1] but is inhibited at higher concentrations [2] and the stomach lining is irritated [3]. Inhibition of the pituitary gland in the brain causes increased production of urine [C]. Vital centres in the brain are depressed [D]. Judgment, self-criticism, perception and motor skills are impaired and sleepiness, coma and even death may occur [E]. Drinkers may change from being carefree to being irritable, disorderly and, finally, comatose [F].

3 Withdrawal symptoms can follow cessation of drinking after a heavy or chronic intake of alcohol. They can be relatively slight, occurring within 8–48 hours of withdrawal, or quite severe, extending over a period of three to eight days.

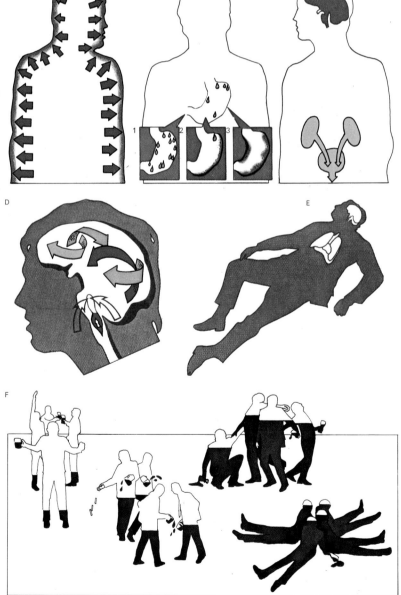

opiates [Key]. New addictive drugs have recently been developed in laboratories.

Addicts who cannot finance their habit (the rapidly developing tolerance to opiates means that ever-increasing doses are required) often turn to crime to finance their addiction. Deprived of his "fix" (jargon for injectable dose), the addict experiences severe restlessness, vomiting, diarrhoea and insomnia, eight to twelve hours after his last dose. Such symptoms last between three and seven days and can cause death. Complications of drug addiction include overdose, liver damage, sepsis and pneumonia.

Dependence on the so-called soft drugs (the barbiturates and the minor tranquillizers) is a serious problem, given the popularity of these drugs in the treament of insomnia [5] and anxiety. Such drugs differ in their ability to produce tolerance and physiological dependence, but their use, depending on the personality and social pressures involved, can in time lead to the abuse of other drugs, such as the opiates.

Illusion, hallucination, altered time sense, distorted judgment, confusion and disorientation are experienced in LSD "trips". There have also been reports of flashbacks – recurrences of the drug effects weeks or even months after the end of a trip. No withdrawal symptoms occur after taking LSD nor on stopping amphetamine intake. Amphetamines [6] produce a rapid onset of euphoria and in large doses can cause a severe paranoid psychosis.

The cannabis controversy

Whether *Cannabis sativa*, a euphoriant and relaxant derived from Indian hemp, is a dangerous drug of addiction or a harmless pleasure remains controversial. The dried leaves of the hemp are termed marihuana, the resin obtained from the flowering tops, hashish. In moderate doses, cannabis can produce a sense of excitement, heightened awareness and well-being followed by a phase of tranquillity and then fatigue.

Drug dependence can be treated by maintaining the addict on a controlled dose of his drug, by replacing the drug by a similar but less potent synthetic or by controlling withdrawal under supervised conditions.

Opium smoking was fostered in China during the 19th century by European traders intent on profit, particularly the British who crushed a Chinese attempt to stop traffic in the drug. By the mid-century, so-called opium dens were sordid features of Chinese cities. Although Europeans were less addicted, opium, laudanum and morphine were legal and common drugs in Europe as well as America in the same period

4 A B C D

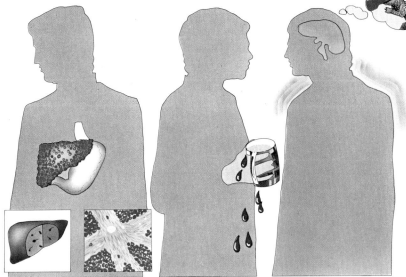
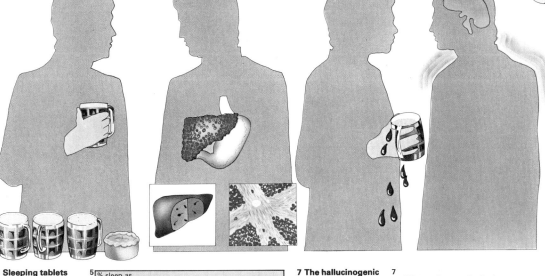

4 People who drink heavily often eat poorly [A] and, in the long term, malnutrition can result. Inflammation of the stomach and disease of the liver (cirrhosis) produce further loss of appetite. In cirrhosis [B], the liver is damaged either by the direct toxic effect of alcohol or by its effect on nutrition. Normal liver tissue is replaced by fibrous scar tissue. Deficiency of vitamin B1 damages the long nerves to the limbs, resulting in peripheral neuritis [C]. A sufferer loses touch sensitivity. Delirium tremens [D], characterized by extreme agitation and visual hallucinations, is a serious withdrawal effect.

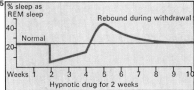

5 Sleeping tablets reduce the amount of REM (rapid eye movement) sleep when dreaming occurs. But REM sleep is higher than normal after withdrawal of the drug, with effects akin to delirium.

% sleep as REM sleep
Rebound during withdrawal
Normal
Weeks 1 2 3 4 5 6 7 8 9 10
Hypnotic drug for 2 weeks

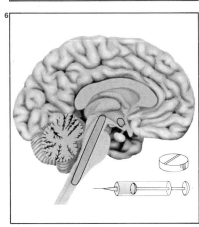

6 The amphetamine group of stimulant drugs acts on the brain in the region of the reticular formation and hypothalamus [yellow]. They can be taken as tablets or injected intravenously to decrease fatigue, increase alertness and lift mood. Tolerance develops rapidly, making progressively larger doses necessary. The euphoria, which follows injection particularly, is short-lived and gives way to depression. Severe paranoid states can occur.

7 The hallucinogenic drugs may produce their effects because they are similar structurally to substances [1] thought to be neurochemical transmitters. Three of the most common naturally occurring hallucinogenic substances are psilocybin, LSD and mescaline. Psilocybin [2] is extracted from the psilocybe family of mushrooms. LSD [3], a highly potent hallucinogen used extensively for therapy by some psychiatrists, was first isolated from a fungus, *Clariceps purpurea*, which causes ergot of rye. Mescaline [4], which has played an important role in the religious rituals of certain Mexican Indian tribes, is obtained from the peyote cactus.

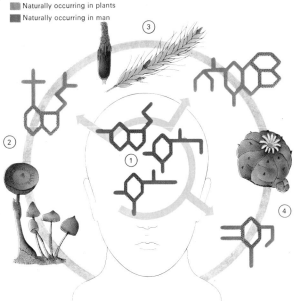

Naturally occurring in plants
Naturally occurring in man

Diseases of the Third World

Health or lack of disease depends, among other things, on quantity and quality of diet; effective control of housing, food standards and water and sewerage systems; and the availability of doctors, drugs and hospitals should a disease occur. This is a measure of the problem facing most of the developing countries that make up approximately 70 per cent of the world's 4,000 million population. All these factors are expensive and presuppose a solid industrial base – something that many Third World countries lack. Hence their health problems are immense: the life expectancy of a newborn baby in such a country is still about 35 years – about the same as it was in Western societies in the fourteenth century.

Insect-borne diseases

The diseases that still affect so many people fall into two main groups – infectious disorders and nutritional ones. The most important group of infections that occur in tropical countries are those spread by insects, which carry disease from person to person.

First among these insect-borne diseases is malaria, still one of the world's greatest killers. It is caused by a tiny parasite that is carried from man to man by the *Anopheles* mosquito [1]. When a mosquito sucks blood from an infected person the parasites are sucked up into the insect's stomach along with the blood. They breed and ten days later their offspring can be found in the mosquito's salivary glands. From then on the insect will inject a dose of parasites into anyone it bites.

The main symptoms of malaria are high fever, headache and violent shivering (rigors). While some species of the malaria parasite are relatively benign others often cause chronic ill health or death. Treatment is with drugs such as chloroquine, but prevention is a better solution. There are three types of prevention: killing the mosquitoes themselves [1]; stopping them from biting people, by using sleeping nets and repellants; and the administration of protective drugs.

Other mosquito-borne infections include various types of filariasis [3], common in the Pacific, the Far East and Africa, and yellow fever, a serious virus disease which occurs in West Africa and South America. For many years now yellow fever has hindered the development and cultivation of much of the South American hinterland. Prevention is the same as for malaria but immunization is an added precaution.

Several other insects are disease carriers. Leishmaniasis [4], dengue fever and phlebotomus fever are all spread by sandflies and cause much suffering and economic disruption, especially in the Middle East. Rat fleas still continue to spread bubonic plague [5] in some Third World countries while the tsetse fly is responsible for trypanosomiasis or sleeping sickness [6]. The rickettsial diseases are spread by various insects.

Lice are the carriers of ordinary typhus, while tick typhus is spread by dog ticks, and scrub typhus (tsutsugamuchi) is carried by rat mites; rat fleas carry murine typhus.

Diseases of insanitation

Almost as important as the insect-borne diseases are those of insanitation. These depend for their spread on the contamination of drinking water or food by human faeces. The linked problems of disposing of excreta and

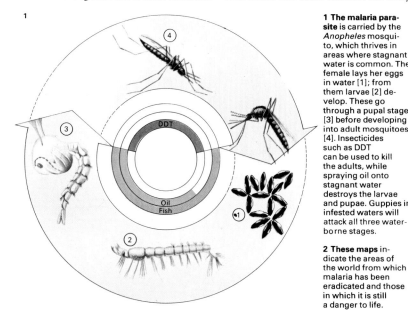

1 **The malaria parasite** is carried by the *Anopheles* mosquito, which thrives in areas where stagnant water is common. The female lays her eggs in water [1]; from them larvae [2] develop. These go through a pupal stage [3] before developing into adult mosquitoes [4]. Insecticides such as DDT can be used to kill the adults, while spraying oil onto stagnant water destroys the larvae and pupae. Guppies in infested waters will attack all three water-borne stages.

2 **These maps** indicate the areas of the world from which malaria has been eradicated and those in which it is still a danger to life.

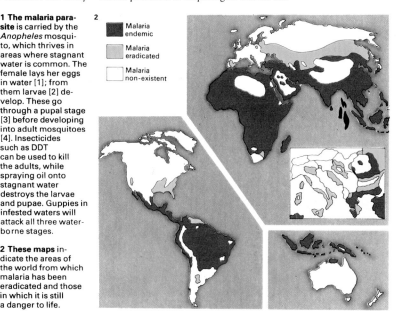

- Malaria endemic
- Malaria eradicated
- Malaria non-existent

3 **Elephantiasis** is one of the most dreadful types of filariasis, which is usually caused by the minute filarial worm *Wuchereria bancrofti*. These worms are injected into the human bloodstream by the bite of an infected mosquito.

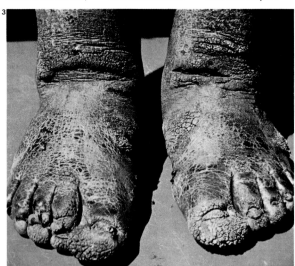

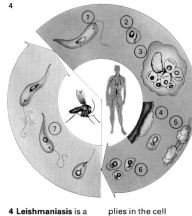

4 **Leishmaniasis** is a widespread disease of tropical countries, caused by a tiny parasite that is spread by infested sandflies. When the parasite enters the bloodstream [1] it changes form [2] and is ingested by a white blood cell [3]. The parasite multiplies in the cell and its "descendants" may attack the skin [4] or lymph nodes [5]. If the victim is bitten by another sandfly, parasites [6] will be sucked up by the insect with the blood and mature into forms that can infect other humans [7].

5 **Bubonic plague** ("Black Death") may still occur, especially in the Indian subcontinent. It is caused by the germ *Pasteurella pestis* [1] which can live in man, in rats and in rat fleas [2]. If a rat has the infection, the fleas that feed on it will ingest the germ. A man bitten by these fleas will probably develop the plague – and may spread the infecting organism to other humans via his breath.

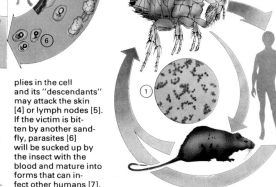

of providing pure drinking water are both difficult and costly to solve.

Infections that are spread by this "faecal-oral route" include many kinds of non-specific diarrhoea as well as typhoid, paratyphoid, cholera, various types of food poisoning, bacillary and amoebic dysentery and probably polio as well. Of this group the disease causing most concern in the 1970s was cholera [7], which was spread alarmingly from its base in the Far East to much of the world in those years. It also penetrated south of the Sahara desert for the first time in medical history.

Other sanitation-related infections to which the Third World populations succumb include hookworm and other worm infestations, which produce anaemia and chronic invalidism; and the tropical disease bilharzia or schistosomiasis [9]. The economic effects of bilharzia, a crippling disease that drastically reduces the ability to work, are severe.

Less important numerically are leprosy and yaws (both spread by prolonged personal contact only, and both now treatable) and rabies which is caught from animals, espe-

cially dogs, and is almost untreatable. The last of the important infectious diseases is smallpox, although this seems, at last, to be well under control and is confined to only six countries in Asia and northeast Africa.

Nutritional disorders

The well-known vitamin deficiency diseases such as pellagra, beri-beri and scurvy are not now of enormous importance in the Third World, partly because it is relatively easy to provide the very small quantities of vitamins needed to maintain health.

Far more serious are outright starvation and the related disease of kwashiokor or protein starvation. Half the world still goes hungry and in areas such as North Africa, where crops have repeatedly been hit by drought, or Bangladesh where floods have destroyed food supplies, millions of people find it desperately hard to obtain any kind of food. As the world's population grows inexorably and as the amount of protein available per mouth gets less and less, it seems that the problem of starvation is likely to get worse rather than better.

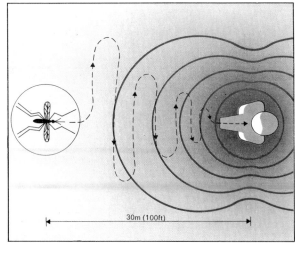

KEY

30m (100ft)

Mosquitoes find their victims in two stages. Carbon dioxide, which humans exhale, first sets them flying. If they contaminate food or water [A]. The victim ingests them and they incubate in his small intestine [B]. They then produce severe, then sense a current of warm, wet air, such as every person gives off, they home in on it; some people are more prone to mosquito often fatal diarrhoea [C]. The patient may survive if his gross dehydration can be corrected with intravenous fluid [D]. If bites than others because they are warmer and wetter. Repellants interfere with the accuracy of the mosquitoes' sensors. such therapy is undertaken promptly (and if no kidney damage has occurred) then full recovery should take place [E].

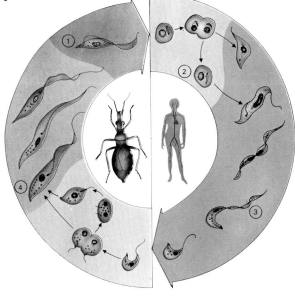

6 Trypanosome infections cause African sleeping sickness and South America's Chagas disease, which often produces fatal heart damage. The parasite [1] enters the blood following a bite by an infected insect such as an assassin bug or tsetse fly; it reproduces [2] and develops into forms that can be ingested by other bugs [3]. In these bugs, intermediate forms [4] develop into parasites to infect more people.

7 Cholera is one of the many diseases that spread because of poor hygiene. Human faeces containing cholera germs

7A   B   C   D   E

8 Insanitary conditions associated with food, water supply or sewage disposal are a sure way to spread disease. This Afghanistan meat market perched over an open sewer, could be a source of infection.

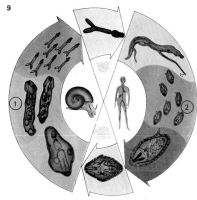

9 Bilharzia (schistosomiasis), named after Theodor Bilharz who identified it in 1851, today affects about 200 million people in tropical countries. It is caused by a minute parasite that spends part of its life cycle in freshwater snails [1] and part in water when it can penetrate the skin of anyone washing or bathing [2]. The invading parasite lays up to 40,000 eggs after developing into an adult. These interfere with blood flow and destroy tissues of liver, lungs and kidneys, reducing energy noticeably.

10 Leprosy has been one of the most dreaded diseases for many centuries – yet it is not highly infectious and efforts to control it are now meeting with great success. Caused by a bacterium (Hansen's bacillus) it is thought to be passed on only by prolonged close contact. It mainly attacks the skin and the nerves and may produce either trivial alterations in the tissues or gross deformities.

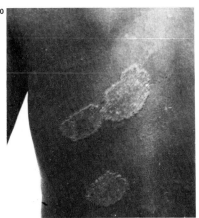

World health

The World Health Organization (WHO) defines "health" as "a state of complete physical, mental and social well-being, and not merely the absence of a disease".

The object of WHO is to try to achieve such a state of well-being for as many as possible of the world's peoples. It may seem at the moment that such a target is far off, when so many millions of people are suffering from disease and malnutrition. But in fact today's pattern of health and disease is in many ways far better than that of past centuries. In Western countries at least, men and women are living longer and healthier lives than ever before – even though "warning signs" in the shape of diseases of "civilization" (such as lung cancer, psychological stress and coronary thrombosis) are now appearing with much greater frequency.

Health problems in the Third World

Even in the poorer parts of the world there have been great improvements in health care during this century. Immunization against infectious diseases has helped to control once-dreadful scourges such as smallpox [7],

and energetic measures against the mosquitoes, flies and ticks that carry disease are slowly helping to defeat killer infections such as malaria, yellow fever and typhus.

The health problems of the poorer two-thirds of the world are still far from being solved, however – a fact easily demonstrated by the mortality figures of various nations [4]. In the less well-off countries there are high mortality rates among the young (in Nigeria over 180 babies out of every 1,000 die in their first year) and survival into later life is still the exception rather than the rule [1].

Several factors account for this situation. One is that the tropical climate of many developing countries encourages the development and expansion of the organisms that cause life-threatening infectious diseases. Another is that providing and maintaining good health care for a rapidly expanding population poses extreme economic problems. In much of the world infant mortality is high simply because there are no proper facilities for looking after mothers and their babies. Those babies who survive infancy may well die in childhood

because of malnutrition. And those who achieve adulthood may well die young because of lack of adequate medical care during illness or maternity. People who are weakened by malnutrition, and who have no doctor within reach, are easy targets for any kind of disease, especially the virulent tropical infections. These diseases need not be confined to the tropics. Today's air travel means that such infections can be carried to all parts of the globe in a short time. Monitoring the spread of such diseases is one of the major challenges to world health.

Organizations to the rescue

It was to deal with all these difficult and complex problems that the United Nations set up the World Health Organization in 1948 [Key]. By the mid-1970s WHO was active in more than 130 different countries and cost $1,000 million a year to run.

The functions of WHO are to act as a clearing house for medical information; to carry out research (especially in the field of epidemiology, whereby the origin and spread of disease is monitored by the Epidemiolog-

CONNECTIONS

See also
106 Diseases of the Third World
110 Community medicine
80 Birth control

In other volumes
302 History and Culture 2
298 History and Culture 2

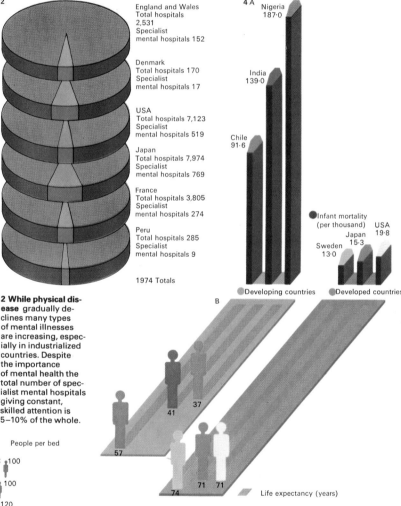

1 Developed and developing countries vary greatly in the availability of health care and this is reflected in the average age of their populations. Comparison of the percentage of the populations of the USA and Papua New Guinea over 75 [A], or of their general age distribution curves [B] shows marked differences. In developing countries with high birth-rates and low life expectancy, the average age of the population is often below 20 and the number of persons living into the 40s and 50s is relatively small. In developed countries most people survive early childhood and the average age is much greater.

1 % of population over 75

A
1·5
0·2 Males
0·1 Females 2·3

USA
Papua New Guinea

B
2·5 % of total population made up by each age
2·0
1·5
1·0
0·5
0 25 50 75
Males
Age

2·5 % of total population made up by each age
2·0
1·5
1·0
0·5
0 25 50 75
Females
Age

2
England and Wales
Total hospitals 2,531
Specialist mental hospitals 152

Denmark
Total hospitals 170
Specialist mental hospitals 17

USA
Total hospitals 7,123
Specialist mental hospitals 519

Japan
Total hospitals 7,974
Specialist mental hospitals 769

France
Total hospitals 3,805
Specialist mental hospitals 274

Peru
Total hospitals 285
Specialist mental hospitals 9

1974 Totals

2 While physical disease gradually declines many types of mental illnesses are increasing, especially in industrialized countries. Despite the importance of mental health the total number of specialist mental hospitals giving constant, skilled attention is 5–10% of the whole.

4 A
Nigeria 187·0
India 139·0
Chile 91·6

Infant mortality (per thousand)
Japan 15·3
Sweden 13·0
USA 19·8

Developing countries   Developed countries

B

57   41   37   74   71 71

Life expectancy (years)

3 A nation's health facilities can be measured by the number of people per doctor and the number of people per hospital bed available. A comparison of several countries provides interesting results. The USSR has more doctors per head of population than any other country in the world; in contrast some tropical countries have only one physician to every 40,000 people. Most developed countries have one bed for every 100 people.

3
People per doctor   People per bed

460 USSR 100
870 UK 100
670 USA 120
2,470 Colombia 400
560 Austria 950
2,330 Formosa 1,040
44,620   2,410
Nigeria
Doctors beds

4 National wealth is the key to good health, as found in developed countries. It can secure balanced diets, good living conditions and expert medical attention. High infant mortality and reduced life expectancy are indicators of poor health and are common in developing countries where such benefits are not easily available. In Sweden, Japan and the USA, fewer than 20 babies in every thousand die during the first year of life. In Nigeria the figure is nearly 200 [A]. Similarly, life expectancy [B] in the same three developed countries is over 70 years, but in Nigeria a newborn baby can only expect (on average) to live to 37. Much of Africa and parts of Asia have similar low life-expectancy levels.

ical Intelligence Service); to administer international sanitary regulations, especially in connection with travel and quarantine; to finance international research programmes including studies of cancer and tuberculosis; and to help individual countries train medical staff and to fight disease [9].

In this last field WHO does much to provide vaccines, antibiotics and other medicines. It also helps to disseminate health education information, which is of great importance in countries where disease and overpopulation are often due to ignorance.

The question of overpopulation now looms large in current thinking on world health. It is generally recognized that there are too many people in the world for its nutritional resources and health care facilities to cope with. More and more people are going to starve to death (or die from lack of medical treatment) unless the earth's rapidly burgeoning population can be limited. Much current world health work is devoted to fertility control research and to teaching people about the urgent need for family planning.

Part of the solution of present problems will lie in increased food production and in this field WHO works closely with one of the UN's other semi-autonomous bodies, the Food and Agriculture Organization (FAO), which has done much to improve world health by research into new crops and new fertilizers and by educating people in better techniques of farming. Unfortunately it seems almost certain that whatever scientific breakthroughs are achieved in this field there will never be enough food to feed even the world's present population.

Hope for the future

If there is to be any hope of a future for the Third World renewed efforts must be made to eradicate the dangerous communicable diseases of the developing world and to prevent the spread of the "civilization diseases" of the West. People must be educated in the ways of good health and more health personnel must be provided. More food must be grown. If all this can be achieved then there is some hope that the advances of this century may be shared by a far wider proportion of the world's population than at present.

The World Health Organization (WHO), whose symbol is shown here, is a semi-autonomous unit within the United Nations Organization. It is based in Geneva, Switzerland, and is active in more than 130 countries, mostly in the Third World.

5 **Cholera** spread in the early 1960s from its "base" in the Far East with alarming rapidity. Some cholera germs had changed their nature and this new variant was called the "El Tor" type. Although cholera has not been seen in most Western countries since mid-Victorian times, by the mid-1970s, as the map shows, it was breaking out in southern Europe, aided by increased air travel.

6 **The spread of rabies** (hydrophobia) across Europe is shown on this map. The actual incidence of rabies is unknown – WHO reports 600 to 700 deaths a year – but the figure is probably far higher. The virus of this terrible disease is spread by dog bites (although even a lick from an infected animal may prove fatal). Foxes and other wild animals can also be carriers of the disease.

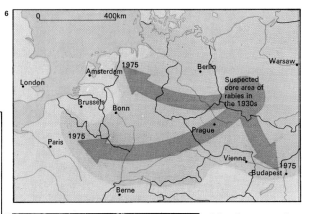

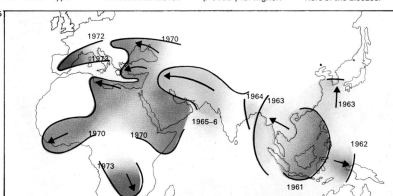

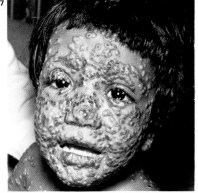

7 **Smallpox,** caused by a virus that is spread by contact with an infected human, was once a world scourge that killed or disfigured thousands of people (even in the most economically advanced countries) every year. Now it has been driven back so nearly to total extinction that many Western countries have abandoned routine vaccination of infants against it. Travellers may still need immunization, however.

8 **The thin, wasted muscles** and protruding bones of these beggars in Calcutta, India, are almost a badge of the developing countries and a sign of the tremendous problems facing them. These problems are exaggerated by the natural disasters (typhoons, flooding and droughts) that seem to plague them regularly.

9 **Regular, small-scale clinics,** like this one at Keneba, Gambia, for fighting malaria, contribute significantly to the control of disease. Particularly in areas of poor transport and few ancillary services such as electricity, this type of centre, tailor-made for local conditions, is more effective than prestigious hospitals of the Western type.

Community medicine

Anyone who has suffered from an attack of food poisoning after a meal in a restaurant, caught influenza from someone on a bus or fleas from a hotel bed or developed athlete's foot after visiting a public swimming pool will know only too well how the health of the individual depends on the health and cleanliness of the whole community.

People cannot function properly when they are ill and in the same way a society cannot function properly unless its members are healthy. As a measure of self-protection, therefore, every community takes steps to promote the health of its members. This is increasingly important as the size of the community increases, both because the results of ignorance, negligence or incompetence can potentially affect many more people more rapidly and because the opportunities for the spreading of disease are so much greater.

The scope of community health

Social medicine has two major functions: preventing disease and, if that fails, curing it. The second includes the building of hospitals and clinics, the training of nurses and doctors

and the financial support of sick people and their families. This aspect of social medicine is relatively expensive, both because of the cost of the facilities needed and because of the loss of productive work. The preventive aspect of social medicine is less dramatic and goes largely unnoticed, yet it affects all of us.

On one level these measures range from the proper disposal of the dead to the myriad regulations, typical of almost all countries today, that control building standards; density of population in towns and cities; the numbers of people that can be carried on ferries or seated in cinemas; conditions for workers in factories; and permissible levels of aeroplane noise, emissions from car exhausts and industrial toxic wastes. The tragedy during the 1950s and 1960s in Minamata, Japan, where unsupervised dumping of paper mill effluent into the sea caused the death or disablement through methyl mercury poisoning of those who ate fish or shellfish caught in the area, highlights the need for community action to ensure public health.

The same vigilance is needed in testing new drugs or similar products which are

brought on to the market before they are thoroughly tested, sometimes with unpredictable and tragic side effects.

A clean water supply

The giant Roman aqueducts that are still scattered across Europe [6] and the ingenious system of underground cisterns or *qanaats* that honeycombed ancient Persia, attest to the vital importance that every community has attached to water. And the linked needs of providing houses with a regular supply of pure drinking water and taking away sewage in such a way that the two do not mix are still major preoccupations of health authorities. For the contamination of water with even tiny amounts of faeces can cause an epidemic of dysentery, cholera or typhoid.

The risk is greatest in developing countries where these diseases are endemic (being highest when an earthquake, typhoon, flood or other disaster disrupts the normal facilities), but can easily occur even in a country as proud of its cleanliness as Switzerland. An outbreak of typhoid in Zermatt in 1963, which affected 313 people and killed

1 The risk of infection from contaminated food is high and so all stages of its production, storage, processing and preparation are carefully regulated and monitored. Imported food is subject to the same standards and is rigorously checked both to ensure adequate hygiene and to exclude any new animal and crop pests or diseases that are not already endemic in the country.

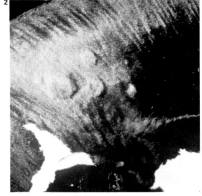

2 Tuberculosis of the bone and brain has been largely eradicated in developed countries over the last 40 years by stringent control of milking herds. Cows are tested every year for the presence of tuberculosis bacilli by injecting them with tuberculin. A positive reaction (swelling under the skin as shown here) indicates they carry the disease and can pass it in their milk to humans.

3 Efforts by restaurants and hotel managements to maintain standards of cleanliness, proper water supply and waste disposal, adequate toilet facilities and staff health are checked by official inspectors.

4 Animals have to be slaughtered in accredited abattoirs both to ensure painless death and to enable the meat to be inspected for the presence of such infections as tapeworms and tuberculosis.

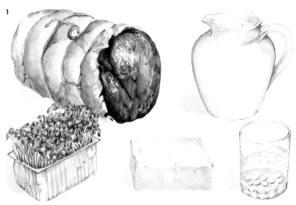

5 Vaccination of cattle against contagious abortion or brucellosis not only prevents the loss of calves but also protects humans from the unpleasant undulant or Malta fever, found in tainted milk.

6 The Pont du Gard, near Nîmes, France, is one of the best known of the Roman aqueducts still standing. The longest built was the 92km (57-mile) Acqua Marcia from the upper Aniene valley to Rome, of which only 11km (7 miles) was above ground.

7 Even fresh vegetables can be a potential source of disease. Watercrass, for example, can carry on its leaves the eggs of the sheep liver fluke (*Fasciola hepatica*), a parasitic flatworm which causes liver "rot". Parts of its life cycle is spent in the freshwater snail so every effort is made to prevent infected snails from entering the beds. Water in well-managed watercress farms is drawn from wells or springs that carry no risk of infection and sheep are kept away.

three before it was finally eradicated, was traced to faulty sewage disposal.

Water has to be properly collected – from springs, rivers or deep bore-holes – stored, purified (usually with chlorine, although ozone is sometimes used) and then distributed. In some parts of the world fluoride is added in minute quantities (1 or 2 parts per million) because experiments have shown that its presence decreases the incidence of tooth decay. Constant tests are carried out at all stages to make sure that no contamination from sewage or other impurities takes place.

The technology exists to clean sewage so thoroughly (and in the process remove the river, lake and sea pollution that is such a feature of industrial countries) that the water can be re-used directly; but the costs involved in these processes are high and, all too often, the priority is regrettably low. There are usually two or sometimes three stages in the cleaning of sewage, and the first stage alone removes about 30 per cent of organic wastes.

A further advantage of an adequate water supply is that it makes a water-carried sewage system possible. Such a system, if linked by proper sewers [10] to sewage treatment works, greatly reduces the danger of diseases spreading, particularly those such as food poisoning, poliomyelitis and dysentery that are transmitted from faeces to food by flies.

Maintaining food standards
All stages of the production of food, from growing and processing to preparation and eating, must be carefully regulated. Techniques for the improvement of food include the addition of vitamins to such common foods as margarine and bread, the pasteurization of milk [8] to prevent the spread of tuberculosis and brucellosis, the inspection of meat to exclude tapeworm infestation [4] and the addition of iodine to salt to combat goitre.

But no matter how high the quality of food it can still become a danger to health if it is not properly handled. Thus there are laws to ensure hygienic conditions in food factories (faulty canning is particularly hazardous, carrying with it the risk of botulin poisoning), food shops and restaurants [3]. The aim is to prevent food becoming contaminated in any way.

KEY

Public wash-houses were common in European cities up to the 1930s. Erected by the authorities, they were heralded as a great advance in hygiene and even of substantial moral benefit to the "poorer" classes. Those built in villages simply provided cold running water and rubbing boards, but later town wash-houses were equipped with large boilers for hot water and even steam-driven spin dryers. Rising wage levels and the advent of cheap electrical washing machines largely removed the need for them and they have been replaced by coin-operated laundries equipped with washing machines, spin dryers and often dry-cleaning machines.

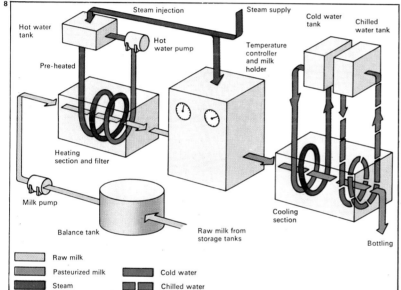

8 Pasteurization of milk destroys the bacteria that cause tuberculosis and undulant fever. In the process, named after the French scientist Louis Pasteur (1822–95), the milk is heated to 62–71°C (144–160°F), which is lethal to most bacteria but not high enough to alter the quality of the milk. As fewer cows carry infectious diseases, pasteurization's main use is in improving the keeping qualities of milk.

9 Unwanted solid refuse can be disposed of by dumping on land or at sea and by incineration. Refuse poses wide environmental problems through its increasing volume and the use of materials that break down slowly.

Steam injection
Steam supply
Hot water tank
Hot water pump
Cold water tank
Chilled water tank
Pre-heated
Temperature controller and milk holder
Heating section and filter
Milk pump
Balance tank
Raw milk from storage tanks
Cooling section
Bottling

Raw milk
Pasteurized milk
Steam
Cold water
Chilled water

10 The provision of adequate facilities for disposing of storm water and sewage in towns is a major health responsibility of local government. Drains and sewers remove such waste for treatment and purification before it is discharged into rivers, lakes or the sea.

11 Many diseases are spread by infected animals or their parasites. Rabies is mainly transmitted by foxes, dogs, badgers and cats. Where there are no barriers to the movement of animals such diseases spread rapidly. The sea is the most effective barrier. Provided all imported animals are kept in quarantine for the incubation period of the organism, any island, like Britain, can keep such diseases at bay.

Preventive medicine

Preventive medicine is the sum of all those measures – general physical check-ups, vaccinations, dental visits and eye tests – that forestall illness either by entirely preventing its occurrence or by catching it so early that it can be easily and quickly treated. Waiting for teeth to ache before visiting the dentist, for example, could well mean extensive filling and even the removal of nerves or teeth.

Care before birth

Understandably, preventive medicine is most in evidence and most widely accepted during pregnancy. Indeed, the process often begins even before conception. At this stage it is often possible to predict the chances of parents with a genetic defect producing a diseased child, and help them to make a decision for or against parenthood. A woman carrying the trait for haemophilia, for example, might decide against bearing a child on being told that any daughters, while directly vulnerable, would have a 50 per cent chance of being carriers and that any sons would have a 50 per cent chance of suffering from the disease.

During pregnancy, an expectant mother makes regular visits to an antenatal clinic so that the progress of the fetus and her own health can be monitored. She is checked to ensure that she is not getting too fat, that blood pressure is not too high, that she is not suffering from diabetes and that she has not contracted syphilis [4] or any of the diseases specific to pregnancy. Particular vigilance is exercised against the metabolic disorder called pregnancy toxaemia.

These antenatal visits also guarantee early action in the event of complications such as narrowness of the birth canal or malposition of the placenta. Monitoring the fetus assures its viability and health and that, at the end of pregnancy, it is in the correct position in the womb for easy delivery.

In recent years a sophisticated new process has been developed to monitor the fetus – amniocentesis [1]. This technique, usually employed when it is suspected that the fetus may have some hereditary defect, involves the passing of a small hollow needle through the mother's abdominal wall into the bag of fluid, the amniotic sac, in which the fetus lies. A little of the fluid is withdrawn and the cells that it contains are grown in tissue culture. These cells are then examined for abnormalities of their chromosomes and chemistry. Not all the abnormalities identified by this method have an ominous significance. However, the discovery of the characteristics of mongolism and Tay-Sach's disease (which causes blindness and severe mental retardation) will force the physician to think of ending the pregnancy prematurely.

Vaccines versus disease

In the early years of life preventive medicine is largely concerned with ensuring proper nutrition and with vaccination against the common infectious diseases of childhood.

The use of vaccines began in 1796 when Edward Jenner (1749–1823) demonstrated the capacity of *Vaccinia* virus to provide protection against smallpox. Today in the temperate countries of the world the objective is to vaccinate all infants during the first year of life against diphtheria, tetanus, whooping cough, poliomyelitis and measles. So effective have these programmes been that, with the exception of whooping cough,

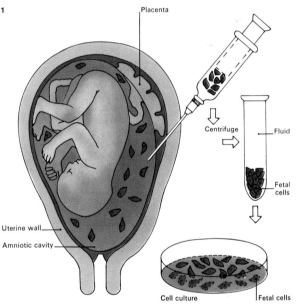

1 Amniocentesis – the extraction by needle puncture of cells from the fluid around the fetus – greatly increases the chance of identifying the presence of hereditary diseases. At present the fetus's sex and some of 40 of the more than 1,600 genetic diseases can be detected after the cells have been grown in culture. The best time for the "tap" is about the 16th week of pregnancy. Problems of using this technique are its delicacy and the risk of long-term damage to the fetus. The possibility of its being used for eugenics or genetic engineering and antenatal sex determination raises questions about the social effect of wide use.

Placenta
Centrifuge
Fluid
Fetal cells
Uterine wall
Amniotic cavity
Cell culture
Fetal cells

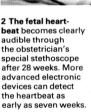

2 The fetal heartbeat becomes clearly audible through the obstetrician's special stethoscope after 28 weeks. More advanced electronic devices can detect the heartbeat as early as seven weeks.

3 Vaccines are produced by artificially cultivating the illness-causing viruses or bacteria. They are collected from infected animals or humans and grown in conditions suited to fast development as here. The micro-organisms produced are washed, centrifuged in a sterile salt solution and killed. The viruses are finally weakened to make them less virulent. A dose injected into the body induces the blood to form antibodies, which combat invasion by the disease.

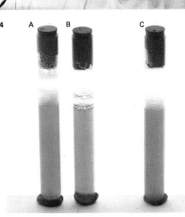

4 Maternal syphilis is a serious threat to an unborn child. The blood of pregnant women is always examined for this infection, usually by the Wassermann reaction. Syphilitic blood inhibits the activity of a substance called "complement", which dissolves specially prepared red blood cells of sheep. These test tubes show a positive control [A], a negative control [C] and a positive test [B]. Fever-causing diseases such as malaria can produce a false positive reaction.

all these diseases are increasingly rare. Tuberculosis has presented more enduring problems, but mass miniature radiography [6, 9] for early diagnosis and the use of antibiotics for treatment have reduced the threat of this once epidemic disease to almost negligible proportions.

In some circumstances extra vaccinations are provided on a more selective basis. German measles (rubella) vaccine is used for adolescent girls who have not had the disease, to prevent the risk of an attack during pregnancy. There are also influenza vaccines, rabies vaccine for the staff of quarantine centres and zoos, anthrax vaccine for veterinary surgeons and hide porters, and a vaccine for hospital workers likely to contract hepatitis as a consequence of handling blood. The length of time for which these vaccines give protection varies. For cholera it is six months, for yellow fever a lifetime.

Preventive medicine for adults

In later years the emphasis in preventive medicine changes from absolute prevention of infectious illness to early diagnosis by

periodical check-up of the illnesses that threaten in middle life.

A physical examination combined with measurement of blood pressure, an X-ray of the chest, an electrocardiograph and an examination of the urine for the tell-tale sugar of diabetes would constitute a typical thorough check on the more common diseases and ailments of middle age [8]. In more elaborate examinations the blood is tested for excess fats and cholesterol.

For women, there are extra hazards associated with their reproductive systems and special clinics are today becoming more common in developed countries. The functions of such clinics include early diagnosis of cancer of the breast and of the cervix. Careful palpation of each breast usually reveals any lumps suggestive of cancer, but more complex methods using soft X-rays and heat sensors are sometimes used [5]. Examination of the cervix is usually accomplished by the cervical smear method in which the cells from the cervix are smeared on a glass slide and examined microscopically for evidence of cancerous changes [7].

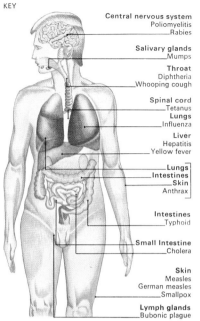

KEY

Central nervous system	Poliomyelitis, Rabies
Salivary glands	Mumps
Throat	Diphtheria, Whooping cough
Spinal cord	Tetanus
Lungs	Influenza
Liver	Hepatitis, Yellow fever
Lungs	
Intestines	
Skin	
	Anthrax
Intestines	Typhoid
Small Intestine	Cholera
Skin	Measles, German measles, Smallpox
Lymph glands	Bubonic plague

The prevention of infectious diseases through vaccination is an important part of preventive medicine. The administration of most vaccines causes the production of substances known as antibodies in the blood of the vaccinated person. These, either alone or in concert with other components of the blood plasma or with the white blood cells, attack invading microbes and so prevent infection. The effects of antibodies are quite specific and those induced by a vaccine intended to prevent one infectious disease give no protection against other diseases. Shown here are the various diseases that can at present be combated by vaccines and the sites of their attack.

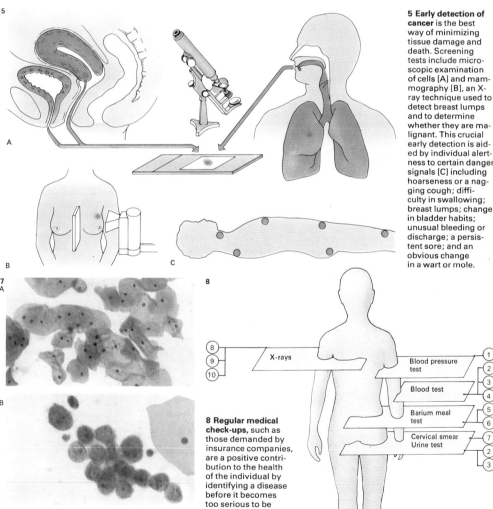

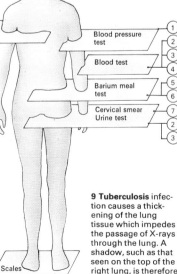

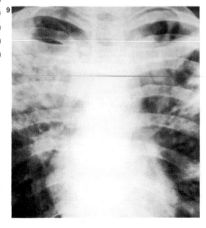

5 Early detection of cancer is the best way of minimizing tissue damage and death. Screening tests include microscopic examination of cells [A] and mammography [B], an X-ray technique used to detect breast lumps and to determine whether they are malignant. This crucial early detection is aided by individual alertness to certain danger signals [C] including hoarseness or a nagging cough; difficulty in swallowing; breast lumps; change in bladder habits; unusual bleeding or discharge; a persistent sore; and an obvious change in a wart or mole.

6 Mass miniature radiography is a cheap way of examining the lungs by X-ray and has been instrumental in the campaign to eradicate tuberculosis. X-rays passing through the patient's chest fall on a fluorescent screen to form a picture similar to that on an X-ray plate. This picture is photographed and the small photographic negative can then be examined by the radiologist.

8 Regular medical check-ups, such as those demanded by insurance companies, are a positive contribution to the health of the individual by identifying a disease before it becomes too serious to be easily cured. Factors watched out for include high blood pressure [1], diabetes [2], nephritis (kidney inflamation) [3], anaemia [4], intestinal [5] and rectal [6] abnormalities, vaginal cancer [7], tuberculosis [8], cancer of the lung [9] and breast [10], and overweight [11].

X-rays
Blood pressure test
Blood test
Barium meal test
Cervical smear
Urine test
Scales

7 Microscopic examination of cells can help detect several different cancers. A smear taken from the cervix during what should be a yearly examination would reveal whether the cells are normal [A] or have undergone cancerous changes [B].

Cells from the urinary tract shed with urine may indicate kidney, ureter or bladder cancer. Lung cancer may be verified by examination of sputum, while oral cancer can often be detected by examining cells scraped from the mouth.

9 Tuberculosis infection causes a thickening of the lung tissue which impedes the passage of X-rays through the lung. A shadow, such as that seen on the top of the right lung, is therefore produced on the X-ray plate. The heart shows as a bulge at the base.

The body's natural defences

A characteristic of all living matter is a certain resilience to disease and damage. The human body possesses a formidable array of such natural defences [Key]. If we are threatened our bodies have many resources at their disposal. These include an early-warning system of muscular reflexes to keep us away from harmful stimuli; frontier defences such as skin; the blood-clotting mechanism plus an emergency regime for dealing with excessive blood-loss; a chemical immunity system to overwhelm or neutralize potentially harmful germs; and mechanisms for the natural repair of damaged bone and soft tissue.

Invading micro-organisms

There are several distinct groups of potentially harmful germs or micro-organisms. They are, in order of size, viruses, bacteria and protozoa and can invade the body by way of contaminated food or water, during contact with other people or animals, through a wound or through direct inhalation in the air we breathe. Although many are harmless – and the presence in the body of certain of these microscopic organisms is essential for

good health – some cause disease.

The three frontiers through which these organisms invade are the skin (and the mucous membrane lining the mouth, vagina and other orifices, which is a derivative of skin) and the respiratory and digestive tracts. The skin has a slightly acid surface and is too cold and hostile an environment for most germs. An additional safeguard is sweat, which contains an enzyme able to break up bacteria. Delicate membranes such as those of the vagina are protected by acid secretions and by the presence of harmless bacteria that subdue disease-producing forms.

At the entrance to the respiratory tract hairs in the nostrils filter out some unwanted particles; others are trapped in the mucus lining the nasal passages and are disposed of by way of the stomach. Farther on, foreign matter is wafted away by microscopic hairs known as cilia which line the trachea and bronchi of the lungs. And patrolling the internal surfaces of the lungs are special scavenger cells that engulf intruding organisms and signal a warning to the immune system. In the digestive tract – possibly the

most hostile frontier of all – germs are destroyed by stomach acidity or by digestive enzymes produced in the intestine.

If these various threshold defences fail, bacteria, viruses and the rest are free to set to work in the body, damaging tissue or releasing toxins (poisonous substances) to cause illness. Yet these germs usually cause disease only if they are allowed to muster in force – a situation often forestalled by the flushing action of various body fluids.

The immune system and tissue repair

When intruders do manage to take hold, they achieve the status of antigens. Now it is the turn of the immune system, the body's own chemical warfare division, to produce the antibodies necessary to fight off infection. A defensive mechanism dispersed throughout the tissues, the immune system consists principally of front-line troops known as lymphocytes. These are particular kinds of white blood cells produced in the bone marrow, thymus and spleen and found in the lymph nodes that are located at intervals throughout the lymphatic system [5].

1 **The blood-clotting mechanism** is the result of a complex chain reaction involving various substances in the blood. Through the interaction of blood platelets, plasma and tissue-clotting factors the soluble protein fibrinogen breaks up into fibrin threads and forms a mesh across the wound. Prior to injury [A] the fibrinogen circulates in the plasma. At the moment of injury [B] the platelets plug the wound while the clotting reaction takes place. Fibrin threads are laid down [C] across the site. Platelets and blood cells become trapped in the fibrin web [D]. This semi-solid mass shrinks, extruding serum (a yellowish fluid) and the clot is formed.

Tissue factor
Plasma factor
Fibrinogen
Platelet
Red blood cell
Fibrin

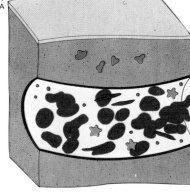

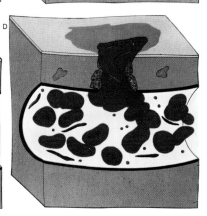

2 **White cells** are important in defence against bacterial infection such as that following entry of a splinter. The most numerous type, polymorphonuclear leucocytes [1], surround infecting microbes [2]

and prevent them from spreading to adjacent tissue. Many polymorphs may be killed in the process and these form the pus. There are more to replace them and eventually the infection is eliminated.

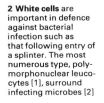
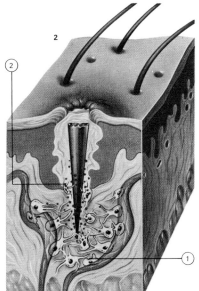

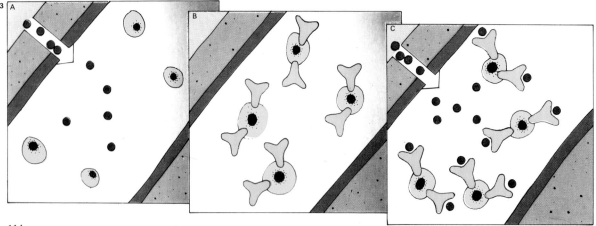

Antigen
Antibody
Antibody/antigen complex

3 **When an antigen enters the blood** [A] it causes a proliferation of lymphocytes (yellow) with specific membrane-bound antibodies [B]. These remain in the bloodstream for prolonged periods. If the same antigen [C] attacks again, the antibodies encounter the invaders and neutralize them.

The oustanding feature of lymphocytes and other disease-fighting cells is their ability to "recognize" specific antigens and to produce precisely the right antibodies to combat them. Antibodies work in different ways and each offers protection against only one disease. Some cause the antigens to clump together, making it impossible for them to spread; some coat the invaders with a substance that makes them more readily ingestible by phagocytic cells such as macrophages; and others neutralize the harmful toxins that are released by antigens [3].

A sequel to victory over infection is the repairing of tissue. Wounds are healed by the arrival of fibroblasts – cells that lay down strands of fibre forming a scaffolding into which new cells grow from the surrounding healthy tissue. Fibroblasts are also mobilized to renovate tissue damage resulting from injuries such as cuts or fractures [6].

Blood clotting and conservation

An important protective mechanism in the case of an open wound is the coagulation of blood [1]. The clotting process depends on various physical and chemical changes but is due ultimately to the conversion of the soluble fibrinogen of the blood plasma into the jelly-like protein known as fibrin. Threads of fibrin from the blood clot and it is this coagulation that arrests the flow of blood [4].

The body also employs techniques to make good the effects of blood loss. If the haemorrhage is severe there is a fall in blood pressure. The severed vessel narrows and there is a tendency for the inner lining to stick to itself. Other blood vessels near the surface of the body become constricted, thus conserving an adequate supply of blood for direction to vital organs such as the heart and brain. Contraction of the spleen may add a pint or more of blood to the general circulation to compensate for the amount lost. Also, since it is not immediately vital to replace red cells, the blood vessels borrow fluid from the tissues to restore volume to the circulation and bring blood pressure back to normal. In this way, with a natural "transfusion" of tissue fluid, a human being can survive the loss of up to a quarter of the total amount of blood that circulates in his body.

KEY

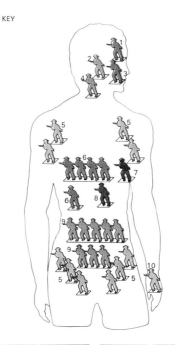

The body fields an array of defences apart from the skin [10]. Lachrymal glands [1] secrete tears to wash away foreign particles. Lymphoid tissues, important in the destruction of germs, include the adenoids and tonsils [2]. Saliva [3] is an effective barrier fluid. Foreign matter is trapped by mucus in the nose and throat [4]. Lymph nodes [5] release protective white blood cells. The liver [6] destroys germs and produces substances vital in blood clotting and tissue repair. A source of defensive white corpuscles is the spleen [7]. Acid in the stomach [8] kills many germs; those reaching the intestine [9] succumb to digestive enzymes.

4 A wound is first sealed by a blood clot [A]. As healing progresses macrophages move in [B] to remove wound debris such as dried blood, bacteria and damaged cells. When the fresh layer of skin grows across the site [C] the remains of the blood clot are sloughed off in the form of a scab.

- Macrophage
- Debris
- Blood clot
- Epidermis
- Dermis

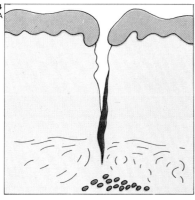
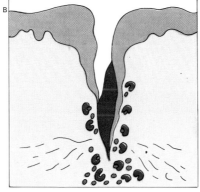
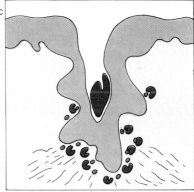

5 The lymph system is a network of lymphatic vessels [1] which collects tissue fluid (the lymph) and conducts it back to the bloodstream at the large subclavian veins [2]. In the process it transports nutrients from blood to cells and cell wastes back into capillaries. Lymph drains through the system but the lymphatics possess valves [3] to prevent backflow. Lymphatic nodes [4] are scattered along the lymph vessels but particularly in the neck, armpits and groin. In the tissue around the nodes micro-organisms are destroyed by macrophage cells, while antibody-synthesizing white blood cells, the lymphocytes, are produced in the lymph nodules [5].

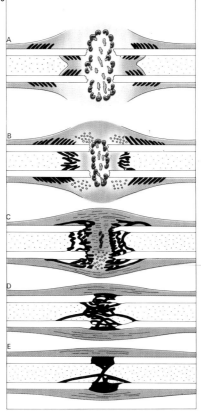

- Macrophage
- Debris — old bone
- Bone
- Bone marrow
- New bone
- Blastema
- Connective tissue
- Cartilage

6 A fracture is the term used for any break in a bone, whether or not the bone penetrates the skin. After an injury of this kind inflammation sets in and a blood clot forms, sealing the ends of damaged vessels [A]. Macrophages invade the site, and remove the wound debris; then after one or two days, long, thin fibroblasts lay down a mesh or grid on which new tissue can grow [B]. A blastema or area of new growth develops and new bone is forged that links the broken ends [C]. Later remodelling strengthens the bone, restoring its true shape [D, E]. At least four to six weeks are required for a fracture to heal completely.

Origins of curative medicine

The practice of medicine in one guise or another predates written history and much of our knowledge of its early forms is therefore a matter of conjecture. It seems likely that medicine did not emerge as a specialized craft until settled communities developed. Then men learned about different plants – which were good to eat; which poisonous; and, presumably, which appeared to alleviate symptoms of the various diseases that were probably as prevalent then as now. Certainly, from the first written records, it appears that familiar ailments were common.

Religion, magic and medicine
Curative medicine was apparently often allied with religion, as it was in Egypt [3]. The administration of medicinal materials was accompanied by complex rituals and incantations and it is quite possible that the psychological therapy of the rituals was as effective as the medicaments, which were often chosen on magical grounds.

One of the basic tenets of magic is the principle of correspondence, known also in medicine as the doctrine of signatures. This magical principle asserts that there are connective links between different things on the grounds of analogy. Plants were seen to have similarities with different aspects of the human condition; so, for example, the flat, mossy liverworts were used until the eighteenth century in the treatment of liver ailments simply because their shape resembled that of the liver.

A related belief concerns the transference of powers of one sort or another between objects [Key]. When a cannibal ate the heart of a strong warrior whom he had slain in battle, he believed that some of the dead man's strength would pass into him. By the same token, mummified flesh was at one time a popular remedy, presumably because anything that had survived death so well would help the eater to do the same.

The converse of this is the passage of bad qualities from the sick person to some other person or object. As much early medicine was based on magical beliefs, it is not surprising that many ailments were interpreted as possession or attack by a malevolent being. Such beliefs may well have prompted surgery, some of the earliest evidence of which is found in trephined skulls from prehistoric sites. Trephining – making a hole in the skull – was an operation performed presumably as a way of letting an evil spirit out of the sufferer's head. Excess pressure in the brain may lead to behaviour that could be interpreted as demonic possession, so this operation may, ironically, have been of value to those who survived the treatment.

In the first millennium AD, victims of "elf-shot" (lightning) in some parts of Europe were buried up to their necks in the ground, so that the malignancy would be transferred to the earth, which would then bind it. In more recent times, rheumatism sufferers slept in contact with young and healthy slaves in order to transfer the rheumatic pains to them. A continuing and equally baseless cure for rheumatism is the wearing of copper bracelets.

Later theories of anatomy and medicine
As civilization developed, many of the early magical practices continued in a distorted form. The Greek division of bodily principles

1 A   B   C

1 Pliny the Elder (AD 23–79) [A] wrote *Historia naturalis*, for centuries a source of information on herbal remedies. Later, John Gerard (1542–1612) [B] and Nicholas Culpeper (1616–54) [C] both produced bestselling herbals. Plants were the first medicines ever used and still form the basis of much of today's curative medicine. Catalogues of plants effective against disease have always been popular.

2

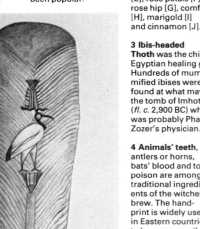

3

2 Herbalists still sell assorted plant materials and claim curative properties for them. Here are sandalwood [A], camomile [B], Chinese herb teas [C], blue mallow [D], kola nuts [E], rose petals [F], rose hip [G], comfrey [H], marigold [I] and cinnamon [J].

3 Ibis-headed Thoth was the chief Egyptian healing god. Hundreds of mummified ibises were found at what may be the tomb of Imhotep (*fl. c.* 2,900 BC) who was probably Pharoah Zozer's physician.

4

4 Animals' teeth, antlers or horns, bats' blood and toad poison are among the traditional ingredients of the witches' brew. The hand-print is widely used in Eastern countries to keep away evil spirits – as the copper bangle is intended to ward off rheumatism. Not all folk remedies are unpleasant: the opium pipe and gin bottle have since been used more for pleasure than relieving pain.

into four "humours" – blood, phlegm, yellow bile and black bile – reflected their philosophical view of the universe as being composed of the four elements – earth, air, fire and water. Long after the alchemists had discovered that matter consisted of more than these four, the humours were still the basis of medical practice. To restore health, it was believed necessary to find treatments that would restore the balance between them. Chinese medicine was based on a similar maintenance of balance, but the principles in which they believed were called yin and yang (negative, feminine and dark; and positive, masculine and light). Drugs were prescribed to restore harmony between these two contrasting principles.

Ancient cures still used

Although the early systems of medicine were not based on scientific premises, not all the cures proposed were useless. Presumably doctors then were observant enough to note when a plant substance was genuinely effective. In some early civilizations there were strict laws to keep the doctor on his toes.

About 5,000 years ago, in Babylon, a doctor who killed his patient was liable to have his hands cut off. Similar penalties for bad medicine existed in ancient Egypt, with the consequence that in medical papyri the doctor is frequently advised, after a description of symptoms to effect a diagnosis, not to treat the patient, but let nature run its course.

From the Indian, Egyptian, Greek and Chinese civilizations, the diseases and cures of the times have come down to us by way of written accounts. In many cases, materials originally used medically have now become widely used for non-medicinal purposes because of their pleasant taste [9]. Examples are rhubarb, tea, coffee and tobacco – the latter once being thought beneficial rather than harmful to health. Some of the very early medicaments are still used in medicine, because the scientific basis for their activities is now understood. What is surprising is the continuing use of many others in which no curative properties have yet been discovered [5] (for example, rhinoceros horn as an aphrodisiac, or ginseng, a herb given in China, as a cure-all).

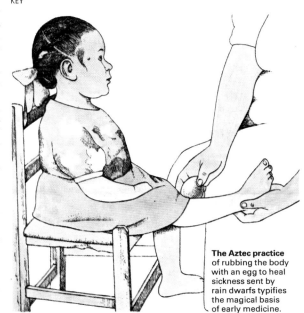

The Aztec practice of rubbing the body with an egg to heal sickness sent by rain dwarfs typifies the magical basis of early medicine.

5 Quack doctors peddling useless patent medicines are a perennial feature of human history. A familiar stereotype is the glib opportunist who toured the Wild West of America in a covered wagon. Mobility was essential so that he could move on before the fraudulence of his cures was exposed.

6 While herbalists' shops are rare in Western countries, in the less developed parts of the world the local medicine man does a thriving trade, compounding remedies from plant extracts in the market-place, with little competition from orthodox doctors. Here one is seen at work in Goulimine, Morocco.

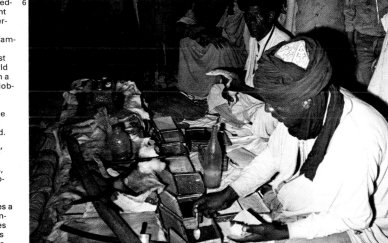

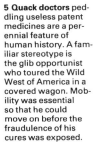

7 Powerful alkaloids that affect the nervous system are contained in henbane [A], deadly nightshade [B] and thornapple [C]. These plants are all members of the Solanaceae family, which also includes the potato. Plant extracts containing these alkaloids have been used widely in medicine for thousands of years.

8 The Andean Indians have chewed the leaf of the coca shrub to combat fatigue for hundreds of years. In 1860, when folk medicine was being rapidly overtaken by science, the leaves were found to contain the valuable local anaesthetic cocaine. Extract of coca leaves was originally an ingredient of the soft drink Coca-Cola.

Coca *Erythroxylon coca*

9 Tea drinking is a traditional Japanese ritual. One reason for the popularity of tea in many countries is that it contains the alkaloid caffeine (also present in coffee), which is a mild stimulant. A variety of herbs are used as medicinal infusions.

Natural remedies

As civilization developed, the stock of folklore about the curative properties of various substances or rituals was gradually classified. This was the first step towards scientific medicine. Once alleged remedies were listed and classified those that stood the test of time could be noted. However, it was not until the nineteenth century that major advances were made in understanding why some plant extracts, for example, had curative properties, for by then chemistry had advanced to the stage at which the molecular structures of individual substances could be identified, so that they could be synthesized in the laboratory and tested therapeutically.

Early natural remedies

Among the earliest pharmacopoeias (catalogues of medicines) was that compiled by Dioscorides (*fl.c.* AD 60), a surgeon in Nero's army. During his travels he made notes about the remedies used in different countries. These notes provided the basis for a tradition that, although largely lost after the fall of the Roman Empire, passed back to Europe from the Arabs in the fifteenth century. At this time pharmacists were known as *aromatarii* because the materials they sold were aromatic substances extracted from plants. Remedies with an animal or mineral base were also used. Paracelsus (*c.* 1493–1541) was as renowned for his cures with mercury as for those that used opium.

At this time there was a continuous interplay of different factors that advanced the science of pharmacy on several fronts simultaneously. Although there were several thousand known "remedies" for various ailments one school of medical theorists still argued that there was, albeit undiscovered, a substance that would cure all ills. This idea of the "cure-all" is in the alchemical tradition of a philosopher's stone that will turn other materials into gold and the alkahest, or universal solvent, that will dissolve all other substances. In their search for such universal substances the alchemists and their followers stumbled on some useful processes. Raymond Lully (*c.* 1235–1315) is credited with discovering how to prepare pure alcohol; he probably discovered how to react this with acid to produce anaesthetic ether.

The availability of pure alcohol was a great advantage in the preparation of tinctures and essences because many of the active principles in medicinal plants dissolved more readily in this solvent than in water. And as the developed European countries discovered other parts of the world, a great variety of new plants became known. These were added to the pharmacopoeias and mixed with other substances in alcoholic solutions.

Isolating the active ingredients

The major problem was to sort out which of the substances were effective and which could be left out. In the eighteenth century when the physician William Withering (1741–99) was introduced to an old country woman whose secret herbal mixture seemed to be surprisingly effective in cases of heart failure, he had to sort through 20 different ingredients to discover that it was the foxglove leaves that were effective. (The fact that they contained the substance digitalis was not discovered until later.)

By the time the New and Old Worlds had been thoroughly explored the number of

Carbon
Oxygen
Hydroxyl
Hydrogen
Naturally occurring in plants

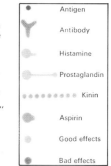

Antigen
Antibody
Histamine
Prostaglandin
Kinin
Aspirin
Good effects
Bad effects

2 Aspirin reduces fever [A] by regulating the brain's temperature control centre [1] and by increasing sweating [2]. It reduces pain by cutting reverberatory impulses at the thalamus [3] and inflammation at joints [4, 5]; but it can cause bleeding [6] and skin rashes [7] with some people. Aspirin reduces inflamation [C] by limiting production of the chemicals prostaglandins and kinins which are released in an "inflammatory cascade" [B] when the body's defences tackle an antigen or foreign body. Histamine is not affected.

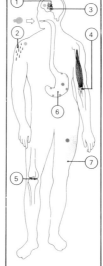

1 Aspirin, the most widely consumed drug in the world, is a chemical substance called acetylsalicylic acid. Plants such as willow (*Salix fragilis*) [1], meadow sweet (*Filipendula ulmeria*) [2] and shallon (*Gaultheria shallon*) [3] contain the related substance salicin. The presence of this pain-reliever was the basis of many old remedies. As salicin is too bitter to be taken internally (the same is true of salicylic acid, the first chemical derivative of salicin), it was not until it was made into its acetyl derivative in 1899 that aspirin [4] emerged.

3 From the roots of *Rauwolfia serpentina*, reserpine, a valuable tranquillizer is obtained. Analysis of its chemical structure has led to the development of more effective synthetic tranquillizers with no damaging side-effects.

4 Quinine is a bitter substance derived from cinchona bark. Brought from Brazil by the Spaniards, it is a remedy for malaria and has only recently been superseded by synthetic compounds.

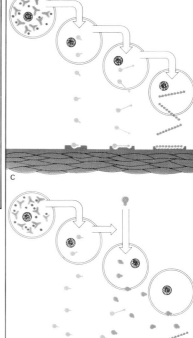

natural plant substances with known pharmacological action was enormous. There was opium from poppies [8], emetine (a remedy for amoebic dysentery) from the ipecac plant, quinine from cinchona bark [4], castor oil from castor seeds and salicin from willow [1]. A number of other plant substances with major effects on living organisms were also known including belladonna, strychnine and curare [7]. All were used as poisons but are now valuable in medicine. But the physiological effects of these substances are so powerful that they must be used in controlled doses. It was only as the active ingredients were isolated that it became possible to use them quantitatively.

Sophisticated techniques

During the nineteenth century chemistry developed rapidly as an exact science and many plant substances were isolated as pure, crystalline compounds. They were all classed as "organic chemicals" and it was not until 1828 that scientists realized it was possible to make such chemicals in a laboratory. In the same year Friedrich Wöhler (1800–82)

synthesized urea from ammonia cyanate and changed the entire emphasis of organic chemistry. Until that time it had been believed that such substances could be produced only by living processes. Nevertheless the chemical structures of many of the pharmacologically useful plant substances are so complex that it is more economical even now to extract them from plants than to make them in the laboratory.

Just how complex some of these substances are was not discovered until well into the twentieth century when sophisticated techniques of molecular analysis were devised. But it is not necessary to understand the structure of a chemical completely in order to modify it chemically. This frequently produces a related substance that will share some properties of the original substance but not others. By trial and error chemists gradually began to improve upon natural remedies, producing analogues of natural compounds that were safer, more effective or had less unpleasant side-effects. Thus began today's pharmaceutical industry which not only modifies nature, but has added much of its own.

Nearly 2,000 years ago a brew made from white willow leaves was recom-mended for gout. To kill pain today we take an aspirin. Now solely a product of the chemical laboratory, aspirin can also be made from willow brew.

5

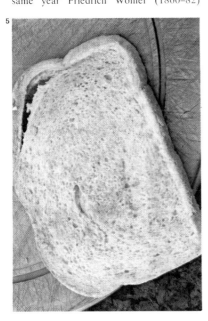

5 Mould on bread is usually viewed with disgust, but it is from variants of one common bread mould that penicillin is derived. This is doubtless why "mould poultices" are an old, effective remedy.

6 Penicillin, which is derived from fungi, is now only one of many antibiotics that kill or inhibit the growth of other micro-organisms. This makes them useful in treating bacterial infections. The normal bacterial cell processes [A] are attacked by different antibiotics [B] which may alter cell membrane structure [1] or inhibit cell wall synthesis [2], protein synthesis [3], energy production [4] or DNA replication [5].

6 A

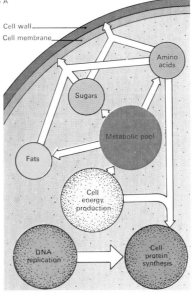

Cell wall
Cell membrane
Amino acids
Sugars
Metabolic pool
Fats
Cell energy production
DNA replication
Cell protein synthesis

B

Amino acids
Sugars
Metabolic pool
Fats

7

7 South American Indians carry pots of raw curare into which they dip their arrows to render them poisonous. This toxic substance is used in controlled doses in surgery for total muscular relaxation.

8 Most derivatives of opium [1] such as morphine [2], heroin [3] and meperdine hydrochloride [5], although extremely effective pain-killing (analgesic) drugs, are at the same time highly addictive. It is possible in the laboratory to pinpoint which parts of the molecule are responsible for the analgesic and which for the addictive effect. On this basis drugs such as codeine [6] and pentazocine [4] can be designed and synthesized. Extensive clinical trials are necessary to confirm their safety.

8

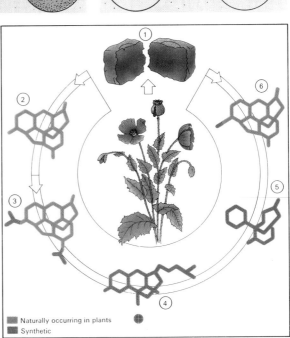

Naturally occurring in plants
Synthetic

Man-made cures

At the beginning of the twentieth century most effective remedies were still derived from plants. Now they are produced in factories following directions worked out in chemical laboratories.

The origins of the drug industry
In the late nineteenth century the nature of chemical structures came to be understood and research chemists, notably in Germany, synthesized hundreds of new organic molecular compounds based on carbon atoms. As a result the idea of a chemically based pharmaceutical industry emerged. Attempts to make synthetic versions of natural compounds on the basis of incomplete knowledge often led to curious results. The first synthetic dyestuff, mauveine, was produced accidentally in 1856 by William Perkin (1838–1907), while he was trying to make quinine.

The result of Perkin's discovery was the start of the synthetic dyestuffs industry. Within a few decades hundreds of synthetic dyestuffs had been manufactured. Ironically, it was from this wealth of new material that

chemotherapy, the modern concept of treating disease with chemicals, came into being. The German scientist Paul Ehrlich (1854–1915) [2A], discovered that certain dyestuffs stained only specific tissues when he treated microscopic specimens with them. From this he conceived the idea that some dyestuffs might selectively and safely destroy the micro-organisms that cause diseases.

Ehrlich tested 500 different dyestuffs on mice that had been infected with trypanosomes – blood parasites that cause sleeping sickness and other diseases. None of them worked. He then tested other compounds with structures similar to the dyestuffs but which also contained atoms of arsenic. His compound "606", or Salvarsan, although ineffective against trypanosomes, turned out to be effective against the bacteria that cause syphilis. It was first manufactured in 1910.

The sulpha drugs and penicillin
Another German scientist, Gerhard Domagk (1895–1964), was responsible for the second great therapeutic discovery to come from the dyestuffs industry. In the early 1930s he

discovered that the red dye Prontosil was an effective bacteriostatic – a synthetic substance that prevents the proliferation of bacteria in the bloodstream. Until that time the most dangerous part of many surgical operations had been the period after the operation when the patient was liable to die from septicaemia – infection from bacteria introduced into the body during the operation.

French scientists soon discovered that only one part of the Prontosil molecule – sulphanilamide – was effective. By the end of the 1930s the sulpha drugs [1] had lowered post-operative mortality dramatically and, in 1938, a bonus appeared in the form of a modified sulphanilamide molecule – sulphapyridine – effective against tuberculosis.

At about the same time the bacteriostatic properties of the penicillin mould, discovered by Alexander Fleming (1881-1955) [2B] in 1928, were being further investigated by Howard Florey (1898-1968) and Ernst Chain (1906–) [2C]. The development of penicillin manufacture during World War II heralded the age of antibiotics. Many micro-organisms were found to contain complex

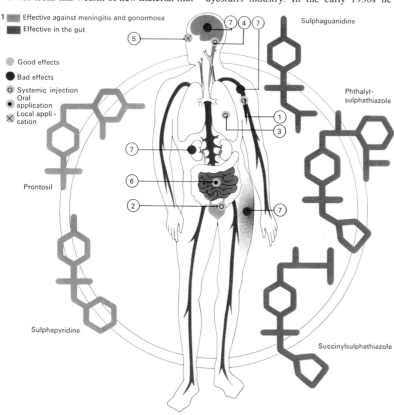

1 Effective against meningitis and gonorrhoea
Effective in the gut

Good effects
Bad effects
Systemic injection
Oral application
Local application

Sulphaguanidine

Phthalyl-sulphathiazole

Prontosil

Sulphapyridine

Succinylsulphathiazole

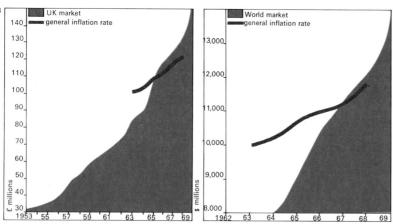

3 UK market / general inflation rate (£ millions) 1953–69

World market / general inflation rate ($ millions) 1962–69

1 The sulpha drugs act by preventing the multiplication of bacteria and so give the body's natural defences the chance to overcome them. The drugs combat blood [1] and urinary tract [2] infections, pneumonia [3], meningitis [4], eye infections [5] and alimentary tract [6] infections such as bacillary dysentery. Overdoses [7] can produce skin rashes, raised temperature, impaired kidney function and anaemia.

2 Paul Ehrlich [A] is regarded as the founder of modern chemotherapy. He believed that individual chemicals could act as "magic bullets" against the agents of infection and his discovery of Salvarsan proved his case. A second era of expansion for the drug industry was based on the accidental discovery by Alexander Fleming [B] that moulds could produce antibacterial substances. Ernst Chain [C], who helped in the development of penicillin, took chemotherapy a stage further by suggesting that semi-synthetic drugs would prove useful.

3 The growth of the pharmaceutical industry during the twentieth century has been staggering. As medical care has spread so the demand for new and better drugs has increased. This demand can only be met by extensive research programmes. Shown are the rising values of drug sales in the United Kingdom and the world.

chemicals that would combat other micro-organisms. Streptomycin was discovered in 1943 and since then dozens of new antibiotics have been discovered and manufactured on a commercial scale [Key, 3].

In most cases it is cheaper to cultivate the micro-organisms by fermentation and then extract the active materials, than to synthesize these materials in a laboratory. A compromise between the two approaches was the development of semi-synthetic penicillins in the 1960s. In 1959 the key fragment of the penicillin molecule, 6-amino-penicillanic acid, was isolated. When this is reacted with various organic molecules it is possible to manufacture penicillins that do not occur in nature. These tend to attack micro-organisms that have developed resistance to natural penicillins, and they are therefore of greater use therapeutically [6].

Natural and synthetic products

The concept of modifying natural molecules has spread dramatically through the pharmaceuticals industry. The contraceptive pill, developed in the early 1960s, emerged as

the result of the chemical modification of natural sex hormones, which had been isolated in the 1930s. The chemical structures found in human sexual and adrenal cortical hormones are also found in slightly different forms in many living organisms. By chemical modification of a starting material obtained from wild yams it has been possible to manufacture the large quantities of oral contraceptives now in demand.

Many other natural products, such as vitamins and adrenaline, have been synthesized in laboratories and some of them are made commercially by synthetic methods. Adrenaline belongs to an important class of chemical compounds called phenthylamines; this also includes ephedrine, an old Chinese plant remedy now used in asthma treatment and in "pep pills". Many completely synthetic drugs such as barbiturates have been discovered by accident. The results of unpredicted side-effects, such as the impairment of limb development caused by thalidomide, mean that rigorous safety checks have to be made and many governments have set up safety committees to test new drugs [5].

The high-speed tableting machine has supplanted the apothecary's crude apparatus. This is a natural consequence of the need for large amounts of drugs and strict quality control.

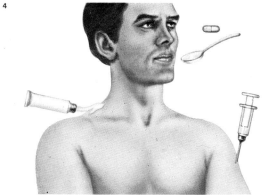

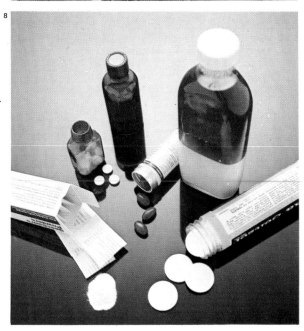

4 Modified forms of penicillin mean that the drug can be given orally, by injection or by local surface application. Many different sorts of penicillin are now available, all with slightly different molecular structures. These slight differences can mean quite large differences in effect. Apart from self-administered substances such as aspirin, penicillin is probably the most commonly used antibiotic the world over.

5 Samples from each batch of every manufactured drug must be sent to analytical laboratories. There checks can be made to ensure that consistent standards of purity are always maintained.

6 Antibiotics have been increasingly used in recent years to keep animals healthy and to combat disease in humans. A disturbing side-effect has been that the bacteria they used to kill are becoming immune.

7 Viruses are resistant to most synthetic drugs. To find a cure for viral diseases such as the common cold, scientists experiment, after artificially infecting a volunteer, with complex naturally occurring defence substances.

8 Drugs can be formulated for oral consumption as pills, powders, solutions or suspensions. Different colours and shapes are used to distinguish the thousands of products now marketed.

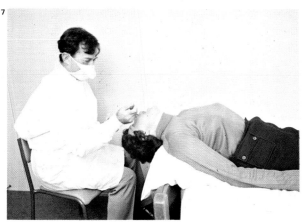

Non-medical healing

Non-medical healing describes any form of healing accomplished without medical or chemical aid, through the manipulation of the life-force – the mental or the spiritual being. Verification of illnesses and cures is difficult in this area, which is thus open to fraudulent practice. But there has been growing interest in the subject recently, with increasing acceptance by the medical profession of the importance of psychosomatic aspects of illness and with the unexplained success of non-medical healing techniques in many cases that had proved to be unresponsive to conventional medicine.

Healers and their methods
The many forms of non-medical healing range through Shamanism (the witch doctors and medicine men of primitive cultures), "laying-on-of-hands", various types of prayer and prayer groups, to meditation and guidance by "discarnate" (previously living) doctors. Also included are placebos (chemically inactive pills or prescriptions that work psychologically, making the patient feel he is taking something therapeutically useful), the

so-called psychic surgery of faith healers in the Philippines [3] and elsewhere, cleansing of "auras" (fields of energy said to surround all living things) and biofeedback.

Among contemporary faith healers, Kathryn Kuhlman in America [1] and Harry Edwards in Britain [2] have helped thousands of sick people by means of fervent prayers and their conviction of their power to heal through a divine or spiritual agency.

Olga Worrall, who with her husband Ambrose conducted a healing "ministry" in Maryland, USA, for more than 40 years, became famous for remarkable results, many of them verified by doctors. She made herself "a channel for God's healing power through prayer". Working recently with scientists in America to investigate the power of prayer, she achieved significant results on the growth of plants, apparently influencing germination of seeds and faster growth (compared with a control batch) from a distance of several hundred kilometres.

Many healers combine laying-on-of-hands with great mental concentration, as do most prayer and healing groups of the

Christian Church. Some churches, such as the Pentecostal, incorporate such "spiritual" healing in their services. Spiritual healing is also an integral part of the New Thought Movement, which includes the churches of Religious Science, and of Christian Science whose healing principle is "scientific prayer". The church of Dr Norman Vincent Peale in New York has become renowned for the healing power of "positive thinking" and The World Healing Crusade, headed by Brother Mandus with its headquarters at Blackpool, England, applies the healing power of God, claiming cures in more than 100 countries.

The transfer of energy
In America, John Scudder, an engineer who is also a church pastor and is allowed entry to hospitals, combines a breathing technique with "magnetic passes" of his hands over the patient's body, together with a psychological approach to alter the patient's preconceptions of sickness. Norbu Chen, originally Charles Alexander of Kentucky, USA, uses techniques learned in Tibet in which he "raises his vibrational level" with chanting

1 **Kathryn Kuhlman's belief** in the healing power of faith in God has drawn thousands to "Miracle Services" at the Andrew Carnegie Auditorium in New York. Working by on-the-spot therapy and prayer, she has been able to help many people with ailments that had not responded to conventional medicine.

2 **Harry Edwards,** the best-known British faith healer, uses a combination of laying-on-of-hands, sympathy and understanding. His public demonstrations of healing have filled the Albert Hall, London. His powers have been attributed to the advice of a "discarnate" doctor as well as to faith.

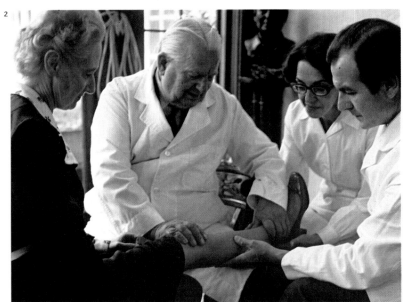

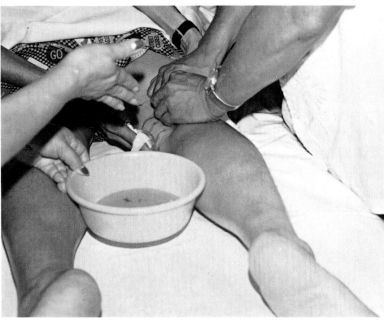

3 **"Psychic surgeons"** in the Philippines appear to operate on patients without instruments, "cutting" into the body with a finger and opening it up with their bare hands to remove organs or tissues they say cause illnesses. Such tissues have been shown to be from animals in some investigations, but many patients have reported cures.

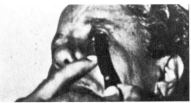

4 **José Pedro de Freitas,** known as "Arigo" (good guy) became famous for successful operations (such as this removal of an eye cataract) carried out painlessly with any handy knife and without anaesthetic. Medical research in 1963 showed that his skill matched that of highly trained doctors.

and meditation until he feels ready to channel the energy of his consciousness to the patient. His results are so remarkable that his work is being studied by a group of scientists in America.

Dr Carl O. Simonton in America and Gilbert Anderson of The National Federation of Spiritual Healers (with 9,000 members) in England both claim promising results in the treatment of terminal cancer through a combination of meditation and laying-on-of-hands and with the patient visualizing the healing of the affected parts.

Biofeedback and acupuncture

Biofeedback, the voluntary control of normally involuntary states, uses the patient's own mind for healing by showing him the results of his own state of mind on an instrument. In some way not yet definable he can cause changes in the reading by altering his state of mind, or vice versa. It is possible, for instance, for people with hypertension to lower their blood pressure. Through deep inner relaxation the patient can relieve tension, headaches, asthma and other symptoms of stress, as well as muscular complaints.

Although the technique of acupuncture dates back at least 5,000 years in Chinese history [5, 6], it first arrived in the West during the 1960s and was brought to prominence only after an investigative visit to China by a group of Western physicians in 1972. Now being used more widely, it involves the concept of 14 main meridians (12 of them bilateral lines or channels which connect organs deep in the body to surface points), each having a point of stimulation as well as a point of sedation. The organs are paired, so that if the heart, for instance, is stimulated, the lung will be sedated.

There are at least 1,000 acupuncture points on the body and possibly many more. When fine steel needles are inserted at an acupuncture point [7] and gently manipulated, they are said to redirect energy flow along the meridians to correct an imbalance. The Chinese called the use of this energy flow "Qi" and it was based on the Taoist yin-yang principle of opposing but complementary aspects of energy which must be kept in balance for health to be assured.

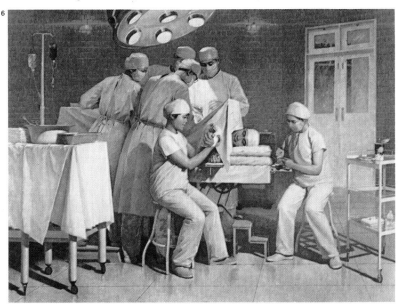

Lourdes became a famous centre of faith healing after a peasant girl, Bernadette Soubirous, had visions of the Virgin Mary by a spring in 1858. Said to effect miraculous cures, the spring has been visited by countless pilgrims.

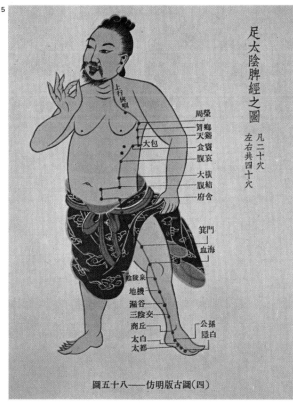

5 Acupuncture as a method of curing illness originated in China more than 5,000 years ago. Metal needles are inserted into the skin at specific points which were mapped out on a bronze figure dating from AD 900 and have been illustrated on many charts since then. The points are located on the 14 meridian channels of which one is shown here in a nineteenth-century version.

6 In China most kinds of surgery are now performed under acupuncture anaesthesia. Patients are aware throughout and feel nothing but numbness; they can often walk away from the operating table.

The development of acupuncture as an anaesthetic began in 1959 when Chairman Mao gave acupuncturists the status of physicians. Before this time acupuncture was a therapeutic procedure.

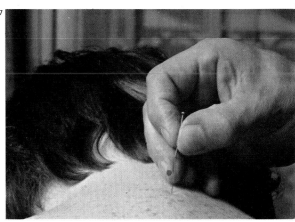

7 Acupuncture procedure involves the rapid insertion of fine steel needles up to 6cm (2.5in) long at acupuncture points, followed by gentle twirling between the thumb and forefinger. This is said to redirect the energy flow along meridians, restoring balance and correcting the disorder. In a treatment for toothache, for instance, the needle may be inserted in the patient's hand, at an acupuncture point between the thumb and forefinger.

8 The use of hypnosis in studying hysteria was pioneered by Jean Martin Charcot (1825–93) at the Salpêtrière Hospital in Paris, where Sigmund Freud was one of his students. Hypnosis is now used in many medical fields, including anaesthesia. In dentistry it can reduce anxiety and in other procedures it can also induce patients to exercise control of such bodily functions as blood pressure and heartbeat. It continues to be used in psychotherapy.

Radiology and radiotherapy

The effects of radiation may be used both for investigating diseases, in diagnostic radiology, and for treating them, in radiotherapy. Much of the early work on the phenomenon of radioactivity was undertaken by Marie Curie (1867–1934) [Key] who, with her husband Pierre (1859–1906) and jointly with Antoine Becquerel (1852–1908), was awarded the Nobel prize for physics in 1903. In 1911 she was also awarded the Nobel prize for chemistry for the discovery of radium and polonium. Her death from leukaemia was probably due to over-exposure to radiation.

Over-exposure illnesses and the long-term effects of atomic bombs point very clearly to the powerful and dangerous effects of radiation waves. By careful and controlled use, however, many of the properties of beta, gamma and X-rays are exploited in medicine for both diagnosis and treatment.

X-ray photography
Most kinds of energy waves are able to pass through some materials and not others. Light waves, for instance, pass through air and also through some liquids (such as water) and even solids (such as glass), but not through walls and doors. X-rays on the other hand can pass with varying ability through body tissues, but can be blocked and absorbed by other substances, such as lead, which is often used in radiology. In the same way that varying amounts of light waves can produce a picture on a photographic plate, so also can X-rays activate such a film [1].

A simple X-ray machine is rather like a camera. An X-ray source is beamed at the part to be examined and focused on to a photographic film, where the image resembles a film negative. An X-ray picture of a hand, for example, shows bones (which stop X-rays quite effectively) as white, the other tissues as greys and the area around the hand (which receives full X-ray exposure) as black. In order to show up other organs within the body more clearly, radio-opaque substances (ones that do not allow X-rays through) are used. Barium sulphate is one such substance and is used in barium meals and enemas to show up the upper and lower bowel respectively [3]. Other substances injected or given by mouth, often containing iodine, are particularly concentrated in certain body organs. This enables better pictures to be obtained of kidneys and bladder by monitoring the appearance of the dye.

Scanning techniques
Pictures obtained by conventional X-ray machines are not as clear as ordinary photographs. Organs within the body are seen superimposed upon each other and the picture is a composite one. To see an organ or part within the body with less overlaying detail the technique of tomography is used [2]. Here the X-ray source, and the detection system or photographic plate, are rotated in an arc round the body with the specific organ as the centre of the arc. This organ then appears relatively still and in focus, whereas other parts are less clear.

In a major advance, this principle has been used in a very sophisticated manner to scan the brain and, using a computer, to integrate the information. The EMI scanner [5] produces a picture of a "slice" through the brain, and can produce a series of such pictures to give a view of the whole brain. By the use of a

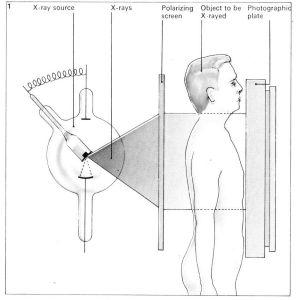

1 The use of X-rays in medicine to photograph the body internally has enormously advanced the diagnosis of disease. Since X-ray waves emerge from an X-ray source travelling in several different planes, they are first passed through a polarizing screen. This organizes or directs them into a single plane and so concentrates them. The body's various organs and tissues (skin, bone, brain, lungs, liver, arteries and so on) absorb varying amounts of the X-rays passing through them. Hence the X-rays striking the photographic plate are of varying degrees of intensity. The result is an image that resembles a film negative.

X-ray source   X-rays   Polarizing screen   Object to be X-rayed   Photographic plate

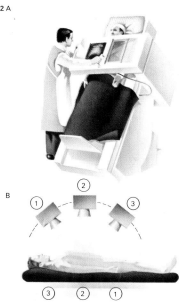

2 Radiology uses X-rays to visualize internal parts of the body as an aid to diagnosis. Various machines are used to produce X-rays of controlled energy levels and to focus the rays and control exposure. A chest X-ray [A] can be viewed on a television screen, stored on videotape or made into X-ray films. To facilitate diagnosis, substances that do not transmit X-rays can be swallowed or injected to provide a means of contrast. An X-ray film is a negative and the contrast appears white. By rotating the X-ray source [B] and taking pictures from several angles [1, 2, 3] it is possible to focus on organs in a particular plane of the body.

3 Barium sulphate is used in radiology to provide contrast in X-ray photography. Internal tissues such as the digestive system and kidneys allow the passage of X-rays to varying extents. But when filled with a substance that blocks the X-rays, they can be made opaque and show up as white. The barium meal, a harmless paste that passes through and out of the intestines unchanged, can be swallowed [1] or injected into the rectum [2]. When swallowed, it outlines in turn the oesophagus or gullet, the stomach and the small intestine [A]. Ulcers, tumours and alterations of function can be seen. Similarly, the colon or large bowel can be photographed [B].

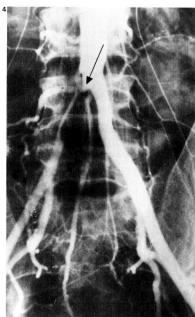

4 Arteriography is a specialized technique of radiology developed to show up the inside of arteries so that thickening and furring can be detected. This tissue damage in arteries is the degenerative disease known as arteriosclerosis. It contributes to coronaries and strokes and to poor circulation in the legs. The arteries are shown up by the injection of a radio-opaque dye which reveals irregularities of the vessel wall. The picture here, obtained after injection of a dye into the aorta, shows the large arteries of the pelvis and upper legs and a local narrowing of the artery (arrowed) that is interfering with the blood flow.

computer, a very much more detailed and precise picture is obtained. This technique is now likely to replace many of the techniques in which dyes or air are injected as contrasts into the fluids bathing the brain, for these latter procedures are longer, more complicated and potentially dangerous. A computer scanning system is more expensive (about US$320,000) than those with more conventional machines, but the X-ray dose for the whole procedure is smaller, comparable to that of a skull X-ray. The EMI Whole-Body Scanner, which produces a scan of a body section that is processed in 20 seconds, is now also in use.

X-rays are not the only type of radiation that can be used to obtain information about the body. Gamma waves emitted from certain radioactive isotopes (forms of certain elements) are also able to pass through body tissues. Small doses of such isotopes can be injected and a measure of the radioactivity taken up by a particular organ may give evidence about it. The thyroid gland, for example, is responsible for using nearly all the iodine taken into the body. If a radioac-

tive iodine isotope is injected into the bloodstream, the activity of the thyroid can be gauged by measuring its rate of uptake of iodine. This can be done by measuring the gamma waves emitted from the region of the thyroid. In addition, scanning the thyroid will give a picture of the activity of each part of the gland. Parts that are over- or underactive may thus be detected.

Treatment of cancer by radiotherapy
Using high doses of isotopes can cause tissue damage. In treating thyroid cancer doses several thousand times higher than for diagnosis are used to take advantage of this damaging effect to kill the cancer cells. This is the principle of radiotherapy, where high doses of radiation are beamed on to tumours to kill them [6]. All living cells are generally sensitive to the effect of radiation, and cancer cells are particularly susceptible. Care must be taken to judge the dosage required and to apply it to the correct area. Some cancers respond better than others to rays of high energy, and for these radiation of a particular energy may be required.

KEY

Marie Curie's work on radium led to

X-ray and isotope techniques for both

the diagnosis and treatment of disease.

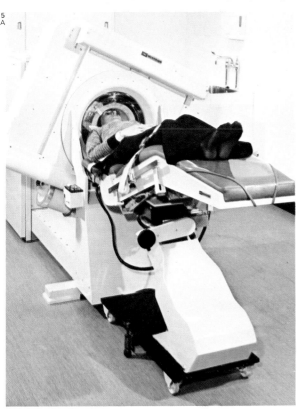

5 A

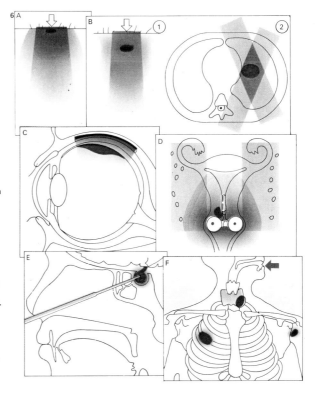

6 **Radiotherapy** uses radiation waves to treat disease. Ionizing radiations (beta, gamma or X-rays) cause changes in cells (particularly when they are dividing) that kills them. This can be used against malignant cells, which are more sensitive to the energy released by irradiation than healthy cells. Treatment of cancer can be by a beam of low-energy rays focused on to the skin [A] or of higher energy [B] focused below the body surface [1], sometimes from more than one direction [2]. A local surface applicator can be used, as in the eye [C] or combined with beam treatment, as in the neck of the womb [D]. Minute radioactive pellets can be placed deep in the body to treat the pituitary [E], or an isotope administered by mouth may be taken up selectively, as is iodine by thyroid cancer, even if the cancer has spread [F].

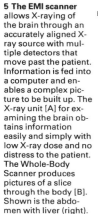

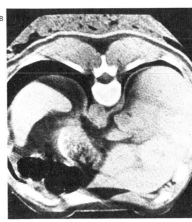

5 The EMI scanner allows X-raying of the brain through an accurately aligned X-ray source with multiple detectors that move past the patient. Information is fed into a computer and enables a complex picture to be built up. The X-ray unit [A] for examining the brain obtains information easily and simply with low X-ray dose and no distress to the patient. The Whole-Body Scanner produces pictures of a slice through the body [B]. Shown is the abdomen with liver (right).

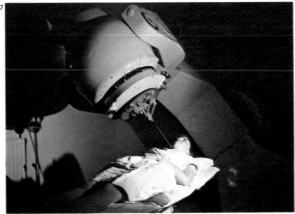

7 To treat a tumour, radiation is delivered by sophisticated machinery with the operators protected by lead screening from the powerful X-ray or gamma-ray source. The patient is also screened so that only a specified area of the body is irradiated for the correct time. Treatment in stages minimizes side-effects. The type and energy of the rays are vital for success and new, very high energy rays can now treat previously unresponsive cancers.

Development of surgery

Surgery is the manual treatment of injuries, deformities and other disorders. Early man probably practised simple forms, such as setting a fractured limb, binding a wound, lancing an abscess or pulling a tooth, and there is evidence that he may have attempted delicate operations with some success.

Advances in anatomical knowledge

Even 10,000 years ago "medicine men" were capable of performing the difficult operation known as trephining – the removal of a piece of skull to cure headaches or madness – and the Romans removed endangered babies from their mothers' wombs by what came to be known as Caesarean section. But generally surgery was limited to simple external operations and excluded anything internal. The sophistication of modern surgery is really the result of discoveries made in the last 300 years.

A surgeon's first need is a clear and accurate understanding of the working of the human body. Until late in the Middle Ages surgery remained a primitive trade (closely allied to that of the barber) because neither doctors nor surgeons had any real knowledge of anatomy. The only way to find out is by dissection of the dead, a process looked upon with horror by Church and state alike. Only in the fifteenth century was this taboo even partially overcome, but from that time onwards doctors were able at last to begin their investigation of the structures that lie within the body. Ambroise Paré, often called the father of modern surgery, was one of the earliest to use this new knowledge [1]. Important subsequent landmarks were the demonstration by William Harvey (1578–1657) of the circulation of the blood and the anatomical discoveries of William Hunter (1718–83) and his brother John (1728–93).

Eliminating pain and infection

Even the most exact knowledge of anatomy does nothing to solve the two major problems confronting the surgeon and, more important, the patient – namely pain and infection. It was the nineteenth century that brought about the great revolution that transformed surgery from an agonizingly brutal and dangerous business into the relatively safe and painless one that it is today. The first advance came in 1844 when Horace Wells (1815–48), an American dentist, used laughing gas (nitrous oxide) when extracting a tooth. Within a few years ether [2] and chloroform were also being used to eliminate pain during major surgery. William Morton (1819–68) pioneered the use of ether (general anaesthesia).

The second factor in the surgical revolution was the advent of antiseptic surgery. Until mid-Victorian times it was commonplace for patients to develop lethal post-operative infection.

The reason for post-operative infection remained hidden until Louis Pasteur (1822–95) showed that it was caused by germs. The first surgeon to grasp fully the implications of Pasteur's discovery was Joseph Lister, who eventually realized that antiseptic chemicals such as carbolic acid could kill germs and protect operation wounds from infection. His technique of placing surgical instruments in antiseptic solution, of spraying carbolic acid in the air during operations and of applying antiseptic

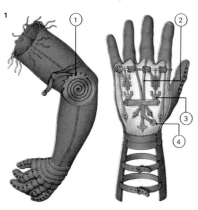

1 French surgeon Ambroise Paré (1517–90) designed this artificial arm. A ratchet and pawl [1] locked the elbow in any position; springs and catches [2, 3, 4] opened and then closed the fingers.

2 The first "purpose-built" operating tables [A] appeared in the 19th century. They were cumbersome but basically the same as modern ones. The ether inhaler [B] replaced whisky or rum as a pain-killer.

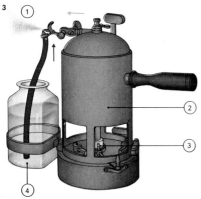

3 The discovery by Pasteur that germs cause infection led to Joseph Lister's introduction of antisepsis – the treatment of wounds with carbolic acid (phenol). Lister (1827–1912) first introduced a hand-operated spray, then a steam model. The steam [1] from the boiler [2] – heated by methylated spirit on a wick [3] – sucked carbolic acid vapour from a jar [4], thus filling the air with an antiseptic mist.

4 Wars accelerate all technologies, not least that of medicine, and the carnage of World War I caused great strides in surgery, particularly in bringing damaged tissues together and keeping patients alive with a saline solution, in spite of heavy fluid losses. Here, in a makeshift field operating theatre, an army surgeon removes a bullet from the arm of a wounded soldier.

5 A constant supply of fresh blood from donors [A] to a blood bank is critical, for it deteriorates after three weeks. Blood reaches hospitals whole (at 4°–6°C [39°–43°F]) or split into its constituent parts by centrifuge. Once separated and dehydrated, plasma can be frozen and kept indefinitely as powder. To increase plasma supplies [B] a donor can give blood 40 times a year instead of the normal three, receiving back his red cells from the previous visit.

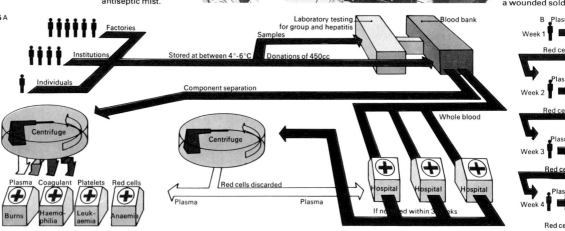
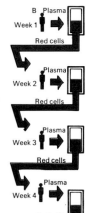

dressing to the stitched incision dramatically reduced the death-rate after surgery [3].

Revolutionary and beneficial as it was, antiseptic surgery began to give way, at the start of the twentieth century, to the modern practice of aseptic surgery, in which the aim is not to kill germs in the operating theatre but to exclude them as far as possible from the patient. Surgeons developed an aseptic (germ-free) technique in which the operation site is made bacteria-free by swabbing the skin with a bactericide before an incision is made. Sterile towels, gowns and masks were introduced and the surgeon and his assistants learned to scrub their hands carefully for a full five minutes before every operation. As an additional precaution sterile rubber gloves, first developed in the USA by William Halstead (1852–1922), became customary.

The 1920s saw the widespread introduction of blood transfusion [5] which made it possible to replace blood lost during an operation and, at the same time, there were refinements in anaesthesia to enable opening of the chest for lung operations.

Shortly after World War II chest surgeons found ways of inserting a finger into the heart in order to clear a partially obstructed valve. Next it became possible to lower an anaesthetized patient's temperature (hypothermia) so much that his heart could actually be stopped for a short time, thus permitting even more delicate operations within it.

Modern innovations in surgery

The advent of the heart-lung machine (which completely takes over the functions of pumping the blood and of breathing) meant that surgeons could take far more time over their complex procedures (with correspondingly better results) and eventually progress to the transplantation of hearts [8]. By the 1960s and 1970s it became possible to transplant other organs (kidneys, liver, lungs and pancreas), although the incidence of failure is still very high. Indeed, surgery and its technology have advanced to such an extent that it is now possible to conceive of a situation in which anyone who has a damaged, defective or worn out organ will be able to get a working replacement, real or artificial, as well as purely cosmetic operations [6].

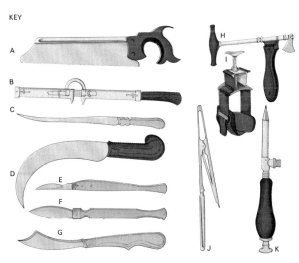

KEY

Surgical instruments of the 18th and 19th centuries looked crude but were effective. Basic amputation saws [A, B], knives [C, D] and scalpels [E, F, G] have hardly changed. Thal's mechanical saw [H], in which the handle was rotated to move the blade, was sometimes used. Today, clamps have replaced tourniquets [I], while forceps [J] now resemble scissors more. The trocar [K] was a sharp, hollow tube used for drawing off fluids or tissue.

6 "Spare parts surgery" has undergone dramatic changes in recent times, making it possible to transplant living organs and fit artificial substitutes within the body. New metals, acrylic and other plastics have helped overcome the problems of rejection, the refusal of the body to accept the installation of foreign materials. Possible spare parts include [1] surgical wig; [2] "Vitallium" skull plate; [3] plastic nose implant; [4] cosmetic acrylic eye; [5] in-the-ear hearing aid; [6] silicon plastic ear; [7] cosmetic plastic ear; [8] metal jaw-bone; [9] dentures; [10] chin implant; [11] Spitz-Holter valve to control fluid on the brain; [12] electronic larynx; [13] shoulder joint replacement; [14] heart valve replacement; [15] heart pacemaker; [16] filter preventing blood clotting in circulation to the lungs; [17] Dacron heart patch; [18] Dacron artery replacement; [19] elbow replacement; [20] Dacron vein and artery graft; [21] metal bone plate; [22] plastic replacement after removal of a part of small intestine; [23] hip joint replacement; [24] wrist bone replacement; [25] finger joint replacement; [26] thighbone replacement; [27] cosmetic plastic testicles; [28] knee joint replacement; [29] plastic artery graft; [30] artificial leg with knee and ankle movement; [31] shin; [32] artificial arm.

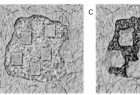

7 In basic plastic surgery, the burned area [A] is cleaned and "postage stamps" of skin from elsewhere on the body are grafted on [B]. Healthy skin grows out until the whole area is covered [C].

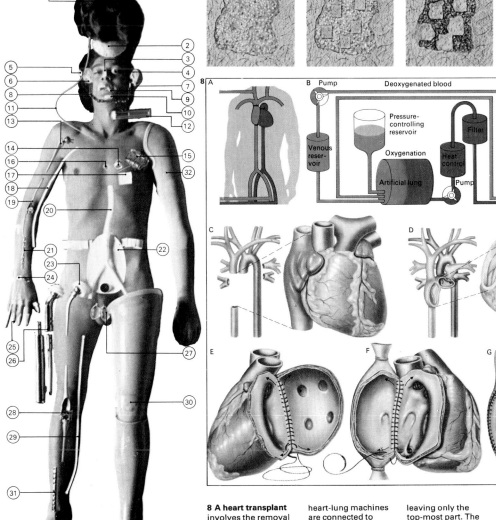

8 A heart transplant involves the removal of a badly diseased heart, replacing it with one from a "donor" – someone who has just died, usually as the result of a road accident. At first, heart-lung machines are connected to both the donor [A] and the recipient [B], to keep their hearts alive. The donor heart is then removed [C] and the recipient's heart is cut out [D], leaving only the top-most part. The donor's heart is now stitched on to this "stump", first on one side [E] and then on the other [F]. Salt solution is then pumped through the heart to clear it of air and the final stitching to the great vessels is completed [G]. The recipient's new heart is now stimulated into beating by means of an electric shock. The operation is extremely rare.

Having an operation

Although advances in modern techniques have made such well-publicized operations as organ transplants and brain operations relatively safe, these are still only a minute fraction of the total number of operations performed each year. There are, for example, only about 500 kidney transplants performed each year in England and Wales out of more than 260,000 surgical operations.

Common surgical operations
To a certain extent operations are a matter of fashion. In the United States circumcision was, until recently, the most common operation. Virtually the entire male population underwent circumcision shortly after birth. In other countries, however, it is becoming quite rare because it is only really necessary in relatively few cases.

In the United Kingdom and many other countries appendectomy (the removal of the appendix) is probably the operation most frequently performed by general surgeons. Other common ones include herniorrhaphy (the repairing of a hernia, or rupture); partial gastrectomy (the removal of part of the

stomach, usually for ulcers); cholecystectomy (gall bladder removal); vagotomy (cutting of the nerves to the stomach, again for ulcers); mastectomy (breast removal); and haemorrhoidectomy (removal of piles).

The general surgeon may spend a great deal of his time removing benign and malignant tumours from various parts of the body and, in many hospitals, be expected to strip out varicose veins. The latter is also in the province of the vascular surgeon, who carries out operations on blood vessels, such as the removal of aneurysms (swellings) in arteries.

Orthopaedic surgeons are responsible for operations on bones, joints, tendons, ligaments, nerves and muscles. The common procedures they undertake include menisectomy (removal of a damaged cartilage from the knee joint) and operations for the relief of "slipped discs" in the vertebral column.

Urological surgeons deal with the urinary tract and the male reproductive organs. One of their main operations is prostatectomy or removal of the prostate gland. Gynaecological surgeons deal with the female reproductive organs and carry out the extremely

common operations of hysterectomy (removal of the womb) and "D and C" (dilation and curettage, or widening of the neck of the womb and scraping of the lining).

Other surgeons specialize in operations on the ear, nose and throat; the most common operation they perform is probably still tonsillectomy, or excision of the tonsils. Considerably less frequent are the delicate operations carried out by eye surgeons, brain surgeons and plastic surgeons.

The procedure for surgery
Most people have some sort of operation (though often only a minor one) in the course of their lives, but few understand very much about what is going to happen to them [1]. On admission to hospital the patient is carefully examined by a house surgeon (a junior doctor working for the specialist) to ensure that he is fit for an operation. The house surgeon may order blood tests or X-rays and, if the operation is to be a major one, may ask the blood bank to put aside some blood that has been carefully "cross-matched" with the patient's blood.

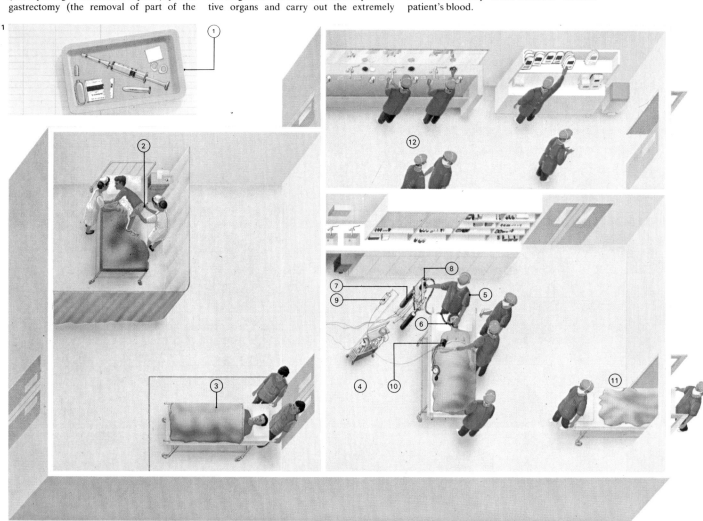

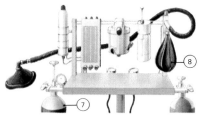

1 Before an operation a pre-medication injection [1] is drawn up into a syringe and administered [2] by a nurse. The "pre-med" takes effect after half an hour or so and then a hospital porter transfers the patient to a trolley [3] and wheels him into the anaesthetic room [4]. The anaesthetist [5] "induces" the patient with an anaesthetic drug [6]. Administering it into an arm vein makes the effect swift, although sometimes the patient is put to sleep by inhaling gases from the anaesthetic machine. This is a movable trolley carrying cylinders [7] of oxygen – which has to be administered throughout the operation to keep the patient alive – and anaesthetic gases such as cyclopropane. Oxygen and whatever gas has been chosen by the anaesthetist are mixed together and they flow through a black rubber bag [8]. When the patient can no longer breathe for himself, this bag is squeezed manually by the anaesthetist or his assistant. After an initial intravenous injection the anaesthetist will continue anaesthesia with gas as the injection wears off. In the anaesthetic room, a "drip" [9] may be set up if it is felt that the patient may need fluid or blood replacement. A blood-pressure

At some time during the 24 hours before the operation the patient is asked to bathe thoroughly and the skin around the operation site is then shaved. For about six hours before, no food or drink is allowed since, at the start of the anaesthetic, such food might be vomited and inhaled into the lungs. An hour or so before the operation a nurse gives the patient a "pre-med" injection, which makes him feel calm and relaxed and dries up chest secretions that might interfere with his breathing under the anaesthetic.

Eventually the patient is taken to the operating theatre suite. Here the anaesthetist gives him an injection into a vein; this produces complete unconsciousness in about ten seconds. The patient is now wheeled into the operating theatre and placed on the table. The anaesthetist places a mask over his face and keeps him unconscious with a mixture of anaesthetic gases and oxygen, often supplemented by injections.

The surgeon, his assistants and the theatre sister take up their positions. They clean the operation site carefully and place sterile towels round it. The surgeon then makes his incision and probes down to the organ or tissue he is looking for, while his assistants use small forceps to close off any bleeding points. The theatre sister's function is to pass instruments and swabs to the surgeon. She is backed up by a team of more junior nurses and by theatre technicians.

Most operations are over in about an hour. At the conclusion, the surgeon carefully stitches up the wound and applies a dressing. The patient is then taken to the recovery room where the anaesthetist (or a nurse) looks after him until he awakens. From there he may go to the intensive care ward, but more usually he will be returned to the ordinary ward, where he will be observed carefully for a few hours.

The recovery period
The patient will spend a few days in hospital after surgery if the operation has been a minor one. Stitches are usually removed about a week after the operation. After discharge from hospital the patient usually sees the surgeon at least once more before being left in the care of his own family doctor.

Obstetrics 20·3%

Nervous system 1·6%

Eye 4·2%

Ear, nose and throat 10·9%

Urinary system 5·0%

Endocrine system 1·0%

Female genital system 13·3%

Heart and lungs 2·6%

Abdomen 17·4%

Male genital system 3·8%

Circulatory system 2·2%

Skin 3·8%

Orthopaedics 8·7%

Breast 2·1%

Upper digestive tract 3·1%

The diagram shows the numbers of operations performed on the body's various systems, expressed as a percentage of the England and Wales yearly total of 260,000.

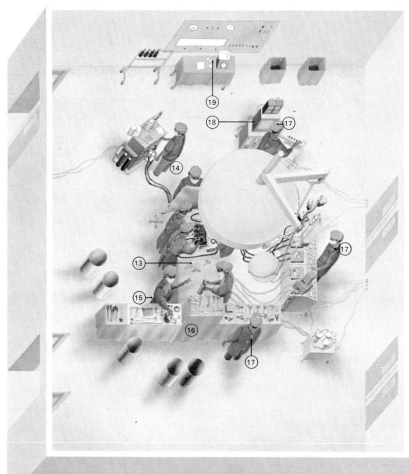

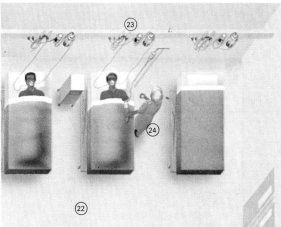

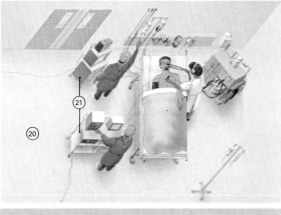

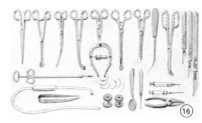

cuff [10] and monitoring devices may be applied to his arm before he is finally wheeled [11] into the operating theatre. Meanwhile the surgeon, assistant surgeons, theatre sister and other theatre staff have been scrubbing up [12] and dressing.

With the patient on the table [13] they commence the operation while the anaesthetist [14] monitors the patient's condition. The theatre sister [15] passes instruments [16] to the surgeon while other technicians and nurses [17] adjust lights, check blood pressure and brain rhythms [18] and put swabs on a scale [19] to check blood loss. After the operation the patient may need to stay in an intensive care room [20] where machines [21] monitor his vital functions. He should

soon be able to move on to a recovery room or a small observation ward [22] which is equipped with oxygen, suction equipment and drip-feed [23]. Here the nurse [24] will keep an eye on him until his pulse and breathing are satisfactory.

Dentistry

The 5,000-year-old history of dentistry reflects not only man's desire to preserve his teeth but also his attempts to avoid and relieve the agonies of toothache. Despite advances made over the centuries, and the highly scientific nature of dentistry today, most people still have dental problems – mainly teeth decay and gum disease. The statistics are alarming, showing, for instance, that in 1968, 36 per cent of the population of England and Wales over the age of 16 had none of their natural teeth.

Dental decay

The teeth are covered by a gelatinous bacteria and food laden layer [2C] called dental plaque. Decay or caries begins with the breakdown of carbohydrate foodstuffs caught in the plaque by bacteria normally present in the mouth. The resulting strong acids dissolve away a protective layer of enamel overlaying the dentine of the teeth [1]. Once the enamel is penetrated, destruction of tooth tissue is rapid.

It is known that decay can be eradicated almost completely by the use of simple and effective preventive measures. These include fluoridation of water supplies, control of dental plaque and advice on diet.

The action of fluoride on teeth, in strengthening the enamel and increasing resistance to decay, was first recognized less than 40 years ago, although it has benefited people for centuries in places where water supply naturally contains fluoride. Since the 1930s, addition of fluoride to water supplies in selected areas at a level of one to two parts per million has significantly reduced the occurrence of dental decay.

The effectiveness and ease of administration of this form of fluoridation make it attractive, although other measures can be used. They include "painting" the teeth with strong fluoride solutions, adding fluoride to toothpastes and mouthwashes, and "fissure sealing" (a technique whereby susceptible parts of the tooth are sealed by the application of a plastic coating).

Preventive efforts must obviously be directed at children, for teeth, unlike other tissues in the body, cannot repair themselves once they have been damaged. When decay sets in, older established methods of dental treatment in the form of fillings, crowns, extractions, bridges and dentures must be used. The repair work done by a dentist aims to remove all decay and to stop further destruction by reconstructing the tooth using a filling material – either metal or plastic [4].

Where dental decay is so far advanced that little of the tooth remains, crowns may be required. This involves removing nearly all of the tooth above gum level and building up in gold or porcelain [5]. Porcelain crowns may also be used to improve the appearance of misplaced or discoloured front teeth.

More serious damage

While decay can continue for a time quite painlessly, it will eventually cause toothache, which becomes acute once the decay penetrates the nervous and vascular tissue (the pulp) in the centre of the tooth. By this stage an ordinary filling may not be enough to save the tooth, for unless the bacterial inflammation of the pulp can be eradicated, an abscess will develop around the root of the tooth and cause its loss. Teeth with inflamed pulps or

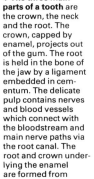

1 The three main parts of a tooth are the crown, the neck and the root. The crown, capped by enamel, projects out of the gum. The root is held in the bone of the jaw by a ligament embedded in cementum. The delicate pulp contains nerves and blood vessels which connect with the bloodstream and main nerve paths via the root canal. The root and crown underlying the enamel are formed from porous dentine.

Enamel
Capillaries
Nerves — Crown
Lymphatics
Pulp
Gum
Dentine — Neck
Jaw
Cementum
Root
Root canal

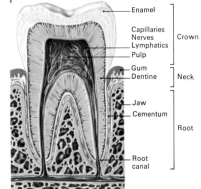

2 Tooth decay (caries) occurs because sugar and other carbohydrates caught in dental plaque (a yellowish film that sticks to the teeth) [C], are turned into strong acid by the fermenting action of bacteria in the mouth. Under the layer of plaque the acid destroys the tooth, first dissolving away the enamel and then eating into the softer dentine, leaving dark, disfiguring stains [B]. Decay [A] starts by attacking the crown [1]. If untreated, it progresses down until it affects the pulp [2]. Once in the pulp, the bacteria begin spreading down the root canal and an abscess develops in the bone around the root, causing pain and swelling [3].

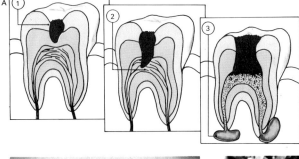

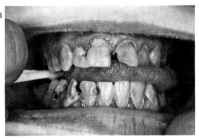

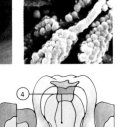

3 Gum disorders are dealt with by periodontology. Gum recession [B] follows infection and inflammation caused by the bacteria in dental plaque. If untreated, this leads to loss of bone [A] around the teeth, which finally loosen and fall out.

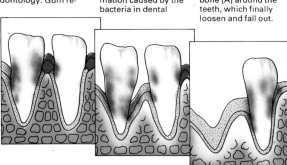

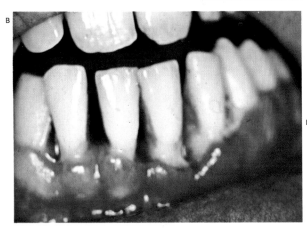

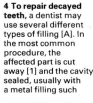

4 To repair decayed teeth, a dentist may use several different types of filling [A]. In the most common procedure, the affected part is cut away [1] and the cavity sealed, usually with a metal filling such as silver amalgam in a back tooth [2] or a white plastic filling in a front tooth. When decay has spread very close to or into the pulp a special sedative lining is placed under the filling to protect the pulp and prevent toothache [3]. If the pulp is badly decayed, a root treatment may be used; all the damaged pulp is removed and the root canal is sealed with a special cement [4].

The effectiveness of good dentistry is shown by a case before [B] and after [C] treatment. Decay and a broken down incisor were repaired using silicate fillings and a crown on the incisor.

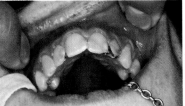

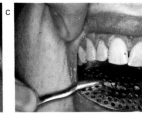

abscesses can sometimes be saved by a technique known as root treatment in which the entire contents of the root canal are removed and replaced by filling material.

As a last resort, a tooth can be extracted under a local anaesthetic given by injection (Where there is an abscess a general anaesthetic is sometimes used.) The tooth is firmly gripped by a pair of specially designed forceps and moved in such a way that the surrounding bone yields gently and the tooth can be pulled out without damage. When wisdom teeth remain buried, or a tooth is badly broken and decayed, extractions involve cutting back the gums and removing some of the jaw-bone surrounding the tooth. Such surgical extractions are usually carried out under general anaesthetic.

When a tooth has been lost it is often desirable that it is replaced, not only for appearance and to make chewing easier, but also because once a gap exists other teeth may have impaired function. Such gaps can be filled with dentures [6] or bridges [7].

As with dental decay, periodontal or gum disease can usually be prevented by conscientious oral hygiene. However, even when gum disease is established, with loosening of the teeth and gum recession, effective treatment is available [3]. The gums are cut back, the damaged tissue removed and the patient told how to prevent recurrence by the regular and thorough practice of oral hygiene.

Specialized forms of dental surgery

One of the more specialized branches of dentistry is orthodontics, which includes the realignment of tooth positions using spring-loaded appliances (braces) within the mouth. Such treatment is most effective in childhood [8]. Orthodontics also extends into the field of general surgery, being used to treat cleft palates that prevent patients from eating and talking normally [9]. Repairing fractures to the jaws and facial bones, correcting facial deformities, and treating tumours within the mouth also require special dental skills.

The scope of dentistry is extremely wide. While one of modern dentistry's concerns is the prevention of caries and gum disease, much can be done to help patients once disease or injury exists.

KEY

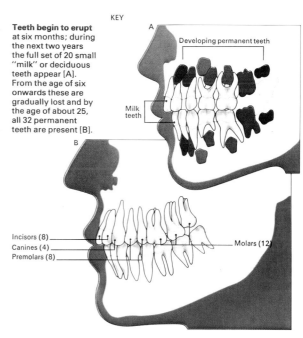

Teeth begin to erupt at six months; during the next two years the full set of 20 small "milk" or deciduous teeth appear [A]. From the age of six onwards these are gradually lost and by the age of about 25, all 32 permanent teeth are present [B].

Developing permanent teeth

Milk teeth

Incisors (8)
Canines (4)
Premolars (8)
Molars (12)

5 Crowns – gold for back teeth and porcelain for front teeth – are used to repair badly damaged teeth. Where decay has made an ordinary filling impossible [A], the tooth is cut down to a peg and the dentist takes an impression so that an accurately fitting crown can be made. This is then cemented on to the remaining tooth stump. Even where a tooth has broken off at gum level [B], it can still be restored by first fitting a gold post into the root of the tooth [C] and then using this as a support for the crown [D]. Crowns are also used for cosmetic purposes to replace unsightly discoloured or misshapen front teeth.

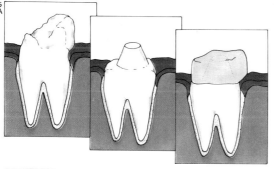

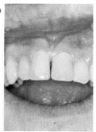

7 A missing tooth can be replaced by a bridge construction [A]. The teeth on either side of the gap are cut down to a peg shape so that gold crowns can be fitted over them. A unit of three crowns, one of which replaces the missing tooth, is then cemented into position. If necessary, a bridge can also be made using two teeth on the same side of the gap [B]. One of the teeth used to hold the bridge was badly decayed [C] and needed root treatment and the fitting of a gold post [D] before the bridging crowns could be made and fitted [E].

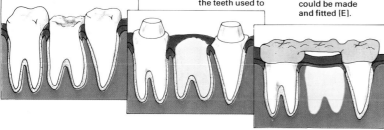

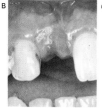
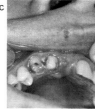
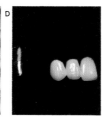
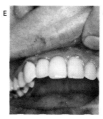

6 A denture may be the only solution if too many teeth in one jaw are lost. Dentures are fitted not only for cosmetic reasons and to allow chewing, but also to maintain sufficient teeth to prevent rapid deterioration of those that remain and imbalance in the muscle action of the jaws, which can cause spasm and pain. The making of an accurately fitting denture is a skilled job. A wax impression is taken of the mouth and a model made to serve as a mould for casting the denture. The extent of a denture varies, as do the materials that can be used to make it. Plastic [A] and metals such as gold or chrome cobalt [B] are common.

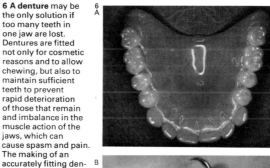

8 Orthodontic braces are used to realign misplaced teeth such as protruding upper incisors. A plate clips into the mouth and stainless steel springs are used to move the teeth gently into the right line.

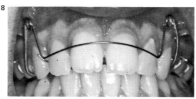

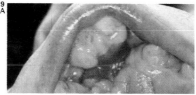
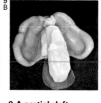

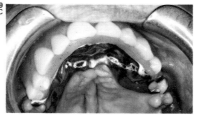

9 A partial cleft palate [A] can be treated by crowning the teeth and clipping a special gold denture [C] to them which seals the gap. A full cleft palate needs a larger denture [B] with a rubber extension to seal the cleft.

131

First aid

First aid is emergency attention given after an accident or sudden illness by someone with or without medical, nursing or other qualifications or experience. Lives are often saved by completely untrained people who simply keep their heads in an emergency, but to give first aid efficiently and safely some form of training is desirable.

In many parts of the world training can be obtained through courses organized by one or other of the voluntary organizations such as the Red Cross (the Red Crescent in Muslim countries). These courses are well worth taking, especially as some of the techniques once thought of as good first aid practices are nowadays felt to be useless or even dangerous.

The principles of first aid
There are a number of important general principles of first aid that everyone should know because they might save another's life. The first and most important of these is succinctly expressed in the famous Latin phrase *Primum non nocere* – "First of all, do no harm". Inexperienced people may unknow-

ingly act unwisely when faced with someone who is injured or suddenly taken ill. They attempt to pour drinks into the mouth of the casualty, even an unconscious one, which is a good way to choke him. Or they try to sit him up or walk him around, when most people who are injured or ill need to be kept lying down. Before giving aid to the casualty knowledge is essential – it might be better not to disturb him at all. And without training heart massage, for example, can be fatal.

Next it is essential to keep calm and to give reassurance. Many people panic at the sight of an injured person and this has a bad effect upon the patient. Keep cool; detail somebody to telephone for an ambulance, stay at the patient's side (unless there is no one else to go for help) and, most important of all, reassure him that help is on the way.

Ensure breathing, stop bleeding
As far as positive action is concerned, it is vital to maintain the airway. Many unconscious people die because the passage leading from the nose and mouth to the lungs is blocked by blood or vomit, by loose or false

teeth or by the tongue falling back into the throat. Deaths from blockage can almost always be prevented if someone is there to size up the situation and take action.

If you are dealing with an unconscious patient act promptly. Immediately loosen any tight clothing around the neck, then take out any dentures and clear the mouth and throat of any material blocking the airway. Finally, turn him into the recovery position [1]. In this the face is turned slightly down (to let any fluid flow out) and the head is bent back to ensure that the tongue does not fall back and choke him to death, which is a real risk if the patient is lying on his back. If his breathing continues to be noisy this is a sure indication that there is still some obstruction that must be removed.

Any heavy bleeding or haemorrhage must be stopped because it threatens life. If a large artery is spurting or even if a major vein is pouring blood any person, but particularly a child, can bleed to death in a few minutes.

Do not put on a tourniquet (a tight bandage above the injury) – this is dangerous. Equally, do not attempt to control the bleed-

CONNECTIONS

See also
84 The causes of illness: 1
86 The causes of illness: 2
94 Diseases of the skeleton and muscles

1 Unconscious patients should be placed in this recovery position immediately to avoid danger from choking. The hidden arm is behind the patient to prevent him from rolling over.

2 The most common method of giving artificial respiration is the mouth-to-mouth system. The mouth and throat are cleared of obstructions [A] such as false teeth and the patient's head is tilted well backwards. The

first-aider holds the nose shut and blows into the mouth [B] until the patient's chest rises, waits for the air to be expelled, then blows in another. The optimum rate is 12 breaths per minute for adults, 20 for children.

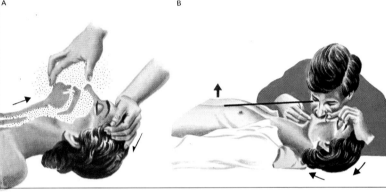

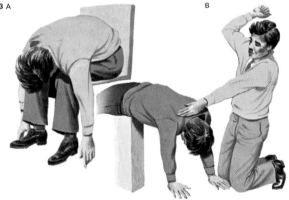

3 Fainting is caused by a failure of the blood supply to the brain. The victim should lie down or sit on a chair, legs apart and head down [A], then sit up after a few minutes. Repeat if necessary. Loosen tight clothing round neck, chest and waist. Lean a choking adult [B] over a table and slap him hard between the shoulder blades. Do not use fingers to remove an object. A child should be held upside down and slapped on the back.

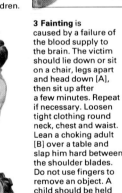

4 The control of a haemorrhage is an important part of first aid. Most nosebleeds can, for example, be controlled by getting the patient to sit up and squeeze the soft part of his nose between finger and thumb for 15 to 20 minutes without interruption [A]. Bleeding from the palm can be effectively controlled by firm and continuous pressure on the place that is bleeding by squeezing a clean pad or handkerchief tightly in the fist [B].

5 Electric shock injury is fairly common. When a person is the victim of shock from a high-voltage cable and has not been thrown clear [A], he should not be approached until the current has been turned off. If a shock occurring at work or

home [B] renders a person unconscious, electrical contact must be broken before any assistance is given, either by switching off the appliance at the mains or pushing the victim away with a dry piece of wood. The next thing is to see if the cas-

ualty is breathing, and if not, to give him mouth-to-mouth respiration. A badly shocked but conscious person should be laid on his back with legs raised [C], but if unconscious, he should be put in the recovery position.

ing by putting your fingers on "pressure points" at some other site. Although nearly everyone seems to have heard of pressure points they are very seldom used in modern first aid practice. The way to stop bleeding is to press firmly – and keep pressing for at least ten minutes – at the place where the blood is coming out [4]. Ideally, it is best to place a sterile cotton wool pad on the bleeding site, but in an emergency a clean handkerchief will do. When a person's life is at stake, use your fingers if nothing else is available.

Shock and its treatment

Look out for signs of shock. People who have been injured or who may have lost a lot of blood, or those who have had heart attacks, may be deeply shocked. The signs are faintness, giddiness, paleness, clamminess of the skin, shallow, rapid breathing and a fast but weak pulse.

Shock can prove fatal and it is essential to act so as to prevent its developing. The way to treat or prevent shock is to keep the patient lying down, preferably with his legs higher than his head (if possible, raise up the lower

part of his body on a rolled-up jacket or similar available object). The patient should then be made as comfortable as possible (loosening any tight clothing) and reassured, for fright increases shock. Do not give him hot, sweet tea or anything by mouth as fluid swallowed may be vomited and, in an injured or unconscious person, then flow back into the windpipe. Equally do not try to heat him up with hot water bottles and heavy blankets as overheating dilates blood vessels of the skin so drawing blood away from the organs inside the body which need this blood in an emergency. These traditional first aid practices are potentially dangerous.

Last of all, no attempt should be made to move a badly injured person unless there is skilled help at hand. Trying to carry someone with broken bones may be dangerous unless you know what you are doing. It is a rule of particular importance where spinal injuries are concerned and these may be present, even if not obvious, in many cases of accident. It is far better to wait until an ambulance man or other trained help arrives – even if this means diverting the traffic.

KEY

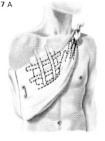
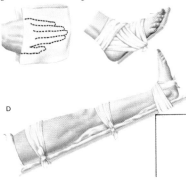

6 Burns caused by dry heat and scalds caused by moist heat are treated similarly. The object is to cool the skin down as fast as possible. Skin damage often occurs after help has arrived because tissues are still at great heat. In the case, therefore, of a burn or scald, smouldering or saturated clothes should be ripped or cut off at once; seconds are vital. The burn should be cooled by immersing

it in cold water [A] for at least ten minutes and the casualty asked to lie down [B], if shocked, with his legs raised. No creams or ointments should be used. All that is necessary is to cover the damaged area with a clean dressing such as a sheet or tablecloth. If the rescuer finds the victim's clothes are still alight he should at once try to smother the flames with a rug or blanket [C] but if this is impractical, pull him to the ground (which helps to keep flames away from the head)

and cover the flames with coats, rugs, cushions, blankets, etc. He should never be rolled over and over. The extent of burns [D] is measured in the percentage chart [1]. If in doubt burns should always be referred to a doctor or hospital. If large areas are damaged (more than 15 per cent in adults and 10 per cent in children) it is necessary to treat the patient for fluid loss. A superficial burn [2] rapidly returns to normal, while a slightly deeper one [3] heals after first going through a

blistering stage. A deep burn [4] that goes right through the epidermis and dermis is still capable of healing if a skingraft is stitched into place. The area from which the graft was taken (hips, buttocks, thighs, inner arm, etc) [5] will soon return to normal.

7 A patient with a fractured (broken) bone should not be moved before expert help arrives. Any bleeding should be stopped and wounds dressed to prevent germs reaching the bone ends. The break should then be immobilized by strapping or bandaging to a splint. For this a piece of wood can be used or, more

generally, the patient's own body. For an injured arm, shoulder or rib [A] a temporary sling is used. A dressing, to protect a fractured finger [B], for instance, and held in place by strapping, can be applied under

the sling. Suspected fractures of the foot [C] and leg [D] can be splinted as shown. It is important that the bandages should be firm but not too tight and that they should not be put over the suspected fracture.

6 D

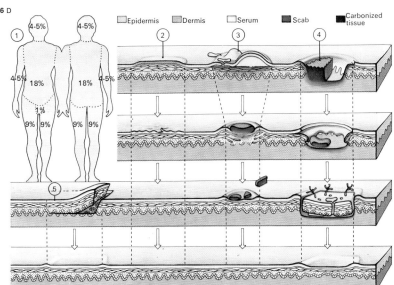

4·5% 4·5%
1
4·5% 18% 18% 4·5%
1%
9% 9% 9% 9%
5

☐ Epidermis ☐ Dermis ☐ Serum ■ Scab ■ Carbonized tissue

2 3 4

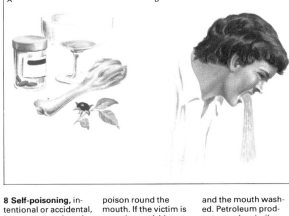

8 A B

8 Self-poisoning, intentional or accidental, is common and can be countered by prompt action. If the casualty is unconscious he should be placed in the recovery position and moved to hospital. Should artificial respiration be necessary beware of any

poison round the mouth. If the victim is conscious ask him what he has taken and look for evidence [A]. If it is corrosive (acid or alkali) do not make him vomit; give him water to drink slowly. Burned lips are a clue; they should be splashed liberally with water

and the mouth washed. Petroleum products require similar treatment. Most cases do not involve either corrosives or petrol products and the correct action is to induce vomiting [B] by forcing two fingers down the throat after giving water to drink.

133

Introduction to mental health

Psychiatry, derived from two Greek words meaning "mind" and "medical treatment", is that branch of medicine devoted to the diagnosis, treatment and prevention of mental ill health. There is no clear dividing line between physical medicine and psychiatry. Psychiatry cannot safely ignore the relationship between bodily condition and mental state, and medicine is becoming increasingly attentive to the manner in which a person's emotional state can precipitate and aggravate physical ill health.

The causes of mental illness

There is never a simple or single cause for mental illness [2]. In practice, a distinction is drawn between predominantly intrinsic or endogenous causes, such as inherited factors, and predominantly extrinsic or exogenous causes, such as physical injury and disease, and mental stress such as bereavement, unexpected financial reverses or the loss of a job. The interaction between a person's basic personality and any physical or emotional stress affects the extent of any subsequent mental reaction. One man or woman may be

able to cope with massive stress quite adequately, whereas another may be overwhelmed by what seems a minor set-back.

Disorders of the mind

The symptoms and signs of mental illness are grouped under a number of headings: disorders of perception; of thought and speech; of memory; of emotion; of the experience of the self; of consciousness; and motor disorders.

Hallucinations are sensory perceptions without any external, objective stimulus and are a common perceptual disturbance in psychotic illnesses and brain disorders. Symptoms of thought disorder include flight of ideas, in which the thought processes are speeded up; perseveration, whereby the patient "perseveres" with a particular response long after a change in his immediate environment has demanded a different response; and thought blocking, in which there is a sudden arrest of the train of thought and the start of an entirely new one.

In certain disorders (obsessional states for example), the patient recognizes that he is compelled to think about certain things

despite his every effort to rid his mind of them. Such compulsion is understood by the patient to originate within him and not as a consequence of some alien activity. In thought alienation, however, the patient experiences his thoughts as being under the control of some external agency. He may believe that others are participating in his thinking or that thoughts are being inserted into or extracted from his head, or that others think his thoughts in unison with him and are aware of his most intimate contemplations.

A delusion is defined as a false belief of morbid origin, and which is an absolute conviction unamenable to reason or contradiction although usually absurd or even impossible. Disorders in the form of thinking are characterized by a fragmentation of the links connecting successive thoughts, and the phenomenon of over-inclusiveness, in which the patient is unable to maintain the boundaries of a concept.

Disorders of memory include disturbances in the registration of information (such as by lack of concentration) and in the retention and recall of material. Where there is a

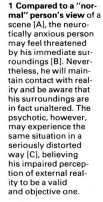

1 Compared to a "normal" person's view of a scene [A], the neurotically anxious person may feel threatened by his immediate surroundings [B]. Nevertheless, he will maintain contact with reality and be aware that his surroundings are in fact unaltered. The psychotic, however, may experience the same situation in a seriously distorted way [C], believing his impaired perception of external reality to be a valid and objective one.

2 Mental health is influenced by widely different factors. Genetic make-up is important; identical twins [A] have a much higher chance of both suffering from schizophrenia than normal siblings. Genetic factors are less significant in neuroses, in which emotional trauma, particularly in childhood [B], is more important. Sexual difficulties are traditionally linked with psychiatric disorders, particularly in adolescence [C]. Social isolation [D], to which immigrants and the elderly are vulnerable, causes depression. Some jobs have attendant psychiatric risks [E]: alcoholism in journalists and seamen; depression in housewives and semi-skilled workers; and psychosomatic complaints in doctors and businessmen. Some physical disorders, like hypothyroidism [F], are associated with psychiatric symptoms and it is difficult to separate them from a purely mental condition such as anxiety neurosis.

pathological memory loss, the gap may be filled with elaborate fabrications, as in alcoholic psychosis.

Emotional disorders consist of variations in the intensity or duration of the emotional response, which may also be inappropriate to the particular situation. The term "affective disorder" refers to a sustained disorder of mood, such as depression or mania, rather than to a transient emotional reaction.

Depersonalization, a disturbance in the experience of the self, occurs when the individual feels himself changed in comparison with his former state. He feels like an automaton and watches his own actions "from outside". In derealization the person perceives the outer world as strange and altered in some significant way. Both these disturbances are experienced on occasions by perfectly healthy people.

Disorders of consciousness are mainly due to physical causes and include alterations in attention and concentration, a slowing in thinking and a lack of direction in thought and action. The patient may be disoriented, may manifest disconnected and incom-

prehensible behaviour or may be delirious.

Motor disorders include lack of initiative, retardation in speech and action and stupor. Major forms of motor disorder seen in mental illness are psycho-motor activity, seen often in manic states; catatonic excitement, characterized by stereotypy (monotonous repetition in speech, mannerisms and movement); and passivity feelings (whereby patients believe their impulses or feelings are controlled by some outside agency).

Major categories of mental illness
There are three major categories of mental illness. The psychoses [1] consist of schizophrenia, manic-depression, paranoid illnesses, the organic psychoses and psychoses associated with physical disorders; the neuroses, consisting of anxiety and phobic states, obsessive-compulsive disorders, hysterical and depressive neuroses; and the personality disorders, which include alcoholism and drug dependence, the behavioural disorders of childhood and a group of anomalies or deviations of personality that are not the result of a psychosis or any other illness.

The blank, emotionless expression of a shell-shocked soldier shows a typical response to overwhelming stress.

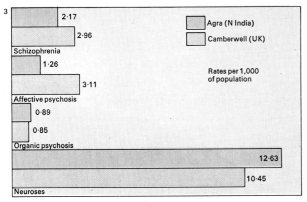

3 | 2·17 |
| 2·96 |
Schizophrenia
| 1·26 |
| 3·11 |
Affective psychosis
| 0·89 |
| 0·85 |
Organic psychosis
| 12·63 |
| 10·45 |
Neuroses

Agra (N India)
Camberwell (UK)

Rates per 1,000 of population

3 Variations in the prevalence rates of psychiatric illnesses between different countries may reflect genuine difference. However, they may be the result of differing diagnostic practices among psychiatrists or of a marked variation in the availability of psychiatric care. For this reason it is difficult to assess whether one country or one environment is more liable to encourage a particular category of mental illness than another. There do, however, appear to be variations in types of neuroses. For instance, the commonest neurosis in Agra is hysteria whereas in Camberwell it is depression.

4 Man shows cyclic variation in many of his functions, as do almost all plants and animals. The most prominent cycles are the circadian or diurnal rhythms [A], which are about 24 hours in length and vary little over a long period (inset clocks represent six weeks). The best known are the diurnal fluctuations in body temperature, steroid secretion and sleep. These rhythms can be disturbed in various ways. In constant darkness they shift forwards so that sleep [B] occurs later and later. Extreme disturbance of sleep can lead to psychotic conditions. Jet travel across international or national time zones [C], where the body is having to readjust its clocks to local time, can produce a number of psychiatric symptoms. The outer clock represents local time; the inner clock the patient's sleep time.

5 Responses to a set of questions may give indications of the presence and degree of psychiatric illness. The questionnaire is designed to elicit different responses from people with different mental illnesses or disorders of behaviour. Questions relate to various areas of behaviour or diagnostic categories, such as hysteria, phobia or thought disorder. Answers are grouped into such categories and the score on each category registered on the graph. The result is a characteristic profile for different mental states. Those shown here are for a neurotic patient [A], a boy suffering from a behaviour disorder [B], and a psychiatric patient [C].

How mental illness has been treated

In ancient times an individual with a disturbed mind was assumed to be influenced by spirits, demons, gods or other supernatural forces. As a result healing practices were generally unsystematic and relied heavily on the power of suggestion.

Madness through history

The theory of the four humours, proposed by the Greek philosopher Hippocrates (*c.* 460–377 BC) and Galen (*c.* AD 130–*c.* 200), introduced a degree of order and influenced medical thought for over 2,000 years. This theory postulated the existence of four key elements in the body (blood, phlegm, yellow bile and black bile). Different diseases and different personality temperaments were believed to be associated with the predominance of one or other humour. An excess of black bile or dry phlegm, for example, was considered the cause of melancholia and this condition was treated with physical methods, such as vapours, baths, diet and emetics.

Arab medicine continued the traditions inherited from Greek medicine and between the eighth and thirteenth centuries a number of asylums for the insane were opened at Damascus, Cairo and Baghdad. In Europe, with the spread of Christianity, care of the mentally afflicted was one of the duties of monasteries, other religious houses and hospitals run by the clergy. The first hospital to be built exclusively for the insane is believed to have been opened at Valencia, in Spain, in 1409.

Paradoxically, this greater care and compassion for the mad among the religious institutions coincided with a tendency among ordinary people to reject them along with paupers, cripples and other social outcasts. Added to this there was medieval Europe's obsession with sorcery, heresy [3], witchcraft and demonic possession. Many violent or dramatic outbursts of insanity were attributed to these "evil forces" and dealt with by torture, imprisonment and often death. Yet the most popular treatise of the time, *De proprietatibus rerum*, written by an English Franciscan friar, distinguished between physical and psychological causes of mental illness and prescribed rest, sedation and music therapy for the violently disturbed.

The seventeenth and eighteenth centuries witnessed a remarkable growth in the scientific basis of medicine and surgery. Theories about mental disturbances, however, lagged behind and the fashion to explain madness in terms of moral flaws, lack of impulse control and the degeneration of personality flourished. Attempts were made, nonetheless, to explain how the mind worked in physiological and chemical terms, even if these did take unusual forms.

Theory and therapy

Franz Mesmer (1734–1815) blamed mental illness on the accumulation of a magnetic fluid in the body that could be removed by special magnetic powers possessed by certain therapists, such as himself.

Around this time, Franz Joseph Gall (1758–1828) claimed to have discovered 27 organs within the brain, each of which was responsible for a particular mental function. The better these organs worked, the larger they were and the more they affected the overall shape of the skull. Phrenology, the art of judging character and mental stability by

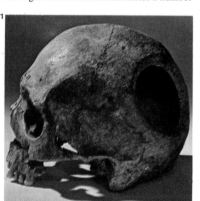

1 **Evidence of trephination,** the removal of a part of the skull probably as a treatment for mental illness, has been found in many cultures dating back at least 10,000 years. It was believed that the spirit causing the mental distress might be released through the opening in the skull. Later the medieval theories of demoniacal possession provided another theoretical basis for such a therapeutic approach to be readopted.

2 **Pilgrimages** to the shrines of special saints were made by the mentally ill during the late 12th century. One of the most famous shrines was that of St John at Ghent. A complex ceremony of offerings, processions, penances and prayers lasted nine days. This led to the founding, in 1191, of the first reception centre for vagrants, paupers and the insane, which was paid for by the citizens of the town.

3 **Insanity** has often been linked with magical practices, sorcery and religious heresy. Girolamo Savonarola (1452–98), the Florentine religious and political reformer, seen here being burnt to death, was almost certainly mentally unhinged. There were few medical authorities, however, who distinguished between insanity and heresy. The court physician of the Duke of Cleves, Johann Weyer (1515–88), was prominent among those who insisted that almost all the so-called witches were elderly women who were psychotic rather than evil or possessed, but neither the Church nor legal experts agreed. It was not until the latter half of the 17th century that the general persecution of witches ceased.

4 **Benjamin Rush** (1745–1813) is generally regarded as the father of American psychiatry. A controversial figure, he was a vigorous advocate of blood letting and the use of mechanical devices in the treatment of mental illness. He is better remembered for his progressive views on the need to combine psychological and physiological approaches in the study and treatment of individual patients.

5 **Philippe Pinel** (1745–1826) was the French physician who on 24 May 1798, with the consent of the Legislative Assembly, removed the chains from 49 insane patients at the Bicêtre Hospital in Paris, ushering in a new type of treatment.

6 **William Tuke** (1732–1822), an English wholesaler of tea and coffee and a philanthropist, founded with his son Henry (1755–1814) the York Retreat (opened 1796). This became an influential early institution for the humane care of the insane.

the shape and features of the skull, became widely popular until it, too, fell into disrepute around the mid-nineteenth century.

Meanwhile the era of confining the insane had dawned and large mental hospitals and private madhouses multiplied. Conditions in these institutions were often appalling and treatments relied on physical methods [7].

Towards modern psychiatry

In time, a number of progressive pioneers removed the chains, and methods of treatment based on kindness, understanding, limited freedom and work began to supplant those elaborate mechanical restraints previously in vogue. At the same time two contrasting trends, both present in amorphous forms for several centuries before, began to crystallize. The first, initiated mainly by German diagnosticians, led to the gradual introduction of an orderly system of classifying mental disease. The second involved the discovery of the unconscious.

The French psychologist Pierre Janet (1859–1947) described a hierarchy of mental functions at the bottom of which he placed automatic functioning, such as in catatonic stupors, and at the top, rational, experienced and conscious activity. Janet's theories, together with the elaborate demonstrations of hypnosis by Jean Charcot (1825–93) at the Salpêtrière in Paris and the remnants of the theories of mesmerism, magnetism and phrenology provided the background to the theories of Sigmund Freud (1856–1939) concerning the unconscious.

The twentieth century has witnessed the development and popularization of psychoanalytical theory, particularly in the United States. At the same time the discovery of the therapeutic efficacy of electrical treatment in severe depressive illnesses, the development of powerful anti-psychotic drugs, such as chlorpromazine, and the slow unravelling of genetic aspects of some forms of mental illness illustrate the progress made in comprehending the biological contribution in psychiatry. The necessity for understanding and effective countermeasures is illustrated by the fact that in Britain one man in every nine and one woman in every six can expect to be treated for mental disorders.

KEY

Bedlam, the popular name for the Bethlem Royal Hospital, became infamous for its ill-treatment of the insane. Public torture was common and a penny tour through the crowded cells was a fashionable pastime in 18th-century London.

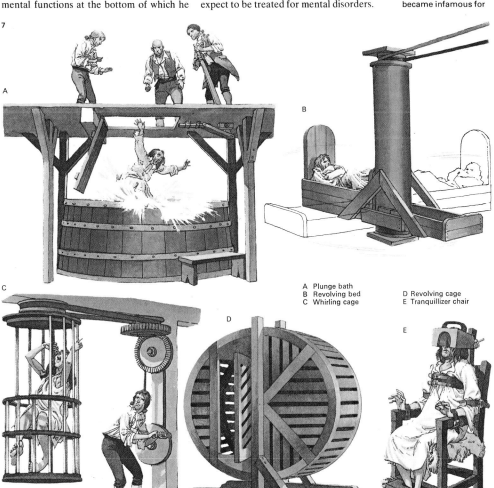

7

A

B

C

D

E

A Plunge bath
B Revolving bed
C Whirling cage
D Revolving cage
E Tranquillizer chair

8
A

8 General paralysis of the insane is a dreaded complication of syphilis appearing some 10 to 25 years after the primary infection. The condition is characterized by severe personality changes, social uncontrollability, impulsiveness, grandiosity and serious mental deterioration. During the latter half of the last century, before the organic cause had been discovered, physicians looked for psychological causes.

The Japanese microbiologist Hideyo Noguchi (1876–1928) [A] put paid to imaginative speculations when, in 1911, he discovered at postmortem the syphilitic spirochete in the shrunken atrophied brains [B] of patients dying from general paralysis of the insane. His discovery led to a resurgence of hope that similar distinct and treatable causes might be found for other mysterious psychiatric disorders.

8
B

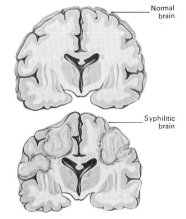

Normal brain

Syphilitic brain

7 Elaborate mechanical contrivances superseded, in the 19th century, the physical methods used in earlier periods – the purges, emetics and bleedings. With the simple intention of subduing violent and manic patients, the most extraordinary contraptions were devised. Leather straps, canvas jackets, muffs and handcuffs replaced the chains. Patients were fastened in wooden chairs in such a way that only the minimum of movement could occur. The induction of sudden and intense fear was believed to have a beneficial effect in mental illness with the result that physicians employed whirling chairs, spinning beds and padded hollow wheels into which patients were strapped and which rotated at speeds of up to 100 revolutions per minute. Such rotation produced vomiting, incontinence, loss of consciousness and, in a number of cases, death. Another cruel device was the plunge bath. The unsuspecting patient was invited to cross a floor which was opened beneath him, plunging him into a bath of ice-cold water. Benjamin Rush, in America, tried to call attention to the need for therapy and the employment of suitable companions to listen sympathetically to patients.

Psychoses

A psychosis is a behavioural disorder typified by a fundamental break with reality and a tortured internal experience [Key]. The major psychosis is schizophrenia. This term refers to the severe mental disturbances, common to all forms of the disorder, in which there is a detachment from the external world and a disintegration of the internal one.

Schizophrenia

Schizophrenia usually, although not invariably, first occurs at puberty. The most common symptoms include incoherent thinking, lack of emotional response, delusions and hallucinations [1]. Thought processses are jumbled and the resultant speech is often incomprehensible and illogical. The schizophrenic may receive moving news without any sign of emotional response. He may smile or appear indifferent when told of some personal tragedy or when confronted by a sombre scene. Delusions are often bizarre and occur with a strength of conviction that renders them beyond reasonable argument. A schizophrenic in a bar, on being offered a drink by the barman, might con-clude that this was a signal to the other cus-tomers that he was a murderer. Hallucinations, particularly involving voices, are also common in schizophrenia.

The cause of the condition is obscure. The role of genetics, in the light of recent studies of twins, is undoubtedly important but it is clear that, although necessary, genetic factors may not be sufficient for the illness to appear. Certain emotional stresses such as family disharmony, physical precipitants such as infection or childbirth, and a number of drugs including amphetamines and LSD, may trigger the illness in predisposed people.

Schizophrenia may be sub-classified as paranoid (the patient has delusions of persecution and is mistrustful of everyone); simple (he is apathetic, withdrawn and uncommunicative); hebephrenic (the patient exhibits silliness and bizarre behaviour); and catatonic (he may sit for hours in one position with muscles rigid, unable to feed or look after himself, seemingly unaware of anything around him). Schizophrenia commonly shows as a mixture of these types. Treatment includes drugs (phenothiazines), social rehabilitation and psychotherapy. The illness usually has serious consequences, and although some patients do recover, permanent damage is the more common outcome.

Paranoia and manic-depression

Paranoia is used by some psychiatrists to refer to mental disorder that occurs when someone has a permanent and unshakeable system of delusions and, at the same time, complete clarity and order in thought, action and will in all other areas. The condition is rare – many cases that begin as paranoia end up showing the characteristic widespread intellectual and emotional deterioration of schizophrenia. However, paranoid illnesses do occur in some solitary and shy individuals who project their own doubts and insecurities on to others, in a number of deaf people, who may misinterpret comments they do not hear clearly as insulting; and in the morbidly jealous.

The other serious psychosis is called manic-depression. As implied, there are two main forms: mania, characterized by excitement, euphoria, grandiose schemes, rapid

1 **Schizophrenia** is characterized by disjointed and episodic thought [A] and inappropriate emotional responses [B]. There is a tendency to delusion with the reading of innocent gestures as malicious signals [C]. Hallucinations and illusions occur as when a peaceful country scene seems oppressive or terrifying [D].

flights of ideas and overactivity; and depression [2]. Patients may suffer from one or other type or from both.

Psychotic illnesses often occur in association with organic diseases. Many different poisons and various kinds of brain damage may affect the mental state. Psychotic symptoms commonly occur in delirium associated with fever, drug intoxications and other causes of brain disturbance. Patients in severe delirious states have illusions and hallucinations of the senses, especially vision. Dislocations of their sense of time and space also help to differentiate the organic psychoses from the so-called functional psychoses, schizophrenia and manic-depression.

Brain diseases and psychosis

The most serious brain disorder often appearing with psychotic features is dementia. This is a diffuse cerebral disease in which there is serious intellectual deterioration and emotional impairment. The brain shows general atrophy, there is a loss of brain cells and specific changes in the cerebral cortex. The psychosis may take various forms

– manic, depressive or paranoid. The manic variety is characterized by pointless activity, silly boasting and a marked overtalkativeness. In the depressive variety, the emotions are rather blunt, there is obvious irritability and hysterical symptoms may be intermingled with hypochondriacal ones. The paranoid variety is most dramatic – there are often elaborate delusions of gases being pumped into rooms, food being poisoned, spies trying to kill the sufferer, and so on.

Dementia occurring before the age of 50 is commonly termed presenile. Pick's disease, in which there is a circumscribed atrophy of the frontal lobes, and Alzheimer's disease, in which there is widespread atrophy of the brain, may be accompanied by psychotic features. Psychosis can also occur in cerebrovascular disease with brain tumours and as a consequence of head injuries. Epilepsy, particularly affecting the temporal lobe, may be accompanied by hallucinations, mood disturbances and paranoid delusions. In the period immediately following childbirth psychoses can occur in predisposed women.

"The Scream"—Edvard Munch, 1893

2 Depression, in psychiatric terms, refers to a condition in which depression is not the only symptom, but is accompanied by insomnia, loss of weight and appetite, diminished libido, feelings of worthlessness and guilt and suicidal thoughts. Some depressive states may have an obvious precipitating factor such as a bereavement or a broken engagement ("reactive" depression). In others, and particularly in patients predisposed genetically to develop manic-depressive psychosis, the mood disturbance often occurs with little or no obvious provocation ("endogenous" depression). In extreme depression the patient may be severely retarded or markedly agitated. He tends to be preoccupied with feelings of guilt and may explain these by concluding that he has committed some terrible crime or sin for which he is being punished. Depressed patients commonly feel a burden to others, particularly to those close to them, and may attempt suicide to relieve not only their own feelings but also those of their concerned relatives and friends from whom they invariably feel cut off and alienated. A delusion sometimes seen in severely depressed people is the so-called "nihilistic" delusion. Influenced by the profound emotional malaise that is central to depression, the patient's mood may become completely negative. He may declare that he has no name, no age, no parents, relatives, wife or children. He may insist he has no head, no chest, no body – indeed, he may deny everything, and resist everything. Treatment in the severe forms of depression is by electrical shock and antidepressants. Psychotherapy is helpful, particularly in reactive depression.

Personality defects and neuroses

The neurotic person has obvious abnormal psychological symptoms, but does not show that sharp break with reality that characterizes the psychotic person. Particular neuroses can and do appear for a short time in otherwise healthy people, but they are usually exaggerated versions of normal experiences. All of us have felt depressed, anxious, fearful at some time – what distinguishes the neurotic is not a qualitative difference from the norm but a quantitative one.

Types of neuroses

Classification of neuroses into various categories is fairly arbitrary but is based on the major symptom present. The main neurotic states identified are anxiety neurosis, hysteria, depressive neuroses and obsessive - compulsive disorders.

The overriding complaint in anxiety neurosis is an excessive fear, often amounting to panic and associated with physical symptoms such as a dry mouth, palpitations and sweating. It may be complicated by a phobia [3, 7], obsession [4] or depression [6].

In the popular mind, hysteria is linked with uncontrollable tantrums and screaming fits. For the psychiatrist the term refers to those symptoms, both physical and mental, such as paralysis, tremor and amnesia, that are caused by a psychological disturbance and aim, unconsciously, at escape from a seemingly insurmountable difficulty or the fulfilment of some need.

Hysteria takes two main forms known as conversion or mainly physical disturbances [5] and dissociation. The most common of the dissociative states is the fugue when someone quite suddenly and without any warning signs wanders off. Such a journey has no plan or destination, is usually accompanied by loss of memory, and often allows the person to avoid some unpleasant situation, such as an embezzlement charge or divorce.

Allied to fugues are the so-called trances and twilight states when a person withdraws and insulates himself from the world. More rare but of the same type is the dual or multiple personality, when a person manifests a number of traits that are opposed to each other and poorly integrated. By a process of dissociation a shy, prudish girl may become flirtatious and seductive. In extreme cases people keep different names and styles for their different personae and when playing one role deny all knowledge of any other.

Anxiety, neurotic depression and hysterical neurosis are the reactions to stress of more or less neurotic personalities. The strongest and best-balanced individual may react with depression to a severe set-back or with anxiety to a major stress. However, most anxious, depressed or hysterical patients are more the victims of their own personalities than of outside events.

Personality disorders

Personality disorders are a group of anomalies or deviations which, although not the result of a psychosis or other illness, are odd enough to upset or puzzle others and sometimes even the sufferers themselves. Such disorders resemble mental illness and may require psychiatric help and understanding but the sufferer is less ill than abnormally developed, in the same way that a physically handicapped individual is not ill but has an abnormal physical constitution.

1 A psychopath displays recurrent antisocial, delinquent, and criminal behaviour in many areas of his life. Data on the prevalence of the disorder are unreliable but it is seen frequently by psychiatrists, usually because of an associated alcohol problem or depression. The family backgrounds of most psychopaths are grossly disturbed. There is commonly a history of parental alcoholism, crimi-nality, separation, divorce or early death. The first signs usually occur in late childhood or early adolescence and are commonly a restlessness, an unresponsiveness to discipline and a tendency to be cruel to smaller children or animals. A disturbed school history, with truancy and academic failure, is frequently found and later the job record is marred by poor performance, unreliability, constant changes and an inability to accept even minor criticism. Psychopaths have great difficulty in maintaining close emotional relationships and their marriages are characterized by infidelity, separation and divorce. Most psychopaths end up with criminal records. Many of them indulge in minor petty delinquency but a small minority commit brutal and callous acts of physical violence.

2 Inadequate individuals show a low physical and mental strength, a lack of resilience and flexibility and an inability to cope with stresses. They are often nervous and dependent in early childhood and their parents are either intensely protective or harsh and rejecting. They are excessively shy, socially and sexually inhibited, egocentric and introverted. They often lead lonely, somewhat joyless and anxious lives in adulthood.

Some personality disorders are due to faulty brain development caused by genetic make-up, injury, infection, poisoning or malnutrition in childhood. Some sex chromosome anomalies also seem to be associated with antisocial behaviour, in particular the so-called XXY male who is unusually tall, aggressive and sometimes of subnormal intelligence. However, environmental factors, including early childhood deprivation, parental quarrelling and severe stress, are also known to play their part.

A number of specific categories of personality disorder, including the antisocial [1] and the immature [2], have been described. The paranoic is sensitive and vulnerable, reacting to everyday experiences with an excessive sense of inferiority and humiliation, and apt to be touchy in preserving what he conceives to be his rights. People who have so-called affective personalities have long-standing anomalies of mood – either a predominantly gloomy and pessimistic attitude to life or the opposite. In contrast the hysterical personality has shallow and changeable emotions and,

although unreliable himself, craves love and attention. The schizoid is notably aloof, shy and reserved, tending to be markedly introspective and eccentric. Finally, the anankastic or obsessive personality is characterized by a strong sense of insecurity, an excessive caution combined with a stubborn inflexibility and a rigid perfectionism.

Sexual problems

Various forms of sexual problems are referred to psychiatrists. Some, such as fetishism (the exclusive derivation of sexual pleasure from inanimate objects) are private anomalies of sexual behaviour. There are others, however, such as paedophilia (the desire of an adult to engage in sexual activity with children) and exhibitionism (the need to expose the male genitals to females) that involve public conduct.

Transvestism (dressing in the clothing of the opposite sex) and trans-sexualism (the wish to be and to function as a member of the opposite sex) are rare but important. These problems are regarded as developmental anomalies rather than as true illnesses.

The Munchausen syndrome, named after Baron von Munchausen (1720-97), is a severe personality disorder wherein a person malingers or consciously feigns an illness to obtain a desired end. A few people repeatedly malinger, going from one hospital to another, often under a variety of assumed names, but almost always telling the same story, faking the same symptoms and submitting to innumerable investigations and operations. Chronic malingerers resemble the Baron, who entertained his friends with extraordinary (and patently untrue) tales of his supposed travels so that he eventually gained the reputation of being an incorrigible liar.

3 Anxiety neurosis may be complicated by a phobia or dread of a specific object or situation. A common phobia is that of "creepy-crawlies", especially spiders. Mildly phobic individuals experience anxiety only when required to handle spiders but the severely phobic can become terrified just by thinking about them and will avoid any situation in which there is the slightest chance of further exposure.

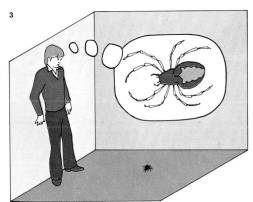

4 In obsessional neurosis the sufferer is occupied with thoughts that do not really interest him and impulses that seem alien. He feels impelled to perform actions which not only give him no pleasure but which he is powerless to stop. The compulsive rituals may be eccentric (eg touching objects a given number of times) or involve repetitive activities (eg cleaning) aimed at neutralizing morbid thoughts (of dirt).

5 A common feature of hysterical neurosis is conversion, a physical disturbance psychologically caused. Symptoms are paralysis, tremor, fits, blindness, deafness and sensory disturbances. Conversion is a primitive mechanism for dealing with difficult situations and evoking sympathy. Hysterics differ from malingerers in that although they may produce illness they are unaware of what they are doing.

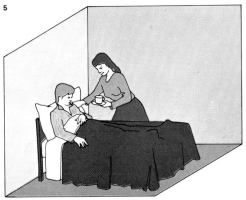

6 Depression is a reaction to the stress or frustration produced by some incident like a loss, grief or disappointment. Depressive neuroses, as a rule, are an exaggerated form of what happens when the average man is temporarily cast down. The severe loss of weight, appetite and sleep, the agitation, delusions and deep guilt feelings seen in manic-depressive psychoses are nearly always lacking.

7 Agoraphobia, a fear of open spaces, is commoner in women and includes panic experienced when out alone, shopping or travelling. Other phobic stimuli are heights, fire, lightning, death and hair. Fears of flowers, water and numbers are rarer, but fears of social situations are common, while children are often afraid of the dark. Snakes, closed places (claustrophobia), cats, spiders, insanity, dirt and mice are other well-known causes of phobia.

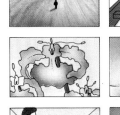

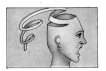

Psychosomatic diseases and retardation

Certain physical disorders such as hypertension and bronchial asthma [3] have classically been designated psychosomatic since the early 1930s. In the intervening years, many other conditions have been similarly labelled including anorexia nervosa, obesity, psychogenic vomiting, some abdominal pains, diarrhoea, torticollis, diabetes mellitus and writer's cramp.

There is little agreement on which illnesses are or are not psychosomatic. Their only common characteristic is the assumption that psychological factors play a major role in their complex and obscure causes. Some critics believe the term to be misleading in that the description of some illnesses as "psychosomatic" implies that the remainder are not. In practice probably few illnesses do not include psychological factors.

Psychosomatic diseases are believed to be the result of a chronic and exaggerated state of the normal physiological expression of emotion, affecting a particular vulnerable organ. If persistent, such an emotional reaction can often produce structural damage.

There has been much speculation about the possible link between certain personality types and susceptibility to psychosomatic disorders. It has been claimed, for instance, that some sufferers from peptic ulcers may be people who unconsciously want to remain dependent. Such a wish is productive of low self-esteem, however, and runs counter to the adult ego's pride and desire for independence. Accordingly, it is repressed and compensated for by aggression and ambition. The resulting inner conflict produces chronic anxiety which in turn leads to overproduction of gastric juices and so to ulcers.

In anorexia nervosa [1] an unwillingness on the part of the patient to develop womanly characteristics, both physical and psychological, with their attendant social and sexual expectations, is thought to lie at the heart of the refusal to eat, the fear of obesity and the desperate attempts to maintain an abnormally low weight level.

Mental retardation

Mental retardation refers to intelligence defects from birth or before full brain development irrespective of the cause. If the IQ (intelligence quotient) of a large sample of people is assessed [7], the same kind of normal distribution curve results as when physical characteristics, such as height or weight, are studied in population samples. IQs between 90 and 110 are considered average. Mental retardation is divided into a number of grades: borderline (IQ 68–85), mild (IQ 52–67), moderate (IQ 36–51), severe (IQ 20–35) and profound mental retardation (IQ under 20).

Such a grading acts as a rough guide to the capabilities of people falling within each range in terms of possible social adjustment and functioning, ability to learn and the acquisition of skills. For example, an adult with profound mental retardation requires nursing care and may achieve only a very limited level of self-care. The moderately retarded adult may be able to work in an unskilled or semi-skilled capacity under sheltered conditions, whereas the person who is only mildly retarded can achieve adequate social and vocational skill for a minimal level of self-support.

Many factors at different stages can cause

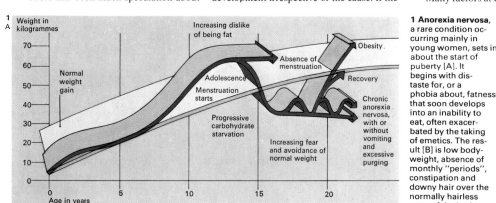

1
A
Weight in kilogrammes

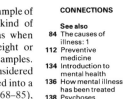

1 Anorexia nervosa, a rare condition occurring mainly in young women, sets in about the start of puberty [A]. It begins with distaste for, or a phobia about, fatness that soon develops into an inability to eat, often exacerbated by the taking of emetics. The result [B] is low body-weight, absence of monthly "periods", constipation and downy hair over the normally hairless parts of the body.

2 The "vapours" were a fashionable psychosomatic complaint of distinguished ladies of the 18th century, characterized by dramatic fainting and a variety of nervous fits, sometimes attributed to over-tight corsetting. Famous physicians of the day treated attacks of the vapours with such "modern" treatments as electric shocks or a course of baths. Impressionable young women were frequent victims.

3 Psychosomatic diseases are believed to be the result of excessive and prolonged exposure of a vulnerable organ to the normal physiological changes that are caused by emotion. If persistent, such an emotional reaction can produce structural damage in the organ affected. Major physical diseases in which psychological factors are thought to play a significant causal role, in addition to modifying the way in which such illnesses develop, include bronchial asthma [1], thyrotoxicosis (resulting from overactivity of the thyroid gland) [2], hypertension [3], coronary artery disease [4], peptic ulceration [5], ulcerative colitis [6] and rheumatoid arthritis [7]. There has also been speculation concerning possible links between certain personality types (for example those who repress anger) and susceptibility to psychosomatic diseases.

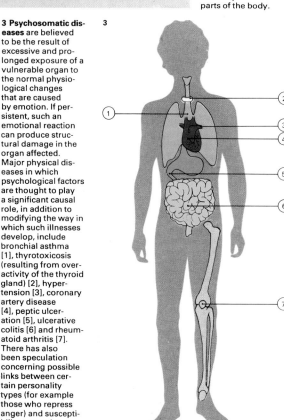

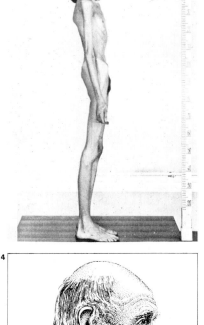

4 Charles Darwin (1809–82), following his journey in HMS *Beagle*, developed a chronic and incapacitating illness, characterized by lassitude, palpitations, headaches, sleeplessness and tremulousness. There has been much speculation concerning the underlying cause of these symptoms. Analysts have seen them as psychosomatic – a psychological reaction to difficulties in Darwin's relationship with his father, who is usually portrayed as a stern and cold disciplinarian.

mental retardation. Some mental defects are the result of genetic diseases. Such conditions include craniostenosis, in which there is premature closing of the cranial sutures (where the bones of the skull are joined) and skull deformity and a number of serious metabolic disorders. Conditions such as mongolism or Down's syndrome are associated with aberrations of the number or shape of the chromosomes.

Uterine factors
Uterine factors in mental retardation include severe dietary deficiency in the pregnant mother giving rise to fetal damage. Virus infections, such as rubella (German measles), contracted by the mother early in pregnancy, can give rise to physical abnormalities and mental deficiency. Irradiation by X-rays or from an atomic explosion may damage the fetus as may certain drugs, such as thalidomide, taken by the mother during pregnancy. Rhesus factor incompatability of a severe nature is also associated with mental retardation in the newborn. Abnormal labour, with prolonged asphyxia and brain

trauma during delivery, and premature birth may also result in intellectual deficiencies.

During infancy and early childhood serious infections, such as meningitis and poliomyelitis, and the rare central nervous system involvement in viral conditions such as mumps, measles and whooping cough, can result in serious mental retardation.

Prevention and treatment
The management of mental retardation can be divided into primary, secondary and tertiary prevention. Primary prevention includes genetic counselling and those medical measures, particularly in the field of obstetrics, aimed at reducing prenatal, natal and post-natal complications. Secondary prevention is achieved by the early identification and treatment of hereditary metabolic disorders. Tertiary prevention involves management of the mentally retarded so that the maximum potential of even the most seriously handicapped can be realized. For the mildly subnormal, special schooling may be required [Key]. For the more severely retarded, institutional care may be necessary.

Mongolism, or Down's syndrome, is a chromosomal abnormality and occurs approximately once in every 600 live births. The overall appearance bears a superficial resemblance to that of members of the Mongol race. The head is small, the neck short and thick and the face has a flat-bridged nose. The eyes are slanted and the hands are broad with short, stubby fingers. Mongols show severe mental retardation.

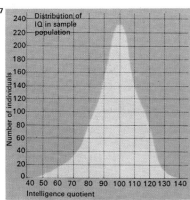

5 Alfred Binet (1857-1911), an experimental psychologist at the Sorbonne, Paris, devised a standardized intelligence test to discover mentally defective primary school children. The test was developed from intelligence scales for the investigation of normal and subnormal children. These were extended to become an intelligence test for the 3–15 age range from which modern tests developed.

6 Extremely withdrawn or moderately mentally handicapped children can be aided by using simple behavioural modification techniques. Normally, the child's solitary play is encouraged or "reinforced" by the teacher's attention [A]. But in this instance he is only rewarded by the teacher's full attention when he joins the group [B]. As a check on the treatment, the original reinforcement is reinstated [C], which at once encourages the child to revert to his former behaviour. The new treatment is started again and the desired behaviour – the child playing in the group – is encouraged until it is firmly instilled [D]. In time the teacher can leave the group to its own devices [E] and the child will remain with it. Such behavioural methods are being extended.

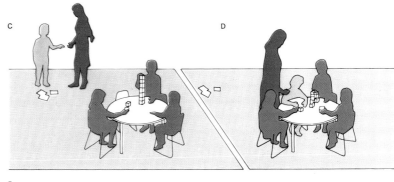

7 The distribution of adult intelligence quotients is from a sample of 2,052 people in the USA in 1958.

8 This example of dyslexic writing illustrates the bizarre character of the spelling disorders in dyslexia, which may be connected with retardation, although not always. Syllables may be omitted or put in the wrong order and sometimes it is not clear where one word ends and another begins.

7

Distribution of IQ in sample population

Number of individuals (y-axis: 0, 20, 40, 60, 80, 100, 120, 140, 160, 180, 200, 220, 240)
Intelligence quotient (x-axis: 40 50 60 70 80 90 100 110 120 130 140)

8

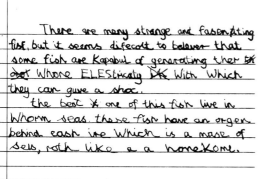

Physical cures for mental illness

Physical treatments in contemporary psychiatry include the use of tranquillizers, sedatives, stimulants and hallucinogenic drugs, electroconvulsive treatment (ECT) and certain psychosurgical procedures.

The use of drugs

A tranquillizer is a drug that induces sedation without loss of consciousness even when given in relatively large doses. The most potent tranquillizers are the phenothiazines of which chlorpromazine is the prototype. Synthesized in a French laboratory in 1950, this drug soon became widely used to reduce severe agitation or excitement of whatever cause. The phenothiazines appear to have selective effects on the delusions and hallucinatory phenomena seen in schizophrenia and manic depression. As a result they have become the drugs used in the treatment of these conditions. A long-acting phenothiazine, fluphenazine, has been developed that can be given by injection every 2-4 weeks and which maintains schizophrenics in a relatively stable state. The phenothiazines are also used to control excitable behaviour in manic states and in confusional states due to physical disease or drugs. They are not effective in the treatment of depression. Side-effects include drowsiness, muscle stiffness and rigidity, skin rashes, excess salivation and, occasionally, jaundice.

The most popular group of minor tranquillizers is the benzodiazepines, of which diazepan (Valium) and chlordiazepoxide (Librium) are the best known. These drugs, are used in the treatment of phobic anxiety, obsessive-compulsive disorders and minor anxiety and tension as well as in the management of the withdrawal symptoms occurring in barbituate and alcohol dependence. Another drug in this group, nitrazapam (Mogadon), is a popular and relatively safe hypnotic, used in the treatment of insomnia.

There are two main groups of antidepressants – the tricyclics, of which imipramine and amitriptyline are the most widely used, and the monoamine oxidase (MAO) inhibitors, of which phenelzine and tranylcypromine (Parnate) are the best known. The tricyclics compare quite favourably with ECT in the treatment of severe depression characterized by suicidal feelings, guilt, self-reproach, ideas of worthlessness, insomnia, weight loss and impairment of libido. The MAO inhibitors are less effective in such states but are thought to be useful in mixed anxiety-depressive states and in depressions believed to be reactions to some obvious environmental stress. They do, however, give unpleasant side-effects (palpitation, sweating and collapse) especially if the patient takes cheese, bean pods, some wines and other foods containing the substance known as tyramine. Lithium salts have been found to be useful in the control of mania, although why they work is not clear.

The powerful hypnotic drugs, the barbiturates, once widely used in the treatment of anxiety, have been superseded by the safer benzodiazepines. Because of their tendency to induce dependence, they are now used only as anti-convulsants in epilepsy. Similarly, the amphetamines, stimulant drugs once used to alleviate depression, have been replaced by the antidepressants and are now rarely prescribed. The hallucinogenic drugs, LSD and mescaline, have been used by some

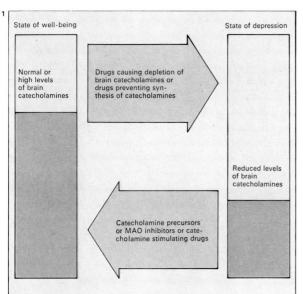

1 The catecholamine theory of depression states that mood change is related to drug-induced changes in the level of chemicals called catecholamines in the brain. The theory is that a feeling of well-being is maintained by a continuous stimulation of certain receptors in the brain by catecholamines, such as adrenaline, noradrenaline and dopamine. Drugs such as reserpine, a potent substance used in the treatment of high blood pressure, deplete brain catecholamine stores and can cause lethargy, apathy and severe depression. In contrast, other compounds have been found that produce an increase in brain catecholamines and hence a marked improvement in mood.

State of well-being
Normal or high levels of brain catecholamines
Drugs causing depletion of brain catecholamines or drugs preventing synthesis of catecholamines
State of depression
Reduced levels of brain catecholamines
Catecholamine precursors or MAO inhibitors or catecholamine stimulating drugs

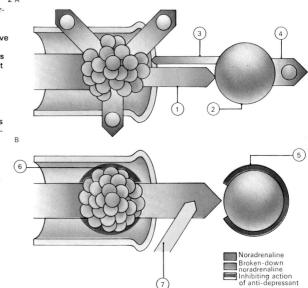

2 A catecholamine neurotransmitter, noradrenaline, is stored in a nerve ending. On stimulation of the nerve [A] noradrenaline is released [1] and exerts its physiological effect on the receptor site [2]. Some noradrenaline is reabsorbed by the nerve ending [3] to be used again while the remainder is broken down [4]. Antidepressants [B] are thought to act by preventing [5] the breakdown and reabsorption of noradrenaline. This increases the amount of free noradrenaline acting on the receptor site. They may also inhibit the breakdown by monoamine oxidase of stored noradrenaline [6]. Other antidepressants increase the site's sensitivity to noradrenaline [7].

Noradrenaline
Broken-down noradrenaline
Inhibiting action of anti-depressant

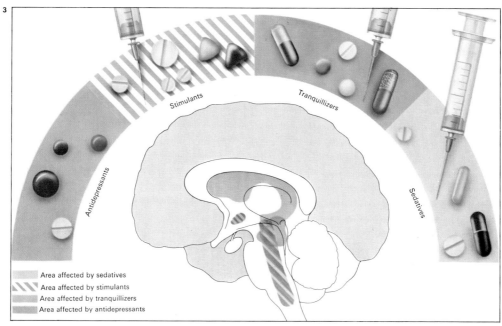

Area affected by sedatives
Area affected by stimulants
Area affected by tranquillizers
Area affected by antidepressants

Stimulants
Tranquillizers
Antidepressants
Sedatives

3 Those areas of the brain believed to be affected by antidepressants, stimulants, tranquillizers and sedatives are shown here. Antidepressant drugs are thought to effect catecholamines stored in the midbrain. Stimulant drugs act on the reticular activating system (RAS) and the hypothalamus. Tranquillizer drugs also act on the RAS and may suppress RAS response in the cortex. They are also thought to affect the limbic system. Sedative drugs such as barbiturates, which are used to induce sleep and as anticonvulsants in epilepsy, have a two-fold action on the brain. They act on the RAS and on the cerebral cortex itself, thereby suppressing brain activity.

psychiatrists in the management of alcoholism and certain neurotic states but the results are unconvincing.

Electroconvulsive therapy
Electroconvulsive treatment was introduced into psychiatry in the late 1930s. It is still widely used in some hospitals and in severe suicidal depressions may be the treatment of choice. The usual course of ECT consists of 6-8 treatments spread over 3-4 weeks. The patient is given an anaesthetic and a muscle relaxant (which paralyses the main muscles and thereby reduces the severity of the convulsion). When unconscious, the patient has two electrodes applied to his head (one each side, in front of and above the ear). An electric current, usually 80 volts with a duration of 0.1-0.3 seconds, is passed between the electrodes producing an epileptic discharge and a convulsion, "modified" by the muscle relaxant and manifested by small twitchings of the facial, hand and feet muscles. The convulsions over, the anaesthetic and relaxant wear off and the patient regains consciousness. There is frequently an associated partial memory loss which is transient, but may take some weeks to clear completely. Studies have revealed that the crucial therapeutic element is the seizure discharge although why this exerts an antidepressive effect on mood is still not known.

Psychosurgery
Psychosurgical operations involve the destruction or removal of normal or apparently normal brain tissue for the purpose of altering certain behaviour. Such procedures are among the most controversial in modern psychiatry. The first such operations were popularized by Egas Moniz [6] during the early 1940s and the so-called "standard" leucotomy (the cutting of certain selected brain fibres) was performed on thousands of chronically incapacitated patients.

The development of potent antipsychotic and antidepressant drugs during the 1950s resulted in a loss of interest in surgical procedures but the development of better operative techniques together with a renewal of interest in brain physiology has recently given new impetus to this field.

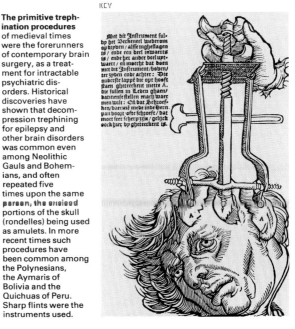

The primitive trephination procedures of medieval times were the forerunners of contemporary brain surgery, as a treatment for intractable psychiatric disorders. Historical discoveries have shown that decompression trephining for epilepsy and other brain disorders was common even among Neolithic Gauls and Bohemians, and often repeated five times upon the same person, the excised portions of the skull (rondelles) being used as amulets. In more recent times such procedures have been common among the Polynesians, the Aymaris of Bolivia and the Quichuas of Peru. Sharp flints were the instruments used.

4 A large multi-centre study of the relative effectiveness of ECT, a tricyclic antidepressant (imipramine), an MAO inhibitor (phenelzine) and an inert placebo was carried out in Britain in 1965. The 250 depressed patients were divided into four treatment groups. Evaluation after one month showed that 71% of those treated with ECT had few or no symptoms compared with 52% for imipramine, 30% for phenelzine and 39% for placebo. This last figure may be taken to indicate the spontaneous short-term remission to be expected in these depressions – they get better automatically.

4

% of each group depressed

100

75 — Phenelzine, Placebo

50 — Imipramine

25 — ECT

0

1 week | 2 weeks | 3 weeks | 4 weeks 1 month

5

5 Ugo Cerletti (1877–1963), the Italian psychiatrist, believed during the 1930s that epilepsy and schizophrenia rarely occurred in the same patient. He suggested that an epileptic convulsion might "protect" the patient against schizophrenia. Accordingly he induced epileptic seizures in schizophrenic patients, some of whom greatly improved. In fact, it is now known that the two are not antagonistic.

6 Egas Moniz (1874–1955), the Portuguese neuro-psychiatrist, impressed by reports that the surgical removal of the frontal lobes in animals eliminated pathological behavioural responses to frustration, suggested that such a procedure might be effective in relieving anxiety in man. His enthusiastic advocacy of such therapeutic procedures led to their widespread adoption by many psychiatrists.

6

7 The technique of leucotomy, as performed in the early 1940s, consisted of making a burr hole in the side of the head above and in front of the ear. A cutting instrument – a leucotome or thin, sharp knife – was swept in an arc longitudinally, thereby cutting the connections between the frontal lobes and the rest of the brain [A]. The same procedure was then repeated on the opposite side of the head. With development of more precise surgical techniques and with modifications in the theoretical basis of psychosurgery, prefrontal leucotomy became less popular. Today, a probe is introduced through a burr hole [B] and is guided to the target area under X-ray

7 A

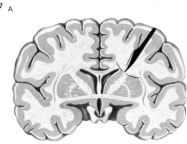

B

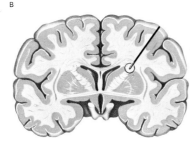

C   Frontal lobe   Cingulate gyrus   Hypothalamus

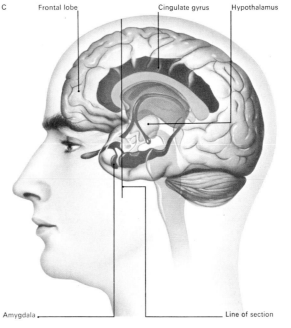

Amygdala   Line of section

control. The area is then destroyed either by the introduction of radioactive seeds through the probe or by the application of intense heat or cold. A favourite target [C] in chronic intractable depression is the inner aspect of the frontal lobe. Destruction of the anterior part of the cingulate gyrus is undertaken in obsessional disorders, of the amygdala in aggressive and hyperactive states and of centres of the hypothalamus in anxiety states and certain sexual disorders. A combination of centres may also be destroyed. Although Egas Moniz, the pioneer of psychosurgery, was awarded the Nobel Prize in 1949, there is still much dispute about the effectiveness of leucotomy.

Behaviour therapy

Many psychologists and psychiatrists, concerned with the causes of neuroses or mental illnesses, believe that they result from emotional problems which, once identified, can also point the way to the cure.

Behaviourist psychologists have a quite different approach. They argue that the cause of, say, a phobia about snakes is irrelevant and that the important thing is to get rid of the phobia by working on it directly. The individual learns his behaviour. It is therefore possible, the behaviourists argue, for him to unlearn or modify his behaviour.

Therapy by conditioning
Behaviour theory states that a neurotic reaction is acquired through a simple process of conditioning – the term used by the Russian psychologist Ivan Pavlov (1849–1936) in his experiments with dogs [1]. What is being conditioned is an emotional feeling of fear or anxiety. Such reactions easily give rise to more complex neuroses. A person with a conditioned fear of dirt may end up severely crippled by his compulsive need to wash off any dirt and with it his fear of contamination.

The behaviourist's answer to such faulty conditioning is to submit it to a process of deconditioning or counter-conditioning. One such method of deconditioning is extinction. In simple extinction the unconditioned stimulus is not repeated when the conditioned stimulus is presented. When applied to Pavlov's original experiments, for example, the bell is rung on a large number of occasions but no food is presented. Gradually the salivary secretion in response to the bell, built up in the earlier experiment, diminishes until it finally disappears completely. Simple extinction is probably responsible for the spontaneous "disappearances" of some neurotic illnesses.

One reason why all neuroses do not so respond may be that, unlike the Pavlovian dog, human beings can avoid things that frighten them and this avoidance, by relieving anxiety, serves as a further conditioning stimulus. For example, a woman with a fear of spiders will run away from any situation in which she may be exposed to these creatures. Her relief reinforces the conditioning process involved so that her phobia

for spiders is strengthened. In such a case, behaviour therapy is required to force the individual to face up to the noxious stimulus.

Other behaviourist techniques
In systematic desensitization [2], an attempt is made to condition an alternative response to the fear-producing stimulus so that the occurrence of one inhibits the occurrence of the other. Anxiety involves the tensing of muscles. By systematically persuading the patient to relax his muscles while asking him to imagine the feared object (for example, a snake) in its least feared state, the therapist relieves the associated anxiety. Relaxation rather than anxiety is conditioned as a response to the imagined snake. Gradually the therapist works up a "hierarchy" of ever more threatening situations so that eventually the subject is able to tolerate the presence of snakes and even handle them.

Flooding [3] has been used with some success in the treatment of certain crippling neurotic states, particularly the severe obsessive-compulsive disorders. This involves making the patient experience his

1 Ivan Pavlov, the Russian physiologist and Nobel prize-winner, proposed the theory of the conditioned reflex as the basic model of mental activity. One of his experiments involved a bell being rung simultaneously with the presentation of food before a dog. Eventually the sound of the bell alone, in the absence of food, caused the dog to salivate. The dog had become conditioned by the experiment to the sound of the bell.

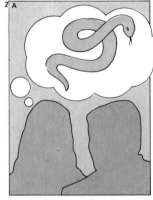
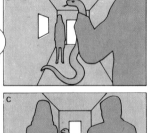

2 Systematic desensitization is a behavioural technique used clinically in treating phobias such as the fear of snakes. A reassuring therapist encourages the patient to think about snakes [A] until she can do so without anxiety. She is then confronted with a distant live snake [B] which is gradually brought closer to her [C]. In the final stages of the treatment the patient progresses from touching the snake through a wire screen [D] to handling it by herself [E] without fear. Thus her fear response has been deconditioned and reconditioned into a fearless response brought about by positive reinforcements.

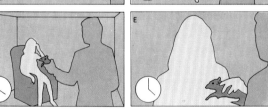

3 The aim of flooding or implosive therapy is to help a patient break the phobic cycle by having her make a deliberate effort to feel and experience her fear without avoiding it. The patient sits in a room while the therapist enters with the feared object, in this case a rat [A]. The rat is taken out of the cage [B] and the patient screams with fear. She becomes progressively more frightened [C], but gradually she calms down due to the extinction of her severe anxiety reaction [D]. Finally she is able to hold the rat [E]. The crucial variable is the duration of exposure, for too early a termination of exposure to the rat increases rather than decreases the patient's great fear of the animal.

4 Unwanted behaviour can be diminished by giving unpleasant stimuli every time the behaviour occurs – so-called negative reinforcement or aversion therapy. A person's desire to stop smoking, for example [A], is reinforced by his doctor's warnings [B]. Under treatment he is given an injection of apomorphine, a nausea-producing drug [C]. Soon afterwards he begins to smoke in a contrived environment in which the "positive" aspects of smoking are overemphasized but almost immediately he feels sick because of the drug [D]. This process is repeated over several days. Later, in a normal social situation [E], the nausea returns whenever he is offered a cigarette [F]. He soon stops smoking.

fear fully until it reaches a peak and then diminishes. Aversion therapy [4] has been used in the treatment of sexual deviations (such as exhibitionism and paedophilia) and alcoholism and drug dependence. The noxious stimulus may be a painful shock, or a nausea-producing or paralysing drug. Following such a course of aversion treatment, the patient is negatively conditioned.

Operant conditioning

Operant conditioning, developed by B. F. Skinner [5] and employed in a number of treatment situations including so-called "token economies" [6], works by means of a mixture of rewards and punishments. If a particular action is consistently followed by a reward ("positive" conditioning), then that particular action is likely to be repeated; if, however, it is followed by a punishment ("negative" conditioning), then the action is likely to cease. It is crucially important for the reward or the punishment to follow quickly upon the behaviour in question. Too long a gap results in a failure by the individual to connect the two and hence the conditioning

quality of the reward or punishment is low.

Behaviourists make considerable claims for their treatments. But while behaviour therapy has a significant role to play in the treatment of neuroses (such as simple phobias) in which there is a single fear-provoking stimulus to be deconditioned, its success rate in more complex states, particularly in the long term, has been questioned. In the treatment of personality disorders behavioural techniques have not to date been very successful, while for the psychoses they have little to offer. In addition anxieties are often aroused concerning the ethics of certain forms of behaviour therapy, particularly aversion therapy. It is felt that behavioural techniques are too much like "brainwashing" or the shaping of a person's character into something that it is not and that this is open to abuse. Behaviourists insist, however, that it is up to the individual to decide whether or not to receive treatment for his problem and that the ethical problems raised by such forms of treatment do not qualitatively differ from those raised by other forms of effective psychiatric treatment.

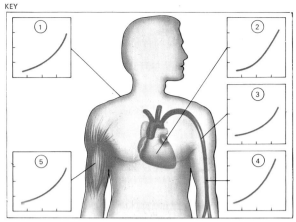

Many treatments of neuroses seek to reduce emotional arousal. One way of assessing objectively the degree of stress or emotion experienced is to measure certain physiological changes during treatment. Under stress, the electrical resistance of the skin to the flow of an electric current decreases and the skin conductivity increases [1]. The heart rate accelerates [2] and there is a constriction of the blood vessels causing a rise in blood pressure [3]. Blood flow through the limbs increases [4]. High levels of muscle tone during stress are measured by electrodes inserted in muscles [5].

5 The operant conditioning technique was first described by the American B. F. Skinner (1904–) in the 1930s. Whereas Pavlov took a hungry animal and paired a previously neutral stimulus (a bell) with the giving of food to an animal, Skinner waited for an animal to behave in a certain way, and then stepped in with a reward such as food, so encouraging the animal to repeat the behaviour.

6 Token economies, a technique based on principles of behaviourist psychology, are being used increasingly in the treatment of the mentally ill and handicapped. Many such patients are socially withdrawn and do not participate in group activities [A]. But whenever they do [B] therapists and nurses can reinforce such positive behaviour by rewarding them with tokens [C]. These are equivalent to money within the hospital and are exchangeable for goods and privileges such as cinema or other entertainment [D]. The patient's social behaviour can be gradually reshaped in this way and lead to a normal life [E] outside the hospital.

7 The patient's immediate social environment can affect his mental state and behaviour, as psychiatrists have become aware in recent years. As a consequence, hospital wards today are no longer grim, large and impersonal, with locked doors and few personal possessions [A]. They have been redesigned so that they are smaller, open and much more social with outside visitors being actively encouraged [B].

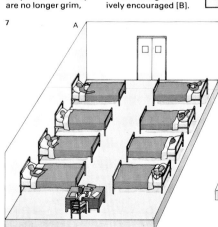

8 William James (1842–1910), a prominent American psychologist and philosopher and brother of the writer Henry James (1843–1916), laid the foundations of behaviourist psychology. His theory that "emotion" is merely the perception of the bodily changes that accompany stress, while controversial, has greatly influenced clinical psychology. This attention on the outward symptoms of anxiety rather than the inward feelings provided an important theoretical framework for the research of the behaviourists.

Psychotherapy

Psychotherapy is the general term for those forms of treatment in which conversation between the therapist and patient is the primary technique. The object of psychotherapy is not just the removal of symptoms, as with physical treatment of a physical disorder, but an exploration of the patient's personality. Indeed, in many cases the removal of symptoms ceases to be as important as the gaining of understanding and insight. To speak of "cure", therefore, as if psychotherapy were a surgical operation, is inappropriate.

Pyschotherapy may be on an individual or group basis, may be interpretative, supportive or suggestive, and may be superficial or deep. By interpretative is meant the exploration by the therapist of something obscure or hidden to the patient. For instance, a psychotherapist interprets a dream by attributing to it some meaning over and above that given to it by the patient. Different psychotherapeutic traditions or schools, Freudian or Jungian for example, may give radically different interpretations of the same dream [1]. By supportive or suggestive psychotherapy is meant that the therapist adopts a reassuring or advisory approach. Deep psychotherapy refers to treatment which concentrates on the patient's early life or on repressed areas of his personality; while superficial psychotherapy centres on more recent and accessible experiences.

Patient and therapist

Not every patient is suitable for psychotherapy. Factors which favour a good response include youth, good intelligence, an ability to self-explore, a capacity to tolerate uncertainty, anxiety and frustration, and verbal skills. Psychotherapeutic techniques are most effectively applied in the treatment of the neuroses and personality disorders, sexual deviations and childhood disturbances. As well as the relatively orthodox, analytically based group and individual psychotherapies, there is marital and family therapy, encounter group therapy, counselling, and the less orthodox "fringe" approaches which include Gestalt therapy, primal therapy, psychodrama and the Japanese Morita therapy.

Until recently, more attention was paid to the particular school of psychotherapy or the therapeutic approach of the therapist than to more personal characteristics. However, it has become clear that certain factors in the therapist's personality, such as ability to empathize, non-possessive warmth, sincerity and open-mindedness, are important.

The most intensive form of psychotherapy is individual psychoanalysis. Psychoanalytic theory postulates that the individual defends himself against pain and anxiety by the use of certain mental mechanisms such as repression, denial and projection. These mechanisms are believed to be characteristically exploited in such neurotic states as hysteria and such psychotic conditions as paranoia. It is these mechanisms that are analysed and broken down in the psychoanalytic interaction.

Approaches to group psychotherapy

The development of group therapy has meant a greater availability of certain psychotherapeutic approaches. The more orthodox groups [2], which may meet daily or

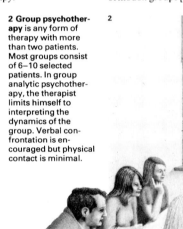

2 Group psychotherapy is any form of therapy with more than two patients. Most groups consist of 6–10 selected patients. In group analytic psychotherapy, the therapist limits himself to interpreting the dynamics of the group. Verbal confrontation is encouraged but physical contact is minimal.

1 Interpretations of dreams vary widely. A young man dreams [A] that he is climbing some stairs with his mother and sister. At the top he is told that his sister is pregnant. For Freud the dream reflected conflict between primitive instincts. This often has roots in a family situation [B]. The stairs represent sexual intercourse, the boy's incestuous wish for his mother [C]. The expected child is an indication of brother/sister sexuality [D]. Jung interpreted the same dream in terms of symbols or "universal archetypes" [E] that exist in all cultures. For Jung, the boy's guilt at neglecting his mother is revealed by his neglect of work (here symbolized by Dionysus in relaxed pose); his sister, a "love of womanhood" (here symbolized by Aphrodite); the stairs his passage through life (here symbolized by the Ages of Man); and the coming child, his rebirth (here symbolized by spring).

3 Encounter therapy is a group approach aimed at restoring spontaneity and involvement in social relationships by the creation of an atmosphere in which members can express their innermost feelings and explore new attitudes. Dramatic use is made of techniques such as body touching and mutual exploration.

weekly, are run under the leadership of a therapist trained in one or other of the established analytical schools. Less formal and orthodox groups have evolved over the years under the general label of "encounter" therapy [3]. These groups tend to be less theoretical; they are relatively unstructured, geared to action rather than analysis and are mainly beneficial to those who do not suffer from severe personality disturbances or mental illness.

An offshoot of encounter therapy is marathon therapy in which the participants and therapists meet continuously for 24–72 hours. Enthusiasts claim that it is a particularly effective treatment for couples with marital problems and for certain neuroses.

Recent developments
Encounter groups often use principles developed by Jacob Moreno, the founder of psychodrama [4]. Janov's primal therapy [5] and Gestalt psychotherapy, with its emphasis on the patient's need to achieve behavioural change through understanding and experience, both involve intense, often exhausting

participation and the acting-out of the individual's emotional turmoil.

Such treatments contrast sharply with a Japanese form of psychotherapy, Morita therapy, in which the patient, under the guidance of a specially trained mentor, undergoes a rigid and ritualized form of re-education that requires his participation in a number of prescribed social activities. Throughout his treatment, he is actively discouraged from baring his soul or analysing his feelings, activities that are believed to lie at the core of all neuroses.

One therapeutic approach that has developed outside the main medical, analytical and behavioural traditions is counselling. Although concepts taken from medicine, behaviourist psychology and psychoanalysis are utilized, the major influence is derived from the American psychologist Carl Rogers and his client-centred therapy. Such an approach emphasizes the growth of an equal relationship between therapist and client, the mutual exploration of problems and possibilities and the rejection of any imposed system of attitudes or values.

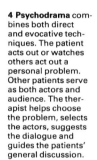

Sigmund Freud (1856-1939)   Carl Jung (1875-1961)   Alfred Adler (1870-1937)

Melanie Klein (1882-1960)   Jacob Moreno (1889-1974)   Carl Rogers (1902-   )

The foundations of psychoanalytical theory and practice were laid by Sigmund Freud. Carl Jung, disenchanted with Freud's teachings, turned for inspiration to philo- sophy, religion and metaphysics. Alfred Adler emphasized the importance of man's inferiority feelings and need for power. Melanie Klein concentrated on the role of primitive infantile wishes. Jacob Moreno developed psychodrama and Carl Rogers advocated the adoption of non-interpretative psychotherapy.

4

5

4 Psychodrama combines both direct and evocative techniques. The patient acts out or watches others act out a personal problem. Other patients serve as both actors and audience. The therapist helps choose the problem, selects the actors, suggests the dialogue and guides the patients' general discussion.

5 Arthur Janov's primal therapy portrays neurosis as a defence against pain and seeks to induce cathartic emotional responses. Direct experience of pain is believed to be curative and so the patient is encouraged to undergo a series of "primals" (angry, fearful, sad or violent outbursts and upheavals).

6

7

6 Family therapy is a method of psychotherapeutic intervention requiring direct and continual emotional involvement between family and therapist. For tactical or practical reasons some family members may be excluded from a few sessions but the focus and benefit is aimed at the whole family rather than individuals.

7 Marital therapy is designed to prevent marital disharmony and breakdown. In the typical joint interview, husband and wife are treated together by two trained therapists, one of either sex. Their relationship is explored in depth and skills derived from psychoanalysis are employed in this kind of therapy.

Human development

Physically a baby is an adult in miniature and physical growth is basically a process of enlargement of bones, muscles, fat and of the baby's internal organs. The human eye may grow as much as 30 per cent in the first five years, but it retains its original structure. Such changes are quantitative. Psychological or personality growth, on the other hand, involves marked qualitative changes. How these come about and what factors are most influential in the establishment of the character of the adult are subjects that have produced widely different interpretations.

Heredity versus environment
Perhaps the area that excites most controversy is the extent to which the individual is the product of his nature (heredity) or his nurture (environment). Heredity comprises two main elements: those aspects of our behaviour that are inherent in, or peculiar to, our belonging to the species *Homo sapiens,* and those aspects determined by our being born of two particular individuals. It is the difference between being a man rather than a lion; and being you rather than someone else.

There are some researchers who detect the presence of common human factors in people all over the world – similarities that range from the universality of religion in one form or another to common patterns of social organization. These imply a common human personality that overrides culture, race and geography. Noam Chomsky (1928–), the American psycholinguist, for instance, asserts that there is in each of us a model or blueprint of language that accounts for our remarkable ability to master complex linguistic rules early on in our life. Nikolaas Tinbergen (1907–), Konrad Lorenz (1903–), and other ethologists (students of animal behaviour) have shown the importance of our animal heritage in crucial areas of human behaviour – such as aggression.

As far as personal heredity is concerned – leaving aside the obvious examples of size, shape and eye colour – there is much (often heated) debate about the extent to which genetic inheritance controls emotional and social development and sets upper limits on intellectual potential.

Opposed to these theories are the so-called behaviourists, headed by B. F. Skinner (1904–), who emphasize the things that each individual is given by being involved in the process of living. This point of view stresses the particular situations and circumstances to which we are exposed. The behaviourists believe that human beings are, if not good, at least innately "neutral" and that the "evil" in people is a product of their environment – poor housing, bad education and other influences of that kind [2].

Is the child active or passive?
Within and around the nature/nurture controversy are several related areas of contention. Does the child actively manipulate his environment, trying to shape it to his skills and abilities? Or is he the creature of that environment? Essentially, both the behaviourists and the Freudians follow the latter approach, while thinkers such as Jean Piaget (1896–) favour the former. But although the behaviourists and Freudians agree that environmental influences are paramount, they differ radically in their interpretations of the springs of human

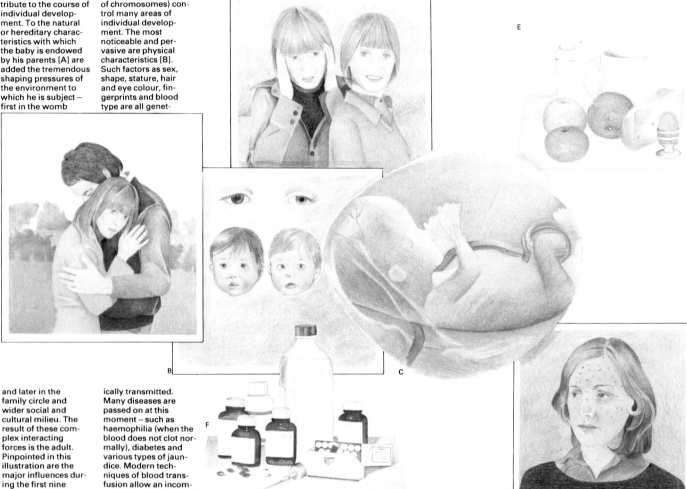

1 Many factors contribute to the course of individual development. To the natural or hereditary characteristics with which the baby is endowed by his parents [A] are added the tremendous shaping pressures of the environment to which he is subject – first in the womb along the 23 pairs of chromosomes) control many areas of individual development. The most noticeable and pervasive are physical characteristics [B]. Such factors as sex, shape, stature, hair and eye colour, fingerprints and blood type are all genet-

and later in the family circle and wider social and cultural milieu. The result of these complex interacting forces is the adult. Pinpointed in this illustration are the major influences during the first nine months of life, from conception to the moment of birth. At the moment of conception, when the mother's egg cell unites with the father's sperm, life begins. In the resulting fertilized egg the bead-like strings of thousands of genes (arranged ically transmitted. Many diseases are passed on at this moment – such as haemophilia (when the blood does not clot normally), diabetes and various types of jaundice. Modern techniques of blood transfusion allow an incompatibility of the child's and mother's blood-types (the Rhesus factor) to be remedied. Psychological traits such as temperament, intelligence and disposition are known to be genetically influenced, but how and to what extent is still not entirely clear.

The embryo is in the womb for roughly 40 weeks [C]. During that critical time, particularly the first vulnerable nine weeks, it is exposed to many factors that can affect normal growth.

The mother's emotional state during pregnancy is important [D]. It is thought that severe or continuing stress can influence the infant's activity level, birth weight and emotional reactions.

Food supply [E], together with the efficient exchange of oxygen and waste products is crucial. Diets deficient in calcium, phosphorus, iron and various vitamins are associated with poor develop-

ment. Many drugs [F] taken by the mother during pregnancy may have an adverse effect on the fetus. Excessive smoking produces smaller babies. Babies born to drug-addicted mothers are also drug dependent.

German measles [G], if contracted during the first three–five months, can lead to a newborn baby with abnormalities in sight, hearing and brain and heart functions. Untreated syphilis may affect the fetus.

personality. If, for instance, a child is afraid of dogs although he has never been attacked by one, the Freudian might argue that the dog's fierceness symbolizes the anger of the child's father. He will say that the child has transferred his fear of his father to the dog. In other words, he will look behind immediate behaviour for a deeper or "unconscious" cause. The behaviourists, in contrast, will seek to relate the fear to an actual incident that led the child to conclude that all dogs are dangerous. Persuading the child to face the cause of his fear will remove it.

Is development continuous?

Whether development is continuous – a process of gradual evolution – or whether it proceeds by marked, step-by-step stages, is also a matter of debate. The sequence of psychosexual development put forward by Sigmund Freud (1856–1939) implies that a child's pleasure drive is successively directed to different parts of the body, beginning with the oral and moving through the anal and phallic to the genital stage at puberty.

Another argument concerns the direction of development. Does it tend towards a final, ideal goal or is it open-ended and indeterminate? Behaviourists favour the open-ended view. Others, such as Lawrence Kohlberg (1927–), believe that there is a desirable endpoint which, when reached, defines and identifies maturity.

Scientific verification for any of these views is hard to come by. In the nature/nurture controversy the nearest approach to laboratory standards are studies using identical twins raised either in different backgrounds after adoption or together in their natural family.

The various views described are extremes but they illustrate the type and the complexity of the problems that confront the student of human development. Shape, size and features, for example, depend not only on genetic inheritance but on such purely environmental factors as nutrition and disease both in the womb and later.

Heredity, culture, socio-economic class and personal experience all play vital parts in the process by which the newborn infant is transformed into a particular kind of adult.

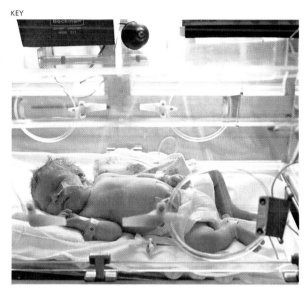

KEY

The development of the newborn baby into an adult is subject to many interlinking influences that stretch back to the moment of conception and forwards to the quality of his environment.

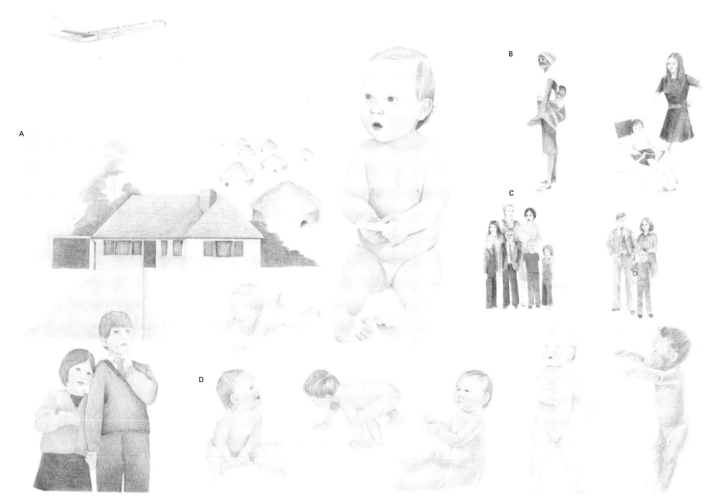

2 The social environment and culture [A] in which a baby is born and reared exerts a potent and decisive effect on the development of his personality. The atmosphere and values of every culture differ and they are what condition individual assumptions and aspirations. Not least of the ways in which a culture affects individual development is through the methods of child-rearing it practises [B]. Size of family [C] and birth order are significant to development. First-born children tend to trust adults more and to conform more to parental models than later-born ones. Studies show that children with older siblings are usually less aggressive and creative as adults. They adjust better to inequities in communal living. The solitary child is thought to have a richer fantasy life. The physical and psychological characteristics with which the baby is born [D] are instrumental in modifying and moulding the environment he grows up in to a considerable extent. Size, sex, temperament and attractiveness are qualities that affect the people he comes into contact with and so alter the ways that they interact with him. Hereditary differences in the senses, muscle mass and sensitivity of the central nervous system are among possible factors that determine why some babies are more responsive, alert and adaptive than others.

Thinking and understanding

Until the early 1950s it was generally believed that a child, at least for the first few months of life, was capable of only dimly appreciating physical sensations. Because a baby's sensory awareness and intellectual ability were supposed to be primitive, he was pictured as existing in a world that amounted to little more than a buzzing confusion.

The many talents of the infant
It now appears that within the first few weeks of life a baby has already learned to focus his eyes, see one image with both eyes and appreciate detail [1]. He can distinguish between different tastes and smells and between sounds of different frequency and loudness.

As a measure of the speed of this learning process, the baby, by the time he is 16 weeks old, is able to predict even the quite complicated trajectory of a ball that he has seen moving but which, during flight, disappears temporarily behind a screen. From birth the baby also actively selects what he looks at or listens to, attending to those things that differ in brightness, contrast, pattern or movement from objects he has seen before [Key].

Adults have an organized picture of the world that takes many things for granted. To arrive at this state involves a long, hard journey for the child. Perhaps the most influential researcher into this process is the Swiss psychologist Jean Piaget (1896–). From years of observation and inventive experiment with his own and other children, he has formulated the theory that cognitive or intellectual growth takes place in a series of four stages or phases. These he has called the sensorimotor [2A]; the preoperational [2B]; the concrete operational [2C]; and the formal operational [2D].

During the sensorimotor stage the infant is constructing a picture of the physical realities of the world by touching, tasting, manipulating and destroying. By these means the building blocks for later thinking are built up. From then on each stage marks a growing ability to think in abstract terms that moves away from the physical reality of actual objects and involves reaching towards an adult grasp of the world. During the preoperational period the child has difficulty in understanding that an apparent change in

an object is not an actual or real change. If two glasses of equal size are filled with liquid and then one is emptied into a taller but narrower glass, the child will maintain that the taller glass has more liquid in it. Later on the child will not be deceived by this, thus revealing an advance in thinking power towards the concrete operational stage.

Stimulating development
At the centre of Piaget's theory is the notion that what propels the child forward from stage to stage is "disequilibrium" – the situation that arises when a child's current picture of his world is shattered by the facts that his observations reveal. A young child will treat a magnet like any other toy, examining its size, weight and taste until he suddenly discovers that, unlike other objects, it attracts metal. This, according to Piaget, will put the child's world into disequilibrium and compel him to revise his picture of it.

A prominent feature of childhood is the markedly self-centred thinking typical of younger children. They assume that others see, experience and think about the world in

1 The acquisition of visual skills is an important part of intellectual growth. The ability to discriminate between lines, patterns, colours, sizes and shapes develops in a number of stages.

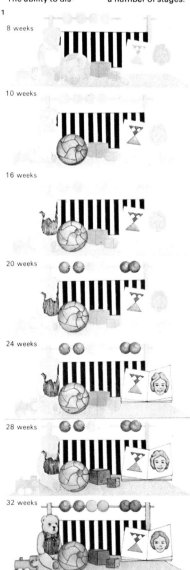

8 weeks
10 weeks
16 weeks
20 weeks
24 weeks
28 weeks
32 weeks

2 Jean Piaget, the Swiss psychologist, has shown that intellectual development, although a continuing process of discovery, can be divided into four main stages. In each stage different methods of understanding the world are used, each with its own logic and consistency, even if it does not conform to strict adult patterns. Piaget states that two processes, which he calls assimilation and accommodation, are essential for this growth. The child either assimilates new information into his existing view of the world, thus filling in more detail, or, if this is not possible because it does not fit into his past experience, he accommodates the new information by revising his way of thinking. The increasing use by the child of accommodation indicates intellectual growth. For the first two years [A], the child is said to be in the sensorimotor stage of development. He understands the world through direct contact with objects and comes to realize that they have a distinct and separate existence. Gradually he learns that they continue to exist even when out of sight. He shows his intelligence through his actions. From two to seven years [B] the child is in the preoperational stage. Thought is more abstract and the image of an object is linked with thoughts about it. The child can also

visualize something without seeing it, but thinks everyone shares his view of the world. Unconscious contradictions are common. From seven to 12 years [C] the child is in the concrete operational stage. He is able to distinguish between present and historic time. He can order and arrange objects according to weight, size and volume and describe a series of actions (like walking to school), carry-

ing the entire sequence in his mind. From 12 years onwards [D] the child is in the formal operational stage. Adolescents are able to use abstract rules to solve problems and they become more preoccupied with thought (and why they are thinking thoughts). They will perceive and explain things in increasingly objective, naturalistic terms. Presented with problems, they will consider solutions and choose the most suitable.

2 A

B

Sometimes God sets him on fire
I can make the sun move, he follows me around. Sometimes a cloud eats him.
Sun
Cloud
Tree

C
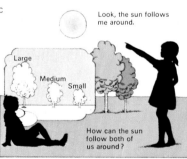
Look, the sun follows me around.
Large
Medium
Small
How can the sun follow both of us around?

D

If the sun doesn't follow either of us, it must be still. Yet it sinks, therefore the earth must move around the sun.
The big tree dropped some seed which has grown into a little tree.

3

3 Classical conditioning arises from "association". A child shows unlearned automatic fear of such things as loud noises. If a child sees a rabbit, about which he has no preformed ideas, at the same time as he hears a

loud noise, he will begin to associate the rabbit with fear. Soon, the sight of the rabbit alone is enough to make him afraid. The Russian physiologist Ivan Pavlov was the first to observe and record such behaviour.

precisely the same way and from the same physical standpoint as they do. However, children are gradually made to realize that theirs is not necessarily the correct or only view. Another change in thinking that takes place during childhood is a decrease in attempts to provide magical explanations. Children more and more try to understand and describe their experiences in terms of physical cause and effect.

Learning involves a number of complex mechanisms. The Russian physiologist I. Pavlov (1849–1936), basing his theories on the results of his experiments with dogs, showed that we learn to connect impressions or bits of information by the process of association. He rang a bell at the same time as he fed the dogs and their mouths started to water. Soon the ringing of the bell by itself stimulated them. This is called classical conditioning [3] and many of our emotional reactions are learned on this principle.

Operant conditioning

The American behaviourist B. F. Skinner (1904–) believes that we learn to do things,

or not to do things, by observing the reactions of those around us. We tend to repeat those actions that are rewarded and avoid those that are punished. Our behaviour is said to be positively or negatively reinforced. This is called operant conditioning [4] and is a decisive factor in moulding human behaviour.

Operant conditioning has been found to exist even in newborn babies. In experiments, babies learned to perform quite complex tasks such as turning their heads twice to the right, three times to the left and then once to the right. The reward offered was the stimulus of a white light coming on for a few seconds. But once the babies had mastered the rules they lost interest in the experiment [6]. This reveals a crucial point about learning and reward: it is essential to find the right motivation before the child will respond positively and so extend his field of knowledge. The pre-school and elementary education system of Maria Montessori (1870–1952) is based on this self-directing drive to learn and on the belief that the child has creative potential and a right to be treated as an individual.

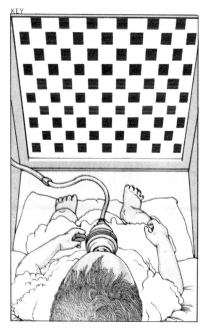

KEY

Recent technical advances in equipment and procedures have made it possible to carry out sophisticated studies even with week-old babies. These involve the use of transistors, video-tape recording, one-way viewing screens and infra-red scans. The apparatus shown here, for instance, tests the baby's ability to respond to a quite complicated learning test. By sucking on the teat he brings the picture on the screen into sharp focus. Results show that the baby soon learns to master the problem but requires the picture to be changed frequently to retain his interest. The findings of such studies have shown that young children have greater capabilities than had been thought.

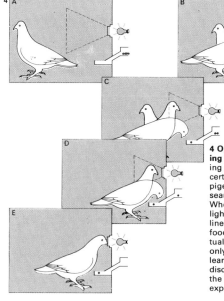

4 Operant conditioning involves rewarding or punishing certain actions. The pigeon in the box [A] searches for food [B]. When it enters the lighted zone [dotted lines] it is given food [C, D]. Eventually it is rewarded only when it learns to peck the disc [E], this being the behaviour the experimenter wanted.

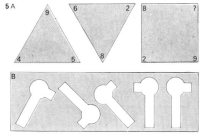
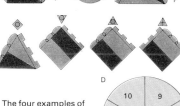

5 IQ tests were first devised in 1905 by the Frenchman Alfred Binet (1857–1911) to select slow learners for special courses. Since then they have been used to measure almost everything of importance in intellectual development. Creative ability, productive thinking and problem-solving were all assumed to be measurable in a

single test. Such tests have recently become the subject of controversy on the grounds that they fail to measure intellectual growth, they do not show ability to learn and are often loaded in favour of a middle-class cultural approach. Attempts have been made to devise culture-free tests that would avoid this bias.

The four examples of "culture free" tests measure ability to see links between numbers and shapes. [A] and [D] are numerical tests, [B] and [C] visiospatial tests.
Questions
A and D: What is the missing number? B: Which is the odd one out? C: Which shape on the bottom row gives the correct answer?

Answers
A: The missing number is three. B: The odd one out is the fourth from the left. C: The pairing shape is the third from the left. D: the missing number is 15.

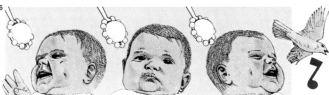

6 Habituation – or boredom – is one of the more important factors in discouraging learning and exploration. It is perhaps best illustrated by the common experience of a child soon getting

bored with some new toy, such as a rattle. Children react positively to new challenges as long as the rewards for success are changed when they show signs of becoming habituated to them.

7 Chaining is the process of putting together related but separate movements into a coherent, co-ordinated act. Many of the things that adults do without thinking – such as the seemingly simple

opening of a door – in fact involve a complex series or chain of actions. The child has to learn these painstakingly by trial and error. Everything we do is similarly pieced together.

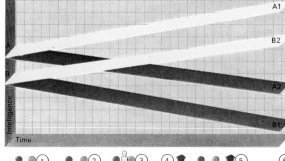

8 Environmental conditions can influence the IQ scores of initially intelligent children [A] and initially less intelligent [B]. This difference may be accentuated [A1, B1] or even reversed [A2, B2].

Many factors have been found to encourage high IQ scores. These include small family size [1], intelligent parents [2] who encourage the child [3], small, well-equipped schools with high teacher/pupil

ratios [4], where the child, teachers and parents are on good terms [5] and a generally active life [6]. Conversely, either sort of child will develop slowly if his environment is poor or deficient.

153

Language development

The acquisition of language by the child is a remarkable achievement that has fascinated both parents and those who study child development. At present the central and fiercely debated issue of language development concerns the degree to which language structure and organization are innate or built into the human mind. At one extreme some researchers such as Noam Chomsky (1928–) maintain that every human being is born with both a natural capacity for language and an already programmed model or pattern in his brain. In contrast the "social learning" theorists, inspired by the work of B. F. Skinner (1904–), insist that cultural elements are crucial to language development.

The beginning of language

There is a close interplay between intellectual and language development. From the start the human infant is active and curious and by the first month of life he closely attends to the speech he hears. (The ability to hear is essential to speech development.) Although he can usually distinguish between different speech sounds much earlier, the child is usually a year old before he is able reliably to produce patterns of sounds that are identifiable as distinct words. Apparently the muscular control required for such a feat is gradually acquired, first through crying and then through cooing and babbling.

By the time he is about a year old the infant communicates his feelings and wants by varying the intonation, stress and frequency of his utterances. Soon afterwards, as he truly begins to become a speaking human being, he explores the way in which speech is organized and discovers the systems of rules (grammar) for putting words together to form sentences.

This breakthrough is dependent on the parallel development of the child's ability to perceive the world efficiently and to reason about it. He must be able, for example, to retain an abstract image of an object that he has seen, heard or touched before he can name it or respond to a name. Only in the second year can he hunt for a named object that is absent.

By the time he is about two years of age an infant can string together two or three words to form rudimentary sentences [2]. He can usually comprehend, as well as express, a variety of fundamental concepts about animate and inanimate objects and about actions and events. He uses such words as "here" and "there", for example, to indicate location and "all-gone" or "all-done" to indicate disappearance or cessation.

At this period, because the child responds positively to a request such as "Please fetch me that coat", parents are often misled into thinking that he understands more than he does. The fact is that conversation, particularly with children, is rich in non-verbal cues such as glances or gestures and it is only in combination with these cues that the child can make sense of the words he hears.

The growth of language

From about the beginning of the third year onwards the child rapidly becomes skilled at comprehending and producing sentences that do not depend on immediate context in order to be understood by others. By about the age of four or five almost all children have mastered the fundamentals of language,

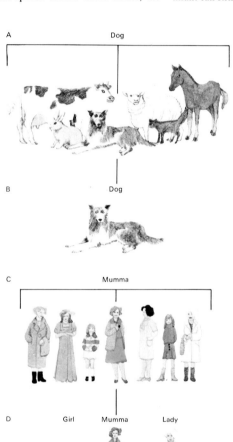

1

A Dog

B Dog

C Mumma

D Girl Mumma Lady

1 Language development in children does not always occur at the same age but most children do show similar steps in language production and comprehension. Typically, a child's first words work as a complete utterance. The word "dog" [A] initially refers to all animals with which the child is familiar. Similarly, "mumma" [C] is used for all females with whom the child comes in contact. The word means "I see a dog, or some other animal" or "I want mumma or some other female". As the child develops, words gain more specific meanings. "Dog" [B] now refers to only one animal, a dog, and "mumma" refers to the child's mother alone and he uses other specific words to refer to all other females [D].

2

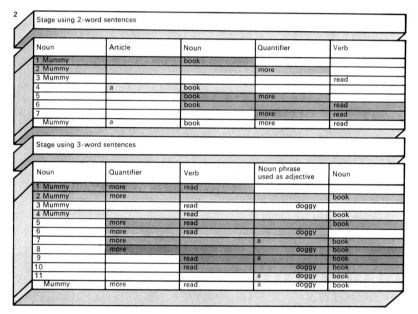

Stage using 2-word sentences				
Noun	Article	Noun	Quantifier	Verb
1 Mummy		book		
2 Mummy			more	
3 Mummy				read
4	a	book		
5		book	more	
6		book		read
7			more	read
Mummy	a	book	more	read

Stage using 3-word sentences				
Noun	Quantifier	Verb	Noun phrase used as adjective	Noun
1 Mummy	more	read		
2 Mummy	more			book
3 Mummy		read	doggy	
4 Mummy		read		book
5	more	read		book
6	more	read	doggy	
7	more		a	book
8	more		doggy	book
9		read	a	book
10		read	doggy	book
11			a doggy	book
Mummy	more	read	a doggy	book

2 Children soon start to make up rudimentary sentences using first two words, then three. At the two-word stage they choose between easy, unrelated words they hear and know are linked with what they want. They can then construct a number of sentences that express their needs. Once they are using three-word sentences they are already using more complex noun and verb phrases and their range of expression is wider.

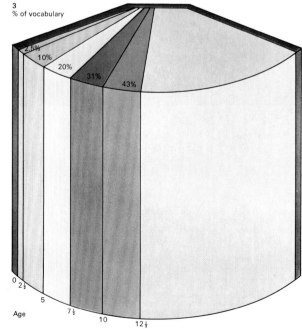

3
% of vocabulary

2.5%
10%
20%
31%
43%

0 2½
5
7½
10
12½
Age

3 The child's vocabulary can be seen as a percentage of his adult vocabulary. The child may use between 1,200 and 2,000 words when he starts school, and be using between 3,000 and 5,000 words at school leaving age. Shakespeare used 29,000 words.

although they may have difficulties with certain sounds, such as the v in "Valentine".

One of the most interesting aspects of language development is undoubtedly that studies carried out in different cultures have shown the universality of the patterns and timings of word use already described. In addition, regardless of the specific language involved, children use similar, non-adult yet nonetheless formal rules in an attempt to impose order on their language almost as soon as they start to talk. They simplify language into telegraphic speech, producing characteristically over-regularized "errors", as in the plurals of nouns (mouses) and the past tenses of verbs (holded).

What parents can do

Language is not only an intellectual achievement but also largely a social product. Parents therefore play an important role in encouraging their children's interest in and practice of language by talking to them as often as possible, answering questions, describing events and by reading. In the early stages, when the infant cannot separate individual items from the whole, it is better to make connections between object and word as clear as possible. "Body" is less clear as a word than "hand" or "finger" since it includes so many different parts.

Exactly what type of language is used between parents and children, whether normal adult or stylized "baby talk", does not seem to be very important and is more a question of current fashion than of objective scientific fact. It seems likely that for the first year simplicity and repetition are useful.

One aspect of language development that often disturbs parents is difficulty in speech production. As many as one to two per cent of schoolchildren, for example, suffer from stuttering, mostly between the ages of six and ten. On average three times as many boys stutter as girls. There are various reasons why children stutter, both neurological and psychological. But whatever the cause it can in turn set up anxieties, especially at school where fellow pupils are quick to pick on and exploit a weakness, and this can set in motion a vicious circle. Should the problem persist it is best to seek professional help.

A child's first words usually consist of only one or two syllables, often repeated. Commonly, they are only a part of a word, for example, "og" for dog. The parents' active interest are an essential ingredient for language development.

4

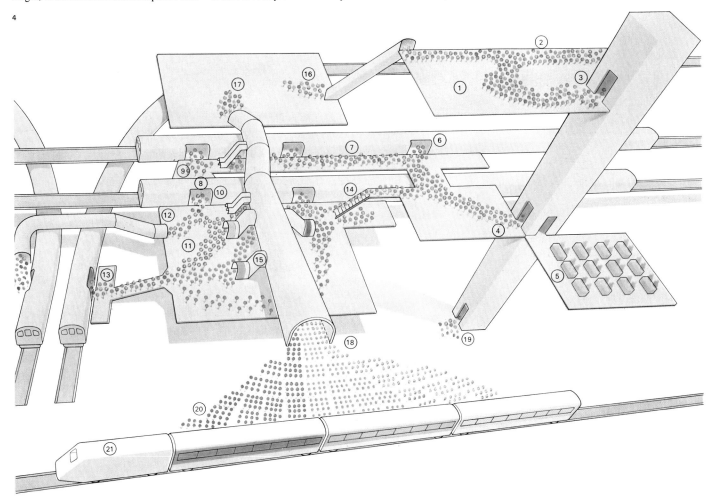

4 The child's rapid grasp of language is extraordinary in view of the complexities involved in making up a sentence. In this diagram the lexicon store [1] is the mind's memory bank for words. Language units assembled here [2] move down a lift [3] and travel to the deep structure, which contains the main rules for language usage. These are the syntax (grammar), the transformational rules and probably the semantics, or meaning. From a departure area [4] monitored by an overall control centre [5] words enter a train [6] to journey around the language system. Between the departure point [7] and arrival point [8] words are given both meaning and syntax. Some words [9] may go on a further journey to acquire more information. Others, which have acquired a grammatical make-up [10], arrive in the transformational area [11]. Here they are sorted into connecting trains according to emphasis, active or passive [12] voice, negative [13] or affirmative moods and so on. Stray words [14] might cause an ungrammatical sentence if they managed [15] to enter the surface structure. Common words go straight from the lexicon store [16] to the surface structure [17], which governs the final presentation of words. Words emerge [18] ready to take their place in the sentence. Those that have fallen through the deep structure [19] emerge as ungrammatical language. The organized features of a sentence line up in the correct order [20]. A sentence [21] is then ready for use in the outside world.

Social development

Being human is not just a matter of having an upright carriage, prehensile hands, a well-developed brain and organs that allow speech. These physiological and biological characteristics provide the "what" that makes up the human animal. But turning the "what" into a "who" requires our learning how to be human, and this is achieved through contact with our fellow men.

The wild boy of Aveyron
In 1799 a wild boy of about 11 years of age was captured in the forests of Aveyron in southern France by local villagers. He had been abandoned soon after birth, but had miraculously survived alone, living the life of a wild animal. His captors, however, found a human being only in the physiological sense. He could not speak, but grunted like an animal. He walked on all fours and seemed devoid of all the usual signs of human emotional responses such as pleasure or affection. A physician, Jean Itard, took the child into his care, determined to teach him some human characteristics. His success was limited. After several years the wild boy had learned only a few words and could just manage to eat properly and keep clean. Little more could be done, for he had missed the vital social and cultural process that psychologists today describe as socialization.

Stages in social development
Social development begins at birth and ends with death, but there is a crucial period in the child's early life during which the "social self" is formed. There seem to be four basic mechanisms that encourage all children in all cultures to become social beings: the desire to obtain regard, acceptance and recognition from others; the desire to identify with or be like those who are admired and loved; the fear of rejection and punishment; and the tendency to imitate. These four mechanisms influence different aspects of behaviour and dominate at different times.

The process begins at birth. The newborn infant has no specific values, attitudes or beliefs: just a propensity to develop into a social being. All theories of personality, regardless of their specific differences, stress the notion that we learn to be what we are because we interact with others, especially those who are accepting, supportive and encouraging, while guiding us according to those rules generally considered to be socially desirable.

In most cultures it is the parents who are the main socializing agents, with mothers being particularly influential because they normally spend more time with the child during the early years of development [1]. We are all affected by prevailing cultural standards, but the ambience of the family we belong to is a more specific and sharply focused influence because it filters and concentrates that culture.

The family's socio-economic class is also crucial because it conditions what is seen as desirable, probable and possible, and so how we come to regard the world – our aspirations, hopes and fears. The immediate depth of the family's influence is altered by other influences such as religious and political leaders and the mass media. The importance of these elements varies from time to time, but throughout, the original lessons learned at home are crucial.

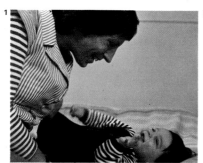

1 The child's early play during his first year primarily involves contact with his mother. He engages little in social play. In the second year, even when in close physical proximity with each other, children are usually quite content to play alone.

2 As children begin to meet more frequently, towards the end of their second year, play often ends in verbal or physical battles with each child demanding the toys and other objects or privileges for himself. This phase lasts for six months or more.

3 Different types of play serve a variety of purposes for the young child. Both solitary and parallel play, involving little or no social activity and an apparent lack of effort to adapt to playmates, are prominent early in development and then decline between ages 2 and 3. At approximately three years, with increas-ing competence in intellectual, social and physical skills, children begin to seek out playmates actively. Their play takes on a more reciprocal and commu-nal quality. In associated group and co-operative play children increasingly come to take into account the feelings, skills and responses of their playmates and others.

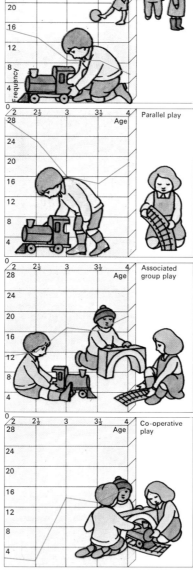

4 A child's popularity [shown by connections between children] or unpopularity [shown by red outlines], in a school class for instance, changes over time although friendships seem more durable than enmities. Research shows that the popular child appears to be more outgoing, more rewarding of others and more likely to seek help and approval than clinging affection.

Between the ages of two and five the child goes through a crash course of learning to make his impulses more social. He absorbs rules about property, the rights and welfare of others and the postponement of gratification of desires. The quality of that period and the way it is handled is critical. Affecting the family situation is the specific interaction between the parents' and the child's different temperaments, and those of his brothers and sisters if he has any.

After the child is five or six, when he begins to go to school, peer groups take on an increasingly important role in many areas of the child's development. Playing with peers provides the child with the opportunity to observe and then practise various skills.

The importance of play is shown by its changing patterns [3]. Up to the age of 18 months or so solitary play is normal, and when children are together they tend to fight [2]. Then a change takes place and there is increasing contact with others in group games that involve and encourage friendly contact, reward-giving, the showing of affection and toy-sharing. That peers imitate behaviour has been shown in many experiments in classroom and laboratory [6]. If aggression is met by resistance it tends to diminish or at least shift focus. Recent studies have shown that when children of four or five see someone giving away toys during a game they are more likely to do the same than those who have not seen a similar model.

Learning sex roles
Biological differences set the stage for sex-role development, and experiments have shown that from a few hours old boys and girls have markedly different patterns of gestures and grimaces during sleep. At the age of two-and-a-half the child is almost fully aware of his or her sexual identity. After this identity has been established it is difficult to reverse. To a varying degree parents convey their own concepts of what is appropriate sex-role behaviour – how one should relate to others if they are male or female and what kinds of work are undertaken by each [5]. Peer contact generally supports the cultural and sub-cultural attitudes originally conveyed by the child's parents.

Adults may scoff at children's games of make-believe, but it has been shown that it is through such games that children explore themselves and others, testing different behaviours and attitudes. In games, too, children learn the rules of society.

5 Boys and girls follow the unconscious behaviour patterns [A, B] presented by their parents almost from birth. In most Western societies males are encouraged to be achievement-orientated, self-reliant and less fearful generally. Girls are taught to be conforming, submissive and responsible. Increasingly, as stereotyped adult roles dissolve, children are exposed to more ambiguous influences [C].

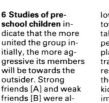

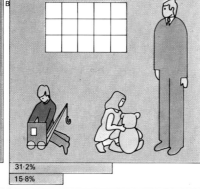

6 Studies of pre-school children indicate that the more united the group initially, the more aggressive its members will be towards the outsider. Strong friends [A] and weak friends [B] were allowed to play with toys which were then taken away by the experimenter, and replaced with less attractive ones. In the resulting frustration the strong friends kicked the experimenter [C]. The weak friends merely called him names [D]. The proportions of time spent in inter-friend co-operation [brown], inter-friend conflict [blue] and aggression to the experimenter [pink] appear below each picture.

46·2%
13·8%

31·2%
15·8%

54·1%
4%
26%

46·9%
10·1%
4%

Type of parent   Effect on child's development

7 Parental behaviour can be affectionate/permissive [A]; affectionate/controlling [B]; hostile/permissive [C]; and hostile/controlling [D]. Each of these types is supposed to produce a different kind of personality in the child: [A] a friendly dominant child with high self-esteem; [B] one more dependent, less friendly, less creative; [C] an aggressive child with poor self-control; [D] a shy withdrawn child with a tendency to guilt feelings and insecurity.

Emotional development

From the moment of birth newborn babies differ widely in level of activity, degree of irritability and general responsiveness to their environment [5]. Today, emotional development – how people come to feel about themselves and others – is seen as being largely the product of early relationships and learning experiences both in the home and in the wider society.

A child's first emotions

The difficulty of gauging just what emotions the newborn child is experiencing is twofold. First, because he cannot communicate in words we have to rely on such non-verbal signs as frowns, grimaces and smiles, which are all open to misinterpretation. Second, our reading of the child's emotional state at any particular moment is strongly influenced by how we would expect to feel if we were in the same situation.

In 1917 the American psychologists J. J. B. Morgan and J. B. Watson proposed the theory that babies are born with three distinct and unlearned emotions – love, fear and rage – and that all others are refinements and elaborations of these. But it is now widely believed that there are no clear-cut distinctions in the emotional responses of newborn infants and that it is only after a month or so that their emotions can begin to be divided into positive and negative. Emotions such as elation, pride, anger and distrust gradually gain coherence over the next two years.

The evolution of the emotions is clearly shown in the child's response to strangers. Until he is about six months old the infant reacts positively to any friendly face. But, as his ability to differentiate between people increases, this indiscriminate response diminishes. By the time he is a year old the child will often react with considerable distress or fear to a stranger.

Fear, pleasure and first relationships

Intellectual development is significant in emotional development. Different fears, for instance, are characteristic of different ages [4], because to fear something implies the ability to hold it in the mind and be aware of its physical and emotional consequences.

Jealousy, especially that prompted by the birth of another baby in the family, commonly surfaces at about 18 months and is at its most explosive for about two years after that. During this period the child may physically assault either the baby or the parents or both. With increasing maturity the outward signs of jealousy become much less direct.

Positive emotions have been much less extensively researched than negative ones possibly because they do not bring problems in their train. Pleasure is widely associated by psychologists with the gratification of instinctual drives such as hunger and thirst. But even the one- or two-month-old infant will show a positive reaction in situations in which the satisfaction of primary drives is not involved. He is receptive to friendly strangers, for example, and also appears to enjoy solving problems for their own sake.

Recent studies with the young of the higher mammals, such as the studies carried out with monkeys in the 1950s by Professor Harry Harlow (1905–) and his associates at the University of Wisconsin [1, 2], and with human infants have focused attention on the complex relationship that develops between

1 **The importance of family life** in developing proper social responses has been shown in experiments with monkeys. Compared to a monkey raised from birth with its mother [1], one kept in total isolation [2] for more than the first six months avoids all contact and appears fearful, clutching himself and crouching. Monkeys brought up with siblings but without a mother [3] lead a normal life but indulge in more hugging than usual. A monkey raised with an imitation mother [4], although behaving normally with her, does not show normal social or sexual behaviour on growing up. If the mother can move [5], the monkey is less fearful.

Avoidance of social contact · Normal social responses

3 **The importance of the bond** between mother and child is seen in the phenomenon of imprinting. After birth the young animal follows and forms an attachment for the first moving thing it sees. This is normally the mother, but any moving object, even a toy, can be imprinted on. The process starts even before birth, through sound. For if an egg that is hatching is placed near a loudspeaker the chick will move towards that.

2 **Comfort and shelter** are by far the most important factors in the development of an affectionate social bond between mother and baby, as experiments with monkeys conducted by Harry Harlow demonstrate. Monkeys raised in cages, separated from their natural mother but with both a wire mesh and a soft, cloth-covered surrogate mother, spent more time clinging to the cloth mother even though it was the wire mother who provided the food [A]. When these monkeys were confronted with a strange object placed near them, like a noisy, drum-beating toy bear, they reacted with great fear [B]. However, when the monkeys could cling to the cloth mother [C] this fear soon subsided. The cloth mother eventually came to provide a reasonably secure base from which to explore. The infant monkey would venture out to investigate novel objects but returned frequently for clinging and contact comfort.

the newborn infant and his primary caretaker (usually, but not always, the mother). The future emotional development of the infant is greatly influenced by this relationship. In animals this is also seen in the way that the young will form a permanent attachment (imprint) for the first moving object that they see, even if it is a human – as in the case of the geese that imprinted themselves on the Austrian researcher Konrad Lorenz (1903–) – and followed him everywhere [3].

Between the ages of one and three the child is particularly vulnerable to emotional disturbance if the primary tie is disturbed for any length of time – as, for instance, when he is sent to hospital, where he is surrounded by a totally foreign environment. This can result in an extremely trying period of crying and clinging and, on occasion, even produces symptoms of depression and despair. Hospitals are gradually changing their once-rigid rules of excluding parents on the grounds that their presence would "disturb" the child. Where prolonged absences are unavoidable, practical experience shows that they ought to be carefully explained rather than ignored

and that the child should be surrounded, as far as possible, with familiar objects.

Erik Erikson, the American psychoanalyst, has stated that it is in his initial relationship that the child must first learn how to receive and give love, and move from complete dependence to increasing independence [6]. Erikson further notes that the child who experiences feelings of security and love in this primary relationship is more likely to develop into a person with a basic trust in the world and in his growing sense of self.

Learning to control emotion
The child must learn how to express and control the needs he feels. For, as with most areas of human behaviour, indiscriminate expression of the emotions is rarely permitted in the developing child. Aggression, a conspicuous feature of childhood behaviour, is an important example [7] of this. In infancy aggression is an immediate and shortlived response to the frustration of wants. But as children learn that violence is not acceptable, aggression and anger become less physically evident and more internalized.

The family environment is crucial in guiding the child's emotional development. A fearful child, for instance, is frequently an unconfident one who feels that he will not be able to meet successfully the challenges he is faced with. Stressing the child's limitations aggravates and prolongs the situation.

4 The nature of children's fears changes as they gain experience. Their emotional reactions to different situations and objects reflect developmental shifts in their understanding and knowledge. Children from four to six years of age are more likely than those below two years to be afraid of animals, bodily harm, dangerous situations, being alone in the dark, creatures of fancy or dreams. These fears depend on imagination and generalization from past experience – unlike those that are more likely to occur in younger children, such as a fear of sudden loud noises and strange situations, people and objects.

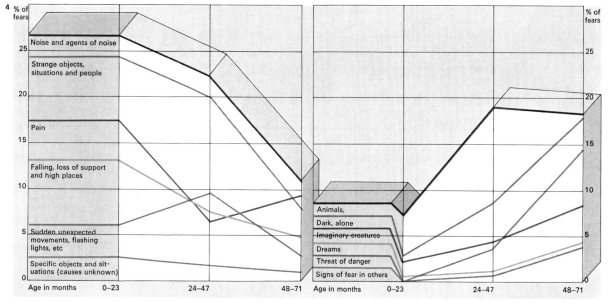

5 Marked differences in temperament or behavioural style are evident in babies only a few hours old. Such characteristics as tempo, adaptability, energy expenditure and mood, in existence from birth, have been found to persist relatively

unchanged into adulthood. Where a child's innate style is compatible with the attitudes and expectations he meets in his environment, his development is likely to be healthy; where it is not there may be behaviour disturbances.

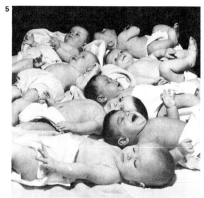

6 Independence is a sense of active mastery of the world, and starts as soon as the infant first discovers that he can influence his surroundings by twisting and turning. Later, crawling, walking and talking all add to his feeling

of self-assertion. Dependence is the need for support in untried, uncertain or unexpected circumstances. Children tend to remain dependent if they are overprotected and not taught to follow some independent urges and activities.

7 Aggressive behaviour in children and factors influencing it have been studied by Arthur Bandura and his colleagues. In a series of experiments in 1963 they tried to determine whether live or filmed models,

acting aggressively towards an inflated rubber doll, had a greater impact on the children's later expressions of aggression in similar situations. Regardless of whether the children saw live or

filmed aggression, they generally imitated the behaviour they had observed. They were also appreciably more aggressive than children with either a non-aggressive model or no model at all.

Moral development

Learning to judge the rightness or wrongness of conduct is crucial to a child's ability to adapt to society. But it is not easy either to give a universal definition of morality or to say how moral values are acquired. Some people have put forward the theory that there is an absolute morality to which everyone should aspire, whereas others adhere to the view that each culture sets its own rules and induces children to conform to them.

Inconsistent values
In the teachings of many religious and ethical systems, morality is sharply defined; right and wrong are presented as polar opposites and the assumption is made that a person who is moral tends to act consistently in accordance with high principles when dealing with other people. But a study conducted during the late 1920s in the United States by H. Hartshorne and M. A. May suggests that most people have flexible moral standards.

Everyone begins as an amoral infant who, as an active learner, acquires his first moral standards from his parents. These early standards are quite rigid and are tied to specific situations. The child often adopts them out of a wish to be obedient and avoid punishment or displeasure. The period between the ages of five and eleven is one of rapid moral development, with friends and teachers becoming significant influences on the standards of conduct adopted – largely through the effort to win their approval. With increasing maturity, obligations to society in general become relatively more important.

Although there are several ways of studying these changes in moral development, one of the most productive has been that of the American psychologist Lawrence Kohlberg. In face-to-face interviews, children are asked to respond to a number of moral dilemmas and the thinking behind their answers is then investigated [2]. Studies carried out in various societies indicate three basic levels in the development of moral reasoning. Kohlberg classified these as premoral, conventional and principled levels.

Steps in progress
Each progressive step is considered to consist of a more complex and balanced way of understanding and thinking about the moral-social world. Premoral reasoning is characterized by primary concern with one's own needs and self-interests. Conventional reasoning includes concern for others and an almost unquestioning acceptance of established authority. Principled reasoning takes into account the welfare of others and reflects self-chosen standards that are based on universal ethical principles. Law is respected but viewed as a human invention that is capable of being changed if circumstances alter.

Higher-level moral concepts are acquired only in late childhood or adolescence, apparently because they require an extensive foundation of intellectual growth and social experience. It appears that levels of moral reasoning may develop in a similar way in all children, regardless of their culture and their religious background [1].

The fact that an individual is able to reason on a principled level does not mean that he will always do so. Different moral situations evoke different responses drawn from all levels. Systems of morality may

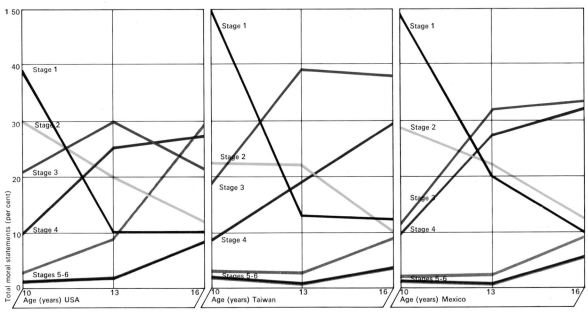

1 **Kohlberg's theory** that moral reasoning develops in rising stages is shown in three different types of society. As the tested children grew older they provided fewer reasons for action at a premoral level [stages 1 and 2] and more at conventional [stages 3 and 4] and principled levels [stages 5 and 6]. The graphs also indicate that moral reasoning seems to develop in a similar way in different cultures, since the pattern of changing values is basically the same in each of the three graphs. The tests were conducted among groups of middle-class boys of various ages living in urban areas of the United States, Taiwan and Mexico.

2 **A child's level of morality** is classified by Lawrence Kohlberg according to the response made to the following story: A sick woman is dying of a cancer from which only one drug can save her. A chemist sells the drug for 10 times what it cost him and the woman's husband has only half the asking price. He tells the chemist his wife is dying, but the price is not lowered. The desperate husband breaks in and steals the drug. Should he have done so and why? Answers both for and against are grouped in six stages. In classifying these it is not the choice of action that is important but the reasons that the individual gives for his action.

Premoral
Stage 1
If you let your wife die because you have not given her the drug, then you will get into trouble. Everyone will blame you for not spending the money to save her and there will be an investigation of both you and the chemist for your wife's death.

Contrary
You should not steal the drug because the chemist will know that you were the thief and so you will be caught and sent to prison. And even if you do get away with it, your conscience will bother you and you will always be worrying that the police will catch up with you.

Stage 2
If you do happen to get caught, you can always give the drug back and then you will not get much of a sentence. It will not be too great a hardship for you to serve a short jail sentence and it will be worthwhile as long as your wife is still alive.

Contrary
You may not get much of a jail sentence if you steal the drug but your wife will probably die anyway before you get out of jail and so it will not do you much good in the end. If your wife dies you should not blame yourself, because it was not your fault that she got cancer.

Conventional
Stage 3
No one will really think that you are bad if you steal the drug, but your family will certainly blame you and think that you are an inhuman husband if you do not. If you let your wife die, you will never be able to look anybody in the face again.

Contrary
It is not just the chemist who will think that you are a criminal; everyone else will as well. After you steal the drug, you will feel bad when you think about the dishonour that you have brought upon yourself and your family. You will not be able to face anyone again.

Stage 4
If you have any sense of honour you will not let your wife die just because you are afraid to do the only thing that can possibly save her. You will always feel guilty that you were the cause of her death if you do not do your duty by her, whatever the consequence for you.

Contrary
You are desperate and you may not know that you are doing wrong when you steal the drug. But you will know that you did wrong after you are punished and sent to jail. You will always feel guilty about your dishonesty and the fact that you have broken the law against stealing.

Principled
Stage 5
You will lose other people's respect, not gain it, if you do not steal the drug. If you let your wife die, it would be because you were afraid, not because you had worked out that it was a wrong act. So you would lose your self-respect as well as that of others.

Contrary
You would certainly lose your standing and the respect of others in your community as well as violating the law if you stole the drug. You would also lose respect for yourself if you allowed yourself to be carried away by emotion and forgot the long-range point of view.

Stage 6
If you do not steal the drug and thus let your wife die, you would always condemn yourself for it. No one would ever think of blaming you and you would have lived up to the letter of the law, but you would not have lived up to your own standards of conscience.

Contrary
If you stole the drug, you probably would not be blamed by other people, but all the same you would condemn yourself because you would not have lived up to your own conscience and the standards of honesty by which you have always lived in the past.

reflect several levels of reasoning. Traditional Christianity, for instance, preaches right actions both because they are in themselves ethical and because the jaws of hell yawn for those who transgress (exemplifying both principled and premoral levels).

Formal moral values

As children develop their own moral reasoning, the way in which parents and friends react to their ideas becomes particularly important. Parents who take seriously their children's opinions on moral issues and discuss them are more likely to have children who reason at a conventional level than at the premoral level.

What parents, friends and others who serve as models actually do has a powerful influence. While words are likely to influence the development of a child's moral reasoning, they do not necessarily result in changes in moral conduct. In general, children copy what others do rather than what they say and are likely to develop consistent moral values only if the actions of influential people in their lives do not conflict with their words.

How a child feels about his own thoughts and deeds plays a pivotal part in his moral development. Typically, a sense of guilt and self-regulation develops gradually and so has an increasingly important role in monitoring behaviour [5]. As this occurs the child starts to regulate his own moral conduct. The child behaves morally principally to avoid his own feelings of guilt.

One of the clearest ways in which parents and others contribute to the development of a child's self-control or sense of guilt is through discipline for disapproved acts. Some research indicates that the most effective discipline is a brief withdrawal of affection combined with a detailed explanation. The latter encourages the child to see things from the perspective of another person and promotes the ability to form independent standards. The timing of discipline may be decisive. It is likely to be most effective if it occurs just before or just as the child begins to do something that is forbidden. When this is not possible, it is helpful if the situation can be carefully re-created by talking with the child and describing the action.

The American researchers Hartshorne and May tested the moral values of thousands of children during the 1920s by means of games and contests in a wide range of situations. They found that if a child cheated in one situation it was not possible to predict whether or not the same child would do so in another situation. They showed too that a child who cheated on one occasion was not necessarily the child who lied then; nor was a child who lied to an adult necessarily the child who lied to a friend. The question in the end is probably not whether children or adults will behave morally or immorally, but rather in what situations they will do so.

3 Our perception of the world is radically affected by our moral values. The sight of a church, for instance, will evoke quite different responses in different people. To a priest [A], it is the sacred house of God, while to an architect [B] it may be merely a construction. An atheist [C] may not distinguish it from surrounding buildings, yet to a Marxist [D] it may represent oppression.

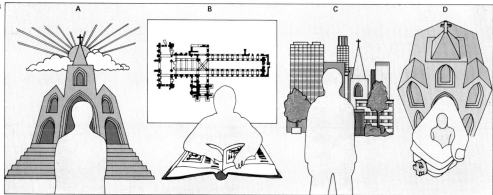

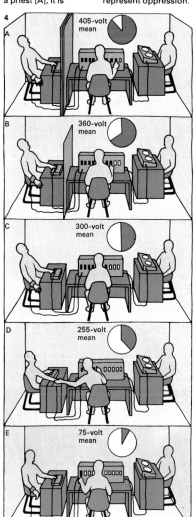

4 Individual scruples and the demands of authority are often at odds. A soldier in uniform will kill but the same man in civilian dress might not. To test response to authority, an American, Stanley Milgram, devised a scheme in which volunteers took part in an "experiment" requiring them to give increasingly severe electric shocks (from 15 to 450 volts) to a protesting man (really an actor) to correct mistakes in a learning test. Authority in the guise of a "scientist" told volunteers they must go on despite a rising crescendo of protest. Where the "learner" was remote [A], 26 out of 40 people went to 450 volts, the mean being 405. Willingness to obey orders declined as the "learner" was brought closer [B, C, D] but it was not until the choice of maximum shock was left to the volunteer that the level fell to 75 volts [E]. Control groups predicted they would disobey rather than cause pain.

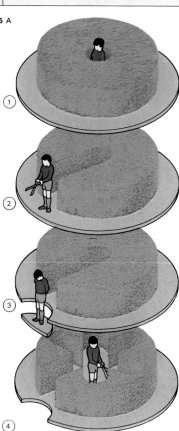

5 The development of a conscience is explained by behaviourist psychologists as the outcome of rewards and punishments [A]. A child starts with no model of how to behave [1]. When he commits a forbidden action [2], the outside world punishes him [3] by at least withdrawing support. Thereafter, in theory, he modifies his behaviour to avoid the mental scar of the action and its result [4]. In contrast, Freudian theory [B] explains guilt as the interaction of the id, ego and superego. The id, present from birth, conflicts with society by trying to satisfy the infant's basic needs. The ego then develops as a mediator to help satisfy these needs. Finally, the superego emerges as internal representative of social values. It inhibits sexual and aggressive impulses of the id while urging the ego to pursue socially approved goals.

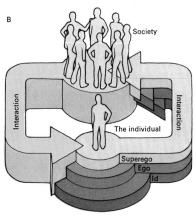

The growing child

Bringing up a child is not like a laboratory experiment – calm and controlled with all inputs carefully measured and predetermined. It is, like all other aspects of being alive, exciting, varied and thoroughly unpredictable and is subject to all sorts of unplanned influences and events, many of them almost totally outside the parents' control. Improvisation within a reasonably worked out framework is the best method of supervising a child's development.

Nature and nurture

Psychologists have explained and clarified the process of development. Their experiments and extensive researches have pinpointed a number of factors that contribute to the physical and psychological growth of the individual from newborn child to adult. The insight they give can be used in building the framework, but parents are largely on their own in supplying the improvisations that are called for.

There is nothing that parents can do about the hereditary factors they pass on to their children – at least not at the moment.

But the behavioural psychologists particularly have highlighted the importance of the learning experiences provided by the child's environment. These both modify and accentuate the child's genetic potential. It has been shown that the quality of the first relationship the child forms with his mother or whoever takes care of him is crucial in setting the pattern of the child's psychological approach to the world in general and personal relationships in particular.

There is also strong evidence that there are particularly sensitive periods in a child's life when the effects of the environment seem to be more readily absorbed and longer lasting. In the first three years, for example, growth in stature is at its most rapid and consequently this is a time when diet exerts most influence on eventual adult height.

If this seems to place a heavy burden on parents for the successful development of their children – and many parents suffer from a feeling that they may not be doing the "right" thing – there is equally strong evidence for the natural and striking durability of children. Certainly, single bad experiences

rarely have a lasting effect. It is a chain of mistreatment that sets up personality problems. The occasional loss of temper by parents is of little consequence and indeed may be beneficial in showing a child that people are not made of glass. Children can weather anger without damage, but not persistent brutality.

Finally, there is the fact, pointed out by the American psychologist Arnold Gesell (1880–1961) and his colleagues, that children brought up in vastly different circumstances grow up to be reasonably similar, fully functioning adults, irrespective of the advantages or disadvantages of their environments. Human growth tends to follow a consistent, natural course.

Changing child care practices

Specific child care practices differ widely both from culture to culture and from generation to generation within a single culture. Some North American Indian babies, for example, spend most of their time strapped firmly to a cradle-board, but they begin to walk at about the same time as chil-

CONNECTIONS

See also
150 Human development
152 Thinking and understanding
154 Language development
156 Social development
158 Emotional development
160 Moral development
164 Adolescence

1

	At birth	Up to 2 months	2-4 months	4-6 months	6-8 months	8-10 months	10-12 months
Body	Almost totally immobile, but with many primitive reflexes, like sucking, foot extension if the sole is tickled and finger grasping.	Able to focus eyes and co-ordinate stare. Can begin to lift chin when lying on stomach.	Can begin to raise head when lying on stomach. Many early reflexes are beginning to be lost. Can turn head and eyes towards speaker.	At 20 weeks baby can open hand for contact with an object. At end of period can grasp things using palm and fingers but not the thumb.	At seven months can sit alone without support for a short time. At end of period can stand with minimum help. Can grasp using forefinger and thumb.	Can crawl by the end of the period and pull himself upright to stand. Better use of tips of forefinger and thumb. Can hold feeding bottle on his own.	Can walk with help. Begins to stop putting objects in mouth. Bowel action starts to become regular.
Senses	Even at three to four days old a baby is able to follow sound from side to side and to tell a buzzer from a bell.	Hearing may be particularly acute for short period, making loud noises painful. Dots and angles on the picture of a face in the face recognition test elicit a smile.	Signs of recognizing things and of losing interest in stimuli. Appreciation of 3-D. *Face recognition test* Needs first eye section alone, then nose to elicit a smile.	Increasing co-ordination between hand and eye. Brings things to mouth to explore. *Face recognition test* Mouth helps on picture. Motion helps to elicit a smile.	*Face recognition test* Attention begins to focus on expression rather than features.	Depth perception is further developed. Child will avoid a visible drop from a surface.	No longer likely to be upset by a sudden unexpected disappearance of an object.
Mind	Reflexes only, though rapidly learns to turn head to left or right to find food in response to either bell or buzzer.	Up to two years is the time to explore dimensions. Baby can't remember existence of object if blocked for more than 15 seconds. This is Piaget's sensorimotor stage.	Can follow trajectory of ball even when it is out of sight. Can tell mother from others but no idea that he has one mother only. Optical illusion of three delights him.	Child now distressed when shown optical illusion of three mothers.	Begins to develop the understanding that hidden or out-of-sight objects continue to exist. Will search for them.	Increasing realization that objects continue to exist even when not seen. Child will search for objects in many different locations.	With first words begins to create and use symbols towards end of period.
Speech	Crying or gurgling are the only sounds.	The first three months is usually the period of maximum crying.	At beginning of period cooing starts; it is sustained for 15-20 seconds. Soon baby responds to human voice and smiles when talked to. Sometimes chuckles.	Cooing becomes more vowel-like and consonants also appear, but all vocalization sounds very different from the mature language.	Cooing begins to turn into babbling. Most common sounds are *ma, mu, dar* and *di*. At end of period repetitions of heard sounds become frequent.	Increased imitations of sounds heard but results inaccurate. Child begins to differentiate words heard by making adjustments to position or behaviour.	First words, often nouns which serve multiple uses (ie "dog" means any four-legged animal). Shows signs of understanding commands, especially with gesture.
Social	Can be divided into active, moderately active and quiet babies.	Has often established a distinct pattern of crying and sociable behaviour. Starts smiling at his mother at four to six weeks. Child is in Freud's oral stage.	Starts to play with objects – rattles, etc – if placed in his hand. Beginnings of differential response to people in his environment.	Beginnings of real differentiation between adults at end of period.	Increasing differentiation between faces. Beginnings of serious play but rarely with other children. Pattern of interaction and support between mother and baby.	Beginnings of preference for play with other people rather than by himself. Receives social support for trying to walk. Starts using imitative gestures.	May show dramatic decrease in excitability and in attention and response when mother is present. Begins to imitate parents' actions (combing hair, smoking, etc).
Emotional	Feelings are almost undifferentiated. Babies are awake or asleep, active or inactive.	Beginning of the differentiation of feelings into positive and negative.	At beginning of period first clearly defined signs of pleasure (smiling, chuckling, etc) and negative emotions.	Pleasure responses to others become selective. Fear and anger begin to emerge separately. Signs of self-satisfaction.	Fear of unknown becomes developed. Attachment to known adults, and relationships become focused. Pleased with reactions from others.	Consciousness of social roles evidenced by timidity towards strangers who are too familiar. Smiles at own image in mirror.	Differences between boys and girls begin to show in way they assert themselves and in readiness to touch toys and others.
Moral	Newborn is amoral, having no values, no attitudes and no beliefs for the first six or seven months.	—	—	—	—	May show signs of withdrawal when admonished for bad conduct.	Understanding of parents' commands remains primitive and tied to immediate situations.

dren from other environments who are painstakingly encouraged to walk.

There are widely differing views – some of them directly contradictory – on many of the most ordinary problems of child rearing. These include whether a child should be breast fed or bottle fed, whether he should be strictly disciplined or never thwarted and whether toilet training is necessary or not.

The question of whether to feed on demand or according to a rigid schedule is a good example of these controversial problems. At the turn of the century it was common to feed babies only when they were hungry. Then it became fashionable to use scientifically determined feeding schedules. The child was fed every four hours regardless of whether he had to be awakened or left crying until the proper time. This technique was thought to be character building because it instilled a sense of discipline and a feeling that he was part of a world bigger than that of his own desires. Then, perhaps as a result of the emphasis that Sigmund Freud (1856–1939) placed on the dangers of thwarting the child's impulses, demand feed-

ing again became popular. Today the question is not considered important, since it has been shown that babies are quite capable of matching their food intake to their needs.

Perhaps as important as anything else to healthy development is the general atmosphere in the family. Some parents prefer an active child, others a passive one. If a child does not conform to his parents' preference, inter-relationship difficulties may follow.

Charting a child's development

There are approximate ages at which a child born at full term might be expected to pass some of the milestones of growing up and these are shown in the chart below. These timings are not a prescription nor are they norms that every child has to achieve to be a successful adult. Boys tend to be physically more adventurous than girls, who start speaking earlier. Allowance must be made for a premature baby particularly in the first year. Instead of beginning to smile at his mother at four to six weeks, for example, a two-month premature baby might be expected to start doing so at 12 to 14 weeks.

It is encouraging for parents who feel they bear a heavy respons-ibility for the development of their child that the results of very different systems of child rearing are so remarkably similar.

2

12-18 months	18 months-2 years	2-3 years	3-5 years	5-7 years	7-11 years	11-13 years
Body Unaided walking at about 13 months. At beginning of period helps to dress himself always interested in feeding himself. Creeps backwards down stairs.	*Body* Control over sphincter muscles begins. Runs but falls often. Climbs stairs holding rail. Girls are half adult height just before two, boys just after.	*Body* Growth more by hormones than by genes. Control over sphincter muscles allows toilet training. Bed wetting increasingly rare. Self feeding improved.	*Body* At end of period girls are slightly ahead of boys in skeletal development. Five-year-old's brain is 75% of his adult weight.	*Body* At six years child's brain is 90% of its mature adult weight. Nerve fibres, etc are almost complete.	*Body* Rise in strength, speed and co-ordination. Fundamental skills advance to those that need instruction and practice – skiing, swimming, etc.	*Body* Girls put on sudden growth spurt and by end of period they average two-and-a-half years ahead of boys. Menstruation begins.
Senses Rapid increase in ability to use characteristic features of objects for identification purposes.	*Senses* Can recognize displacement even though he has not seen object moved.	*Senses* Shows ability to distinguish between primary body parts and features; developing skill at drawing them.	*Senses* Child finds it difficult to distinguish between open and closed letters. Confuses b and d, p and q. Ignores straightness, so confuses D and O.	*Senses* Focuses on significant features of object or event. Increasingly able to discriminate between letters of the alphabet. At end of period can begin to read.	*Senses* –	*Senses* Increasingly perceives non-obvious relationships among apparently unrelated events and objects.
Mind Up to seven years increasingly links image of object with thought about it. Cannot imagine that dimensions stay the same. This is Piaget's pre-operational stage.	*Mind* Thinks everyone has same view of object as he has (ie when he covers eyes and can't see you, you can't see him). Can't describe journey, but can find his way.	*Mind* Progress in ability to classify many sorts of objects according to one or more shared characteristics.	*Mind* Although child realizes apparent change in object is not real change (water to ice), unlikely to understand how or why this is so.	*Mind* This is Piaget's concrete operational stage. Child can use concepts and rules, but deals almost exclusively, at the beginning at least, with here and now.	*Mind* Can now conserve and see parts of the whole. He can represent a series of actions and can arrange objects according to size, weight, etc.	*Mind* Child able to think around a subject, testing many solutions in his mind. Use of abstract rules to solve problems. This is Piaget's formal stage.
Speech Repertoire of 3-50 words. No fuss at not being understood. Vocabulary divided by sex (man-lady) and size (man-boy). Speech is telegraphic – "all-gone-food".	*Speech* Repertoire of more than 50 words at end of period. Can string two words together (noun and qualifier). Grammatical use of articles, plurals, etc. Passive confuses.	*Speech* Over period child able to put 3 or 4 words together. Structuring of language begins, use of tenses, plurals, order, etc.	*Speech* Sentences become longer and more complex. Increasing grasp of basic grammar, like active – passive.	*Speech* Masters irregular endings (ie mice rather than mouses). Vocabulary becomes more complex and school-related.	*Speech* By end of period has almost complete mastery of complex grammatical rules.	*Speech* Ever widening vocabulary of more technical and specialized words. Greater range of concepts and terms to describe phenomena.
Social At end of period fighting and jealousy over toys emerges. Signs of guilt at not living up to socially approved rules. Pleased with particular events or people.	*Social* At end of period solitary play declines. Parallel play with little contact between children in one room is more common. Start of Freud's anal stage.	*Social* By 30 months will be helping with domestic chores. Co-operation between children playing together increased. End of Freud's anal stage, beginning of phallic.	*Social* Beginning of importance of peers in shaping behaviour. Identification with same-sex parent is at its strongest.	*Social* Turns to peers rather than adults. Play teaches social roles and individual limits. End of Freud's phallic stage, beginning of latency.	*Social* Social interactions strongest with the same sex; minimal social interaction between sexes.	*Social* The sharing of possessions, feelings and plans is still primarily with same-sexed peers. End of Freud's latency period, beginning of genital.
Emotional Fear of strange objects, people and noises declines. At end of period shows jealousy over newly arrived babies.	*Emotional* At end of period fear of imaginary creatures and dark begins. More sensitive to ridicule.	*Emotional* Boys express more physical assertiveness and anger; girls express less of these qualities. Attempts at more independence.	*Emotional* Increased satisfaction at accomplishing a self-set task. Rising sensitivity to feelings and responses of others.	*Emotional* Child actively engaged on testing self-image.	*Emotional* Becomes less physical and more verbal in relationships with peers.	*Emotional* Alternating periods of withdrawal and gregariousness. Sexual feelings start to emerge.
Moral Behaviour up to six or seven years is shaped by threat of punishment or satisfaction of needs.	*Moral* Child's comprehension of parents' moral code consists of simple good/bad distinction.	*Moral* Child likely to show substantial guilt-like reactions, though still situation-specific.	*Moral* Still highly egocentric but beginning to show marked guilt and self-regulation.	*Moral* Child's judgment of "right" changes. Respect for, rather than fear of, law.	*Moral* Child does things either to win approval or because the law says so.	*Moral* Actions are guided according to universal ethical principles.

Adolescence

The beginning of adolescence, the period between the ages of 12 and 20, is marked by physical changes brought about by the sex hormones [1]. These physical changes are accompanied by psychological changes. Before adolescence, most children are primarily interested in their own sex. Boys form gangs with boys, girls whisper and giggle with girls. As adolescence advances, each sex becomes more interested in, and more tolerant of, the other. But both boys and girls tend to continue to have attachments to older members of their own sex whom they admire. "Crushes" upon teachers, or the hero-worship of sportsmen or pop stars [Key], also serve a valuable function; for such people act as models that growing boys and girls can imitate, thus learning how to become fully adult members of their own sex [4]. In this way, older people from outside the family can serve as substitutes for the parental models that governed early childhood.

Physical changes in early adolescence
Early adolescence is a time when both sexes may have anxieties about the changes taking place in the body. Girls are sometimes self-conscious about their figures or bothered by menstrual irregularities [2]. Boys often worry about their developing genitals, believing themselves less well endowed than their contemporaries. Because both boys and girls develop at different rates and different ages, those who are late developers easily become anxious that there is "something wrong" with them. This anxiety is often masked by hypochondriacal fancies and many adolescents seek reassurance that they have not got tuberculosis, cancer or another complaint.

In both sexes one unpleasant side effect of the influx of the sex hormones is the development of acne. This unsightly skin complaint, which can affect the chest, back and face, increases self-consciousness especially if it appears on the face. Sufferers not only feel, quite rightly, that contemporaries will find them less attractive, but also sometimes believe that acne is a sign of, or punishment for, their early sexual experimentation. Acne is one adolescent complaint that requires expert medical attention.

Another is the obesity that often afflicts both girls and boys at this period of their lives. Obesity should never be regarded as just "puppy fat" that will be outgrown. For one thing, it often becomes permanent. For another, it is sometimes the result of compulsive overeating, a sympton of anxiety and depression. Like acne, this is a physical disability that is often neglected and it is important that it should be corrected.

Social awkwardness and exhibitionism
Adolescence is in any case a time of social awkwardness. To be neither child nor grown-up is a difficult situation. Some adolescents become intensely shy, hiding themselves away from social contacts. Others become aggressively exhibitionist, flaunting the start of emancipation with outrageous clothes and outlandish hairstyles [5].

Now that adolescents in Western cultures have more money they have become a mass market and the target of advertisers. Many advertisements exploit the adolescent's natural anxiety about physical appearance and social acceptability. The result is that many adolescents overspend on clothes,

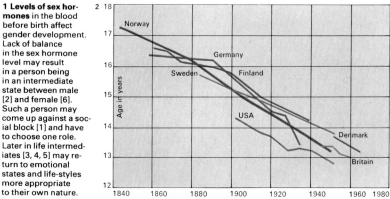

1 Levels of sex hormones in the blood before birth affect gender development. Lack of balance in the sex hormone level may result in a person being in an intermediate state between male [2] and female [6]. Such a person may come up against a social block [1] and have to choose one role. Later in life intermediates [3, 4, 5] may return to emotional states and life-styles more appropriate to their own nature.

2 Over the last century menstruation has begun at an increasingly early age. All girls should be prepared for its onset by being given a straightforward account of its function and purpose. If this is properly done, menstruation can come to be looked forward to as a sign of being grown-up rather than as a "curse" to be concealed and moaned over. Usually most girls experience little discomfort.

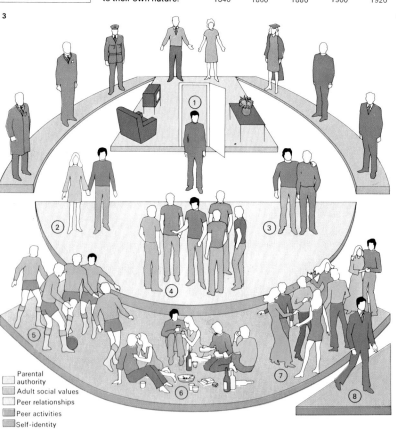

3 Adolescence is a period when the individual tries to develop a sense of his own identity and define his position in the world. As a result, parental values [1] are often rejected and acceptance by peer groups becomes of great importance. Through increasing experimentation with the opposite sex [2] and strong attachments to boys of his own age [3], the adolescent widens his social world and learns social skills. His dress and attitudes often change to conform more closely with the group [4]. By participating in social activities such as games [5], talking, drinking and smoking [6] and parties [7], the adolescent learns to adopt different roles in different social situations. The individual passes through this learning process, which is an aid to his maturity, before joining the adult community [8] and accepting the adult world and its responsibilities.

Parental authority
Adult social values
Peer relationships
Peer activities
Self-identity

4 Adolescents want to leave childhood behind and become "grown-up". It is natural for them to find heroes and heroines on whom they can model themselves and about whom they can have fantasies. Society does not take enough trouble to make use of the idealism of adolescence by providing models of the right kind. Very often the adolescent heroes are themselves hardly more than teenagers and may lend a spurious glamour to such activities as drug-using.

make-up, deodorants and hairdressers, feeling that they must match social expectations that have been artificially implanted in their minds by the media.

Search for identity in the young adult

Adolescence is essentially a time of search for identity [3]. Uneasily poised between childhood and adulthood, the adolescent finds his central problem in the question "Who am I?" In primitive societies, the transition from being a child to being grown-up is generally clearly marked by initiation rites, often of a painful kind. But when these ordeals are over, the individual at least knows where he is and exactly what is expected of him. In Western societies, although confirmation ceremonies such as bar mitzvah [7] may serve something of the same function, the adolescent has no clear picture of his role in society. This is partly because while some adolescents leave school at 16 and have to go out to work, others stay in the more "childish" position of being a student until well into their twenties.

In complex societies, it is much more difficult to define roles in terms of age, since different things are expected of different levels of intelligence and background. At what age should adolescents be allowed to vote, to drive cars and motor cycles, to marry, to have bank accounts? Different countries have different rules [8] and within the same country an adolescent may be put in charge of a potentially lethal machine on the roads, yet not allowed any political say in the running of his country or home town. Deprived adolescents who feel themselves to be undervalued or unwanted turn to hooliganism, a problem in all "advanced" cultures. If they cannot make themselves felt in constructive ways, they may consider violence an acceptable alternative [6].

Aggressive self-assertion is an inescapable part of adolescence for most of us. For how is an adolescent to define himself unless he has some standards against which to rebel, some way of demonstrating that he is an individual in his own right, a different person from his parents? Discovering identity necessarily implies discovering difference; and differences between the adolescent and the parents are inevitable.

KEY

5 Clothes are one of the most obvious ways of expressing individuality, rebelling against parental standards and demonstrating that the adolescent is now ready for sexual encounters. Hence many adolescents dress in ways that are exaggerated and that emphasize their sexuality. Girls use elaborate make-up and boys wear jeans that emphasize the genital bulge.

6 Rebellion against parents, teachers and other authorities is a necessary part of growing up. But often it goes too far. Vandalism, gang-warfare and the affrays that occur at football matches are predominantly teenage affairs. Western society is conspicuously inefficient at providing enough constructive outlets for normal adolescent energy and physical aggression.

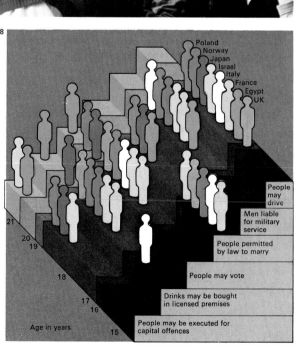

7 Bar mitzvah, one example of a ritual marking the transition from child to adult, is reached by a Jewish boy of thirteen. All the commandments, such as the binding of phylacteries (shown here), symbolizing the Covenant, become incumbent on him. Confirmation and initiation rites in simple cultures and boarding schools are other examples.

8 The age at which adolescents can take on adult privileges or burdens varies from country to country as the chart shows. Teenagers are often confused because they are expected to behave like children in one setting and adults in another. Moreover the expectations are quite different in different social and intellectual strata.

Adolescence: sex and independence

Sex and independence are indissolubly linked. Sex is the biological force that compels an adolescent to look for a mate and the incest taboo ensures that this mate comes from outside his or her own family circle. The taboo on incest, which operates in almost every culture, has more than one kind of significance. First, it removes a dangerous source of disharmony within the family by excluding the possibility of sexual rivalry. Second, it links different families and thus encourages social cohesion. Third, it spurs the adolescent to become independent by forcing him or her to look for sexual experience outside the family circle. Those adolescents who, for one reason or another, stay at home may remain dependent and immature. Lying behind these social and psychological considerations is the biological possibility that incestuous unions may concentrate genetic defects in the family and produce serious mental or physical abnormalities.

Sexual experimentation

The sexual experimentation of adolescents, which often causes parents such anxiety, is in fact a biological necessity that is as much a stage in the development of the adolescent's independence as it is a means of discovery in the sexual field itself [3].

During adolescence boys and girls may experiment with both solitary and mutual masturbation, and the latter may be with their own sex or the opposite one. Transient homosexual encounters are common and should not be taken as meaning that the adolescent is permanently orientated in a homosexual direction. Nor, in most instances, should such encounters be thought to have any lasting effect. Experts believe that those who become permanently homosexual do so partly because of genetic predisposition and partly because of emotional influences during early childhood.

The changes of partner so characteristic of adolescence are also biologically and psychologically healthy. How are adolescent boys or girls to learn what their real preferences and requirements are if they are given no choice and no opportunity at all to experiment with different partners? Parents may be content if their son "goes steady" with, and then marries, the girl next door; but psychiatrists who have seen many such marriages break up in middle age are more confident of the future security of a marriage when both boy and girl have had more experience with different partners.

The need for privacy

The essential privateness of sex is also valuable in encouraging independence in the adolescent. Parents are often over-anxious because their adolescent children do not confide in them and some go as far as to read their children's diaries and letters. Other parents become hurt and try to force their children to tell them everything, as they used to do when they were much younger. But secrets are a necessary part of growing up and should be respected as part of the drive towards independence.

Most adolescent secrets are in any case harmless and those that are not are often best dealt with by adults other than the adolescent's parents. Privacy is the right of every human being, whatever his or her age; forced confidences lead only to resentment. On the

2 James Dean (1931–55) epitomized the often bitter confrontation between the adolescent and adult worlds in his films *East of Eden, Rebel without a Cause* and *Giant*. He revealed the mental turmoil of a youth attempting to come to terms with himself and the conventions of the adult world, and his inability to communicate this inner struggle. As the champion of alienated youth, he highlighted in his roles the gap between the generations and reminded his audience of the problems of self-doubt, frustration and resentment that youth suffers on the way to adulthood. In his own life he exemplified the common gulf that exists between parents and their children at this stage in life.

1 Parents complain when their children are small that they see too much of them. When the children become teenagers, their parents complain that they hardly see them at all. It is of course perfectly natural for teenagers to spend most of their time with their contemporaries. There is bound to be a divergence of interest between parents and teenagers, leading the teenagers to form a sub-culture of their own. The group bond comes partly from shared interest and partly from a need for reassurance. Adolescents are notoriously unsure of themselves and it is comforting for them to find others as uncertain as they are.

3 "The Kiss" by Gustav Klimt (1862–1918) typifies the romantic aspects of sexual involvement. Sexual potency in the male is at its height during adolescence and the urge to find sexual release is a crude, insistent force that often expresses itself in coarse and vulgar ways. However these rather crude passions do not preclude more complete emotional involvement, which is commonly expressed by concern, tenderness and an unselfish preoccupation with the needs and wishes of the beloved. "Commercial sex" (such as prostitution and pornography) does not afford any opportunity for the development of emotional involvement. Adolescent interest in pornography is common, but it does not affect the normal capacity to form deep, loving relationships.

other hand, adolescents also need to be understood and so may feel the need to confide in some adult – though preferably not in their parents. It is usual, if a little disconcerting, for parents to find the adolescent children of their friends and acquaintances much easier to get on with and much more polite than their own teenage children.

Adolescents typically find themselves torn by conflicting impulses. Their wish to be private vies with their need to be understood. Their wish to run their own lives is opposed to their need to ask advice about many aspects of life that they have not yet encountered. In practical terms it is difficult for them to be independent if they have to ask adults for money. And managing their own finances is a problem if they are not familiar with the intricacies of budgeting and opening an account at a bank or post office. It is difficult to maintain secrecy about a social engagement or "date" if advice and reassurances about what to wear and how to behave are desperately needed. Dependence upon parents imposes unwelcome restrictions and often results in frustration, but independence

involves equally unwelcome anxiety. Only tolerance on both sides, from parents and adolescents, sees people through this difficult stage without incessant conflict [1, 2].

In place of parents

A possible solution to this inevitable and necessary tension between parents and adolescent children is the organization of society and family life in such a way that adolescents always have some known and trusted adults outside the family to whom they can turn at any time for support and advice. As an adult it is much easier to be tolerant and understanding towards the adolescent who is rebelling against parents when one is not directly involved. In the extended families of some cultures the adolescent will often be friendly with a number of adults who are concerned about him but who are not his parents. This probably makes life much easier for both parties. In the days when godparents really counted, they could play the part of adviser and confessor, and there is a case for resuscitating the position of the godparent.

KEY

The story of Romeo and Juliet, members of two warring families in Verona, who defied the conventions of their society, expresses the importance of love as a force for independence.

4 Romantic and cynical attitudes to love have been conflicting elements in society since time immemorial. The romantic view is expressed in this nineteenth-century Indian picture of a girl yearning for the return of her lover, while Mozart's opera *Cosi fan tutte* is a tale of mutual deceit. Two girls whose lovers pretend to go into the army take substitutes who turn out, ironically, to be their own lovers in disguise.

5 Opposition to the Vietnam War became a focus for young people's dissent in the 1960s and 1970s not only in the USA but all over the world. Perhaps prompted by the USA's apparently imperialistic and aggressive role in the war, there was evasion of compulsory military service on an unprecedented scale and widespread peace demonstrations that were joined by returned veterans, as here.

6 Greater affluence and a longer period of education have contributed to the emergence of an alternative mass youth culture. The wide use of illegal drugs by Western youth typifies both its affluence, independence and defiance of adult authority.

7 Clothes are a sure pointer to the profound social change that has occurred in this century. In the early 1900s adolescents seem to have emulated their parents' mode of dress. Now the reverse is true.

Adulthood: marriage

Throughout most of human history marriage, even when it is monogamous, was socially less concerned with sexual companionship than with kinship – the movement of men and women of marriageable age between and within extended family groups, the acquisition of kin and the disposal of property [1].

The romantic ideal and its flaws
In Western cultures this pattern of marriage began to change with the general acceptance of the romantic idea of marriage in the nineteenth century. In theory at least this affected to ignore dynastic [2] and financial considerations (as some marriages always had done). Instead it substituted complete mutual absorption of the kind common among people much in love – the difference being that this "peak experience" was supposed to be lifelong. In practice, this change was slow to take place and financial position and class were still of fundamental importance. A governess, for example, as a member of the middle class who had to earn money was not eligible for marriage to a gentleman. Since she usually would not con-

sent to marry below her "station", she remained a spinster.

The romantic concept was an advance psychologically in many ways. Its strengths were that it recognized the woman as a person with equality of choice, that it based marriage on love rather than policy [3] and that it put more value on the sexual relationship. (Again, however, it took a long time for the theory to become fact and women have only recently lost their submissive stereotype).

The romantic concept worked well as long as the variety of social relationships provided by the older kinship system continued to be present. Its inherent weakness, however, is that the isolation sought by lovers becomes burdensome for married couples and for their children who are grossly overexposed to each other's society.

This rather negative side of the development of romantic marriage in the twentieth century has determined the forces for further change, although many have found satisfaction in the opportunities for development and mutual respect provided by the

romantic-monogamous pattern. Other factors placing strains on the existing system are a drastic reduction in early mortality, leading to longer marriages, and the advent of reliable birth control. In societies where population growth is static, the recreational and relational aspects of sex become much more important than its reproductive use. Thus the compelling ethical and practical argument in favour of marital permanence – that children need a stable, two-parent setting in order to grow up undamaged – can be countered by not having children.

Serial polygamy and adultery
Romantic marriage has reacted to these processes by a change of practice, most marked in Scandinavia and America but reflected even in Catholic countries where religion reinforces the institution. The professed attitude of the marriage ceremony, the law and convention is one of exclusive, lifelong monogamy. The actual practice of many couples is to indulge in serial polygamy [5], where marriages ending in divorce occur one after another (since "marriage" is still

1 Dynastic considerations were uppermost in Eleanor of Aquitaine's (c. 1122–1204) marriage to Henry II of England (1133–89). Women were important both as sexual partners and mothers and also as ambassadors between clans and bringers of inheritances. Henry, who had just succeeded to the Duchy of Normandy through his mother, won control of much of southwest France by this fateful marriage.

2 The abdication of Edward VIII (1894–1973) in 1936 to marry a divorced American, Mrs Wallis Simpson, was the epitome of the romantic concept of "giving up all for the woman you love".

3 "Arranged" marriages, here portrayed by the English painter William Hogarth (1697–1764), are evidence of the practical side of marriage. Wealth, power and influence are traded off against each other or exchanged for youth and beauty.

4 Divorce rates are rising as the law becomes less obstructive (in response to public acceptance) and also as people become less tolerant of difficult home situations. Women's increasing ability to achieve financial independence has re-

moved one barrier to successful separate existence. To what extent this signals the decline of marriage is unknown since far more Victorian marriages ended with an early death than modern marriages are ended by divorce.

Number of divorces per 1 million inhabitants
(graph showing USA, Denmark, Japan, Norway, UK from 1900 to 1970; values 0, 500, 1,000, 1,500, 2,000, 2,500)

required to sanctify intercourse). Adultery may be tolerated provided it is not "serious", is for sex only and does not come too blatantly to the notice of the other spouse.

Secondary relationships of this kind are not a new feature of monogamy and were equally common in upper-class Victorian England, or in Catholic countries where divorce was barred. In the modern scene, however, adultery may be changing its role – chiefly through a growing dissatisfaction with the idea that monogamy must be an emotionally exclusive relationship with secondary relations thereby devalued or deprived of humanness and openness.

Contemporary marriage is, accordingly, in a state of flux. Most couples still enter marriage with the expectation of exclusive lifelong monogamy and start a family. But since divorce has been made less difficult many couples resort to it as a way out of marital difficulties [4]. On the other hand, the mutual-sufficiency, mutual-ownership stereotype of marriage is being changed as a growing percentage of younger couples practise trial marriage to test compatibility. (This old expedient in rural areas was used to test fertility or the economic viability of the couple as farmers.) People's expectations, too, are becoming more realistic. There is less of the "happy-ever-after" literature of marriage, which played a part in miseducating young people.

Contemporary marriage

A legally recognized primary relationship seems unlikely to be supplanted as the focus of most man-woman relationships and there is no sign that the popularity of marriage is declining. But the tendency in some societies is for it to become more open, with childbearing recognized both legally and socially as a chosen life-style that imposes limits on the freedom to dissolve a relationship. With this exception a far wider range of life-styles is becoming accepted – and acceptable without guilt or blame [6, 8]. The moving of sex *per se* out of the field of "morals" and the substitution of morals based on responsibility, especially to children, represents a logical outgrowth of the change in the importance of the family.

Marriage has been a pragmatic alliance throughout most of history, particularly for country people like those depicted in *Peasant Wedding* by Pieter Bruegel (c. 1520–69). Families formed primary economic units in such villages. Rural communities, being more inter-dependent than urban ones, have coped better with the potential isolation of wedded couples in romantic marriages where "love" is the foundation.

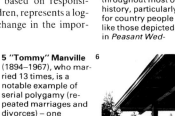

5 "Tommy" Manville (1894–1967), who married 13 times, is a notable example of serial polygamy (repeated marriages and divorces) – one answer to the problem posed by romantic marriage. The most frequent casualties are children, whose experience may be as disturbing as its Victorian equivalent – the death of a parent and coming of a step-parent.

6 Experimentation with different lifestyles is a feature of youth culture, which values unpossessive love and rejects ownership as an index of love. It also compensates for shortage of kin by eroticizing friendship as do such peoples as the Eskimos.

7 Polygamy (marriage with several spouses at the same time) is common in many societies, especially in Asia and Africa. Usually it takes the form of one man with many wives (as here in New Guinea), though the reverse does occur (depending on the definition of husband) among Eskimos, Aborigines, groups in India and Tibet and others. The number of wives a man may have usually depends on economics, though it may be limited by law (Muslims may have only four) or reserved for kings and nobles (King Mtessa of Uganda is said to have had 7,000). Sometimes all the wives have similar status; in other groups the first or principal wife is above the others.

8 Marriage between homosexuals, instituted particularly in the United States, is the most recent form of marriage ceremony. It is a response to the gradual acceptance by society of the possibility of legitimacy of viable and long-lasting relationships between homosexuals who wish to share their lives and often their property. For them, as for any other couple, marriage is an outward and visible sign of their inner commitment to each other.

Adulthood: from 20 to 30

The decade from 20 to 30 is hard to characterize because there are marked differences between the roles of men and women, between the married and the unmarried and between those with children and those without. Ideally, it is a time of hope and increasing confidence. With the tribulations of adolescence behind them, and many exciting possibilities ahead, young men and women start the road towards maturity impatient to prove themselves and certain that they will do better than their seniors.

The bonds of matrimony

Yet for the newly married with a young and growing family financial difficulties are common and, particularly for women, the adjustment to a life centred round the home is difficult to make. Whereas a husband can escape to his work, his wife is often tied all day to the house. For many mothers this leads to a drastic loss of confidence and of self-esteem. Current research indicates that one-third of working-class women with children under six who are at home full time are clinically depressed.

Some women especially those whose parents have kept too tight a rein on their daughters' independence, find the sudden release from control and the new responsibilities of running a house and budgeting difficult to cope with. Some even develop phobias about going out alone and shopping and remain confined to the house because they cannot face the burdens of adulthood.

Both husband and wife may find themselves torn between the attention their partner expects and demands and their own wish to remain loyal to the group of friends with whom they spent time before getting married. It takes time for couples to form new patterns of relationships in society as a pair; and as a result some newlyweds, feel isolated and cut off from family and friends.

For many newlyweds money becomes an important issue for several reasons. During adolescence young people, at least in Western society, are often still living in their parents' home and yet have begun to earn their livings. Before marriage, they have few financial responsibilities and become accustomed to spending what they earn on them-

selves without thought for others. After marriage, their financial horizons become sharply restricted. The couple may soon acquire a home of their own, but they may have the prospect of many years of regular repayments on the house or apartment and on the furniture to fill it. A home means privacy and comfort; but it also means a succession of bills to be paid.

The arrival of children

The situation may become more complicated after the arrival of children because they add to expenditure and because after the birth of the first child many women give up their jobs. This reduces the family's income and at the same time promotes a feeling of dependence in the wife. The fact that the money she spends is earned by her husband may seem to her to be irksome and degrading, particularly as she makes such a contribution to the quality of the home life – a contribution that some women think is taken for granted or undervalued. This can become a focus for resentment and bickering. Equally, some men feel that the responsibility of being the

CONNECTIONS

See also
172 Adulthood: from 30 to 40
174 Adulthood: middle age
306 Uses of leisure
290 Changing patterns of crime

1 The average single man [A] in Western European countries spends a quarter of his leisure time participating in physical recreation. The most popular pastimes are swimming, football, table tennis, athletics, tennis and rugby football. Once married he adopts more sedentary pursuits centered on the home. The arrival of children further consolidates the family unit. Salaried workers [B], who are often under training in their early twenties, on average earn less than their manual-worker counterparts although they have greater earning prospects in the future. Men between the ages of 20 and 30 committed almost a quarter of all crimes in 1974 [C]. However male teenagers were guilty of 26 per cent of all offences, among which car and petty theft were most common. The greatest killer [D] of 20 to 30 year olds is the motor car, causing 30 per cent of all deaths, while suicides account for 10 per cent. Cancer is the major medical cause (leukaemia and cancer of the testes particularly); heart disease ranks second.

Age groups	% of all crimes committed, by this age group
20–29	23
30–39	
40–49	

	% of all deaths, in this age group
20–29	0·6
30–39	
40–49	

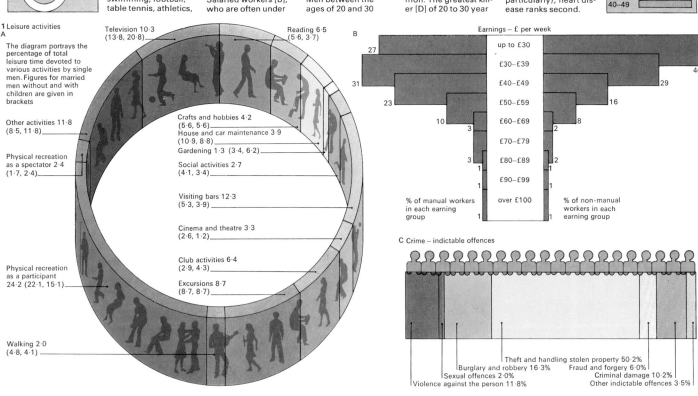

1 Leisure activities

A

The diagram portrays the percentage of total leisure time devoted to various activities by single men. Figures for married men without and with children are given in brackets

Television 10·3 (13·8, 20·8)
Reading 6·5 (5·6, 3·7)
Other activities 11·8 (8·5, 11·8)
Physical recreation as a spectator 2·4 (1·7, 2·4)
Crafts and hobbies 4·2 (5·6, 5·6)
House and car maintenance 3·9 (10·9, 8·8)
Gardening 1·3 (3·4, 6·2)
Social activities 2·7 (4·1, 3·4)
Visiting bars 12·3 (5·3, 3·9)
Cinema and theatre 3·3 (2·6, 1·2)
Club activities 6·4 (2·9, 4·3)
Excursions 8·7 (8·7, 8·7)
Physical recreation as a participant 24·2 (22·1, 15·1)
Walking 2·0 (4·8, 4·1)

B Earnings – £ per week

% of manual workers in each earning group		% of non-manual workers in each earning group
27	up to £30	40
31	£30–£39	29
23	£40–£49	16
10	£50–£59	8
3	£60–£69	2
	£70–£79	
3	£80–£89	2
1	£90–£99	1
1	over £100	1

C Crime – indictable offences

Theft and handling stolen property 50·2%
Burglary and robbery 16·3%
Fraud and forgery 6·0%
Criminal damage 10·2%
Other indictable offences 3·5%
Sexual offences 2·0%
Violence against the person 11·8%

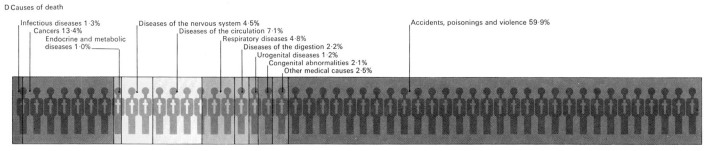

D Causes of death

Infectious diseases 1·3%
Cancers 13·4%
Endocrine and metabolic diseases 1·0%
Diseases of the nervous system 4·5%
Diseases of the circulation 7·1%
Respiratory diseases 4·8%
Diseases of the digestion 2·2%
Urogenital diseases 1·2%
Congenital abnormalities 2·1%
Other medical causes 2·5%
Accidents, poisonings and violence 59·9%

sole supporter of the household is a heavy burden. The classic role of the man as breadwinner can be as constraining and unwelcome as the role of the woman housewife.

In addition children impose financial strains and there is evidence that they may also precipitate problems of communication, particularly for low-income families. While husband and wife both have jobs they share many preoccupations. But as the wife becomes absorbed in her work in the home and the husband in his work outside, these common experiences largely disappear and there is often little to talk about of mutual interest. For many couples this is an important factor in marital disharmony

The constraints of the period

Western societies do not make a clear enough demarcation between being a child and being a grown-up, with the result that the adolescent often does not know what society expects of him. This problem continues in the decade 20 to 30. A particular difficulty is that although at the peak of their powers young men and women find difficulty in expressing their full potential. In a society that is still controlled by the middle-aged they cannot usually gain any position of power. In the medical profession, for example, a doctor aiming at specialist training may not have completed that training by the age of 30 and may not have attained a position of eminence until middle age has overtaken him. In societies where hunting and fighting are the main occupations of younger men, maximum potential and maximum achievement coincide. In Western society youth often lacks the opportunity to use its ability.

This dilemma may be particularly acute for women. The pressure that society exerts on them to measure their success in terms of the children they produce is at variance with their growing wish to participate in the world on an equal footing with men. Their traditional goals clash uncomfortably with their new ones. As a symptom of this there is a shortage of job opportunities and of such facilities as day-care centres for pre-school children suited to the needs of women who wish to combine having a job outside the home with caring for their children.

Women aged between 20 and 29

2 The most popular forms of physical recreation [A] for women between 20 and 30 are dancing, swimming, tennis, table tennis, horse riding and ice skating. The amount of leisure time spent on these activities declines with marriage and the arrival of children. As with men, activities then centre on the home and hobbies and craftwork such as knitting and sewing. The earning potential [B] of manual and non-manual women workers varies little as few non-manual workers undergo long training. There is still a wide difference between the earnings of men and women. Women aged between 20 and 30 [C] commit less than a fifth of the number of crimes committed by men of the same age group (girl teenagers have a higher crime incidence). Shoplifting comprises 50 per cent of offences but violence is increasing. The death rate [D] for women of this age is less than half that for men and deaths due to accidents are considerably lower. Relatively more women die of leukaemia, brain cancer and certain heart diseases.

Age groups	% of all crimes committed, by this age group
20–29	35
30–39	
40–49	

Age groups	% of all deaths, in this age group
20–29	0·3
30–39	
40–49	

2 Leisure activities

A

The diagram portrays the percentage of total leisure time devoted to various activities by single women. Figures for married women without and with children are given in brackets

Television 10·7 (14·2, 19·8)

Reading 7·5 (6·1, 4·0)

B

Physical recreation as a spectator 1·7 (1·7, 1·0)

Other activities 6·7 (9·5, 13·6)

Crafts and hobbies 11·6 (18·6, 20·7)

Gardening 0·3 (2·4, 3·1)

Social activities 7·0 (11·2, 7·0)

Visiting bars 2·9 (2·2, 2·2)

Cinema and theatre 5·6 (2·8, 1·5)

Physical recreation as a participant 28·0 (15·4, 9·8)

Club activities 5·3 (3·2, 1·8)

Excursions 7·0 (8·4, 7·7)

Walking 5·7 (4·3, 7·8)

Earnings – £ per week

	% of manual workers	Earnings	% of non-manual workers	
	9	up to £15	5	
	24	£15–£19	22	
	32	£20–£24	28	
	19	£25–£29	20	
	9	£30–£34	14	
	4	£35–£39	6	
	1	£40–£44	2	
	1	£45–£49	2	
	1	over £50	1	

% of manual workers in each earning group

% of non-manual workers in each earning group

C Crime – indictable offences

Burglary and robbery 3·0%
Violence against the person 4·7%
Theft and handling stolen property 74·8%
Fraud and forgery 9·7%
Criminal damage 4·2%
Other indictable offences 3·6%

D Causes of death

Infectious diseases 1·9%
Cancers 21·6%
Endocrine and metabolic diseases 2·4%
Diseases of the nervous system 5·5%
Diseases of the circulation 11·5%
Respiratory diseases 7·6%
Diseases of the digestion 3·0%
Urogenital diseases 2·1%
Complications of childbirth 2·5%
Congenital abnormalities 3·4%
Other medical causes 3·8%
Accidents, poisonings and violence 34·7%

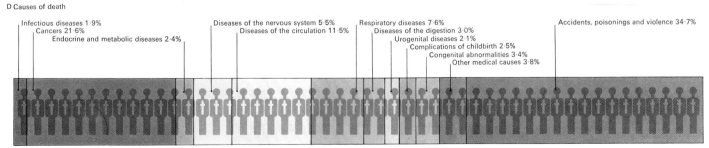

Adulthood: from 30 to 40

During their thirties, most young people become firmly established in their adult roles. The direction in which a man is going and the probable range of his achievement will be settled, although surprising changes of direction can still occur. Equally, a woman will usually have decided on the balance she would like to strike between career and family. A married couple will know how many children they want and their social life will have been established, although this is largely dependent on the age of their children. Many older people look back on this decade as a time of particular happiness. The uncertainties of youth are past; the probems and upheavals of middle age are still in the future. Asked what age they would like to be, many older people answer: "Thirty-five."

Friendship and love

Although friends can be made at any age, by 35 to 40 most people will have a number of established friends, people whom they know so well that they can be completely at ease and relaxed with them. This is partly the result of having overcome the social anxieties of earlier decades and partly because many friendships do not fully mature for some years after the first encounter.

Nevertheless, this decade may present problems. Married couples usually find that the initial delight of being "in love" has worn off. Being in love is a curious condition, which Sigmund Freud (1856–1939) called "the psychosis of normal people". Most young people carry within their minds a somewhat idealized picture of the opposite sex, derived in part from parents and in part from the images presented by television, the cinema, novels, advertising and so on. When they fall in love, the beloved person seems to correspond exactly with this idealized picture. Psychologists would say that the idealized image is "projected" upon the other person. Living with another person day in, day out means, however, that each becomes aware of all the ways in which the image and the reality fail to correspond. Real people are not like creatures of the imagination but are human beings with faults and foibles as well as the attractions that first gained their partner's interest.

Being in love, therefore, must be replaced by learning to love the other as a real person. For some this is not an easy transition since they miss the thrill of being in love and may sometimes become disillusioned or bored with the person they once adored. This is the prime reason for what is called the "seven-year itch", the tendency, especially in men, to look for sexual partners outside the marriage. Infidelity, a prime reason for divorce, is now recognized as a symptom of marital disharmony rather than a cause, and transient infidelity ought to lead to a re-examination of the marital relationship rather than to its immediate break-up.

Psychological maturity

In Western culture, the years from 30 to 40 are commonly regarded as those in which "maturity" is reached. But this prized maturity is hard to define. Both men and women reach their physical peak in their early twenties or earlier. Yet often psychological maturity seems to be a goal that recedes the nearer a human being approaches it.

Men aged between 30 and 39

1 The average married man in Western European countries spends 60 per cent of his leisure time [A] pursuing activities in and around the home. Television, gardening, decorating the house and maintaining the car are main forms of relaxation. Time spent in physical recreation drops to only 10 per cent and favourite sports include swimming, dancing, fishing and golf. Going out in the evening is of minor importance. Earning comparisons [B] show non-manual male workers earning on average 22 per cent more than their manual counterparts, reflecting the earning potential of qualifications gained earlier. Far fewer crimes [C] are committed by men in their 30s than by those in their 20s. Destructive crimes give way to more offences of other types but only some kinds of sexual crimes rise absolutely. Male deaths [D] in the 30–40 age group are virtually the same as for younger men. There is a drop in deaths as a result of accidents. Heart disease and cancer of the lung, brain or testes are the major medical causes of death.

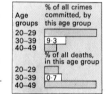

Age groups	% of all crimes committed, by this age group
20–29	
30–39	9.3
40–49	

	% of all deaths, in this age group
20–29	
30–39	0.7
40–49	

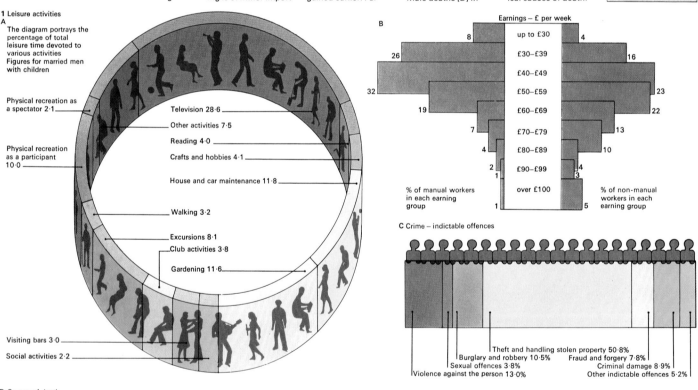

1 Leisure activities

A
The diagram portrays the percentage of total leisure time devoted to various activities
Figures for married men with children

Physical recreation as a spectator 2·1

Physical recreation as a participant 10·0

Television 28·6
Other activities 7·5
Reading 4·0
Crafts and hobbies 4·1
House and car maintenance 11·8
Walking 3·2
Excursions 8·1
Club activities 3·8
Gardening 11·6

Visiting bars 3·0
Social activities 2·2

B

Earnings – £ per week

% of manual workers in each earning group		% of non-manual workers in each earning group
8	up to £30	4
26	£30–£39	16
32	£40–£49	23
19	£50–£59	22
7	£60–£69	13
4	£70–£79	10
2	£80–£89	4
1	£90–£99	3
1	over £100	5

C Crime – indictable offences

Theft and handling stolen property 50·8%
Burglary and robbery 10·5%   Fraud and forgery 7·8%
Sexual offences 3·8%   Criminal damage 8·9%
Violence against the person 13·0%   Other indictable offences 5·2%

D Causes of death

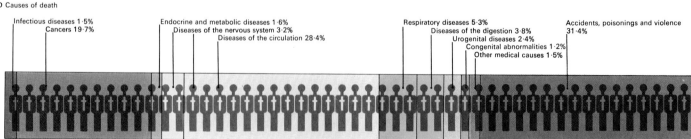

Infectious diseases 1·5%
Cancers 19·7%
Endocrine and metabolic diseases 1·6%
Diseases of the nervous system 3·2%
Diseases of the circulation 28·4%
Respiratory diseases 5·3%
Diseases of the digestion 3·8%
Urogenital diseases 2·4%
Congenital abnormalities 1·2%
Other medical causes 1·5%
Accidents, poisonings and violence 31·4%

What is maturity in psychological terms? Since man is a social being, maturity of personality cannot be defined except in terms of relationships. A person without relationships with other persons cannot be defined as an entity, just as it is meaningless to ask what size an object may be if there is nothing with which to compare it. Although there are many attributes of maturity, one that is obviously important may be singled out: the ability to make fruitful, loving relationships with other people on equal terms, without either being dominated or dominating. This achievement implies an acceptance of the other person as he or she is, without any wish to alter, to direct, or to submit; a recognition of the other person as a separate entity and therefore of oneself as a separate entity also.

Importance of self-criticism
Maturity also demands that a person be realistic, without abandoning the capacity to use imagination. It implies self-control, but combined with the ability to "let go" when this is appropriate. It implies having arrived at some fairly coherent point of view about the universe and the place of man within it, but excludes dogmatism and fanaticism. It implies firmness without rigidity; love without infatuation; decisiveness without being dictatorial; tolerance without a facile permissiveness.

In fact, to be mature in every sense is beyond the reach of most, though this ideal state may be approached more nearly by the habit of vigilant self-criticism. Few human beings in modern cultures become mature by the age of 40. Indeed the extension of the human life-span beyond the reproductive period has necessitated the extension of psychological development beyond middle age well into the second half of life.

Statistics [1, 2] show that people in the 30 to 40 age group are usually healthy, socially stable and tend to spend much of their spare time within the family circle. The material rewards of having gained higher educational qualifications become more evident. In both physical and psychological terms, adults of this age are most likely to store up future trouble by developing a false sense of complacency.

Women aged between 30 and 39

2 Married women in the 30 to 40 age group [A] spend little time on physical recreation, apart from dancing. As with men, more than 60 per cent of leisure time is spent in the home and going out is

confined mainly to places the children appreciate. Television, reading, knitting, sewing, other handicrafts and gardening are the major activities. Earnings of women in non-manual work [B]

rise in the 30 to 40 age group and average 33 per cent more than for women in manual work whose earnings are no higher than for the 20 to 30 manual group. Women's crime rates fall in the 30s [C]

compared with the 20s. Petty thieving accounts for 80 per cent of offences, of which shop-lifting, the only offence committed more by women than men, is 55 per cent of the total. The death rate of women in their

30s rises faster [D] than it does among men. Cancer of the breast, ovary or cervix, together with cerebrovascular and rheumatic heart disease, are by far the main causes of death in their group.

Age groups	% of all crimes committed, by this age group
20–29	
30–39	2·0
40–49	

	% of all deaths, in this age group
20–29	
30–39	0·4
40–49	

2 Leisure activities
A
The diagram portrays the percentage of total leisure time devoted to various activities. Figures for married women with children

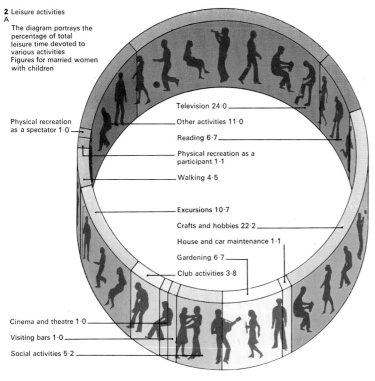

Physical recreation as a spectator 1·0

Television 24·0
Other activities 11·0
Reading 6·7
Physical recreation as a participant 1·1
Walking 4·5

Excursions 10·7
Crafts and hobbies 22·2
House and car maintenance 1·1
Gardening 6·7
Club activities 3·8

Cinema and theatre 1·0
Visiting bars 1·0
Social activities 5·2

B
Earnings – £ per week

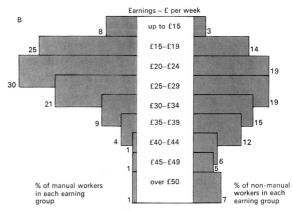

% of manual workers in each earning group		% of non-manual workers in each earning group
8	up to £15	3
25	£15–£19	14
30	£20–£24	19
21	£25–£29	19
9	£30–£34	15
4	£35–£39	12
1	£40–£44	6
1	£45–£49	5
1	over £50	7

C Crime – indictable offences

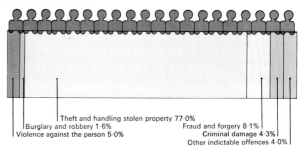

Theft and handling stolen property 77·0%
Burglary and robbery 1·6%
Violence against the person 5·0%
Fraud and forgery 8·1%
Criminal damage 4·3%
Other indictable offences 4·0%

D Causes of death

Infectious diseases 1·5%
Cancers 37·1%
Endocrine and metabolic diseases 1·8%
Diseases of the nervous system 4·1%
Diseases of the circulation 18·8%
Respiratory diseases 6·3%
Diseases of the digestion 3·7%
Urogenital diseases 2·6%
Complications of childbirth 1·4%
Congenital abnormalities 1·8%
Other medical causes 2·3%
Accidents, poisonings and violence 18·6%

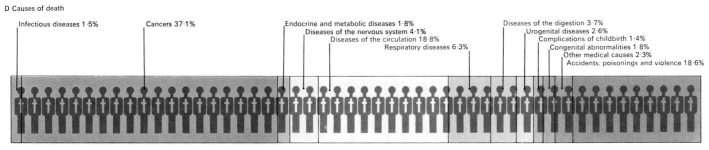

Adulthood: middle age

In modern culture there is a growing awareness of middle age as a period of change and development. It was once assumed that by the age of 40 both men and women were so fixed in their ways that little change could be expected and to an extent this is borne out in statistics [1, 2]. But it is increasingly recognized that middle age can and should be a time in which new interests are developed and new departures undertaken.

Mid-life crisis

Emotional disturbances in middle age are so common that psychiatrists often refer to the "mid-life crisis". This was once thought to be the result, at any rate in women, of the "change of life" or menopause. While it remains true that a few women do experience unpleasant physical symptoms at this time and may become irritable and moody, modern treatment with hormones, although controversial, is able to alleviate most of the symptoms. Moreover, the widely held belief that after the menopause women lose most of their sexual desire is demonstrably false. In fact the opposite is often true, since many

women experience an increase in desire when pregnancy is no longer a possibility.

Although men often go through a mid-life crisis, this is no longer generally believed to be the result of some male equivalent of the menopause. It has been shown that diminution in sexual desire and performance is very gradual in most men, so that the emotional upheaval through which some pass cannot be attributed to any sudden decline in sexual potency. What then is the cause of the mid-life crisis, and what forms does it take?

The name itself suggests one reason. At the mid-point of life, both men and women are at the stage when they should have realized their ambitions. If things have gone well, the traditional goals (for a man, of establishing himself in a job or profession; for a woman, of rearing a family) will have been achieved. However, many women feel there is nothing left for them once their children are no longer dependent and that years of child-care and housework have left them unqualified for work outside the home. And many men regard the effort they have put into "getting ahead" as misplaced.

Inevitably, not all of a person's dreams will have been fulfilled. Some become depressed because they have to come to terms with the reality of what their life is, and abandon some of the hopes of youth. This is a time when a restless dissatisfaction afflicts many people, often showing itself in transient infidelities, increasing consumption of alcohol and changes of occupation. People of this age often feel that the future has nothing to offer but decline into old age and death.

Creative turning-point

A study of the lives of creatively gifted men and women shows that their work often undergoes a change of style at the mid-life period. Some, like the English novelist George Eliot (1819–80), do not begin their real creative production until this time in their lives. Others, like Ludwig van Beethoven (1770–1827), demonstrated an increase in profundity that marked off their late work sharply from what had gone before. Sigmund Freud (1856–1939) published nothing of striking originality until he was 39; and the work that he continued to regard as

Men aged between 40 and 49

1 Leisure activities
[A] of average middle-aged fathers in Western European countries continue to centre around the home, although children are likely to be more independent. Gardening now takes precedence over the house. Less energetic sports are taken up and watching sport is very popular and often shared with sons. Earnings of male manual workers [B] show a drop in middle age because of declining productivity but the non-manual worker increases his earnings and his relative prosperity. Crime rates [C] for men aged 40 to 50 show that they commit fewer impetuous offences than younger men. Most of the crimes in this age group are committed by professional thieves. Men of this age group are responsible for only 4.8 per cent of crimes. Typical causes of death [D] in an affluent society – heart disease and cancers – account for more than two-thirds of all deaths of middle-aged men. Mortality rates in general begin showing a rise between the ages of 40 and 50.

Age groups	% of all crimes committed, by this age group
20–29	
30–39	
40–49	4.8

	% of all deaths, in this age group
20–29	
30–39	
40–49	2.1

1 Leisure activities
A

The diagram portrays the percentage of total leisure time devoted to various activities. Figures for married men with children

Physical recreation as a participant 6·8

Television 30·6
Other activities 8·6
Physical recreation as a spectator 2·4
Reading 2·8
Crafts and hobbies 3·2
Walking 3·0
Excursions 7·2
House and car maintenance 10·8
Club activities 5·2
Gardening 13·6

Cinema and theatre 0·4
Visiting bars 3·4
Social activities 2·0

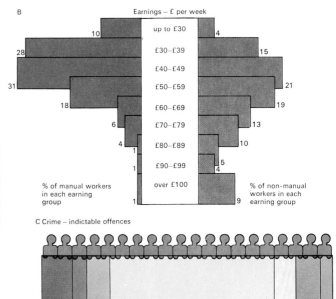

B

Earnings – £ per week

% of manual workers in each earning group			% of non-manual workers in each earning group
10	up to £30	4	
28	£30–£39	15	
31	£40–£49	21	
18	£50–£59	19	
6	£60–£69	13	
4	£70–£79	10	
1	£80–£89	5	
1	£90–£99	4	
1	over £100	9	

C Crime – indictable offences

Theft and handling stolen property 56·3%
Burglary and robbery 8·0%
Fraud and forgery 7·5%
Sexual offences 4·9%
Criminal damage 8·1%
Violence against the person 10·5%
Other indictable offences 4·7%

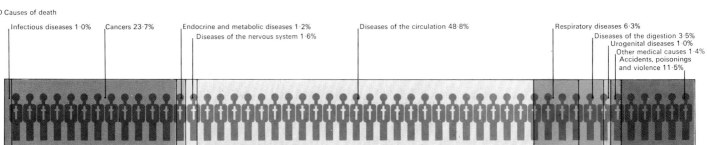

D Causes of death

Infectious diseases 1·0%
Cancers 23·7%
Endocrine and metabolic diseases 1·2%
Diseases of the nervous system 1·6%
Diseases of the circulation 48·8%
Respiratory diseases 6·3%
Diseases of the digestion 3·5%
Urogenital diseases 1·0%
Other medical causes 1·4%
Accidents, poisonings and violence 11·5%

his most penetrating. *The Interpretation of Dreams*, did not appear until he was 43.

Creative geniuses show in their work that they suffer from the same conflicts as other people. Their ability to cope with this, sometimes by changing their style, is something that others might emulate.

Discovering new goals

Often the mid-point of life is a time to rediscover interests and aspects of the self that have been dropped because there has not been enough time to pursue them. Most middle-aged people have had enthusiasms in adolescence that they were unable to follow up – for painting, music, literature, gardening, bee-keeping or bird-watching. The great Swiss psychologist Carl Jung (1875–1961), who specialized in the treatment of middle-aged patients, said that culture was the goal of the second half of life, and that what we needed were schools for 40-year-olds. It was a perceptive and perhaps prophetic suggestion. Middle-aged people sometimes fear that they cannot learn anything new, but this is wrong. At this time,

most people are more realistically aware of their strengths and their weaknesses, and are much more able to apply themselves systematically to whatever they undertake.

The great mistake is to think that in middle age it is too late to continue searching for anything new. For a while, psychologists thought that the human being was always driven by a need to discharge tension and achieve the peace that comes when instincts are satisfied and hungers temporarily assuaged. Now it has become clear that such an idea is quite inadequate. The brain works best when it is given a variety of different, novel stimuli. Shut a man up in a darkened, sound-proof room with nothing either to worry or disturb him, and what happens? He has hallucinations and begs to be released.

New problems are a vital part of living, and if they do not exist it is necessary to invent them. Men and women who have achieved, at the mid-point of life, the goals of youth, must find new problems to wrestle with, and other interests to engage them. Discovering these fresh challenges is largely a matter of having the right attitude.

Women aged between 40 and 49

2 Leisure patterns [A] among middle-aged women show a slight fall in television watching and handicrafts. The extra time is spent in and around the home, in gardening or in reading. With a

greater degree of freedom from children there is an increase in physical recreation, especially dancing. Women engage more in social activities than men. Earnings of female manual workers [B]

are maintained at much the same level in middle-age as they were in the decade 30 to 40 but they are still consistently less than a man receives for the same work. Few crimes [C] are committed by

middle-aged women. Shop-lifting and other petty thefts account for an overwhelming percentage of their offences. Drunken driving is the only offence more common than among younger women.

The death rate [D] among middle-aged women climbs steeply. Cancers and heart disease cause more than 70 per cent of deaths at this age. Of these, breast and cervical cancers predominate.

Age groups	% of all crimes committed, by this age group
20–29	
30–39	
40–49	1·4

Age groups	% of all deaths, in this age group
20–29	
30–39	
40–49	1·4

2 Leisure activities
A
The diagram portrays the percentage of total leisure time devoted to various activities. Figures for married women with children

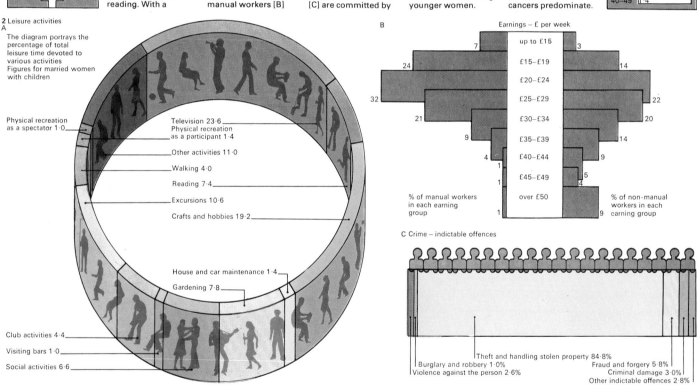

Physical recreation as a spectator 1·0

Television 23·6
Physical recreation as a participant 1·4
Other activities 11·0
Walking 4·0
Reading 7·4
Excursions 10·6
Crafts and hobbies 19·2

House and car maintenance 1·4
Gardening 7·8

Club activities 4·4
Visiting bars 1·0
Social activities 6·6

B

Earnings – £ per week

% of manual workers in each earning group		% of non-manual workers in each earning group
7	up to £15	3
24	£15–£19	14
32	£20–£24	22
21	£25–£29	20
9	£30–£34	14
4	£35–£39	9
1	£40–£44	5
1	£45–£49	4
1	over £50	9

C Crime – indictable offences

Theft and handling stolen property 84·8%
Burglary and robbery 1·0%
Violence against the person 2·6%
Fraud and forgery 5·8%
Criminal damage 3·0%
Other indictable offences 2·8%

D Causes of death

Infectious diseases 1·0%
Cancers 47·1%
Endocrine and metabolic diseases 1·3%
Diseases of the nervous system 2·6%
Diseases of the circulation 25·0%
Respiratory diseases 7·0%
Diseases of the digestion 3·0%
Urogenital diseases 2·1%
Other medical causes 2·3%
Accidents, poisonings and violence 8·6%

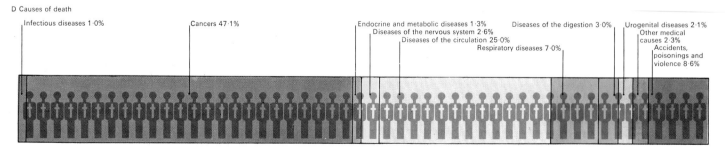

Ageing and longevity

The rate at which human beings age has not changed since prehistory. Civilized man lives longer [2, 4] because his chance of survival to the age of natural death has improved, not because ageing itself has been postponed. Ageing is the term for the process that causes organisms to become more liable to die the longer they live. With time, the capacity for self-repair and resistance to damage decreases and the force of mortality increases [Key] at a rate characteristic of each species, leading to a characteristic life-span.

Longevity, size and inheritance

The life-spans of animals [1] tend to correlate both with their size and, more closely, with excess brain capacity (as calculated by the information capacity carried in the nervous system) in relation to size. The human life-span is often quoted as the biblical "three score years and ten", although certified ages up to 112 are documented.

In several localities there is a popular belief, recently widely publicized, that people live to great ages. The best known of these are Hunza-land, Abkhaz, in the Georgian Republic and Vilcabamba in Ecuador. Abkhazian life-spans, some of them up to 140 or 160 years and more, have been much investigated and much disputed. They can neither be accepted nor rejected out of hand. The Vilcabamban records, better authenticated than most, have been under study. There are claims of high ages elsewhere chiefly in mountain areas where they are linked with "simple life", isolation from disease and an absence of reliable documents.

If such longevity is true, the causes are likely to be in the first place genetic, with social and possibly dietary factors playing a part. Longevity in humans is heritable, but less so than is stature. What is inherited is not so much longevity as the absence of factors for "shortevity". In most known genetic systems, hybrid vigour (the inheritance of a varied genetic repertoire) outranks any other heritable factor in increasing life-span.

Most multicellular organisms age, the probable exceptions being those, such as large trees and sea anemones, in which all of the component cells propagate clonally by splitting into two exactly similar cells. Ageing occurs in mixed organisms such as mammals, which contain both renewable and unrenewable cells, and in organisms (such as certain worms and adult insects) in which no cell division is possible.

The process of ageing

The nature of the ageing process or processes is unknown. One question to be settled is whether in man and other mammals the accumulation of damage is sited primarily in non-dividing cells such as neurons; in dividing cells, which may develop faults with successive divisions; in structural materials such as collagen (a protein in fibrous connective tissue); or in the programming of our physiology. No one of these processes can yet be pinpointed as the "cause" of ageing.

Fixed cells accumulate both viruses and waste materials and are irreplaceably lost with time, although the widely cited notion of massive brain cell loss with ageing is erroneous. Dividing cells accumulate mutations and errors but, except when they escape from normal controls, as in cancer, countervailing mechanisms operate to censor

1 The only mammals comparable in longevity to man are the elephant and the rhinoceros. Small rodents live one year in the wild but may survive two to four years in captivity. Dogs seldom reach more than 20 years, cats possibly longer. About 30 years is the maximum age for most medium or large mammals. Whales are the only possible exception, although the evidence is confused by the difficulty of interpreting the zones laid down in the wax earplug. If, as in tree trunk rings, these are annual, ages of 80 years may be reached by some species. Birds live longer than mammals. Small species, normally annual in the wild, may live 12–20 years in captivity. Large flying birds (parrots, geese, ravens) may be capable of living 60–70 years. The fish with the longest potential lives are sturgeon (100 years). Animals are listed with their known life-spans in years. [1] seahorse 1–2; [2] mouse 3; [3] guppy 5; [4] guinea pig 7+; [5] queen bee 7+; [6] silverfish 7; [7] large beetles 5–10; [8] earthworm 5–10; [9] swallow 9; [10] bats 10–15; [11] sponge 15; [12] rabbit 12; [13] sheep 15+; [14] frog 12-20; [15] starling 19; [16] giant spider 20; [17] dog 24+; [18] seal 20-25; [19] cat 27+; [20] cow 30+; [21] snakes, lizards 25–30+; [22] oyster catcher 27; [23] pigeon 35; [24] lion 30-35; [25] newt 35; [26] toad 36; [27] zebra 38+; [28] chimpanzee 39+; [29] ostrich 30–40; [30] horse 40+; [31] hippopotamus 54; [32] carp 50+; [33] lobster 50; [34] pelican 40+; [35] mussel 50–100; [36] goose 47+; [37] crocodile 50-60; [38] sea anemone 60–70; [39] Indian elephant 77+; [40] cockatoo 70–85; [41] golden eagle 80; [42] sturgeon 80–100; [43] man 112+; [44] vulture 117; [45] big tortoise 100-150.

mismade cells. It was once believed, from faulty experiments of the French biologist and surgeon Alexis Carrel (1873–1944), that cells grown outside the body were "immortal", being capable of indefinite division. Recent work indicates that this is not so, but that normally they are able to undergo only some 50–60 doublings before the accumulation of defects in their chemical machinery kills them. Major changes also occur with time in the body's immune defence mechanisms. The probability here is that divergence in cell structure causes some cells not to be recognized as "self" and so they are destroyed.

Ageing in most animals can be readily modified by the simple process of limiting calorific intake [5]. This limitation need not be so severe as to affect growth; feeding rats two days in three has little effect on growth but increases life-span by 60 per cent or more. Such observations suggest that the rate of ageing in man could be retarded.

Since mechanisms in the brain probably monitor dietary intake and serve as controllers of cellular ageing rates, the study of

these, rather than of lifelong dietary modification, seems a likely course for research aimed at making it "take 70 years to reach 60". Human life can probably be prolonged little by overcoming single diseases, with the possible exception of arterial disease; a man dying at 40 usually has one cause of death, a man dying at 90 may have 15.

Towards longer life

Lifetime dietary or other experiments on man or long-lived animals could be useful but this kind of experiment is in practice confined to rats and mice. High priority in research on ageing is therefore given to tests of short duration. Such tests were first applied to Hiroshima survivors in an attempt to find whether ageing is accelerated by radiation. Some animal studies designed to discover whether anti-radiation drugs delay ageing have prolonged life in mice, but it is likely that changes in food intake or liver chemistry have played a part. A technique to alter the human ageing rate is however almost certainly attainable [3], and much research is being done under the title of gerontology.

KEY

"You start dying the moment you are born" is a favourite saying caught in this

artist's impression of man assailed by death. If we dodge the rocks, arrows,

blunderbuss or rifle of death, the bridge itself stops halfway across the river.

3 Many interlinked factors probably contribute to the ageing process. Certainly no single common factor has been isolated. Consequently the re-

search that is being carried out in several countries is aimed at testing many different areas that are thought to be implicated. Listed below are some

of the most important. Generally tests are carried out on such animals as mice and positive results that can be applied to humans are years away.

Agents	Effects	Examples
3 Antioxidants	Prevent cell damage by groups of atoms called free radicals. Many antioxidants affect appetite (so cutting down food intake) and liver enzymes	BHT (butylated hydroxytoluene), selenium, ETQ (ethoxyquine)
Radioprotectants	Prevent chemical changes that are similar to those produced by radiation	SH (sulphydryl) groups
Immunosuppressants	Prevent the body's immune system from declaring war on the body itself	Azathioprine (Imuran: used in prevention of organ transplant rejection), ALS
Cell or lysosome stabilizers	Prevent cell damage caused by enzyme leakage within the cell itself	Prednisolone (synthetic steroid, induces same effects as adrenal gland), aspirin
Enzyme inducers	Cause the liver to produce enzymes that affect the body's metabolism	DDT, phenobarbitone, BHT, Dilantin
Antiappetants	Reduce food intake and thus restrict calories	BHT
Hormones	No one hormone controls ageing, but many are able to affect it via different physiological processes	
Patent medicines	Most are wholly inactive although high claims are made	Gland extracts, embryonic cells, queen bee jelly, etc

2 Human survival curves show the effect of improvements in living conditions on the lengths of time that people could expect to live in different cultures and at different times. The shape of the curve

is revealing. A rapid plunge at the start, as in Stone Age man (an estimate), Mexico (1930) and India (1930) indicates high infant mortality. The top line is the theoretical optimum, a situation in which almost

everyone lives to be about 70 years old. The curves for modern industrial nations are approaching this point. Current research on ageing is aimed at achieving the optimum, where active life is pro-

longed to the limit. The object is not to produce a population that will live to the age of 120, but to develop one that will retain its vigour up to a natural life-span of about 80 years.

Theoretical
England 1900
USA 1930
England 1970
USA 1900
England 1930
USA 1970
Japan 1930
Mexico 1930
India 1930
Stone Age

4 Life expectancy at birth has risen markedly in industrial countries in the last 100 years, due to better nutrition, advances in medical science, smaller families and improved health and hygiene measures. Total births of boys normally exceed those of girls by about 6 per cent. But stillbirth and mortality rates at almost all ages are higher for men, so typically women live longer and outnumber men in the total population (by 106 to 100 in 1971 in the UK).

Females
Males
War periods

England and Wales

5 Overeating seems to be one cause of ageing, judging by a series of experiments carried out on rats. One group was raised on a restricted but nutritionally adequate diet, others from the same litter were fed normally. After 24 months the normally fed ones [A] had effectively reached old age after only two-thirds of their lives. The "retarded" ones [B] when re-fed were as fit at 39 months as the normal rats were at 24. The survival curves reflect this difference.

Females
1 Reduced feeding
2 Normal feeding
Males
3 Reduced feeding
4 Normal feeding

177

An active old age

In pre-industrial societies the old were considered to be repositories of wisdom. Because they had seen it all before, and had practical experience of how recurring situations had been effectively dealt with, they were a valued and valuable resource to the community. In societies characterized by large, extended families of mixed ages the old had vital functions, not only as guardians of experience and lore but also as those who taught children and watched livestock.

Cultural prejudice towards the aged
In many contemporary societies families are smaller, their range of activities has contracted and jobs have become more specialized. Money has taken the place of skills and general functions have been usurped by specialists. Increasingly, in societies that have made youth an obsession, the old have found that ordinary dignities and privileges as equal human beings have been denied them. In the East both life and death are considered to be part of the whole life cycle and so the older a person gets, the wiser he is thought to be [4]: in the West attitudes

are different. The tendency, particularly since the 1960s, has been to downgrade the importance of experience, to question the authority of those with seniority in professional or political life and to inflate the value of youthfulness.

While physical changes in ageing, such as greying, wrinkling and muscular weakness, are undeniable [1, 2], much of the picture of "old age" in Western cultures is based on social attitudes. Old age is a role imposed by a convention that assumes "the old" to be infirm, unemployable, ineducable, asexual and dependent. All of these assumptions depend on old people playing the expected role. Too often people equate retirement from a job with retirement from life and treat others and themselves accordingly. But many of the supposed disadvantages of age have been shown to be imaginary [3] or to arise from the expectation that they will occur. Although old people are not so readily aroused, sexual function persists lifelong in normal individuals of both sexes. The rate of learning a foreign language for the first time is identical in 15- and 80-year-old subjects if

those with impaired brain circulation or other overt diseases are excluded. The "dependency" of the old, like the deficiency once ascribed to black people or to women, is a product of folklore and prejudice, not a fact.

The existence of a large population of aged is really a twentieth-century problem, since until recently most people did not live past 65. In America, for example, only three per cent of the population was over 65 in 1900; today the figure is ten per cent and it may soon approach 20 per cent. This is a pattern that is evident in all societies with zero population growth and it is now precisely these societies that exclude older persons from full participation and cultivate negative attitudes to their usefulness and worth.

Planning for the aged
There are some signs that the body of prejudice towards older people may change before long. Even without the probability that the rate of ageing can eventually be slowed, so that it takes, say, 70 years to reach the equivalent of 60, society is beginning to move towards the acceptance of two succes-

1 **Certain physical changes** are characteristic of the ageing process. Hair greys or whitens as a result of hereditary factors. On the head, both men and women start to lose hair, although the process is more marked in men. Facial hair may increase, especially in women, but body hair thins or disappears. Height is reduced as spinal discs atrophy, leading to a sagging posture. Vision weakens at about 45 and a white or grey semicircle may later develop around the iris. Changes in the inner ear cause a loss of high-tone hearing. Smell and taste also become less sensitive. In men, a bulbous nose may develop due to faulty gland action. Dilation of blood vessels gives a spidery effect at the root of the nose. Teeth that are not looked after will be lost, leading to shrinkage of the jaw line. Under the chin and in the breasts, soft skin sags as a result of loss of the elastic protein collagen. The top layer of skin thins out all over the body so that blood vessels stand out, especially on the wrists. Skin also wrinkles and becomes discoloured. Senile freckles on the back of the hands and warts or raised red blood vessels on the body are common. Hand grip weakens and women may develop a swelling of the top finger joints.

The heart is less able to respond to extra work as its valves grow more rigid and the body is more prone to illness because blood-making capacity is reduced. Artery walls thicken. Lungs function less well due to reduced respiratory muscle power and distortion of the rib cage. The lungs' air spaces enlarge and breathlessness increases. Cellular units in the nervous system do not reproduce, so that nerve cells decline in number and the brain may lose bulk. The functions of the gastric juices, kidneys, spleen and pancreas deteriorate in effectiveness. Ovaries cease to function at the menopause. In men the testes atro-

phy, although sperm formation may continue to a late age. The prostate, a male gland, enlarges, tending to cause bladder trouble. Muscles lose bulk and become weaker. Bones grow lighter and more porous. Weight-bearing joints, particularly knees and ankles, may swell and become deformed. Osteoarthritis, a degenerative joint disease, is also common in spine and hips. Stiffness and loss of flexibility are increased by calcification in cartilage and ligaments. The shins may become tender and irritated as a result of the dryness of the skin that comes with old age. Corns and callouses grow thicker.

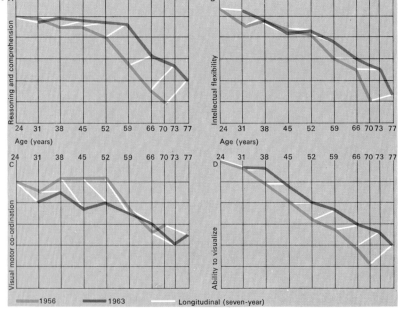

2 **Behind the outward signs of ageing,** several physiological processes [A, B] start to decline.

3 **The popular belief** that intelligence declines with age is challenged by recent research. When adults of various ages were tested at the same point in time the aged fared worse [red and blue lines]. But generational differences may account for this because tests at different times on the same people revealed another picture [white lines]. In reasoning and comprehension [A], intellectual flexibility [B] and ability to organize visual material [D] the same adults did better in 1963 than they had in 1956. Only their visual motor co-ordination [C] declined.

1956   1963   Longitudinal (seven-year)

sive life-styles, with a renewal of education in middle and late life and a recognition of "senior citizens" as a political force.

Many people of both sexes already undergo an identity crisis very similar to a second adolescence at about the age of 45–50 in which they reassess their goals and achievements. This is because they can expect to live to the age of 70 to 80. Our ancestors dreamed of rejuvenation [5, 6,] but seldom lived long beyond 70. With the dissipation of myths about the inability of older people to learn or acquire new skills, senior education aimed at updating skills may be given higher priority. Destructive retirement practices which expect the retired to be idle, and legislation penalizing pensioners who earn, may come under increasing attack.

Ageing does involve an increase in the incidence of illness, which may impair learning power and efficiency. Planning to meet the needs of old people must accommodate both the vigorous who want to remain fully engaged and a small but increasing proportion of those unable to do so, who need social and medical support. A similar range of options is needed in housing; some older people like to live in a mixed community, others with their contemporaries. The aim of social planning should be to make society "age blind" as well as colour blind so that status as a person is no more affected by age than by sex or race. While age inevitably lessens physical efficiency it does not as inevitably bring helplessness, dependence, foolishness or senility in its train.

Psychological problems

Of all the psychological problems that face the aged, the most difficult are perhaps isolation and the need for adaptability. In addition to physical handicaps and the diminishing number of social roles that are commonly available to the old, the aged have to cope with loss of the familiar – career, standard of living, surroundings and, above all, people. The gloomy search through the obituaries for yet another friend who has gone is bad enough; the death of a spouse is often a blow from which the aged do not recover. The ability to come to terms with these changes is vitally important.

Creative power in old age has been demonstrated by many musicians, writers and artists, including Pablo Picasso (1881–1973). His longevity as a painter was surpassed only by that of Titian (c. 1487–1576), who produced some of his finest work in his 80s. Arturo Toscanini (1867–1957) was a vigorous conductor until the age of 87 while at the same age, Konrad Adenauer (1876–1967) was Chancellor of West Germany.

4 A Taoist believes that life can be prolonged by attaining the state of *hsien*. This can be reached by a combination of respiratory, dietary, sexual and gymnastic techniques. Man must inhale deeply when the earth breathes in, during the day. Grain and meat are impure so consumption is regulated and while sexual intercourse is not forbidden, it involves a loss of vital body fluid.

5 "The Fountain of Youth", by Lucas Cranach (1472–1553), is one of the many works that deal with the theme of rejuvenation. The idea that man could restore his youth by bathing in a magical river or spring occurs in many cultures. The Hindu Pool of Youth (700 BC) and the Hebrew River of Immortality are among the earliest examples. Alexander the Great supposedly searched for such a spring and Juan Ponce de Leon (1460–1521) was seeking it when he sighted the Florida peninsula unexpectedly in 1513. Rejuvenation was not thought impossible as snakes seemed to be reborn by shedding their skin and the mythical phoenix was reborn from fire.

6 Satirizing hopes of rejuvenation, an eighteenth-century artist envisaged a windmill that could grind up old women and reconstitute them as ladies of fashion. The most serious attempts to find an elixir of youth were made by alchemists who sought to gain power over nature through science. One of the basic themes of their search was that man could become eternal through association with things eternal. It was this theory that led Chinese alchemists to recommend that food should be eaten off golden plates. For in this way the patient would take in a little of the metal each time he ate and since gold was incorruptible he would gradually achieve a higher state of being.

7 Plastic surgery is a drastic means of "rejuvenating" the face that has grown in favour in recent years. Sagging and wrinkled skin, brought about by a decline in the elasticity of muscles, is stretched back from the face and neck by a surgeon, who then removes the excess and stitches the gap beneath the hairline. The process leaves the skin tauter but often has to be repeated.

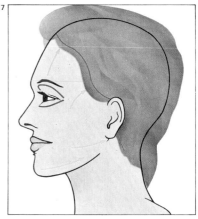

Death, grieving and loss

From the moment of birth each of us begins to develop a model of the world in our imagination. This model includes everything that we know, or think we know, of the world around us; it includes our own bodies and minds insofar as we can view these ourselves; it includes the things and the people we know, and our plans, expectations and hopes [1]. Sometimes, events will occur that invalidate a part or parts of our world model – the unexpected loss of a loved one, for example – and we may be faced with the need to abandon many of our assumptions and painfully to rebuild the world it has taken so long to create. The experience of a change of this kind has been termed a "psychosocial transition" or crisis period.

Reactions to change and loss

Periods of crisis are both time and energy consuming and they follow a more or less consistent sequence [2]. At first, particularly if the change is sudden or massive, there is likely to be a stage of shock, denial or disbelief. The individual is unable to take in the reality of what is happening and tends to behave as if no change were occurring at all.

Before long, however, realization begins to dawn and there is a period of striving in the course of which the person bitterly tries to recover the world that he or she is losing and to preserve the old model. Repeated frustration causes him or her to abandon the struggle and a period of disillusionment, apathy and despair sets in. Finally, little by little, the appetites of life return and new beginnings are made so that the final phase is one of reorganization and recovery.

Death and bereavement

The most devastating and overwhelming type of change is possibly that which occurs when a previously healthy individual develops a fatal illness. Death implies so radical a change that not only is the dying person likely to avoid facing the prospect of his or her own death but everyone around tends to collude in the pretence that the illness will disappear. Remarkably, given emotional support and the relief of pain and other symptoms, many people do eventually express a wish to be told the full facts of their illness. They will often

arrive at a stage of acceptance in which they fully realize the implications of the illness and can reorganize and enjoy to the full the life that remains [6, 7].

Coping with death implies the possession of a philosophy of life that includes death as a valid part of the world model. In modern Western society this is rare and death has taken the place of sex as a taboo topic. Consequently the dying and the bereaved are sometimes deprived of human relationships at the time when they need them most.

It would be fair to assume that once a person has died his or her troubles are over. For those who remain, however, a fresh stage of adjustment may be just beginning. They may be helped or hindered in this adjustment by the rituals and social events that, in all societies, mark the transition from life to death. The survivor is faced with the need to undertake a major change in his or her model of the world. It is hard to deny the fact of death for long and soon the phase of shock is replaced by the pangs of grief – characterized by intense pining.

Some people report an illusory sense of

CONNECTIONS

See also
204 Types of ritual
212 Myths of autumn
214 Myths of winter

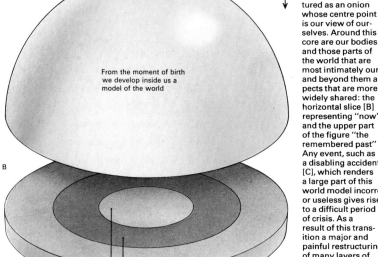

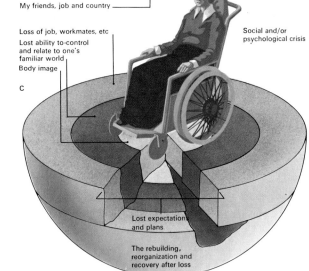

1 A

Time

B

C

From the moment of birth we develop inside us a model of the world

My body
My wife, car and house
My friends, job and country

Loss of job, workmates, etc
Lost ability to control and relate to one's familiar world
Body image

Social and/or psychological crisis

Lost expectations and plans

The rebuilding, reorganization and recovery after loss

1 The model of the world that we contain within us can be pictured as an onion whose centre point is our view of ourselves. Around this core are our bodies and those parts of the world that are most intimately ours, and beyond them aspects that are more widely shared: the horizontal slice [B] representing "now" and the upper part of the figure "the remembered past" [A]. Any event, such as a disabling accident [C], which renders a large part of this world model incorrect or useless gives rise to a difficult period of crisis. As a result of this transition a major and painful restructuring of many layers of the onion is necessary.

2 Death, the most fundamental period of transition in our lives, is a crisis that many people in Western society face unprepared. This lack of preparation is reflected in the way that, all too often, relatives, doctors and nurses try to hide the fact of death and maintain the pretence that a dying person is not very ill [A]. When a person dies those closest to him pass through the grief [B] which for a while will disorganize their lives and make it hard for them to support each other. If they face up to their grief and receive encouragement and understanding from others they will be able to cope better with the successive losses and griefs of old age [C].

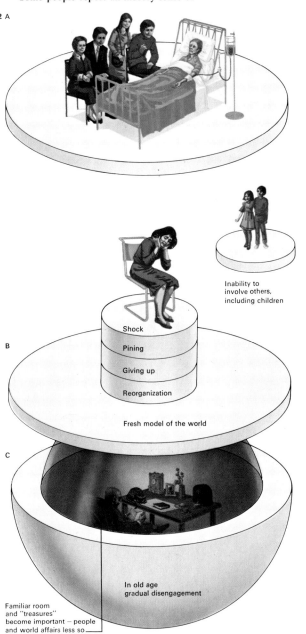

2 A

B

C

Shock
Pining
Giving up
Reorganization

Fresh model of the world

Inability to involve others, including children

In old age gradual disengagement

Familiar room and "treasures" become important – people and world affairs less so

the continuing presence of the dead person that may be retained for many years, but in time the facts are accepted and, if all goes well, the bereaved person discovers a fresh identity. In other types of loss, such as that following divorce, there may be a similar sequence of reactions but these are more likely to be complicated by anger and self-reproach.

Disablement and old age
Illness and accident may give rise to physical disablement. The patient is then faced with the need to grieve not only for the lost ability but also for those aspects of life that went with it. Thus the person who loses a limb grieves for the job that he or she can no longer carry out, for the lost sports and other activities and for the future prospects that seem to have disappeared.

Disabled people need time and emotional support if they are to come through the grief that must be expressed for the real losses they have undergone. They may be reluctant to learn the skills appropriate to their disability until this process of realization is complete.

Old people tend to suffer a succession of bereavements and disablements. They often attempt to cope with these by narrowing their horizons and disengaging themselves from many of the people around them. But at the same time they become vulnerable to changes in the small environment that is all they have left. It is important to recognize that an old person's room and its familiar clutter of objects are much more important than they would be to a younger person.

Periods of grief and loss are times of danger but they are also times of opportunity. Those who come through them may emerge more mature and secure than before. Those who succumb by excessive avoidance or by "caving in" – may find future losses even more hard to take. The loving support and care of close friends and relatives remains the most valuable consolation available to those facing death or loss. Modern society suffers from an increasing embarrassment about the expression of feeling. Spontaneous gestures of affection are the gifts sometimes hardest to give but most valued of all because they tell the bereaved that he or she is not alone.

The expression of grief in the human face reflects a deep-seated need to cry. This has a "signal" function and evokes support in others. But knowing that it is useless to search for the lost person, efforts are often made to inhibit grief, with varying success and inner tension.

3 The rituals attending death have usually been seen as support to the dead, but most of them also have a psychological function for the living. The provision of food, drink and familiar possessions, as in this Bronze Age Danish burial, comfort the survivors as well.

4 In the former Indian tradition of "suttee" the widow of the dead man joined her husband on the funeral pyre. In this culture she was treated as one of the familiar possessions that accompanied him into the next world. It was also assumed, perhaps, that life for the survivor would, or certainly should, be intolerable.

5 Societies vary greatly in the ritual expression of mourning. In this Chinese funeral the mourners wear white and the colourful coffin seems to reflect the hope of a contented future for the deceased. The sarcophagus bears the inscription "good luck" in Chinese characters. But this does not prevent grief from being expressed and professional wailers may be employed by the family to give social sanction to the shedding of tears among relatives and friends.

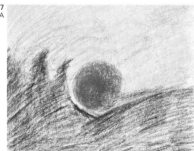

6 In facing the prospect of his or her own death a person can come through the grief and become free to enjoy the life that remains to him or her. This woman, knowing full well that she was within a few days of death, was yet able to enjoy a ride in the garden of a hospital.

7 These pictures were drawn by a woman with terminal cancer. Her feelings when she first entered hospital [A] contrast sharply with her outlook a few weeks later [B]. The pictures reflect a remarkable transformation from all-consuming pain to a balanced, peaceful frame of mind, even though staff made no attempt to conceal from this patient the facts of her fatal illness.

Questions of life and death

Man has always asked questions about what he ought to do, where his duty lies and what rights he possesses or grants to others. He is a moral being because he cannot evade choices. Since these choices are often difficult to make, man has evolved systems of principles or morals – usually as expressed by a religion or philosophy – in order to help him make his choices and measure his actions. In this way he has some concrete and shared idea of what good and evil mean.

The question remains, particularly in a largely secular society, as to whether these principles are absolute – given, as it were, by some external and superior force – or whether they are man-made. Fyodor Dostoevsky (1821–81) asserted that "If God does not exist, everything is permitted", but it may be that our principles represent a consensus of rules that make social life possible.

Attitudes to moral systems can be classified broadly into two types. Moral principles (for example, "Thou shalt not kill") can be formulated and then applied whatever the consequences and regardless of the specific individual case. (Few if any values, however,

have been so applied – men have been encouraged by many churches to kill in war, for instance.) In contrast to this "absolutist" approach to morality, a "situationist" or "utilitarian" approach to moral values is more usual today, whereby the particular circumstances of each case are considered.

Science and morality

There is nothing really new about any present-day moral problems, but they are now perhaps more pressing than in previous ages because modern science has greatly increased the scope and range of the consequences of man's decisions.

This is most obvious in modern warfare. Combatants are now far removed from each other and, in the process, distanced from the consequences of their actions. War is also no longer simply a combat between armies; the humane distinction between combatants and non-combatants has been obliterated. This raises the question of whether there can be any "innocent bystanders" in modern conflicts. War has become total, limited only by the fear of universal destruction.

The problem is at its most extreme with the atomic and hydrogen bombs [1]. Many people have questioned the development and production of these bombs, let alone their use, both because their destructiveness outweighs any justification for their use and because of the probable long-term effects of radiation. Does the end of "saving" lives justify the horrific means of attaining that saving? Even the peaceful use of nuclear power carries the risk of lethal pollution of the atmosphere for many generations. Behind the practical problems of using it peacefully is the moral one of whether it is possible to balance the benefits, however great, against the risk, however small and remote it may in fact be.

Dilemmas in medicine

Advances in medical science pose problems in unusual guises, from the morality of using animals in experiments [4] to the question of exactly when "spare parts" for transplant surgery [3] may be removed from "dead" donors in the interests of another human life.

Another area of medical progress –

1 Atomic weapons raise serious moral problems for man. In 1945, the first two atomic bombs totally destroyed Hiroshima and Nagasaki in Japan. The US president, Harry S. Truman, explained that if the bombs had not been dropped, the war would have dragged on for 18 months longer at a probable cost of two million Japanese and one million American lives. Ending war swiftly and saving lives are unarguably moral intentions for a head of state, but this situation may have been a prime example of the way moral dilemmas defy reduction to practical terms. The "equation of suffering" on which Truman based his argument does not necessarily provide a moral justification for his action.

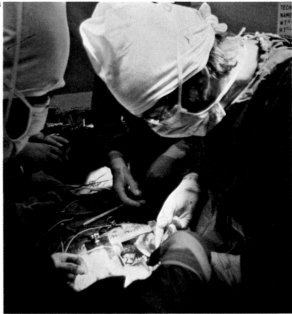

2 The use of gas in World War I was abandoned not for moral reasons, but because it proved to be an unreliable weapon. Other chemical methods have since been used, although germ warfare has not. In many instances, although by no means all, there seem, fortunately, to be "natural sanctions" that set limits to human behaviour.

3 The first heart transplant was made by Christiaan Barnard (1922–) in 1966. Although a great medical feat, the problem remains whether large sums and rare skills should be used to save a few lives while less spectacular operations are neglected. However, without such pioneering operations medical knowledge cannot advance.

increased control over reproduction – has given new dimensions to one of the most intractable of moral questions: who has the right to decide in matters of life and death? The broad question of social responsibility and individual rights is also relevant to another area of birth control – "genetic engineering" and eugenics. For the scientist could eventually be faced with the morality of "improving the breed" at the expense of a couple's right to produce children.

The morality of preventing life – by sterilization, prohibition of reproduction on medical grounds, or abortion [5] – is closely linked with that of taking life. Voluntary euthanasia (mercy killing), for example, in the face of intolerable and incurable pain or because life has been reduced to vegetable existence, has been claimed as a fundamental human right. In such cases the conflicting moral arguments of the "absolutists" and "relativists" are thrown into sharpest relief.

Sometimes moral problems arise because the state wants to impose its will on the individual. Even if it stops short of putting down undesirables [Key], has society the right to order the sterilization of a mentally retarded girl whose children are likely to be handicapped? Should it reform habitual criminals by aversion therapy or, more radically, by a brain operation that makes them irrevocably docile even if the result is a pacified robot rather than a human being? The possibility of such methods being used by an unscrupulous or frightened state in order to control political dissidents is a daunting one.

The roots of morality
All moral traditions, both religious and secular, find such radical measures repugnant because these traditions make man the norm of morality and hold that one may not abuse the dignity of another person without harming oneself. The whole of humanity is, ultimately, interdependent and interrelated, and men are responsible for each other and for future generations. In borderline situations there will always be conflicts of duties, but love and respect for other human beings, truthfulness and honesty in relationships provide, if not final answers, a starting-point on which to base any moral system.

KEY
Capital punishment, here summarily administered in Manchu China, raises the fundamental question of whether life should be taken for any reason at all. Is it an effective deterrent? What sort of crimes should it be imposed for: treason, premeditated murder, acts of terrorism? Because the moral and practical arguments are so difficult and provoke such emotion most societies invest the power over life and death not in the individual but in "justice", "God" or "the people".

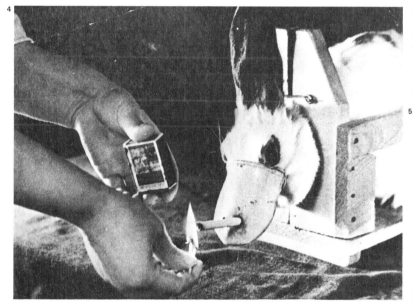

4 Animals are still widely used in medical research. The justification is that the lives and health of humans are more important than those of animals; that it is better to find out the ill effects of drugs or smoking on rabbits than on humans. Does this justify the torture of animals? The balance between benefit to man and intolerable harm to animals is difficult to establish morally.

5 Human characteristics are obvious in a 12-week-old fetus. Its genetic coding has already dictated the colour of its eyes and the pattern of its fingerprints, but it is still totally dependent on its mother for oxygen and nourishment. These conflicting factors invite opposing arguments: either that the fetus is simply a part of the mother's body; or that the fetus is to be considered as a human life. Abortion thus becomes, by the first argument, justified if used to preserve the health of the mother or prevent the suffering of the child; and by the second, an indefensible act of murder.

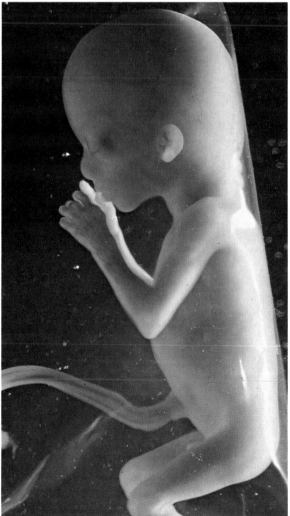

6 Law and order are necessary in all societies, but force should, in theory, be used only to prevent a major breach of the peace. The aim of these riot policemen, once they have to be called in, should be to disperse the crowd without causing injury. The right of assembly and peaceful demonstration is basic to all free societies and the erosion of that right one of the first signs of repression.

The idea of the person

One of the distinguishing marks of humanity is that whereas other living creatures merely act, man can reflect on his ability to act. He is (after he has ceased to be an infant) a creature capable of self-consciousness and since each individual has his own consciousness, each one is unique, not only in appearance and bodily structure, but also in possessing a continuous thread of conscious being which is not shared in its entirety with any other human being.

Self-consciousness

In the Bible story of the Garden of Eden [1] the acquisition by man of self-consciousness is symbolized by his eating the apple and falling from grace. The basis of this story is that self-consciousness is far from being an unmixed blessing, since it makes man liable to be self-conscious on occasions when it would be more enjoyable to be spontaneous, and liable to be led by his own individual interests into actions that others may condemn and that may make him feel guilty. As a result, man's capacity for self-awareness also gives rise to morality and the emotions of

guilt and remorse, from which other animals appear to be free. Another facet of the story is the belief that man was once innocent, incapable of crime, and free from knowingness and guile; hence the religious idea of man's original sin and primal fall from grace. There is also the hope that at some time in the future, either in this or another world, man will be able to achieve a state of blessedness in which consciousness is purged of self-consciousness and selfishness.

Man's unique capacity to be conscious of himself and of his relation to the world around him, which enables him to know about and control his environment to an extent unparalleled by other animals, has resulted in an awareness not only of his individual being but also of the fact that this being is only transient. Hence man's unique interest in his past, which leads him to create myths and write histories, and in his future, which leads him to create memorials that will, he hopes, survive death [4]. He is also driven to pursue fame and to crave descendants who will carry on his line and his name into the remote future. Religion, history, patriotism,

nationalism, ancestor-worship, time-honoured ceremonials – all seem to derive at least some of their vitality and appeal from the fact that they help to decrease man's disquieting awareness of the fact that he personally is not immortal and must face his own death.

Loneliness and self-awareness

Since each individual's consciousness is unique and unshared in its entirety with anyone else, each man's recognition of himself as an individual carries with it the knowledge that he can be lonely. Even if he has companions now with whom he can share his experiences, circumstances could arise in which he would be alone and isolated [5]. The more any individual becomes aware of his own individuality and uniqueness, the more aware he will become of being alone and on his own in the world.

In our present civilization, in which education and culture set high value on the achievement of self-awareness and the development of individuality, and in which society is conceived of as a collection of separate individuals and not as an organic com-

1 The fall of man as seen by the German painter Lucas Cranach (1472–1553), begins when God sternly enjoins Adam and Eve not to eat of the Tree of Knowledge. When they do eat the forbidden fruit, after being tempted by the serpent, God expels them from the Garden of Eden. This story symbolizes man's loss of innocence, his discovery of shame, and his acquisition of a moral sense.

2 Man differs from other animals in being able to ponder on himself and recognize his reflection in a mirror. According to some psychologists, the invention of cheap mirrors (c. 1700) has had a profound effect on the consciousness of man by making him aware of himself as a visible object.

3 Man's ability to see himself as unique and separate from all others makes it possible for him to perform actions that cut across his natural bonds of affection. In the Bible this is symbolized in the story of the slaying of Abel by his envious brother Cain, who was the first tiller of the soil. This followed God's preference for Abel's offerings. In consequence, God put a mark on Cain's forehead to ensure that nobody would kill him and so shorten his punishment for taking away a life that God had given.

4 Man's awareness of his mortality impels him to create objects that will be remembered after his death. Those with wealth and power often build triumphal arches, palaces, pyramids, or cities by which they hope to be remembered forever. The Taj Mahal, shown here, is an example of this striving for immortality, but in one respect it is exceptional. It was built to immortalize not its builder, Shah Jahan (1592–1666), but his favourite wife, Mumtaz Mahal. The mausoleum took 20,000 men 11 years to complete.

munity, techniques of communication have become an important object of scientific study. For example, the state of being in love [6], in which the boundaries between selves seem to be miraculously dissolved, has become the most eagerly desired form of experience. In other, less self-aware cultures people rarely fall in love and when they do it is held to be a threat to social life.

Many people feel that their self-awareness does not include the whole of themselves and that what they do is also directed by other mysterious forces. Sometimes these other forces are conceived to be inside the self but inaccessible to consciousness. These forces have in the past been described as spirits or demons; in contemporary psychological language they are regarded as instincts which impel the individual from within and which he can never completely know or control.

Man, on this view, is like the rider of a horse of which he has only imperfect knowledge and control. At times the rider and the horse may go along smoothly together, sometimes the rider may succeed in forcing the horse to go in a direction in which it is reluctant to go, but sometimes the rider may get carried away by the horse. In other words, sometimes we behave spontaneously and feel at one with ourselves, sometimes we are able to successfully practise self-control, and sometimes we are carried away and do things that we do not really want to do.

Fate or destiny

Where these other forces are conceived to be outside the self, they are either viewed in the abstract as fate [7] or the influence of the stars, or more personally as gods who directly interfere with and control the destiny of individual men. More rationally, the nature of man may be seen to be determined by historical, social forces which mould his nature and way of life from without [9].

One of the great divides is between those who believe that man is impelled by forces within himself and are attracted to psychological explanations, and those who believe that man is moulded by external circumstances and are attracted to historical and sociological explanations.

KEY

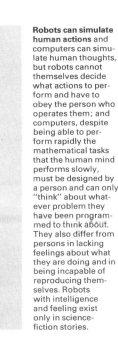

Robots can simulate human actions and computers can simulate human thoughts, but robots cannot themselves decide what actions to perform and have to obey the person who operates them; and computers, despite being able to perform rapidly the mathematical tasks that the human mind performs slowly, must be designed by a person and can only "think" about whatever problem they have been programmed to think about. They also differ from persons in lacking feelings about what they are doing and in being incapable of reproducing themselves. Robots with intelligence and feeling exist only in science-fiction stories.

5

5 "The Ancient Mariner" painted by Gustave Doré (1833–83) depicts loneliness – the awareness of isolation from others. But in this case it is not the loneliness of someone whom chance has separated from his fellow men, but of someone who has committed a crime for which he believes there can be no forgiveness – the slaying of the albatross.

6 When human beings believe themselves to be in love, they lose their sense of isolation which comes with individual consciousness. The boundaries between the two individuals who are in love are dissolved and yet a feeling of completeness remains. Modern Western man is unique in assuming romantic love to be a common experience.

6

7

7 The Moirai or Fates of Greek mythology presided over the birth, destiny and death of man. Clotho spins the thread, Lachesis decides how long it is to be, and Atropos cuts it. Even the gods could not alter such destiny.

8

8 Innocence, supposedly childhood's greatest gift, is associated with lack of knowledge (particularly, in the Victorian mind, sexual knowledge), and lack of self-consciousness and guile. It should perhaps be equated with more positive values and attitudes.

9 Jean-Jacques Rousseau (1712–78), French philosopher and novelist, believed that man is born free and good but is corrupted by the artificiality of society. His political ideas encouraged the French Revolution and his educational ideas still influence teachers.

9

185

Aspects of human nature

There are two contrasting ways of looking at human nature [Key]. One conceives man to be impelled by forces, drives and instincts acting on him from within; the other conceives him to be moulded by environmental, historical and economic forces impinging on him from without. Philosophy, psychology, history and sociology are to a large measure attempts to define precisely what these internal and external forces are.

Aspirations and instincts

Philosophers and psychologists who have attempted to define the internal forces that actuate man can themselves be divided conveniently into two groups: those who believe man has innate aspirations and those who believe that he has inborn instincts. The former, Platonists or idealists, hold that man is born with innate ideas or ideals and that his prime motive and reward in life is to realize these ideals [1, 2, 7]. The latter, Aristotelians, materialists or realists, hold that man is born with innate cravings, passions or instincts that demand enactment or satisfaction. According to the former, there exists somewhere, perhaps in the mind of God, a perfect Idea of Goodness or Beauty or Truth, which each individual has an innate aspiration to become or achieve. According to the latter, the physical nature of man impels him to behave in ways that ensure his survival and the continuation of the race – these impulses usually being called instincts.

Although these two ways of looking at human nature are most easily stated as though they were incompatible opposites, so-called dualist philosophies attempt to reconcile them by asserting, for instance, that the spiritual part of human nature aspires towards realization of the ideal, while the physical part is driven by instinct and the pursuit of pleasure.

However, since the Renaissance, and more particularly since Charles Darwin (1809–82) and Sigmund Freud (1856–1939), this dualistic view of human nature has been abandoned by the non-religious in favour of the rationalist view that even the "spiritual" aspects of human nature have evolved from man's instincts and are, in the last resort, pleasure-seeking and not the effects of divine purpose and grace. Freudian psychology [5], which asserts that all "higher" mental activities are functions and "sublimations" of infantile sexual and aggressive drives, is the outstanding modern example of a theory of human nature based on the assumption that man is ultimately a pleasure-seeking organism. The idealist view of human nature is represented by the work of the psychologist Carl Jung (1875–1961).

Love and hate

Biologists and many psychologists assume the existence of two instincts or groups of instincts: the self-preservative [3] – hunger, aggression and fear – and the reproductive – sexual and maternal impulses. Some psychologists, including Freud, abandoned this straightforward classification in favour of the idea that the two basic instincts are love (or sex) and hate (or aggression).

An interesting sidelight has been shed on the view that man has innate ideals by the animal psychologists, or ethologists, such as Konrad Lorenz (1903–) and Nikolaas Tinbergen (1907–). They have produced evi-

1 Captain Lawrence Oates (1880–1912) walked out into the snow to certain death, so as not to be a burden to other members of Robert Scott's expedition to the South Pole (1911–12). His selfless act of heroism exemplifies the ideal of noble self-sacrifice – in Horace's words a "sweet and noble" thing. Despite Oates's heroism, the expedition ended in disaster and all the remaining members died in a fierce blizzard before reaching safety.

3 The fate of the 147 people who escaped on a raft from the shipwrecked French frigate *Meduse* when she sank in the Atlantic in 1816 epitomizes the triumph of the instinct of self-preservation. After 12 days only 15 remained to be saved (here portrayed by Théodore Géricault [1791–1824]). The rest had been thrown overboard or eaten by the strongest survivors. In such circumstances life becomes, in the words of the philosopher Thomas Hobbes (1588–1679), "nasty, brutish and short".

2 According to the ascetic ideal, not only the renunciation of all pleasures, but even the mortification of the flesh, by fasting and whipping, may be virtuous and pleasing to God.

4 In the Indian tradition, as shown by this temple sculpture, high value is placed by some sects on the body and on eroticism, in striking contrast to the ascetic, anti-erotic ideal which has until recently pervaded the Christian West.

dence of the existence in animals of at least one instinct that is group- or species-preservative and that has been associated with man's "finer feelings". This is the instinct, in some social species, for an individual animal to shelter or protect from attack the group of which it is a member, often at the cost of its own life.

Behaviour patterns
In some scientific circles the word "instinct" has become unfashionable, having been replaced by the concept of "behaviour patterns", some of which are innate and some of which are learned. According to "behaviourist" psychologists, such as Ivan Pavlov (1849–1936) and J. B. Watson (1878–1958), and learning theorists, such as B. F. Skinner (1904–) and H. J. Eysenck (1916–), human nature should be explained without using subjective concepts such as mind, or assuming the existence of inner forces of any kind, but solely in terms of observable patterns of behaviour.

Those philosophers, historians and economists who believe that human nature is

largely moulded by external forces can again be divided into two groups: those who stress man's physical environment and those who stress social forces [8]. The former emphasize the importance of geography and climate in moulding character and explain national differences by reference to them – maintaining, for instance, that the English character is partly determined by the fact that Britain is an island and has a variable but in no way extreme climate.

Those who stress social forces are either empiricists or ideologists: the first maintain that individuals or groups are moulded by the history of the particular society in which they have grown up; the second, the ideologists, include those Nazi historians who saw history in terms of the predestined rise of the Aryan race to world domination, and contemporary Marxists and communists who see history as a struggle between social classes, which will inevitably end in the victory of the working class. Sceptics and empiricists view these and other theories of history as attempts to impose order and pattern on events that have neither order nor pattern.

The two contrasting views of human nature are summarized here: the first emphasizing man's innate faculties – instincts and forces driving him from within; the second, the environmental forces of climate, geography and social structure acting on him from outside. According to idealists, man strives to achieve some moral or religious perfection, but according to materialists he strives rather to satisfy his innate instincts: to survive and to procreate.

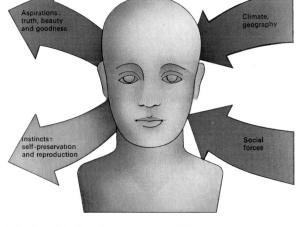

5 Psychoanalytical theory describes how the newborn child [A] does not differentiate between himself and the world. Then [B] he learns to differentiate himself from his parents and other central figures. With the development of sexual pleasure zones [C, D] the child passes through oral, anal and genital stages, followed by a period of latency before puberty and then the adult genital stage. The mind [E] is organized in three parts. The id contains primitive drives such as sex and hunger which the ego attempts to satisfy without incurring the world's hostility. The superego represents values and conscience instilled by parents and society.

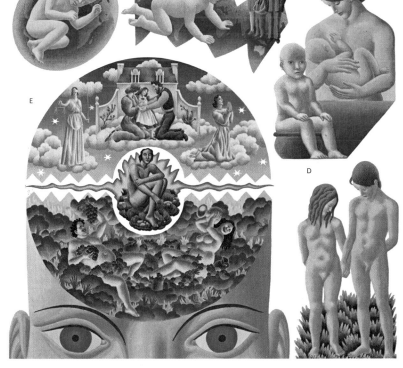

6 Konrad Lorenz has shown that most animals display submission signals that end continued aggression by a stronger animal against a rival of the same species. Man also has submission signals such as avoidance of eye contact. One explanation for the fact that man does kill his own species is that the development of weapons has diminished his ability both to show and to respond to such signals.

7 John Bunyan's Pilgrim's Progress (1678) is an allegory of spiritual progress. It tells of the pitfalls and battles that a Christian faces in his journey to the Celestial City.

8 Differences in class and income play an important part in forming personality. The "Two Nations", a phrase coined by Benjamin Disraeli (1804–81), have gradually grown closer together, particularly since World War II.

187

Classifying personality types

Although each individual is unique and in some essential respects unlike any other person, people do, of course, resemble one another in such characteristics as physique, character, temperament and susceptibility to disease. Since time immemorial attempts have been made to classify people into groups on the basis of these resemblances.

The four humours and modern versions
For centuries the most generally accepted classification of human beings was the "humoral" theory propounded by the Greek physician Galen (c. AD 130–c. 200) [1]. According to Galen there were four different types of temperament – sanguine, phlegmatic, choleric and melancholic – each being produced by the preponderance of one or other of the fluids or humours that could be found in the body. If blood predominated, the individual possessed a sanguine or optimistic temperament; if phlegm predominated, he was phlegmatic or unexcitable; if yellow bile predominated, the individual possessed a choleric temperament and was quick to anger; and if black bile predominated, he

had a melancholic (nowadays, depressive) temperament.

The humoral theory of temperament was generally believed in by both doctors and the general public until well into the nineteenth century and has left its mark in descriptive terms such as sanguine and choleric. Although no one today seriously credits that blood, phlegm and bile play any role in determining temperament, it has, nonetheless, had a curious revival in recent years. According to the British psychologist H. J. Eysenck (1916–), whose ideas are based on statistical research, human personality can be measured by two different yardsticks, one measuring stability-instability, the other measuring introversion-extraversion, the latter being terms made popular by the Swiss psychiatrist Carl Jung (1875–1961). Extraversion is the tendency to be "outgoing" and interested in what is going on outside oneself; introversion is the tendency towards introspection and interest in one's own responses to outside events.

If both yardsticks are used [3, 4], four groups of people emerge – unstable intro-

verts, unstable extraverts, stable introverts and stable extraverts. Here, says Eysenck, the four personality types correspond fairly precisely to Galen's four temperaments – unstable introverts resembling Galen's melancholics, unstable extraverts resembling his cholerics, stable introverts resembling his phlegmatics and stable extraverts being sanguine. Further research by Eysenck suggests that most neurotics are melancholic unstable introverts, that most criminals are choleric unstable extraverts and that most healthy citizens have natures that are either phlegmatic or sanguine.

The remarkable aspect of this system of classification is the way in which it bridges an apparently unbridgeable gap between classical and medieval ideas and modern scientific thought. Eysenck's classification differs, however, in that Galen asserted that one was melancholic, or phlegmatic, or choleric, or sanguine while Eysenck proposes that a person can be two, but not three of these at the same time; for instance, if one is marginally unstable, one is either melancholic-phlegmatic or choleric-sanguine, depending

1 Character and temperament were for centuries thought to be the result of the four humours. These paintings present a medieval view of the sanguine [A], phlegmatic [B], choleric [C] and melancholic [D] man. Any imbalance between the humours leads to what are now called neuroses.

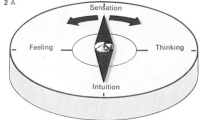
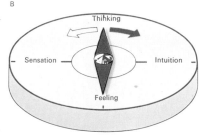

2 Carl Jung's compass of the psyche has four points. For a sensation type [A] intuition is least developed, for a thinking type [B] the feeling side is least in evidence. (Feeling means the ability to evaluate experience without need to analyse.) As there is overlap in everyone, one of the qualities on either side may be as highly developed as the dominant one.

4 The answers given to a series of questions on an Eysenck personality inventory help to locate the individual's place on the extravert-introvert and neurotic-stable dimensions.

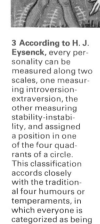
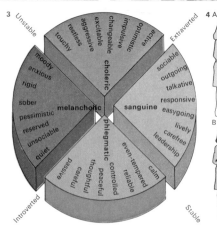

3 According to H. J. Eysenck, every personality can be measured along two scales, one measuring introversion-extraversion, the other measuring stability-instability, and assigned a position in one of the four quadrants of a circle. This classification accords closely with the traditional four humours or temperaments, in which everyone is categorized as being either sanguine, phlegmatic, choleric or melancholic.

Eysenck wheel labels:
Unstable — restless, aggressive, excitable, changeable, impulsive, optimistic, active (Choleric)
touchy, moody, anxious, rigid, sober, pessimistic, reserved, unsociable, quiet (Melancholic)
Extraverted — sociable, outgoing, talkative, responsive, easygoing, lively, carefree, leadership (Sanguine)
even-tempered, controlled, reliable, calm, peaceful, thoughtful, careful, passive (Phlegmatic)
Introverted / Stable

4 A 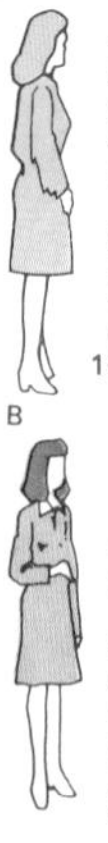
Extravert traits
1 Impulsiveness
2 Hyperactivity
3 Dominance
4 Attention-seeking
5 Verbal expressiveness
6 Gregariousness
7 Leadership
8 Friendliness
9 Active helpfulness
10 Conscientiousness

B
Introvert traits
1 Irritability
2 Emotional fluctuation
3 Suspiciousness
4 Self-consciousness
5 Withdrawal
6 Submissiveness
7 Depression
8 Calmness

on whether one is introverted or extraverted.

Although Eysenck's theory of personality types is based on massive research, it is not universally accepted, largely owing to the fact that it leaves unresolved the question as to whether personality is innate and unchanging or the result of individual development and experience. For if the latter were so, it would be possible for a person's personality type to change over the years in response to changed circumstances, so reducing the value of assigning anyone to a particular type.

Other personality types

The Eysenck-Galen theory of personality and temperament is far from being the only one. Jung, in his book *Psychological Types*, divided people into eight groups, four groups of extraverts and four of introverts, the four divisions being in each case thinking, feeling, sensation and intuition [2]. Here also the classification has limited value since Jung held that we are unconsciously of the opposite type to that of our consciousness and that the fully integrated or "individuated" person transcends types. Freud was not a classifier

but some of his followers have classified people according to the kind of mental illness their personality most resembles – and which they would presumably develop if they ever did fall mentally ill.

There is also an extensive medical literature that attempts to correlate physique, character and predisposition to both mental and physical illness, but its theories have as yet achieved little acceptance.

Reactions to classifications

People vary enormously in their emotional attitude to classifications of man since some people are fascinated by them and others find them morally offensive. Any classification that really worked would be of great practical value in medicine and vocational guidance. If it were proved, say, that tall, lightly built people were susceptible to one range of illnesses and were introverts, while short, heavily built people were prey to another range of illnesses and were extraverts – the German Ernst Kretschmer (1888–1964) did claim this [8] – the task of doctors and counsellors would be much easier than it is.

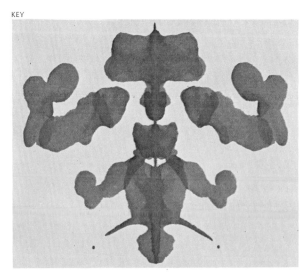

The Rorschach test, invented by Hermann Rorschach (1884–1922), invites a person to describe what he can see in ten standardised, symmetrical ink blots, his answers being used as evidence of his fantasies, his personality, his psychological state and even his intelligence.

5 Thematic apperception tests are used particularly to measure personality improvement during psychiatric treatment. They comprise a series of 20 ambiguous pictures and drawings about which the subject has to make up dramatic stories. He is asked to indicate what the people are thinking and feeling, the events that have led up to the situation and the outcome. A man standing by a lamp-post, for example, might be seen as waiting to meet his girl friend before they go out for a drink [A], or as just having had an argument with his fiancée and being shut out of her house [B]. The stories are then analysed, for it is assumed that the subject identifies negatively or positively with characters that he makes up and so will reveal something about his own inner impulse conflicts and sources of threat.

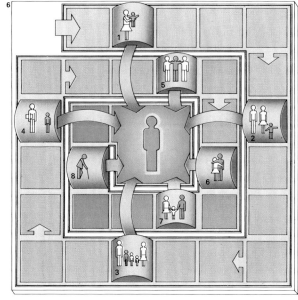

6 E. H. Erikson, the American psychiatrist, has theorized that we go through a series of psychosocial developments. Each stage is characterized by an identity crisis that strengthens the ego or self.

1 The development of hope in infancy.
2 The emergence of will, and the determination to have independence and free choice with the exercise of self-restraint.
3 A secure family gives purpose and initiative to pursue valued goals.
4 The development of competence, casting aside infantile inferiority.
5 The adolescent develops fidelity and loyalty to the self and others.
6 The crisis of young adulthood — love — encourages mutual devotion.
7 Adulthood brings responsibilities and teaches care.
8 In old age wisdom grows.

7 Thin, average-build and fat people, according to the American psychiatrist William H. Sheldon's theory of somatotypes, are endomorphs, mesomorphs and ectomorphs respectively. People tend to prefer objects that are similar in shape to themselves.

8 Three distinct types of people, according to Kretschmer, are the pycknics (broad, fat, short and tending to melancholy); asthenics (thin and tending to the schizoid); and athletics (muscular and intermediate in personality).

189

Freedom and individuality

The question of freedom involves a compromise between the rights of the individual and the rights of society. Each individual is unique, in that he or she is in some respects unlike any other person. But society binds individuals together on the basis of the things they have in common, such as domicile, language, race or religion.

In liberal, democratic societies individuals possess moral and legal rights deriving from recognition of and respect for their uniqueness; limits are set on the right of the state to interfere in the lives of individuals and on the extent to which individuals are allowed to encroach on one another's freedom. In Britain, for instance, each individual possesses the right to privacy in his own home. The police are not allowed to insist on being granted entry without permission from a magistrate and have to give good reason for seeking it. Also other people may be guilty of trespass if they enter without the occupant's permission.

In totalitarian countries the rights of the individual are not respected in the same way and the police may enter someone's home whenever they wish; they may also make an arrest without giving a reason. In some countries such as South Africa, one class of persons, the whites, are recognized as having rights while another class, the blacks, have more limited rights [3].

A similar division existed in Europe before the abolition of slavery [2]. In the classical civilizations of Greece and Rome there were two classes of human beings, a slave-owning class whose members were regarded as individuals with civic rights and a slave class with few rights, the members of which could be bought and sold.

Although Christianity has always held that all human beings are individuals and unique in the eyes of God, it is only in the last 200 years or so that there has been any widespread practical recognition and acceptance of human individuality, regardless of colour, religion, race or way of life.

Freedom and dignity

The concept of individuality is closely linked with that of freedom, the right of each individual to pursue his own happiness in his own unique way, with a minimum of interference from the state or from others [Key].

The fact of individuality forms the moral basis for the idea that all human beings possess a dignity that must be respected. Since each individual is unique, he is irreplaceable and no other person can be a true and complete substitute for him. Therefore he has a value that derives simply from his being himself. As a result, he has a dignity deriving from his individuality and uniqueness, rather than a price deriving from what it cost to produce him or what it would cost to replace him. As Immanuel Kant (1724–1804), the German philosopher, put it: "A thing has a price if any substitute or equivalent can be found for it. It has dignity or worthiness if it admits no equivalent."

Individuality as a scientific fact

The idea that each member of the human race is unique is not only a legal and moral assumption but also a scientific and biological fact. The police can identify an individual by his fingerprints since no two people have exactly the same pattern of loops and whorls

1 Jews destined for Auschwitz, victims of the most callous disregard for the dignity of man ever witnessed, were collected and herded into cattle trucks. In 1940 the Nazi Government of Germany, which had from the beginning persecuted the Jews, instituted the policy of the Final Solution – the mass extermination of the Jews in the gas chambers and execution yards of Auschwitz, Belsen, Buchenwald and other camps. Millions of Jews – as well as gypsies and others – died as a result of this policy, which the Nazis tried to justify by claiming that the Jews were a "lesser race", not fully human, and so had forfeited human rights.

3 Apartheid in South Africa separates black from white and gives greater privileges to whites. In public facilities there is a rigid division between those reserved for whites and for blacks.

2 Slavery, by which a person is bound to work for a master without recompense or freedom, is as old as civilization; Egyptians, Greeks and Romans, all used slaves. The discovery of America gave the slave trade a new impetus. The Spanish, Portuguese, British, French and Dutch all imported slaves from Africa to work on plantations. Britain abolished slavery in its possessions in 1833; in the USA, it was abolished after the end of the Civil War (1861–5).

on their fingertips. It seems probable that the human voice is similarly unique to each individual. Phonetic recordings show voice patterns that are characteristic of each individual and are unaffected by the language being spoken or the accent being adopted. It appears, too, that the arrangement of genes on the chromosomes in the nucleus of each cell in the body is unique for every individual.

Furthermore, since human beings have more genes than any other animals, the possible range of genetic differences is greater than in any other animal. Man's wide range of innate differences is enhanced by the variety of social circumstances in which he grows up and which mould his personality. Man's unique capacity for symbolic thought, expressed in language [6], creates complexities of psychological "development" that do not exist in any other creature. Since all humans have differing bodies, grow up in different environments and have minds that develop in differing ways, all are essentially different and unique; and it is only in the unreal world of dreams that a person could ever meet someone exactly the same as himself.

Identical twins come nearer than anything else to being exceptions to the uniqueness of man; hence the uncanny effect they can create, and the myths and superstitions that exist about them. But even identical twins have different fingerprints and birthmarks. Within a few weeks, they have recognizably different personalities as well.

Sexual reproduction
There are creatures (such as ants and bees) in which nothing analogous to individuality exists; individuality is ultimately an effect of sexual reproduction. In insects and plants that reproduce asexually, one generation may be exactly like the previous one and many plants are not individuals in the literal sense of the word since they can be divided, as they are when a cutting is taken; what was originally one plant can become two or more separate, but not individual, ones. These groups are called clones: the idea of producing human clones – regiments of identical men analogous to worker bees – remains a horrifying if remote possibility discussed by science, but existing only in science fiction.

OURS...to fight for

Freedom of Speech *Freedom of Worship*

Freedom from Want *Freedom from Fear*

The right of every person in a democratic society is to be treated as an individual and to be free. In a speech made some months before the United States entered World War II in 1941, President Franklin D. Roosevelt (1882–1945) made what has come to be regarded as a classic statement of democratic principles – envisaging a world founded upon four essential freedoms: the freedom of speech and expression, the freedom of every person to worship God in his own way, the freedom from want and the freedom from fear. These ideals were represented in a series of paintings entitled "The Four Freedoms" by the American artist Norman Rockwell (1894–   ) in 1941.

4

4 The Coldstream Guards, Trooping the Colour in London, exemplify the ability of humans to suppress one of their most important qualities – individuality – and achieve a rigid uniformity. This machine-like discipline and obedience implies at least a partial surrender of uniqueness and personality which, although in the interests of efficiency, may be dehumanizing.

5 The seeming contrast between the regimentation of the guardsmen and the relaxed informality of this group in an Amsterdam square masks a similarity. Nonconformity can itself take forms that are almost as rigid in their conventions as those of troops.

5

6

6 In the story of the Tower of Babel, here painted by Pieter Bruegel the Elder (c.1520–69), God punished His people by "confounding their language". Language is both a powerful source of social cohesion and an expression of the individual's personality. Each person's use of language is unique.

7 The story of the Good Samaritan symbolizes the need for people not to pursue their own ends selfishly but to be aware of responsibility to others in need. Jesus told of a good Samaritan who helped a man left half-dead by thieves after supposedly more worthy people had "passed by on the other side". Modern "Samaritans" notably offer help to potential suicides.

7

Body image and biofeedback

Body image is the physical aspect of our personality, the image we believe we present to the world. We are so used to experiencing our bodies as if we were observers sitting inside our own heads [Key] that when the letter E is drawn on the forehead of a person whose eyes are closed he commonly reads it as a figure 3 – that is, he sees it from behind.

Views of the self-image

Our view of ourselves may be wrong. We may "see" ourselves as bigger, smaller, less or more attractive than we are and such mistakes can in turn affect our bodies. A timid person fails to "walk tall", for example, and the process of "seeing" our bodies can almost certainly alter functions by subtle processes, sometimes predisposing our bodies to malfunction or disease.

The body image grows gradually with child development and misconceptions can grow with it. The trans-sexualist, for example, finds sexual identity to be at variance with the male or female body he or she possesses. In an obese person overeating may arise from a false body image either at the physical level or, more often, at the self-valuing level. Moreover humans love to manipulate the body image symbolically. They are the only animals that can choose their own "plumage", and they may do so through decorating [7], deforming [2, 3, 4, 6] or surgically altering their bodies, or by using clothing to signal mood, life-style, status and other things. Some clothing affects physical stature (such status symbols as crowns increase apparent height; shoes are a more common way of achieving this). Other clothes emphasize or de-emphasize features that a particular culture values; genital bulge may be exaggerated [5] or hidden in an exaggerated way, and women's bodies have been moulded into many biologically unnatural shapes by addition and disguise [1].

Control through biofeedback

Self-image can affect the body physically, for example, when the posture of a timid person leads to backache. A significant recent development is the recognition of the extent to which the body can also be manipulated favourably, giving a person control of physical processes normally regarded as uncontrollable. This is called biofeedback.

Body functions have long been divided, both popularly and by scientists, into the "voluntary" and the "involuntary" – those we monitor and control, such as muscular movements of the limbs, and those that are largely self-regulating, such as bowel movements, heart rate and blood pressure. The latter are affected by an individual's state of mind, but usually on an unconscious level. It is now generally recognized that a great many of these so-called involuntary processes can be controlled by "willing" in exactly the same way as we can initiate movement of a leg or a finger, although less precisely. Yogis, with their traditional awareness of the body and its numerous functions (which they have achieved through meditation), have recognized this for centuries and can, to some degree, regulate blood pressure, heart rate and other such processes at will. This helps to explain how some yogis can survive being buried underground for several days.

In recent years simple machines have been developed that provide a short cut to

CONNECTIONS

See also
236 Talking without words
266 The individual in society
184 The idea of the person
194 Meditation and consciousness
122 Non-medical healing

1 **Fashion** has dictated that women not only alter their clothes regularly, but also their shape. They have followed trends set by "society" fashion designers with and without the aid of such extensive undergarments as the whalebone corset and the padded bustle. Variously, shoulders, busts, waists and bottoms have been emphasized at different times. The first 50 years of this century witnessed a remarkable series of revisions. The century started with the celebrated S shape [A], switched to the boyish bottomless, waistless, bustless style of the 1920s [B] and then changed in the 1940s to the broad-shouldered "New Look" [C].

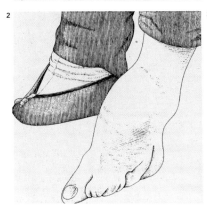

2 **Small, delicate feet** are still universally admired in women. In China, in the past, it was customary to bind young girls' feet to limit their growth. This ensured a higher marriage price at the cost of deformity and dislocation of the foot bones.

3 **The giraffe look,** traditionally admired by some tribes in Burma and east Africa, is achieved by adding year by year to a collar of brass rings. If the collar is later removed the girl's weakened neck may not support her head.

4 **Alteration of the natural shape** of children's heads in an attempt to improve the aesthetic qualities of the skull has been practised by some parents since the time of ancient Egypt. The pliability of the cranium in childhood makes this possible. In different cultures the head has been elongated, broadened or flattened, initially by pressing by hand and then by a variety of head-boards. In France, even as recently as the last century, such head deformations took place. Although elongated, broadened or flattened heads were thought beautiful another aim was to "shape the brain" so that the child would develop bravery, moral uprightness, sincere feelings and intellectual capacities.

such body awareness. These so-called biofeedback machines [9] record changes in a person's physiological processes by moving a needle, lighting a bulb or causing audible clicks. The subject, by playing "hotter, colder" with the display (attempting to extinguish the light, move the needle or make the clicks louder), can acquire the knack of regulating the hitherto "involuntary" function concerned.

The biofeedback machine has been shown to work for a wide range of involuntary processes including control of brain alpha rhythms, blood pressure, heart rate and skin blood flow. In fact it appears that almost any involuntary motor process that can be recorded by the machine can be controlled voluntarily.

The medical uses of biofeedback have not yet been established but are potentially vast. Biofeedback control of brain rhythms has already been used to fight drug-resistant epilepsy by warning patients of the onset of an attack. The chief potential of the technique is as a drugless method of modifying symptoms such as hypertension, the physical expression of anxiety, and other psychosomatic disorders for which drugs have hitherto been heavily prescribed.

Character armour

The importance of body image, including its effect on the body and its control mechanisms, was first stressed by the Austrian psychologist Wilhelm Reich (1897–1957), who saw the posture and organ functions of the individual as a kind of armour ("character armour") that could reflect anxieties, rigidities and negative feelings towards the body and the self, especially in its sexual context. Reich based a form of therapy on the removal of these rigidities, claiming that this would improve both bodily function and self-esteem.

The manner of holding, using and "seeing" our bodies is now regarded as part of a definite language of communication called "body language". This goes beyond ordinary gesture. The whole physical ambience of a person's body is a means of communicating with, and influencing the responses of, other people.

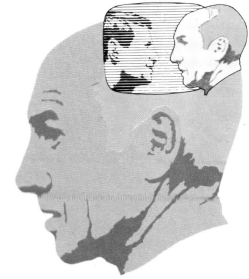

We experience the outside world at second-hand and even experience our bodies as if we were observers sitting inside our heads looking at a TV screen. This is called "body image".

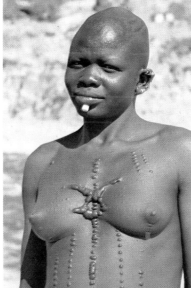

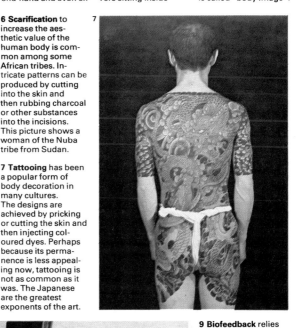

6 Scarification to increase the aesthetic value of the human body is common among some African tribes. Intricate patterns can be produced by cutting into the skin and then rubbing charcoal or other substances into the incisions. This picture shows a woman of the Nuba tribe from Sudan.

7 Tattooing has been a popular form of body decoration in many cultures. The designs are achieved by pricking or cutting the skin and then injecting coloured dyes. Perhaps because its permanence is less appealing now, tattooing is not as common as it was. The Japanese are the greatest exponents of the art.

5 Male manipulation of the body image is most strikingly exemplified by the codpieces once used to accentuate masculinity. Lodovico Capponi wears one in this portrait by Bronzino (1503–72).

8 Distorted demon masks of intricate design, like this one from Sri Lanka, were used to terrify enemies in battle and, in religious ceremonies, to symbolize the forces of evil and darkness.

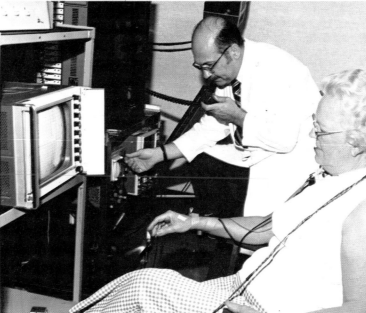

9 Biofeedback relies upon the minute electrical impulses produced when the body performs involuntary processes. These electronic potentials can be picked up by means of electrodes that are placed on certain parts of the anatomy according to the particular body function to be treated. The signal is then amplified, filtered and rectified so that it can be monitored in the form of an oscilloscopograph, a digital display or a series of sound frequencies. The patient then concentrates upon the level of activity and tries to reduce it below a certain point. By this method the intensity of a headache, and such functions as blood pressure and heart rate can be controlled.

Meditation and consciousness

Meditation is an ancient inward or spiritual discipline dating back at least to 1000 BC when it was practised in India. Meditation involves deep thought or contemplation, often helped by postures, sounds or images that concentrate the mind [1]. The object of meditation is not to heighten reasoning powers but to allow intuitive perception of spiritual truth, usually by excluding distracting physical sensations or intellectual activity. Meditative techniques have also been used, particularly in Western countries during the past 20 years, as a means of relieving mental and physical tension.

Altered states of consciousness
Mystics of all major religions believe that God must be approached by inward illumination rather than through reason, logic or outward forms of worship. As a way to enlightenment, meditation has been used in Western as well as Eastern religions for many centuries. But whereas it has a minor place in Christianity and Islam, meditation lies at the very heart of Hinduism and its great offshoot, Buddhism. In most forms of Hinduism the aim of meditation is to discover the inner self as a means of purification. Yoga, the most widely known Hindu technique, is based on the theory that by training both body and mind, centres of psychic power can be tapped.

Buddhism, dating from the sixth century BC, is a more self-effacing religion, teaching that with strict discipline and study a happy few can reach a state of pure consciousness transcending body and mind in which the ultimate reality, nirvana (nothingness), can be understood. Tantric Buddhism evolved more accessible methods of achieving this perfect wisdom and spiritual freedom through the practice of certain rituals and by using body positions (mudras), sacred phrases (mantras) and universal designs (mandalas). Other Buddhist sects use a variety of methods to increase concentration. These range from contemplation of human virtue or misery to gazing at images or lights.

Drugs, herbs and alcohol also have a long history of use as aids to inward vision, sometimes in combination with meditation. In the 1960s many young people in the West, who felt alienated from the materialism of their own society, turned to drugs or to Eastern meditative techniques in an attempt to expand their consciousness and achieve a sense of spiritual harmony. Hallucinogenic experiences with drugs such as LSD often had adverse effects. But meditation seemed to produce physical benefits that could be measured, at least on a psychological level.

Physiological changes
In 1960 studies made on Zen Buddhist priests showed that distinct electroencephalographic EEG brainwave patterns [2, 3] were associated with states of meditation. Practised meditators slipped quickly into an "alpha" brainwave rhythm, indicating a state of serenity and peaceful alertness later bordering on the "theta" rhythm of trance [5]. The priests and monks who acted as test subjects could maintain this state of deep, yet alert, rest for hours without entering the "delta" rhythm or cycle of full sleep. More experiments showed that meditation was associated with other physiological changes including decreases in heart rate and oxygen consumption (about 20 per cent lower) and

CONNECTIONS

See also
224 Religion and the plight of modern man
192 Body image and biofeedback
52 Touch, pain and temperature

1 Five levels of consciousness are generally associated with meditation. Although these can be described only in subjective terms, many mystics have attempted to put them in ascending order as a means of guidance to students. At the first level the body is relaxed and the mind stilled. Ordinary wakefulness is then transcended by a state of deep rest but alertness in which the mind is more intensely aware of itself. This state of personal awareness is transcended by cosmic consciousness, a sense of illumination beyond the boundaries of self and time. The next stage is God consciousness, or the sense of inseparability of creator and creation. Finally, at the highest level of all, consciousness itself is transcended in a unity with the whole, or oneness. The basic position for meditation is the lotus of yoga in which the legs are crossed, the back is erect and the hands lie loosely on the knees [A]. Westerners may modify this to a comfortable upright position on a chair with open eyes gazing straight ahead [B]. Practised yogis achieve difficult postures [C] in the belief that control of the body by the mind is the first step to salvation. Sexual forms of tantric yoga are based on positions that symbolize unity [D]. Beautifully designed mandalas [E] provide aids to concentration.

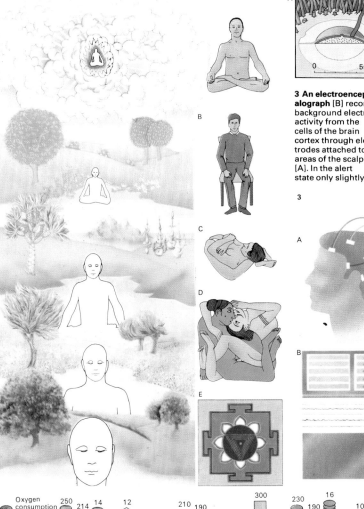

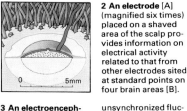

2 An electrode [A] (magnified six times) placed on a shaved area of the scalp provides information on electrical activity related to that from other electrodes sited at standard points on four brain areas [B].

3 An electroencephalograph [B] records background electrical activity from the cells of the brain cortex through electrodes attached to areas of the scalp [A]. In the alert state only slightly unsynchronized fluctuations are detected [C]. A dominant alpha rhythm with regular waves appears in an eyes-closed state of relaxation [D]. With the onset of sleep slower, more irregular waves occur [E]

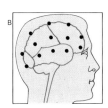

and as sleep deepens [F] large delta waves appear [G] with periodic irregularities [H], apparently related to dream activity. The EEG patterns of epilepsy [I] provide a striking comparison.

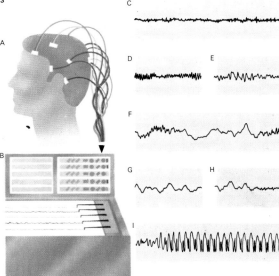

4 Oxygen consumption in cc
Carbon dioxide elimination in cc
Respiration
Blood lactate milligrams/100cc
Skin resistance milliohms
Critical level

250 214 14 12 50

Per minute Sitting

210 190 4

Per minute Meditating

7 300

230 190 16 10 250

Per minute Sleeping

4 A variety of physiological changes takes place during meditation. There is a marked decrease in the rate of respiration and of consumption of oxygen, a decline in the concentration of blood lactate (linked with relaxation of muscle activity in rest) and a rapid rise in the electrical resistance of the skin. In sleep, skin resistance normally rises but not so much or at such a rate. There are distinct physiological differences between sleep and meditation.

reduced output of carbon dioxide, indicating a general slowing of metabolism [4].

When control subjects, unpractised in meditation techniques, adopted the cross-legged "lotus" position of Eastern meditation and were wired to the same instruments, they registered nothing but "beta" rhythm, the state of ordinary wakefulness. But in the early 1970s further studies at the Harvard Medical School indicated that the physiological state associated with meditation could be achieved without years of training or self-discipline. In 1972 a report was published in *Scientific American* on tests carried out on 36 people using a simplified technique known as transcendental meditation (TM). The subjects had all achieved, after only a few sessions of training, similar EEG brainwave patterns to those of meditating monks.

Benefits of meditation

The discovery that meditation could be instrumentally measured created interest among both scientists and psychologists. The effects of achieving certain EEG patterns cannot be equated with the objectives of spiritual meditation and physiological changes convey little information about the mind, but the significance of the Harvard results is that meditation, even at a simple level, appears to have benefits for societies with a high degree of stress in daily life. Since Maharishi Mahesh Yogi [7] brought the TM technique to the West in 1957, it has become a popular form of relaxation particularly in America where some large claims have been made for its social as well as its individual benefits [8]. The TM technique involves two daily sessions of 15–20 minutes during which the subject sits in a comfortable position with his eyes closed and repeats a sound without a special meaning chosen to keep his wandering mind in tow. No intense concentration or special discipline is required nor is any specific religious belief or philosophy. More rigorous yoga techniques have also attracted many Westerners willing to study and practise meditation and sometimes to travel to the East in search of gurus (teachers) and ashrams (retreats). The question of whether there is a special area of the mind that can be reached by such methods remains open.

The practice of meditation takes different forms but, as with this Buddhist monk, it is usually aimed at achieving a higher plane of consciousness in which physical needs are negated and the spirit of man perceives essential truth. To aid transition to this plane some Eastern methods of meditation use an object such as a mandala – a geometric design often symbolizing the cosmos. Other methods include forms of rhythmic breathing and concentration on a point between the eyes or on a mantra, the most famous being "Om", a derivation of the Sanskrit "Aum", said to be the sacred word of power. Mantras are often phrases that sum up ancient wisdom.

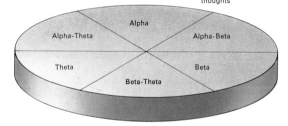

5 Alpha-Theta state. Inner stillness. Decreased self-awareness. Pleasant feeling of freedom

Alpha state. Feeling of transcendence. Alert, calm, elated. Sense of insight

Alpha-Beta state. Wakeful, calm, creative, perceptive. Sensitive, investigative, intellectual thoughts

Theta state. Dreamlike trance state. No self-awareness

Beta-Theta state. Disturbed and negative. Feeling of unhappiness

Beta state. Subject to anger, excitement, tension

5 States of feeling may possibly be linked with differing amounts of alpha, beta or theta rhythms as shown in EEG brainwave patterns. Any interpretation of these relationships is necessarily highly subjective, but a generalized chart can be based on research with numbers of meditators including both Zen monks and yogis of long standing and more recent practisers of transcendental meditation.

6 An Indian fakir demonstrates his ability, in the meditative stare, to negate the pain caused by lying on a bed of thorns, showing how the mind can be trained to gain control of the body.

7 Maharishi Mahesh Yogi (surrounded by some of his followers) is the best known of a number of Eastern gurus who have influenced Westerners to take up the practice of meditation during the past few years. A physics graduate of Allahabad University, India, the Maharishi studied with yogi master Swami Brahmananda Saraswati for 13 years before teaching transcendental meditation in the West, chiefly in America.

8 The benefits of meditation are evaluated somewhat differently in the East and in the West. Eastern mystics see relaxation as a prerequisite for spiritual insight. In the West more emphasis is placed on physiological or social advantages. Considerable claims are made for transcendental meditation by those who practise it. TM researchers report improved physical and mental co-ordination, reduced depression and irritability, fewer stress disorders and a statistically measurable drop in anti-social behaviour. One TM chart links falling crime rates in 12 American cities with the effects of meditation by 1 per cent of the population and compares this with rising rates in matched cities.

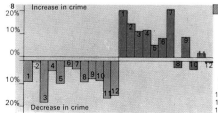

Cities with 1% meditating	% change	Control cities	% change
1 Chapel Hill, NC	9·3	1 Rocky Mount, NC	20·2
2 Ithaca, NY	0·6	2 Poughkeepsie, NY	14·4
3 Lawrence, Kan	18·4	3 Lafayette, Ind	11·1
4 Bloomington, Ind	4·5	4 Columbia, Mo	11·2
5 Carbondale, Ill	9·9	5 Marshalltown, Iowa	5·0
6 Iowa City, Iowa	2·5	6 Oshkosh, Wisc	8·3
7 Ames, Iowa	3·6	7 Norman, Okla	20·8
8 Boulder, Cal	9·1	8 Fort Collins, Cal	3·2
9 Santa Cruz, Cal	7·9	9 Monterey, Cal	8·5
10 Santa Barbara, Cal	8·8	10 Costa Mesa, Cal	3·9
11 Los Altos, Cal	16·4	11 Claremont, Cal	1·4
12 Davis, Cal	15·2	12 Pleasant Hill, Cal	0·01

The occult

Man has always felt the presence of hidden forces around him, secret sources of knowledge and latent powers within himself. The occult describes his beliefs about this mysterious "other world" and the means by which he has tried to contact it. Occult beliefs have varied from culture to culture, changing with the advance of science. But although science explains that gravity makes an apple drop from a tree it does not attempt to say why the apple struck a particular man passing beneath. Occult lore has been concerned less with explaining events than with trying to predict, forestall or induce them.

The historical background
For the occult believer, the supernatural dimension includes a variety of gods and demons that were – and sometimes are – thought to influence weather, crops and procreation. Men once believed their very survival depended these powers. Linked with this was the almost universal belief in life after death and in spirits and ghosts.

Recognizing the power of suggestion or auto-suggestion, holy men such as the Magi (Persian priests from whom the word "magic" comes) devised rituals and symbols to concentrate the mind. Pagan cults based on some of these rituals survived the growth of major religions such as Christianity, but from about the fourteenth century occult practices began to be condemned by the Church as evil. Persecution of the Albigensian sect, which held that the material world was created by the Devil, led to witch hunts in continental Europe, Britain and later America. Witches were said to take any form they pleased, fly at lightning speed by broomstick, change humans into beasts and create an infinite variety of havoc with magical potions and incantations.

Torture and burning of those identified as witches created a climate of hysteria that actually encouraged sorcery and which lasted until the eighteenth century. Tales of people turning into wolves (lycanthropy) and dead bodies remaining fresh by preying on the living (vampirism) flourished, along with belief in less malign spirits such as fairies and elves. Prayers, rituals and talismans were used to invoke assistance or ward off bad luck. A residue of folk superstitions such as "touching wood" remains today and exorcism is still occasionally practised to drive "devils" out of a person said to be possessed.

Prophecy and fortune-telling
Predicting the future has always been a common preoccupation ranging from inspection of the sky or of animal entrails to interpretation of apparently random patterns of coins, cards, dice or sticks as in the ancient Chinese book of wisdom, the *I Ching* [7]. The most notable of European seers was Nostradamus (1503-66), a French physician and astrologer who wrote more than 600 obscure verses that can be interpreted as a remarkably accurate forecast of the French Revolution and of some other major events. By the eighteenth century more bizarre, older methods of reading the future such as kephlomancy (the crackling made by a burning donkey's head), hydromancy (the noise of running water) and onychmancy (reflections in a virgin's oiled fingernails) had fallen into disuse. But the use of Tarot cards [9] and many other traditional methods [3]

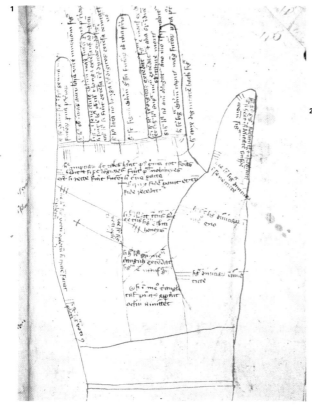

1 **Palmistry,** the attempt to infer human characteristics and foretell the future from the lines of the palm, is an art of great antiquity. In the 15th century the main features of the palm were related to the planets and signs of the zodiac. The art is sometimes called cheiromancy after an expert modern seer, the Irishman Count Louis Hamon, who called himself "Cheiro".

2 **Raising a spirit** was among occult feats claimed by John Dee (1527–1608), the most celebrated English psychic experimenter of the 16th century. With his friend, Edward Kelley, Dee recorded long conversations with various angels. An inventor and astrologer, Dee was a shrewd politician and was an adviser to Elizabeth I (1533-1603). He inspired scholars to study astrology and alchemy.

3 **Seeing man as a microcosm** of the universe is a Greek concept from which sprang a number of occult arts including metoscopy – the interpretation of lines on the brow or moles on the body. In this 17th-century print the subject's characteristics were inferred from a series of circles, perhaps implying a reflection in man of the planets' circular paths round the sun. Adherents of metoscopy believed lines and moles were "stars of the body".

continued. Some gypsies made a business of interpreting cards, dreams, crystal balls, tea leaves or the palms of hands [1].

The mid-nineteenth century brought a revival of interest in the age-old belief that spirits of the dead could sometimes be seen or heard. The idea that ghosts were lost souls trapped between this world and the next was given impetus after the invention of photography when exposed film sometimes showed what appeared to be ghostly forms [6]. Spiritualists believed a psychically sensitive person, called a medium, could go into a trance state and, while in this state, receive messages from the departed.

The Spiritualist movement
The Spiritualist movement spread after the teenage Fox sisters in America in 1848 claimed that their home was besieged by rapping noises and that objects fell or were hurled off shelves as if pushed by a poltergeist, a mischevious spirit. Evidence that they had "communicated" with this spirit was later discredited, but a number of mediums soon emerged who appeared to have

paranormal abilities. Among them were Helena Blavatsky (1831-91), who founded the Theosophical Society, and Daniel Home (1833–86) who impressed European royalty and baffled sceptics with phenomena ranging from sudden temperature drops to floating tables and the elongation or levitation of his own body [5]. It became fashionable to hold seances – meetings for the purpose of contacting the dead. Answers to participants' questions were either spelled out on an ouija board or given in a "yes" or "no" form according to the number of raps or tilts on the table. Exposure of a number of charlatans who exploited the gullibility of the bereaved led to declining interest in spiritualism

During the twentieth century, however, both in the East and the West, scientists are slowly turning their attention to the explanation of "psychic" and "clairvoyant" forces. The result of their studies will, perhaps, be the exposure and full explanation of some of those mysterious elements which, according to James Jeans (1877–1946), make the universe begin to seem more like a great idea than a great machine.

KEY

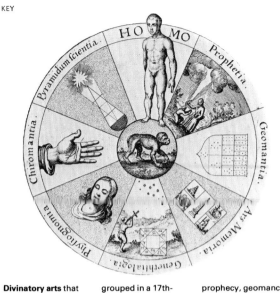

Divinatory arts that tried to glimpse man's fate through occult means were grouped in a 17th-century diagram by Robert Fludd (1574–1637). They included prophecy, geomancy, astrology, physiognomy, palmistry and pyramid science.

6 "Spirit" photographs were produced by some Victorian mediums but most of them could be easily faked by parlour magicians. Several mediums were caught using assistants or various mechanical methods to produce the illusion of a spirit substance called ectoplasm. Spirits have been more shy of infra-red photography in the dark.

4 A planchette was a popular device at the end of the 19th century for producing "automatic writing", supposedly prompted by a spirit. Hands were placed on a free-moving board that held a pencil.

5 Levitation was one of the most astonishing feats performed before critical audiences by Daniel Home, a medium whose apparently occult powers were neither fully tested nor proved fraudulent.

7 Fortune-telling of the kind practised in China before 1949 was usually based on the use of coins or sticks like those on this Canton street stall. The classic *I Ching* method dates back to 1000 BC.

8 Jeane Dixon, an American clairvoyant, used a crystal ball to predict several electoral defeats and victories and the deaths of Dag Hammarskjöld, Marilyn Monroe and Robert and John F. Kennedy.

9 The Tarot pack contains 78 cards in four suits – Wands, Cups, Swords and Pentacles. The picture cards carry symbols that in some cases date back to ancient Egypt and are among the oldest known to man: the Sun, the Moon, the Lovers, the Devil, the Tree of Life. Each suit has an underlying theme. Chosen and laid out by one of several systems, each card is said to moderate and influence its neighbour. Successful Tarot readers use much intuition.

Parapsychology

Parapsychology is a term used for scientific research into extra-sensory perception (ESP); that is, the apparent ability of a person to communicate with some other entity, or to become aware of something, without using the normal sense organs.

Classification of phenomena

Psychic phenomena such as these are usually divided into four kinds [Key]. Telepathy is communication between one mind and another in some unknown way. It involves the sending or receiving of messages, thoughts or feelings and is often called "thought transference". Clairvoyance is the awareness of some event, object or person not known to anyone else, such as the presence of a letter in a secret drawer or a fire in an empty house. Precognition is the ability to foretell future events. Psychokinesis (mind-movement or PK) occurs when a person causes physical objects to move or fall without apparent contact with them [6].

Other odd or unexpected events such as the alleged seeing of ghosts, voice communication with "the dead", the produc-

tion of "ectoplasm" (foam-like substance supposedly taking the shape of a disembodied person) and other inexplicable manifestations of the "beyond", are not generally part of parapsychological research.

The first attempts to examine claims about paranormal events scientifically were initiated in 1882 when the Society for Psychical Research was set up in London by a group of interested intellectuals headed by Henry Sidgwick (1838–1900), later Knightsbridge Professor of Moral Philosophy at Cambridge University. The society was concerned with investigating and classifying anecdotal evidence. It was not until some 50 years later that J. B. Rhine (1895-) at Duke University, North Carolina, began the first controlled experiments into extra-sensory perception.

Parapsychology and J. B. Rhine

J. B. Rhine's initial work was the study of clairvoyance by means of written records of so-called spirit communication and by tests with mediums. He hoped to find confirmation for the existence of disembodied spirits.

By 1934, despite exhaustive work with mediums, including Eileen J. Garrett (who was later to found the Parapsychology Foundation in New York), Rhine felt there was no irrefutable evidence possible and turned his efforts to a duller but more tangible form of research with a system of card-guessing experiments using Zener cards [3]. These are a pack of 25 cards, bearing five different symbols such as a circle or a cross, each symbol having its own colour. To test clairvoyance, for example, the subject of the experiment tries to name the colours or symbols on the cards, one at a time. No one knows what symbol or colour will come up.

The laws of chance would produce five correct answers out of 25. If the subject consistently scores higher than that it could be said that he has some clairvoyant ability.

Rhine's experiments indicated that there were people who clearly had extra-sensory perception. He published his findings in *Extrasensory Perception* (1934), a book that aroused both considerable interest and a great deal of criticism over the mathematical validity of his tests. Rhine responded by

1 Levitation of chairs is achieved by a medium in this contemporary artist's impression of a seance in Germany in the late 19th century. The well-dressed people around the table indicate the fashionable interest in psychic phenomena at that time. Although early experimental research was concentrated on the alleged powers of a medium, charlatanism soon turned parapsychology to the laboratory.

2 Joseph Banks Rhine, pioneer of scientific enquiry into psychic phenomena, is seen here in his laboratory engaged in research into psychokinesis (PK). The machine spills dice randomly onto the board, and the subject tries to influence which side of the line they will land on. Any discrepancies between the statistically expected results and those of this and similar experiments are ascribed to extra-sensory powers.

3 The Zener cards are the basic equipment for card-guessing experiments. They have clear geometric designs (a square, circle, star, cross or waves) and are used as a pack of 25. A score of more than one in five is considered "above chance". Seen here [A] are the results of experiments conducted over several years by Rhine. The cards can be mixed in three ways: they may be shuffled by hand by the ex-

perimenter; they may be put in a box and rotated for a predetermined length of time; and they may be run in a machine [B] that rotates them for an arbitrarily chosen length of time. In all cases the subject must guess what order the cards will appear in after they have been shuffled. In modern laboratories at parapsychology institutes the cards may be shuffled and randomly sorted for selection by electronic means.

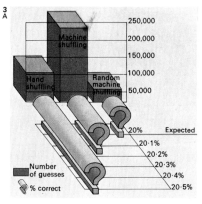

250,000
200,000
150,000
100,000
50,000

Machine shuffling
Hand shuffling
Random machine shuffling

20% Expected
20·1%
20·2%
20·3%
20·4%
20·5%

Number of guesses
% correct

4 At the dream laboratory of the Maimonides Medical Center, New York, experiments are carried out on an isolated sleeping subject with electrodes attached to her scalp to show the rapid-eye movements (REM) of dreaming. An experimenter concentrates on a randomly selected art print in an effort to communicate it by telepathy from his mind to that of the sleeper. The subject is awakened periodically to describe her dreams and the results are independently analysed.

tightening the controls of his experiments.

New Frontiers of the Mind (1937), Rhine's next book, became a best-seller and his statistics were pronounced valid by the American Institute for Mathematical Statistics. A separate parapsychological laboratory was then established at Duke University.

More and more varieties of tests followed and during the 1940s Rhine was no longer trying simply to prove the existence of ESP, but looking for the various reasons and conditions that favoured its production. He discovered that mood and attitude were important factors, as was the relationship between experimenter and subject, and that extremely low scoring was just as indicative of ESP as extremely high. Circumstances, environment and many other effects on the subjects' psychic responses were closely studied and the results then computed.

Debate about ESP
From the 1950s onward there followed more specialized and sophisticated experiments involving an increased use of electronic instruments for both randomizing and computerizing the test material [7]. Psychokinesis, in particular, was being studied at this time.

In 1969, after long, hard resistance, parapsychology won a place in the American Association for the Advancement of Science and since that time the exploration of ESP has become associated with many other fields of research under the inclusive title of "psi". Parapsychology conferences are now regularly held in many countries.

While the various researchers into parapsychology seem to have advanced some way towards establishing the fact of the paranormal, they have not yet explained what it is or what range of energy or force engenders it. Only about ten per cent of psychologists believe in the fact of ESP; ten per cent dismiss it totally and the remaining 80 per cent believe that more evidence is necessary before they can wholeheartedly accept its existence. The main problems are the difficulties of verifying findings that have often been established by those most eager for positive results and of effectively ruling out the possibility of cheating.

KEY

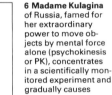

Psychic phenomena or "psi" are of four kinds: telepathy, clairvoyance, precognition and psychokinesis (PK). PK is subdivided into PK-MT (PK on moving things); PK-LT (PK on living things); and PK-ST (PK on static things).

5 Ingo Swann, a gifted psychic, took part in experiments to test out-of-body perception. Specially drawn art "targets" were laid on a platform placed high above his chair and well out of sight [A]. When he was in a "relaxed state of mind" he began to draw what he could "see". The results were compared [B] and analysed. Ingo Swann completed many of these drawings with remarkable accuracy.

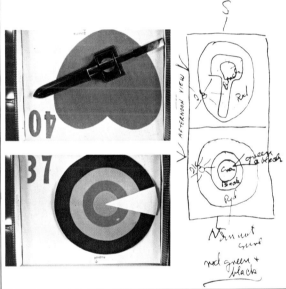

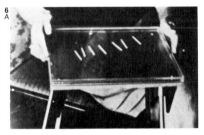
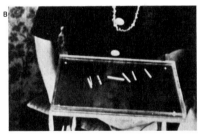

6 Madame Kulagina of Russia, famed for her extraordinary power to move objects by mental force alone (psychokinesis or PK), concentrates in a scientifically monitored experiment and gradually causes matchsticks to move [A], alter their direction [B] and finally bunch together [C]. In other experiments she has moved objects of many different materials and caused both compass needles and cases to rotate.

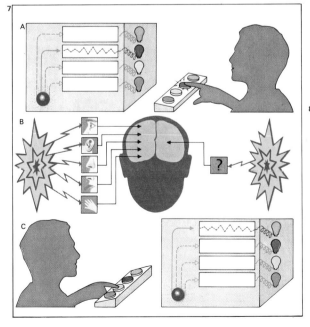

7 We normally assume all information received by the brain comes through the five senses, but many experiments point to the existence of extra-sensory perception [B]. In recent trials four coloured lights were lit randomly and subjects had to indicate, by pressing a button, either which would light up [A] or which would not [C]. The results of both tests after more than 83,000 trials were higher than statistically expected.

8 Metal bending by some mental power has attracted a great deal of attention from the public demonstrations given since 1971 by the Israeli Uri Geller. He appears to be able to bend a variety of objects including nails, spoons, keys and clock hands, apparently just by mental force. Many others have claimed this power and they can be classified according to their individual strengths. In the highest category there is an unusual preponderance of children.

Astrology

Astrology states that the Sun, Moon and planets play a vital role in human affairs. For centuries, astrology was virtually inseparable from the study of the universe which we now call astronomy. In technically advanced countries the influence of astrology has waned and it has been dismissed as worthless or fraudulent. But it is still highly valued in some Eastern countries. There has also been a more general revival of interest in astrological lore and the possibility of a link between it and biological rhythms detected by scientific investigation, which is now being more seriously and systematically pursued.

The aims of astrology

Columns in some newspapers and magazines foster an impression that astrologers attempt to predict the future. But true astrology has little to do with "what the stars foretell". Serious astrologers claim to do no more than indicate trends which may be averted – or promoted – by taking appropriate action.

Astrologers base their deductions on the apparent positions of the bodies of the Solar System and use the stars only as reference points. In drawing up astrological charts or horoscopes the Earth is taken as the central point. This geocentric view is reasonable enough since astrologers can make their observations only from the Earth. No modern astrologer really believes the Earth to be the true centre of the universe.

Because the planets (and the Moon) move in roughly the same plane, they seem to keep to a certain region of the sky, making up a belt known as the Zodiac [Key]. The Zodiac is divided into 12 equal sections or signs, each named after a constellation. These constellations are Aries (the Ram), Taurus (the Bull), Gemini (the Twins), Cancer (the Crab), Leo (the Lion), Virgo (the Virgin), Libra (the Scales), Scorpio (the Scorpion), Sagittarius (the Archer), Capricornus (the Sea-goat), Aquarius (the Water-bearer) and Pisces (the Fish). The names have no significance, and neither have the star-patterns themselves, except as reference points. Moreover, the "vernal equinox" or First Point of Aries – the point where the apparent path of the Sun, or ecliptic, cuts the celestial equator – is no longer in the constellation of Aries as it was in 900 BC; it has shifted into the adjacent constellation of Pisces. But this so-called "precession" makes no difference to the astrological signs.

When casting a horoscope, astrologers work out the positions of the Sun, Moon and planets at the exact time of an individual's birth; for this it is also important to know the place of birth. The celestial pattern that emerges is supposed, in ways unspecified, to determine the personal characteristics of an individual born under its influence rather like the Moon's effect on the tides. Although nobody would claim that all Librans, for instance, are similar in personality, some statistical studies indicate general trends in line with astrological lore.

Historical background

Western astrology may have arisen in Mesopotamia. The earliest-known planetary tables date from the mid-seventh century BC. Early Babylonian astrology was not directly personal. It was concerned with large-scale events such as the advent of wars, floods and eclipses and with their possible effect on the

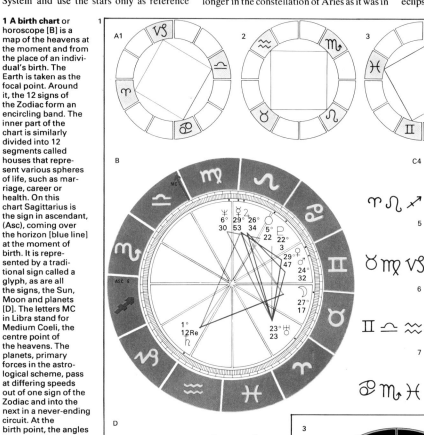

1 A birth chart or horoscope [B] is a map of the heavens at the moment and from the place of an individual's birth. The Earth is taken as the focal point. Around it, the 12 signs of the Zodiac form an encircling band. The inner part of the chart is similarly divided into 12 segments called houses that represent various spheres of life, such as marriage, career or health. On this chart Sagittarius is the sign in ascendant, (Asc), coming over the horizon [blue line] at the moment of birth. It is represented by a traditional sign called a glyph, as are all the signs, the Sun, Moon and planets [D]. The letters MC in Libra stand for Medium Coeli, the centre point of the heavens. The planets, primary forces in the astrological scheme, pass at differing speeds out of one sign of the Zodiac and into the next in a never-ending circuit. At the birth point, the angles formed with the Earth are measured in degrees [figures]. When there are specific angular relationships between them they are in Aspect [green, red and orange lines]. From the Aspects, astrologers interpret the subject's potential personality and motivation. A predominance of planets in certain signs is significant as each sign is assigned one of three qualities [A]. The quality of enterprise belongs to Aries, Cancer, Libra and Capricorn [1]. Taurus, Leo, Scorpio and Aquarius are fixed or steadfast [2]. Gemini, Virgo, Sagittarius and Pisces are mutable or adaptable [3]. The signs are also allotted one of four elements [C]. Aries, Leo and Sagittarius correspond to Fire [4], Taurus, Virgo and Capricorn to Earth [5], Gemini, Libra and Aquarius to Air [6] and Cancer, Scorpio and Pisces to Water [7]. Signs with the elements of Fire and Air are masculine or positive; those of Earth and Water feminine or negative.

D

♈ Aries	♎ Libra	☉ Sun	♄ Saturn
♉ Taurus	♏ Scorpio	☽ Moon	♅ Uranus
♊ Gemini	♐ Sagittarius	☿ Mercury	♆ Neptune
♋ Cancer	♑ Capricorn	♀ Venus	♇ Pluto
♌ Leo	♒ Aquarius	♂ Mars	⊕ Earth
♍ Virgo	♓ Pisces	♃ Jupiter	

2 A baby's first cry is usually taken by astrologers to signify the moment of birth. Accurate timing is crucial, as shown by this medieval woodcut of astrologers charting the celestial bodies at that particular moment. If the time of birth is not known an astrologer will use the planetary positions as at noon on the birthday as an approximation. This, however, will restrict his conclusions.

3 The rulership of the signs springs from the fact that traditionally all the planets of the Solar System have had special relationships with certain signs of the Zodiac. The Sun and Moon ruled one sign each, with Mercury, Venus, Mars, Jupiter and Saturn ruling two. This remained the case until the discovery of the so-called "modern planets" (Uranus in 1781, Neptune in 1846 and Pluto in 1930). Astrologers studied many thousands of birth charts to discover what effect they would have. Eventually the new planets were allotted joint rulerships with the "old". Uranus now shares the rulership of Aquarius with Saturn, Neptune the rulership of Pisces with Jupiter, and Pluto that of Scorpio with Mars. Modern astrologers consider the effect of Uranus on Aquarius to be stronger than that of Saturn, and that of Neptune on Pisces stronger than that of Jupiter. Pluto's influence is still a matter of debate.

king, who embodied the affairs of the state and its well-being. The Mesopotamian tradition may have been transmitted to Egypt and through the Middle East to India and thence to China and the rest of Asia. In about the fourth century BC, the Greeks began recasting astrological lore in terms of their own traditions. It was they who popularized a method of working out individual destinies based on the moment of birth. Ptolemy (c. AD 90–168) is credited with the first astrological textbook composed in the West – the *Tetrabiblos*. The planets, houses and signs of the Zodiac were rationalized and set down in a way that has changed little since.

With the splitting up of the Roman Empire, the Arabs became the chief exponents of astrology and astronomy. To calculate their horoscopes and other charts they needed to know the positions of the stars and movements of the planets with great accuracy. This led them to draw up tables of planetary motion, together with star catalogues, which surpassed anything that the Greeks produced. But after the rediscovery of the Hellenic tradition in Europe

during the fifteenth and sixteenth centuries astrology was ranked as one of the foremost sciences in all European universities.

From Copernicus to the space age

A gradual ebbing of interest began with the great astronomical discoveries made from the time of Nicolas Copernicus (1473–1543), who showed that the Sun, not the Earth, was the centre of the planetary system. The invention of the telescope was followed by the work of Isaac Newton (1642–1727) whose book *Principia* laid the foundations of modern astronomy. But although the Earth was no longer regarded as all-important in the universe as a whole, the basic principles of astrology remained unchanged. Newton and Copernicus were profoundly interested in astrology.

Recent technical developments are bringing greater understanding of the universe and of its influence on living things. In an age when men have been to the Moon and sent messages out towards the planets it may be significant that interest in astrology is reviving rather than passing into total eclipse.

KEY

The Zodiac [1] with its symbolic signs and the apparent path of the Sun [2] circles the Earth, which is shown at the centre of a cutaway sphere.

4

4 The 12 houses occupy the central section of a birth chart. They are associated with everyday activities, some of which are listed below. The influence of the planets and signs that fall within a particular house is focused on those activities. If a house has no planet within its segment of the chart this does not mean that the sphere of life that the house is normally concerned with is of no importance to the subject. In this case the astrologer will consider the Zodiacal sign that is on the cusp or starting point of the house. Each of the houses has its partner across the chart, the first relating to the seventh, the second to the eighth and so on. The houses are believed to show a relationship between the Zodiacal signs and the turning of the Earth on its axis. The symbolic nature of the houses was originally worked out by the Greeks.

1st house:
personality
health
temperament
7th house:
emotional life
business partners
marriage

2nd house:
possessions
worldly resources
income
8th house:
inherited money
life forces
insurance

3rd house:
family ties
education
communication
speech
9th house:
further education
long-distance travel
languages

4th house:
home
family commitments
start and end of life
10th house:
career
social standing
personal image

5th house:
creativity
love affairs
children
11th house:
societies
friends
intellectual pleasures

6th house:
physical well-being
subordinates
work
hobbies
12th house:
service to others
escapism
the unconscious

6 Ptolemy [A] (AD 120-80), the last great astronomer of classical times, who charted 300 new stars, was chiefly an astrologer. Paracelsus [B] (1493-1541), a pioneer of medical chemistry, was interested in the interrelationships between man and the planets. Louis de Wohl [C] (1903-61) was employed by the British to predict what astrological advice was being given to Hitler.

5 Astrological medicine was, until the eighteenth century, an important part of conventional medical practice. The various parts of the body were regarded as being under the rulership of specific signs and planets, which were also associated with specific diseases. More recently, astrologers have stressed relationships between the signs and the glandular and nervous systems. They hold that this relationship is often one of polarity; an Arian, for example, may be affected by ailments of parts of the body ruled by the opposite sign, Libra. Astrological anatomy thus divides the body according to the ruling signs and their opposites.

5 ♊

Gemini-Sagittarius
Gemini rules the lungs, nerves, arms and shoulders. Geminians are, therefore, prone to break collarbones.

♌

Leo-Aquarius
Leo rules the heart, spine and back. This makes Leos prone to heart attacks.

♎

Libra-Aries
Libra rules the kidneys. Any disturbance of the usually well-balanced Libran way of life, through accident or argument, is soon reflected in a serious kidney upset.

♏

Scorpio-Taurus
Scorpio rules the sexual organs. Scorpians are the most highly sexed of all the signs and, with their tendency to do nothing by halves, any frustration or suppression of this sexual energy can lead to unpleasant or cruel behaviour.

♑

Capricorn-Cancer
Capricorn rules the knees, bones and teeth. Orthopaedic and dental troubles are accordingly relatively common, as is any disease, such as rheumatism, that limits movement.

♓

Pisces-Virgo
Pisces rules the feet. Pisceans do not respond well to drugs so these must be administered carefully.

♈

Aries-Libra
Aries rules the head and as a result Arians often suffer from headaches. Aries also controls the supply of adrenalin.

♉

Taurus-Scorpio
Taurus rules the throat and neck, which makes Taureans vulnerable to colds and chills.

♋

Cancer-Capricorn
Cancer rules the stomach and the alimentary canal. Indigestion and ulcers are fairly common, giving Cancerians a reputation for delicate health.

♍

Virgo-Pisces
Virgo rules the nervous system and intestines. Virgoans tend to worry so suffer from indigestion.

♐

Sagittarius-Gemini
Sagittarius rules the liver, hips and thighs. Being active by nature, Sagittarians stagnate if they do not get a considerable amount of physical exercise. They tend to put on weight easily, which in women particularly goes to the hips and thighs.

♒

Aquarius-Leo
Aquarius rules the circulation, so Aquarians suffer from varicose veins and hardening of the arteries.

201

The meaning of ritual

Ritual involves behaviour that cannot easily be explained by what the participants hope to achieve. This indirect connection between actions and results is a striking aspect of ritual, whether seen in the elaborate courting dances of birds [1], in great state occasions such as the 14 July parade in France [Key] or in formal greetings [3, 4]. It is not immediately clear how the dancing of grebes, for instance, facilitates copulation or how the movements and words of a witch doctor ultimately can achieve a cure.

Human and animal ritual

Ritual behaviour is displayed by both man and animals. The common factor is that the elaborate way in which an action is performed is as important as, or more important than, the action itself. Equally, in both human and animal ritual repetition is essential. Rituals follow an intensely detailed programme that rigidly specifies the movements and sounds that should be made and their order of performance, so that the ritual is always the same.

Although the rituals of animals and men share these characteristics, they nevertheless differ fundamentally. Most animal rituals are instinctive [2], while human rituals have to be learned. Birds whose courting involves complex rituals, for example, display the same kind of male or female behaviour throughout the species. But in man, ritual surrounding a similar occasion varies extensively from one social group to another and also within groups; some tribes, for example, prescribe certain ritualistic roles for initiated youths that are entirely different from the roles taken by the uninitiated.

Human ritual is of two rather different kinds. On the one hand it may be simply an aspect of an activity that is not itself ritual — a cultural "frill". For example, in all cultures the essential activity of eating is accompanied by ritual embellishments that are given great importance and which are called table manners. On the other hand some entire activities are "rituals". In the masonic ritual, for example, the freemasons express their system of morality by allegorical acts, in which the tools of a working mason are used as symbols.

The nature of ritual can be seen most clearly by comparing the everyday use of language, gestures and signs with their use in ritual. Ritual language is of a different type from everyday language: it tends to be more formal and is often archaic. Indeed rituals are often carried on in a totally different language from the normal, such as Latin, Hebrew or Arabic. Ritual speech is more ponderous or may take on a singsong character. Most commonly, ritual uses singing and chanting rather than plain speaking. Ritual gestures are also characteristic. They tend to be both stylized and expressive, often verging on dance as in military or some religious rituals; indeed dance is a common element of the most sacred and important rituals [5].

Religious and secular forms

A third characteristic of human ritual is the use of objects as symbols [6] displayed on special occasions or used to give meaning to certain actions. The flags of military parades are an obvious example; more complicated but no less typical is the symbolic use of water in baptism or the use of animals and plants in

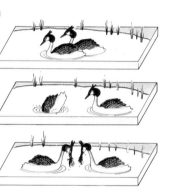
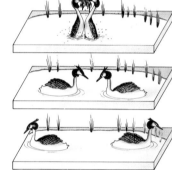

1 Animal ritual is more fixed than human ritual. These great crested grebes have not had to learn their complicated courtship movements from watching other birds but know what to do instinctively. The similarity with human ritual lies in the fact that the sequence of actions is predetermined, repetitive and only indirectly related to the end in view: in this instance, mating

2 Many animals, such as the red hartebeest, found in southwest Africa, establish dominance in the herd by regular stylized combats; the boundary between ritual and purposeful behaviour is often indistinct.

3 Raising hats and shaking hands are examples of the sort of ritual that pervades all human life, sometimes unconsciously. Elaborate procedures of greeting and leave-taking are characteristic of human societies throughout the world. In one sense, they appear meaningless because they seem to convey no information. But the significance of such rituals – once they are established in a society – appears when they are omitted and cause grave offence.

4 Maoris in New Zealand rub noses in a greeting ritual that performs much the same function as the French custom of kissing on both cheeks. One of the interesting aspects of human ritual is that although the use of ritual on certain occasions such as greeting is extremely widespread, the particular form it takes can vary considerably from culture to culture. Animal ritual, by contrast, shows little variation within a species.

5 Zulu war dances, and many other tribal dances, have strong elements of ritual. The arts of both theatre and dance originated in religious rituals and many secular rituals also involve dancing or its equivalent: highly stylized gestures or words following in fixed order according to recognized rules and often performed in unison. The use of the body in a more stylized way is characteristic of many forms of ritual behaviour.

fertility ceremonies to represent the various forces of nature.

Activities that are associated with ritual behaviour have some remarkably similar aspects in all cultures, whether in remote parts of New Guinea or in New York, and they fall into two categories. Rituals such as initiation ceremonies, church services or invocation of ancestors are clearly of a religious character. Others, such as formal government receptions for visiting dignitaries, are clearly secular. Often, however, it is not clear which element dominates. The Chinese New Year ceremony [9], for instance, combines secular elements marking the passage of the seasons and the settling of debts with more religious aspects such as ancestor worship and purification.

Greetings are perhaps the most common secular rituals. When people meet, in all cultures, they usually have to perform a series of set phrases in a given sequence, accompanied by equally formal gestures and postures. Greeting rituals are significant because they are a convenient way of establishing or acknowledging personal relationships and

giving order to social life. They express the degree of familiarity between people, or their differences in rank. It is not surprising that the greater these differences, the more elaborate the ritual aspects of greetings, so that a monarch, for example, is greeted with more ritual than a friend. What is begun by a greeting is continued by other ritual aspects of social behaviour, such as rules of etiquette and politeness, and is completed by another ritual sequence – of leave-taking.

Defining relationships

Even relationships between equals may have ritual aspects, although these take various forms. In Madagascar, for instance, brothers in-law go through standardized, obligatory routines of joking with each other and fixed, reciprocal insults. It is hard to see the practical value of many social rituals, but they help to define relationships and give a sense of predictability. Instead of every new social encounter having to be treated differently, ritual offers routine forms of behaviour so that people and circumstances do not have to be evaluated instantly.

A ritual such as the laying of a wreath to commemorate the nation's dead unites the whole community. It is characteristic of all ritual that it stresses the importance of the group rather than of the individual. The carrying out of some rituals, even after they have lost their significance for many who take part, has led many to reject ritual as meaningless.

6 A 15th-century stained glass window represents the four evangelists symbolically rather than pictorially. Such symbolism depends on social conventions. Rituals of all kinds characteristically refer to objects, people, events and emotions or abstract concepts by symbols that often seem to be chosen in a quite arbitrary way. Here, John is represented by an eagle, Matthew by an angel, Mark by a lion and Luke by an ox. Another example is the use of a lamb to represent Christ, although the symbolism of this is clear. The reason why symbols and ritual usually go hand in hand is complicated. Ritual uses stylized language, song and dance in a generalized way to evoke wide agreement. Symbols contribute to the remoteness and vagueness of ritual while adding depth.

7 Right- or left-handedness is given special ritual meaning in many cultures.

The left hand is associated with evil and darkness, the right with good and the light. Use of one or the other is specifically prescribed for some actions.

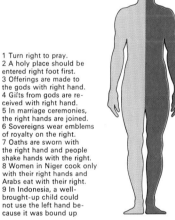

1 Turn right to pray.
2 A holy place should be entered right foot first.
3 Offerings are made to the gods with right hand.
4 Gifts from gods are received with right hand.
5 In marriage ceremonies, the right hands are joined.
6 Sovereigns wear emblems of royalty on the right.
7 Oaths are sworn with the right hand and people shake hands with the right.
8 Women in Niger cook only with their right hands and Arabs eat with their right.
9 In Indonesia, a well-brought-up child could not use the left hand because it was bound up

1 If greedy spirits of the dead need placating by a gift it is always given with the left hand.
2 Sinners leave the church by the left door.
3 Guinea Coast tribesmen believe they have the power to kill by putting the left thumb in an enemy's drink.
4 Left-handedness marks a possible sorcerer.
5 Some Christian saints are supposed to have refused as babies to suckle their mother's left breast.
6 In the Arabic world, toilet cleaning is done with the left hand.
7 Left is the hand of perjury, treachery and fraud

8 The coronation of Queen Elizabeth II of England in 1953 was an example of the use of elaborate ritual to transform a person from one role to another. Wedding and consecration ceremonies serve a similar purpose. Ritual is a means of formalizing and acknowledging social position.

9 The dragon procession mounted at new year by the Chinese community of San Francisco, USA, is one of countless ritual festivities that mark the passing of years or of seasons. Rituals have always been used to organize time and to attempt to give meaning to its passing by dividing it into units given special significance.

Types of ritual

Ritual colours all aspects of life from the most commonplace to the most sacred and is present in differing degrees in all known cultures. One function of ritual is always to give the appearance of order, security and meaning to the unpredictable sequence of events that characterizes human life.

Another major function of ritual is to define situations and people's roles in them according to set routines. This is clearly shown in such rituals as initiation in tribal societies, marriage and installation ceremonies. Anthropologists call these rites of passage because they are rituals that involve the public declaration of a change of status of an individual [2, 3, 4].

Rites of passage

The number of rites of passage and their significance vary widely from culture to culture. An East African Samburu man [1] would traditionally go through complicated rituals at birth, at initiation, on becoming a warrior, at marriage, on becoming a junior elder, on becoming a senior elder and finally at death, whereas a European Christian

would probably go through elaborate rituals only at baptism, confirmation, marriage and death. The latter might, however, also go through less universal rituals associated with entering a club, a school or an association or on the occasion of becoming a public official such as a judge or a mayor.

However varied such occasions might appear, rites of passage usually follow a similar pattern: first they consist of a phase marking separation from the old status, then an intermediate phase, thought to be particularly significant, then a phase marking acceptance of the new status. In the traditional European marriage, for example, the stag party marks separation from the old status, the central phase consists of the marriage ceremony and acceptance of the new status is represented by customs such as the groom carrying the bride across the threshold of her new home.

The same general features are found again and again in the three stages of rites of passage. Separation is marked by ritual acting out of the status that is being lost – being a single man with "the boys", for

instance. This is often accompanied by ritual violence, especially the revenge of the group towards the individual who is in some sense betraying them by leaving. A parallel violence from the group that is being joined may take the form of an endurance test. The intermediate stage is always a period set apart from ordinary life and is sacred, solemn and is sometimes dangerous. The final period of reintegration is usually happier expressing relief that the transition is complete.

Seasonal rituals

Rituals that mark certain times of the year and the passage of the seasons have been called rites of intensification because they intensify general group solidarity. Nearly everywhere the beginning of growing seasons especially for plants and animals on which people depend for food, is marked by rituals associated with fertility and rebirth. Similarly, rituals accompany the harvest. Sometimes the beginning of the harvest and gathering of the first fruits is chosen as the occasion for ritual, as in much of Asia. Alternatively, rituals may mark the end of

CONNECTIONS

See also
202 The meaning of ritual
222 Methods of worship
108 Death, grieving and loss

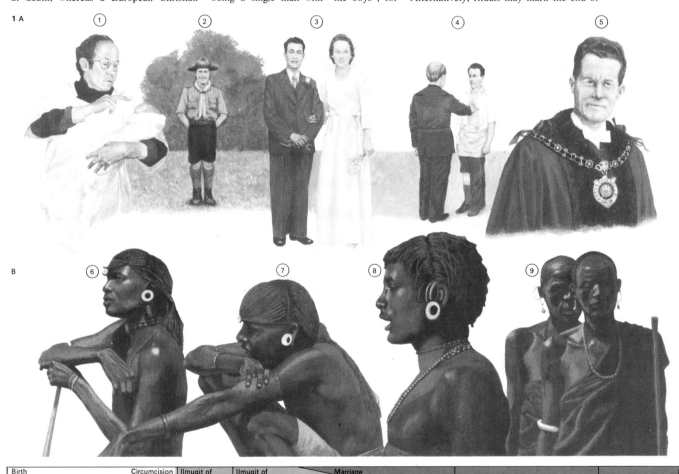

Birth		Circumcision	Ilmugit of the Arrows	Ilmugit of the Name	Ilmugit of the Bull	Marriage	Ilmugit of the Milk and Leaves			Burial
Age in years			Junior moran 15	Senior moran 20			Junior elder 30	Firestick elder 40	Senior elder 55	

1 Rites of passage, which mark the movement of an individual within society, vary widely between communities and cultures. In most Western societies [A] people do not pass through any set pattern of stages. A male may be baptized [1], confirmed

or married [3], or become a boy scout [2], an elector, a member of a club, a church warden, a freemason [4] or a mayor [5]. Many of these stages involve little ceremony and none is obligatory for all members of the society. In contrast to this

pattern, in a society such as that of the Samburu [B] of East Africa every member of the tribe goes through each stage, which is accompanied by elaborate ritual. The birth of a son will be marked by the gift of cattle. Until he reaches puberty, the

boy will grow and braid his hair [6]. At the age of 15 he is circumcised and joins a group of other youths as an initiate. A month later he becomes a junior moran (warrior) by going through the Ilmugit (ceremony) of the Arrows during

which he slaughters an ox and vows to his mother to keep certain laws. To mark his status he first wears red ochre [7] and then cuts his hair [8]. When the group of youths reach the age of 20 they attain senior moran status through the Ilmugit

of the Name. This is the most important ceremony and is repeated a month later. A senior moran may father children but may not marry. The Ilmugit of the Bull, at about 26 is the first of a number of ceremonies that lead to marriage and

elderhood, culminating in the Ilmugit of the Milk and Leaves at the age of 31. Once a man is an elder [9] he moves steadily upwards within this social order as new groups of boys are introduced at the initiate stage.

the harvest, as in the harvest festivals of Europe. All these rituals embody the same idea – the public recognition of natural changes in the lives of people and seasons and the activities associated with them.

Other common occasions for rituals arise when a community feels that order has been disturbed and needs restoring. The most universal examples are rituals of purification. Throughout the world there are rituals to restore purity after contact with death, often involving a ceremony of washing or stepping over fire. Many of the rituals that follow battles also contain this element. Rituals are common in many countries to purify women after childbirth, after menstruation and even after sexual intercourse. Perhaps nowhere are purification rituals so elaborate as in India [5] where, for example, high-caste Hindus have to purify themselves after involuntary contact with any one of innumerable polluting substances.

Other important restorative rituals are those involved with curing. Disease, whether physical or psychological, is a threat to the order emphasized by ritual, so particularly

complex rituals are needed to bring back a patient to the proper state. For example, among certain tribal peoples of South-East Asia disease is thought to involve loss of the patient's soul and prolonged rituals act out the process of recapturing it and making the sick man whole once more. In some religions, restorative rituals of this type are believed to be regularly necessary. Because disease or sin are thought of as inevitable, "curing" is always needed. Christian, Judaic and Muslim worship may be seen as examples of restorative ritual in religion.

Rituals of sacrifice

A recurring theme of the most holy rituals is sacrifice [6], a form that important rituals take throughout the world irrespective of culture or religion. Sacrifice is basically the giving up of something, whether the killing of an animal or a human being, or material offerings. By this means a short period of communication is established with the supernatural or life-giving power. In restorative rituals pollution, disease or sin can then be taken away so that order may be restored.

Ritual, in one sense of the word, is an element of almost all human actions in the form of symbolic elaboration. But there are occasions when this symbolic elaboration predominates and the relationship between means and ends is different from that found in everyday behaviour. Here, in High Mass in Jerusalem, the Last Supper is re-enacted so as to include the congregation in Christ's sacrifice, but no attempt is made at realism. Instead, the ceremony involves gestures laid down by church tradition and symbols whose power is allusive. These elements of symbolic enactment and the establishment of authority characterize ritual.

2 Christian baptism is an example of one of many similar ceremonies that, in some communities, mark the social beginnings of the individual in the world. Such ceremonies nearly always occur at a different time than the person's actual birth. In nearly all cultures of the world, the passage of man through life and the bodily changes that accompany this are marked by ritual.

3 The circumcision ceremony in Madagascar parallels baptism in a different culture. Here, the banana tree symbolizes the bountiful fertility of nature that the ceremony bestows on the child.

4 A funeral procession in Bali is an instance of the elaborate rituals that surround death, expressing separation of the dead from the living and defining the social status of the dead person.

6 The supreme religious ritual, whether in tribal ancestor worship or Christianity, involves a sacrificial gift that establishes a moment of communication between men and gods. The most potent gift is often the life of an animal, as seen here with the sacrifice of a camel by Turkana tribesmen in Kenya. This may take away a god's displeasure from the sacrificer or alternatively enhance his life-giving power.

5 Ritual bathing by Hindus in the holy waters of the River Ganges is typical of rituals that are seen as having a restorative function, purifying the participants of their failings and so allowing them to regain their proper place in the scheme of the universe. The same symbols often recur in different cultural traditions. Water is a symbol of purification and incorporation for Hindus, as it is for Christians in the baptism ceremony.

The meaning and function of myth

Myths are tales or traditions that seek to explain the place of man in the universe, the nature of society, the relationship between the individual and the world that he perceives, and the meaning of occurrences in nature. Today we tend to draw a sharp line between facts that can be proved scientifically and beliefs and ideas that cannot be so proved. These latter are often lumped together and dismissed as "imagination" or "inventions" or "myths". This spurious contrast between myth or fantasy on the one hand and "hard facts" on the other conceals and distorts the value and significance of myths as guides to life.

Myth, science and religion
Myths are found in every part of the world and, despite their bewildering variety, share certain common characteristics [1, 3, 4, 8]. These similarities arise because men everywhere face the same basic problems and ask the same questions. They want to know why they are what they are, why nature behaves as it does and how cause and effect are linked. It is human nature to seek for

meanings as well as causes for everything that arises in consciousness.

Although science has now answered many of the "how" questions, the reasons "why" – man's relation to the cosmos and the nature of the life-force within him – remain basically unanswered and unanswerable.

Myths have in common with religions the fact that they both offer reasons as well as causes; both the "how" and the "why" of the universe. By comparison with many world religions, however, most myths are less concerned with direct guidance. They contain an implicit moral, but their main aim is not to impose it. They are "just so" stories concerned with explaining the unquantifiable aspects of existence and deal with both common human experiences and with the supernatural.

From the mythological standpoint the world we experience directly is not the only world. The phenomenon of birth can be understood as a physical process but that does not exclude it also being regarded as a supranatural event (for example, as a reincarnation). Indeed most people will

admit, if they are truthful, that they actually experience life on two levels – the scientific and the mythological. But in our increasingly literal-minded, science-dominated society it is only in extreme situations – where the rational fabric of society breaks down – that our mythological consciousness surfaces.

The mental processes behind mythology
The step-by-step logical thinking required for the acquisition of scientific knowledge is slow and laborious. It is much easier to arrive at conclusions by comparisons and analogies where "just as . . . so in" are the key words.

Myths explain the phenomena of nature, for instance, by drawing parallels between simple, known things and those that are harder to grasp. Fire has something in common with the sun, the source of heat and energy. Gold is shiny and resembles the sun in colour. It also does not rust with weather and therefore signifies immortality. So out of common physical characteristics, symbolic equations are made and one thing takes on the qualities of another.

As the egg originates life, so the world

CONNECTIONS

See also
208 Myths of spring
210 Myths of summer
212 Myths of autumn
214 Myths of winter
216 The nature of
    religion

1 **Fantastical creatures** play an important part in many myths, probably because they appeal to man's need for the unusual and to his awe of the forces around him. The fire-breathing Chimera [9], for instance, expresses the power of the volcanoes it inhabits. The creatures and events of myth may also be shadows of actual events. Hercules' struggle with the Hydra [4] may mirror the draining of the marshes by some former king; while the Centaurs [15], half-horse, half-human, may have originated with the famous horsemen of Thessaly. Others pictured are Hippocampus [1], the Mermaid [2], the winged Lamassu [3] of Babylon, the Gryphon [5] from Asia, the Satyr [6], Sirens [7], Medusa [8], the Sphinx [10] from Egypt, Minotaur [11], Pegasus [12], Cerberus [13], the Unicorn [14], La Tavasque [16], and the Dragon [17].

2 **Myths create gods in the form of man.** This anthropomorphism was at its height in ancient Greece where the gods expressed in extreme form human qualities such as beauty, anger and love. Shown here is part of the Greek pantheon. Hermes [A] was the messenger of the gods, Apollo [B] was patron of music and Zeus [C] was the ruler of all the gods. Athene [D] was protector of Athens and guardian of the crafts. Demeter [E] was a fertility goddess, often associated with corn.

3 **The snake or dragon** that eats its own tail forms a circle signifying the cyclical nature of all things (a common mythological theme). This illustration from a 15th-century alchemical text bears a colour-coded message; green (faded) for the beginning, red for the goal of alchemy. Tail-eating creatures are found in myths from all over the world. A Japanese map shows a tail-eating snake whose movements under the earth are said to be the cause of earthquakes.

was created out of an egg [5]. Mountains are often inaccessible and inspire awe, as do the beings that man credits with having more power than himself. So the proper habitation for the gods may be a mountain, such as Mount Olympus, abode of the Greek gods. Thunder and lightning inspire fear as do outbursts of anger; hence a man killed by lightning must have offended Zeus, the chief Olympian [2]. Sometimes one characteristic is made use of in the equation, sometimes another. As thunder ushers in rain, so, where rain is scarce, thunder may symbolize fertility. Rivers, trees and animals, all have their characteristics, expressed in terms of such human values as cunning and fertility, destructiveness and courage.

The necessity of myths

But myths do not just explain why man and the world in which he lives are as he finds them. This view of mythology would be inadequate. Image-making is one of the most distinctive human characteristics. The telling of myths becomes a vital necessity not simply to placate or propitiate the suprahuman powers, but to stimulate the very same creative and spiritual gifts that made man invent his myths. Without meaning and purpose beyond the satisfying of the daily physical necessities no man and no culture can flower. By the same token man needs an understanding of his defeats and victories, of birth and death, in order to stave off the despair that the vagaries of fortune and the complexities of life might induce. There is, accordingly, a myth to answer almost every mood and question. There are myths of origin or creation, of fertility and heroism, of resurrection and immortality.

Myths are timeless and perpetual in the sense that man's need to live in harmony with his nature by means of guidelines (which we now call psychological rather than spiritual or religious) is as great as ever. Myths provide a bridge between outer "realities" and the hopes, wishes and fears of our dreams. They provide man with comfort and support. In them he may find a play area in a world which would otherwise be fearsome, unbearable, dull or frustrating. Such play is as necessary as our daily work.

This 1790 Italian map shows the stars visible from the Earth's Northern Hemisphere, grouped together by constellations whose symbols date back to a time when man's observations were influenced by the concepts of Greek mythology. The periphery is formed by the Zodiac. The Zodiacal signs indicate months.

4 This 13th-century Hindu wheel at Konarak, India, symbolizes the circular nature of the universe. The unending cycle of birth and death is a stock element in mythology. The couple, male and female, will always be there creating new generations. The idea of no birth without death signifies the eternal cycle to which Hinduism and Buddhism subscribe in different ways; the former aiming at the perfection of life, the latter at its renunciation.

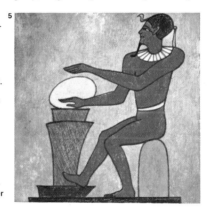

5 One version of Egyptian creation myths has Ptah of Memphis shaping the world, in the form of an egg, on a potter's wheel. But it was also he, the creator of all things, who dwelt in the egg while it was in the primeval waters. The image of the world egg appears in other mythologies; thus Phan-ku, in Chinese mythology, holds the egg of chaos, composed of the yin and yang (female and male) symbols, out of which he was born.

6 This Roman relief of Leda and the swan depicts one of the love stories in which Zeus couples with mortal women. The transformation of gods as well as mortals into animal form is a common mythological theme. Swans are held to be sacred birds in many places, and killing them is believed to bring misfortune.

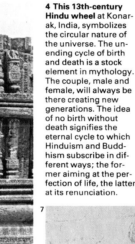

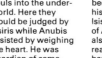

7 Anubis, the jackal-headed Egyptian god, like the Greek god Hermes, conducted souls into the underworld. Here they would be judged by Osiris while Anubis assisted by weighing the heart. He was guardian of cemeteries and may have been deified so as to prevent jackals from devouring the dead. After Osiris had been dismembered by his adversary Seth, Isis enlisted the help of Anubis, who was also the embalmer, to reassemble her husband's body, so creating the first mummy.

8 Hermaphroditic statues, such as this ancestral Nomo figure from Mali, are found the world over as part of the common preoccupation with fertility. Fertility rites may take the form of a sacrifice to ensure new life, or, as here, be fundamentally magical; bisexual figures can fertilize themselves, so guaranteeing offspring. The worship of such statues was expected to result in the fertility of land, animals and humans. Some carvings from Africa and Oceania are divided, showing a male and a female half.

207

Myths of spring

Myths can be divided either regionally, corresponding to the centres of civilization from which they originate, or according to their major themes. As some themes are common to all regions, albeit with varying emphasis, it is convenient to take one – the motif of the four seasons – and look at its mythological counterparts. Creation and hero myths (also called epics) correspond to the spring phase, representing the beginnings of mankind on earth.

The myths of creation

Creation myths deal with the origins of the world and presuppose that in the beginning there was something uncreated. This is usually presented as the Abyss or Chaos, vast and dark like Egyptian Nun, the primordial ocean in which lay the germs of all things and all beings before the creation.

The creator is invariably regarded as divine. But in some traditions, notably the Judaeo-Christian, he is a non-human abstract "Father" and is eternal. In others, such as the Greek or Egyptian, the emphasis is intensely biological. This has two conse-

quences. First, there are several versions, just as there is more than one method of procreation in the animal kingdom. Second, the ruling deity lives under the challenge of rivalry and death. (One explanation for the succession of gods that characterizes almost all mythologies is that it reflects invasions of an area by people with other gods, who have to be set above those of the conquered.)

The Egyptian god Atum (later known as Re) is usually described as a human male, but he is also referred to as bisexual, "that great He-She". Nevertheless, this ruler felt lonely and desired a companion. Atum created by masturbation the first creatures – Shu, who was male and represented air, and Tefnut, who was female and represented moisture. In another version he spat out Shu and Tefnut. To both he gave a vital essence, Ka, which may be regarded as the soul. This compares with the Jewish version of God breathing His divine life into Adam. In an earlier Egyptian myth, Atum is a serpent living in the dark waters of the Abyss, his outer coils forming the limits of the world. In subsequent versions, the creator was a mongoose which

killed the snake (that is, himself), then a primeval goose and then an egg. All this took place in darkness, before heaven and earth were separated and before light was created (the reverse order from the account of the Creation in the Bible).

This Egyptian creation myth demonstrates two general principles. First, a mythology consists of several layers, comprising older and later versions which may co-exist and form an amalgam. Second, myths of various regions share common elements [2], for instance the breath-and-soul-giving and the fashioning of man out of earth.

The origins and functions of the hero

With the beginnings of the world and cosmos accounted for, man has to explain himself and his culture. How did he learn to make fire how to fish, hunt, rear domestic animals, cultivate the earth, discover medicines and, later, develop a complex culture?

Inventions of such extraordinary importance for man's survival were ascribed to cult heroes who had obviously been endowed with unusual talents. These heroes were

1 The giant Ymir was the first living being in Scandinavian myth; he was born from the melting ice and suckled by the cow Audumulla. After dying his body became land, his blood the seas, his skull the heavens, his bones mountains and his hair trees.

2 Trees play a large part in mythology, from the Indian Asvattha or tree of life and knowledge to this Scandinavian world tree Yggdrasil. This represents the entire world as a tree. The branches reach to the sky, the roots go down to Niflhel, the underworld. Near the root gushes the fountain Hvergelmir, the source of the rivers, while the disc-shaped middle world is encircled by a snake.

3 The phallus or lingam is a holy motif in India. The female counterpart is called yoni. They symbolize all creative energy – antagonistic yet co-operative forces of sex; father heaven and mother earth.

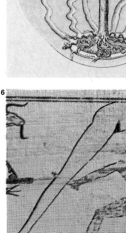

4 Izanagi and Izanami were brother and sister and the last of seven generations of Japanese gods. Standing on the floating bridge of heaven they created the island of Onokoro by stirring the ocean with a celestial lance. When they descended to the land, Izanagi walked round it from the left, Izanami from the right. When they met, Izanami spoke to her future husband first, expressing her pleasure at their sexual differences. So did Izanagi, but he was angry that the woman had spoken first. However, they produced two children; but as they turned out to be a leech and a foam island, they were both disowned. When the gods were consulted they said that the disaster was due to Izanami's mistake of having spoken first. So they went back and performed the ritual correctly and so produced the many islands of Japan. This myth illustrates the importance of ceremonial in Japan and suggests how the sexes should relate.

5 Tangaroa, a creator and sea god, appears in many Polynesian myths. In Tonga and Samoa he existed alone above a vast expanse of water and then threw down a stone that became land. His bird messenger planted a vine, but it rotted and in the

decomposing matter a swarm of maggots became men and women. In the Society Islands he is pictured as existing in an egg-like shell that revolved in space with no sky, land or sun – a parallel with the womb and a common mythological idea.

6 The separation of the Egyptian goddess Nut (sky) from her brother Geb (earth) is an example of the theme of separation from a previous state of existence in creation myths. Their embrace was so close that their father Shu (air) had to help. Inter-

estingly, the Egyptian cosmos differs from most others in making the earth male and the sky female. The daily birth of the sun out of Nut (also identified with Hathor or Isis, all versions of the great mother goddess of Egypt) is the next step in the creation myth.

generally the illegitimate sons of gods who were persecuted by their fathers' offended spouses. The heroes' deeds and inventions benefited man both materially and spiritually. They served as models for people who had to live and struggle, suffer defeats, enjoy some triumphs and die. Thus each epic forms a cycle analogous to the rise, zenith and sinking of the sun each day.

The dawn (or birth) phase of the hero's life, well illustrated in the myth of Hercules [11], foreshadows the aims and objects of his existence. In Hercules' case these were the preservation of life by saving a land from an oppressive ruler, the founding of a civilization and the renewal and assertion of man's spiritual potential over a lowly and purely instinct-driven life [9,10].

Trickery and stealth

Another version of the opening phase may be seen when the heroic deed is accomplished by stealth (notably trickery or theft) [8]. As usual, the ruling powers are offended, but despite the punishment they inflict the deed is done and mankind has progressed one step

nearer equality with the gods.

Among the best-known hero-thieves in Western civilization is Prometheus, who stole the fire that had been the privilege of the Olympians. The significance of fire was not only that man could now cook formerly inedible foods and keep warm. Fire also gave him light – an analogy for an increase of human consciousness. Zeus punished Prometheus by chaining him to a rock, where his liver was eaten in day-time by an eagle but at night it grew again in readiness for the next feast. Zeus also offset the benefit that Prometheus had brought to mankind by a devious and malevolent device : he had the beautiful Pandora (All-giving) created and sent to Epimetheus, Prometheus' stupid brother, who married her. She it was who took the lid off a box out of which flew all the diseases and sufferings that afflict mankind. Comparison with the paradisaical story of the Fall – embracing the view of woman as the dangerous temptress – is obvious. To overcome her threatening power was one of the hero's tasks. This done, woman could become his indispensable helper.

Spring, the time when the fields are prepared for the seed, para- llels the myths of the creation of the earth. In myths the plough is portrayed as phallic, while the furrows represent woman.

7

7 According to an Australian myth, daylight is created when the morning star is blown into the sky by the east wind. Observation of the night sky gave the Aborigines the idea of time as an eternal cycle.

8

8 Maui, a hero from Oceania, achieved his deeds through trickery. He lassoed the sun to give man a full day and stole fire (like Prometheus) from the gods. A clown, he expresses man's need to poke fun at the gods.

9

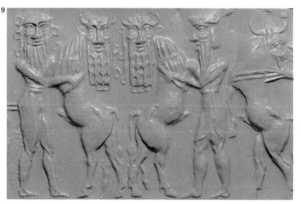

9 Gilgamesh, a hero of Babylonian myths, fought and overcame monsters. In this seal, c. 2200 BC, he and Enkidu, who is still close to the animal stage from which heroes evolve (reflecting the evolution of man from brute creation), fight a bull that was sent against Gilgamesh by Ishtar after he had repulsed her. Among the hero's functions is the conquest of instinct-driven nature and the establishment of civilization.

10

10 St George, seen here in a 17th-century Ethiopian painting, may have originated in an historical figure who lived in Palestine in the 3rd century. His task, shared with other heroes, is to overcome evil in the form of the dragon and free the maiden. The Greek hero Perseus likewise rescued Andromeda from a dragon after killing the dreaded Gorgon.

11 Hercules, here by Antonio Pollaiuolo (c. 1429–98), rescued the Greeks from many dangers. The Lernaean Hydra was ravaging the country. Every time one of her heads was cut off, two sprang up in its place. He solved this by burning them with red-hot brands. By dipping his arrows in her blood, he made them poisonous.

11

Myths of summer

Summer, with the sun at the height of its power, is understandably associated with images of the hero's achievement and female receptiveness. The myths of this season embody the idea of the union and the fruit of sky and earth on the sexual and reproductive planes as well as the spiritual.

Woman as goddess

There can be little doubt about mythology having been told and recorded from a man's point of view. Woman is regarded as the second sex, a newcomer to creation and a definitely inferior and possibly evil one at that. Myths describing the transformation and psychological development of the individual rarely have a heroine as the centrepiece. The story of Eros and Psyche is one of the few. Woman is usually shown to play diametrically opposed roles in mythology – as a source of life, on the one hand, and as a dangerous temptress and ruthless destroyer on the other.

Given that on one level myths reflect human emotions and attributes, there are several possible explanations for this equivocal attitude. One is that women are seen as different sexually, and intolerance for what is different is a consistent characteristic of mankind. When this "other" is also desired, then it may be loathed and feared as well as loved and idealized. Another explanation stems from a baby's experience of its mother. This earliest and most formative relationship see-saws (erratically as far as the baby is concerned) between a warm, protective love and anger and punishment. Consequently an ambivalent attitude towards women develops. Women and the goddesses who represent them come to symbolize the dread aroused by the unpredictable hazards of man's life.

The mother goddesses round the world are seen equivocally as givers and takers of life, as personifications of the earth, as creators of animals and vegetation and as goddesses of love, marriage and maternity. They appear, with some or all these characteristics, under many names: as Kali (India), Inanna (Sumeria), Ishtar (Babylon), Astarte or Anat (Canaan), Aphrodite, Demeter and Artemis (Greece), Cybele and Venus (Rome), Isis (Egypt), Ma (Anatolia) and Freya (Scandinavia). Their rites range from the decorous to the orgiastic and sometimes include temple prostitution.

The hero and women

The hero's encounters with women reveal man's attitude towards women and also embody lessons about how man comes to terms with the conflicting urges in his nature. There is an initial theme of separation from the dangerous (or incestuous) union with the mother. Subsequently, woman, in the form of the fair maiden or the king's daughter, becomes a hard-won prize, either as the goal of the hero's quest or as a helper, inspiring him to accomplish his mission, as Ariadne did when she fell in love with Theseus.

He had come from Athens to King Minos of Crete to pay the yearly tribute of seven youths and seven maidens whose fate was to be devoured by a monster bull, the Minotaur, which lived in the labyrinth. Without the thread that Ariadne gave him, he would not have found his way out of the labyrinth where others had perished. But in terms of his

1 Surviving myths almost invariably portray the supreme diety as male. But extremely old objects such as this "Venus" (c. 20,000 BC) suggest that earlier the earth – the most revered of the gods, giver and taker of life – was represented as a woman and mother. This crude figure is hardly beautiful, unlike the Venus of the Romans yet it suggests an awe-inspiring strength.

2 Diana of Ephesus (not to be confused with the Roman huntress Diana) was probably of ancient Asiatic origin, although the Greeks found her temple and worship established in Ionia. Her prodigious power to suckle infants is portrayed by her many breasts. Multiplication of an attribute to suggest prowess is seen also in the Hindu pantheon whose gods often have several arms, legs or even heads.

3 The Virgin Mary took on many of the qualities of the mother goddess, but none of her fickleness or anger. She is more closely associated with the birth and death (pietà) of Jesus than any other part of his life. Leonardo da Vinci creates a mysterious effect by placing her in a strange, almost unearthly grotto. The cave and spring are classic mother symbols, the former representing the womb. The interpretation here has been elevated to a spiritual plane and away from the biological, which is usually at least implied by the presence of Mary's husband, Joseph, and of the animals.

4 Female figures, often holding children, are commonly kept in Bakongo ancestor cult shrines in Angola, to honour the founders of the family. From the viewpoint of union and fertilization, birth is the harvest; from the viewpoint of the new-born infant the cutting of the cord separates one state of existence from the last. The cycle leads from original unity through separations and initiations to final reunification. Familiarity with the phenomenon of birth does not destroy a sense of awe.

5 Medusa was the only one of the three Gorgons, the daughters of the ancient sea gods Phorcys and Ceto, who was mortal. Her head (from a 3rd-century AD Roman mosaic) was covered with snakes in place of hair, and struck such terror in the beholder that it turned him to stone. Perseus, armed with the cap of Hades to make him invisible and guided by Athene, succeeded in cutting off her head by sighting it in a polished shield and striking it with a sickle.

psychological development, his triumph seems premature. Although he carried Ariadne off, she got little thanks for her love and he abandoned her on a small island. Easy victories had spoilt him, so it is not surprising that he had further trouble with women. His capture, marriage and subsequent repudiation of the Amazon Antiope led to the invasion of Athens, as did his treatment of the young Helen, whom he carried off from Sparta. This story may be interpreted as a cautionary tale. Manhood is not easily acquired and without it woman remains a threat, a devouring, dangerous creature, whether in human form or, as often classically portrayed, as an evil monstrosity, such as Medusa, that needs to be slain [5].

Other elements of this theme are contained in one of the oldest epics recorded, that of Gilgamesh. The story of his partnership with Enkidu, a wild, barely human creature who had been captured while he was being seduced by a temple prostitute, their killing of a ferocious giant and the "bull of heaven", and Enkidu's death at the hands of the angry gods, contains among other themes

allusions to an incomplete development of heterosexuality. In psychological terms, the message is that fear of the opposite sex is overcome not by contempt (killing the monster), not by tributes and worship, but by the granting of equal status to the essential otherness of the opposite sex, recognizing its common humanity.

The pendulum of mythology
Mythological thinking, like man's nature itself, may be compared to a pendulum that swings between the earthy or biological [8] and the more abstract or spiritual [3]. Myths surrounding fertilization and childbirth show this wide spectrum particularly well. Zeus coupled with Leda in the form of a swan. When he fertilized Danaë (who gave birth to Perseus, slayer of Medusa) he did so as a shower of gold leaking through the roof of her prison. As the pendulum swings back, man's mythological thinking has matured and he accepts images of ordinary sexual intercourse, although perhaps under somewhat unusual circumstances, as the symbolic union of opposite qualities in general [6].

Summer is a time of ripeness and the full power of the sun's energy. Wine harvesting is associated with Dionysus, god of wine, who represents the intoxicating power of nature.

6 Marriage is a social institution not much honoured in mythology (especially not by Zeus) apart from occasional praise of marital fidelity, as in the story of Penelope waiting for Ulysses to return. Sexual union, however, is exalted in widely differing cults and rituals. In this 16th-century alchemical version of "the sacred marriage", king and queen, sun and moon join together, their dissimilar elements dissolving and merging in the water.

7 The branches of the Tree of Jesse, in this 12th-century French version, support his descendants, David, the Virgin Mary and Jesus. Like Sarah before her, Anna, mother of Mary, was barren for 20 years until "the Lord took mercy". The symbol of a tree as axis of the world is not unusual. Adonis, the Greek god of vegetation, was said to have been born from the trunk of a myrtle.

8 The yogic posture known as Yab-Yum (Father-Mother), seen in this 18th-century Tibetan bronze, unites the Hindu male god or "absolute reality" with his female counterpart, Shakti. No greater contrast can be imagined than that between this and the denial of sexuality in Christian tradition, although Mary is often called the consort of God. The female emanation of the Hindu god is called Lakshmi (in the case of Vishnu) or Parvati (in the case of Shiva). As the latter, she is portrayed as a beautiful young woman sitting close to her husband. The imagery of sexual union is also delicately expressed in a linear design of interlaced triangles which is called Shri-Yantra.

9 Pan, son of Hermes, with the legs, horns and beard of a goat, was a phallic shepherd-god of the pastures and woods who made the flocks prolific. He was worshipped originally in Arcadia and his cult spread throughout Greece in the 5th century BC. His revels, depicted here by Nicolas Poussin (1594–1665), were similar to those of Bacchus, the god of wine, called bacchanalia. Pan frolicked with nymphs at night but liked to frighten travellers, hence "panic".

Myths of autumn

Autumn, when the fertility and vigour of summer give way to the death of crops and the failing strength of the sun, is associated in mythology with the dying god or hero [2, 5, 7], the destructive power of the mother goddess [3, 6] and the death or peril of the earth and the creatures it supports.

The cycle of the seasons
The Greek story of Demeter and her daughter Persephone shows how myth tries to answer the central questions of life – in this case why the earth annually loses its fertility and nature apparently dies.

The god Zeus was both the father and uncle of Persephone. Without Demeter's knowledge he promised Persephone to his brother Hades (king of the underworld). While the girl was gathering flowers in the fields of Nysa, the earth suddenly opened and Hades carried her off. When she learned what had happened, Demeter, angry with Zeus, left Olympus and as a result, since she was the goddess of fertility, the earth was barren and nothing grew. Famine would have destroyed all creatures had not Zeus sent

Hermes to fetch Persephone back. Hades consented, but he had already given her a pomegranate to eat. Because she had eaten in the underworld she would have to spend a third of the year below the earth. The rest of the year she could spend happily reunited with her mother, who consequently allowed the earth to bear fruit again.

In this myth of fertility and death, Persephone in Hades is the seed corn in the ground; Persephone rejoined with her mother is the sprouting seed that nourishes man and animals. To mark the annual cycle, to make sure that it continued and to propitiate the goddess, a festival, the Eleusinia, was celebrated in Athens, as harvest festivals are in other parts of the world.

In addition to explaining the succession of the seasons, myths are concerned to explain shorter cycles such as the rising and setting of the sun [1] and the phases and eclipses of the moon. In Indo-European myths the orbits of the sun are often interpreted in terms of a horse and chariot. Surya, the Hindu sun-god, for instance, drove across the heavens in a flaming chariot, as did the Greek Helios and

the Slavonic Dazhbog. A Nordic version explains that the sun and moon move because they are being pursued by devouring wolves.

In India the moon represented the cup from which the gods drank Amrita, the elixir of immortality, and its eclipses were due to the monster Rahu. When the gods first extracted Amrita, by churning the Milk Oceans, Rahu stole the first sip. Vishnu immediately cut off his head, which began to pursue the moon ravenously. Eclipses happened when Rahu succeeded in swallowing the moon, but because he had no stomach the moon reappeared and the chase across the heavens was resumed.

Myths of the flood
Floods as periodic, world-destructive events occur so universally in myths that they may reflect actual events, although it is probable that many local inundations were interpreted as world events. The bare bones of these myths are the same. A great flood drowns all the inhabitants of the world with the exception of one man or family whose escape in a boat is made possible by advance warning.

1 The rising sun in Egyptian mythology is symbolized by the scarab beetle Khepri, here being lifted out of the primeval waters (from an 1150 BC papyrus). The *Book of the Dead* tells how the soul, in its journey through the underworld, reaches the divine solar barque where it can ask the god freely about the reasons for all the apparent disharmonies it has met during its lifetime. The scarab's habit of providing larval food by laying its eggs in a ball of dung (regarded by the Egyptians as a symbol of the world) made it seem eternally self-creating. In its cyclic decline, the sun-god was known as Re-Atum, or "the completed".

2 Mithraism was practised as a religion in Asia Minor centuries before Christ and was taken up as a cult by the Romans about AD 75 as a competitor to Christianity for some 200 years. An important element is the bull sacrifice demanded by the sun-god, Mithras, who is also the bull – slayer and slain. The sacrifice marks a state of transition similar to seasonal changes. It ensures fertility and purifies the human soul. The bull suffers; even the god has to avert his eyes. But from the various parts of the newly sacrificed animal a whole renewed cosmos is miraculously created.

3 As life feeds on life, so the beautiful young maiden Parvati changes into the death-dealing Hindu goddess Kali, seen here dancing on her husband, Shiva. What the goddess has bestowed, she will take away. She is the Black One, suitably adorned with a necklace of skulls. Often, she is shown brandishing scissors as well as a sword with which to cut the thread of life. Dishevelled and wild-eyed, with large fangs, protruding tongue and blood-dripping hands, she nevertheless embodies Shiva's dynamic energy, even when dancing on his corpse. The word Kali is the feminine for "time" and is a reminder of the brief life-span of everything in existence.

4 This Aztec mask, covered in turquoise, is believed to represent Tlaloc (or the Mayan Chaac). Although not a major deity in the Aztec pantheon, as the god of rain he was of considerable importance in the dry, hot climate of Mexico. His cult was the most horrible of all. For the festival in his honour, the priests sacrificed babies and young children which they afterward cooked and ate. if the children cried it was a sign of rain to come.

Eventually the gods are appeased, the floods recede and life reappears.

But there are interesting variants. In the Hindu version Manu, unlike the Hebrew Noah [8], was the only survivor because he had been warned of the flood by a fish [9]. When it was all over he felt lonely and wanted a wife. She was duly created by the gods who made her out of Manu's sacrifices of sour milk, butter and curds.

In the Mesopotamian *Epic of Gilgamesh* the survivors of the flood, Utnapishtim and his wife, had similar experiences but their relationship with the gods was not as personal as that between Noah and Yahweh. For one thing Utnapishtim was warned only by a subterfuge of Ea, lord of the waters and wisdom, who by doing so gave away a secret plan of the council of the gods. Nor was there afterwards, any promise by the gods that such a disaster would not happen again.

The coming of death
It is not surprising that death, the ultimate mystery of life, should be a universal theme in mythologies. Death is consistently seen as an intruder, not existing at the beginning when humans renewed their lives repeatedly as did snakes their skin or the moon in its phases. Usually, death appears as a result of an error, as a punishment, or by agreement.

The idea of death as an error often centres on a message that goes astray. In Africa, for instance, God sends the chameleon to tell the first men that they are to be immortal. But because it dawdles it is overtaken by the lizard, who is the messenger of death.

Death as a punishment (often because of a woman's fault, as in the biblical story of Adem and Eve) is a common motif. The Algonquin Indians of North America, for instance, hold that the Great Hare gave man immortality in a parcel that he was forbidden to open. His wife, being curious however, looked in and let immortality fly away.

Death by agreement appears in some parts of the world. A myth of the Greenland Eskimos states that in the beginning there was no death but also no sun. One old woman insisted that if it was impossible to have one without the other, it would be better to have both, as without light, life was worthless.

KEY

Mythological themes of death, mourning and dismemberment are summed up in a medieval scene of the end of a day's hunting with the death of the stag and trees in autumnal colours.

5 The cycle of life and death does not spare the gods of Scandinavian mythology, who are destroyed by monsters at the end of their rule. Here Odin is being devoured by the wolf Fenris. The twilight of the gods reflects the onset of the Nordic winter. But the gods are avenged by their sons. The wolf is slain by Odin's son and a new generation of men and women arise from the world tree, Yggdrasil.

6 Odysseus's encounter with the Sirens (from a Roman mosaic) during his return home after the fall of Troy is typical of threatening myths about women. The Sirens were beautiful maidens whose singing so enchanted sailors that they swam ashore and died miserably. Odysseus filled his rowers' ears with wax and had himself bound to the mast so that he could hear the sweet singing without diving overboard to his death.

7 The Greek myth of Actaeon being killed by hounds reveals a goddess in savage mood. Actaeon was hunting a stag when he caught sight of Diana (or Artemis) bathing with her maidens. In anger, she transformed him into his own quarry to be torn to pieces. It is superficially the story of a virago punishing a man's lust. But the stag was a sacred animal and Actaeon's ritualistic dismemberment perhaps implied an autumnal sacrifice to ensure the next harvest.

8 This multi-storied ark, from an 18th-century Ethiopian text on Noah, carries all the creatures needed to repopulate the world when the Flood finally subsides. Central to the flood myth is a warning to mankind not to be too proud.

9 Vishnu, the great Hindu god, can take on whatever incarnation (*dvatas*) is required. Three of the best known are in animal form: fish, tortoise and boar. During the Flood, man's ancestor, Manu, was saved by a fish whose life he had spared (Vishnu in disguise). The motif of a grateful animal saving the hero's life is well known in Western fairy tales and is perhaps as universal as the story of the Flood among the various myths of destruction.

213

Myths of winter

Death as ultimate finality is unacceptable to most human beings. Myth's function is, therefore, to explain that life in its known form must come to an end as inevitably as the arrival of winter [Key] but also to point to a future that is not accessible to our senses. It is in this void of unknowability that myths are most often employed.

Continuation or transformation?
Seeking to avoid the inevitable, man has created many myths about life-preserving remedies such as magic potions, elixirs of immortality or rejuvenation. In the *Epic of Gilgamesh*, for instance, the hero plunges into the cosmic sea with stones tied to his feet. At the bottom he finds the prickly herb of immortality, plucks it, cuts loose the stones and surfaces. But his triumph is short-lived, for while he is bathing in a fountain a serpent eats the herb. Shedding its skin periodically the snake becomes a symbol of rejuvenation and immortality while man has to come to terms with his mortality.

Other semi-divine beings have tried to cheat death and its messengers. The Polyne-sian trickster Maui even tried to kill the goddess of death herself. With his friends the birds he crept up while she was asleep, intending to crawl into her body between her thighs, kill her and escape through her mouth. At first he seemed to succeed but when a wagtail saw only his legs sticking out, it could not contain a chuckle which woke the goddess. She closed her thighs and her womb became his tomb.

A touching motif that reappears in myths of several regions is the attempt to rescue a loved one from the clutches of the underworld. Thus Izanami, who according to Japanese myth, had with her husband Izanagi, created the world out of the ocean, dies giving birth to fire. Izanagi, disconsolate at the loss, follows her into the Land of Darkness. He finds her in a castle and persuades her to return, but she delays because she has already eaten food there (as did Persephone in a similar Greek myth). Impatiently Izanagi uses a light and sees that she is already in an advanced state of decay. So angry is Izanami at being seen in a humiliating state that she tries to kill Izanagi. There is a chase and Izanagi barely manages to save himself.

A variation on this story is found in the Greek myth of Orpheus who, finding that Eurydice has died of a snake bite, decides to follow her to Hades. Such is the quality of his music that the torment of the damned stops and Eurydice is permitted to return with him on condition that he does not look back at her until they reach the upper world. His anxiety gets the better of him, however; he looks back and she is lost. (The fatal mistake of looking back is echoed in the biblical story of Lot's wife, who looked behind her on the way from Sodom and was turned into a pillar of salt.) The moral of these myths seems to be that man must learn to accept the inevitability of separations in life, of which death is the ultimate and most irrevocable.

Beyond death: heaven and hell
Man's difficulty in accepting death as final is reflected in the universal theme of a world after death [1, 3, 6]. In many traditions this is somewhere on the earth, often in the west (Eden [8] is an exception) and separated by water from the known world, such as the

1 After death, the Egyptians believed, life continued in the underworld. Each person had a double, or Ka, representing the divine essence. Here it is shown as a strange creature with human head and falcon's body. Although Ka is as spiritual as any Christian soul, a concrete explanation was needed for its ability to fly across the underworld with the corpse to which it belonged.

2 One Aztec myth depicted four destroyed worlds with the present one in the centre and suggested that the human race had been wiped out in earlier times because it had been too self-opinionated. If men became too proud, the present world would itself be destroyed by means of an earthquake. The ending of a universe and the beginning of a new era was predicted in many other traditions. Graeco-Roman mythology spoke of a descending order of world eras from a "golden" to an "iron" age, the baser metal indicating a progressive worsening of the human condition.

3 The other world, according to the Egyptian *Book of the Dead,* was a realm in which the blessed dead lived much as they had always done, only in a state of more perfect happiness, farming the Elysian fields. Osiris, the supreme ruler and judge of the dead, is here attended by his wife-sisters, Isis and Nephthys. His insignia are agricultural. There were no spiritual occupations or heavenly choirs as in Semitic religions.

4 The myth of Jonah and the whale represents a Hebrew variation of the theme of death and resurrection. On the surface it is the miraculous story of how the Lord sent a whale to Jonah when he had been cast out into the sea and of how he lived three nights and days in its belly before being delivered to dry land. At a deeper level it is an allegory of how Jonah atoned for disobeying God's command and was delivered from guilt.

5 Implications of sacrifice underlie the Hebrew story of how Samson's hair was treacherously shorn by Delilah. A connection between cutting the hair and losing physical strength is widespread and some Fijian chiefs ate a man as a precaution before cutting theirs. But the theme may also imply a transition from outer to inner strength, in keeping with the sacrifice of hair by nuns and orthodox Jewesses as a preparation for sacred marriage.

Celtic Avalon, the Greek Islands of the Blessed and the American Indian Happy Hunting Grounds. In some it is below the earth, such as the realms of Tumbuka (Malawi) and in yet others it is in the sky as are the Judaeo-Christian heaven and the Buddhist and Hindu paradises.

Some realms admit all the dead regardless of their merits while others restrict entry to those who have earned it. The Greek Hades, for instance, accepts the souls of all who are ferried across the River Styx by Charon provided they have the necessary fare. But in the Judaeo-Christian religions the soul is assigned to heaven or hell according to divine judgment of the person's life on earth, in the same way as the hearts of the dead are weighed by Anubis in Egyptian myths.

But merit is not always measured in moral terms and "heaven" often reflects inequalities on earth. In the Leeward Islands only aristocrats are sent to "sweet-scented Rohutu" while commoners go to "foul-scented Rohutu". The mansions of the sun were open only to the Incas and nobles of Peru, while the Norse Valhalla was the prize

of the mighty in war. Even the Christian heaven is not always gained by a pure life. Jean Calvin (1509–64), the Swiss theologian, held that salvation was through arbitrary divine choice. Generally "heavens" are portrayed as beautiful parks or gardens filled with earthly delights, places of eternal youth and freedom from want.

The end of the world
A final mythological theme is that of the end of the world and a return to chaos. The gods imposed order on the world and they may well revoke their patronage. It is for this reason that festivals are celebrated, rituals performed and sacrifices made. But almost every mythology envisages a time of eventual destruction heralded by wars, famines, floods, hurricanes and earthquakes.

Traditions as separate as those of the Mexican Aztecs [2] and the Indian Hindus and Buddhists envisaged several world ages characterized by decreasing moral standards and piety. The Aztecs believed that when the last age had finished the world would be consumed with fire and everything destroyed.

In dead midwinter, nature rests under a blanket of snow, no | sap rises and trees are being felled. But life lies underground | in the seed corn and burns in the Promethean fires of man.

6 In Chinese mythology, hell was run like a well-ordered bureaucracy, reflecting the importance of administrative efficiency in China. Here, Yama, king of the Seventh Hell and supreme master of the | law courts, dispenses a form of justice in which exact punishment is prescribed for each offence. Misers, for example, and dishonest mandarins have to swallow melted gold, while cannibals and dese- | crators of graves are chased by demons into a river. Investigation of souls takes place in the first of ten courts which assigns each soul to one of 18 hells designed to fit various crimes.

7 Shou-Lao, symbol rather than god of longevity, holds a golden peach. These ripened only once in 3,000 years in a celestial garden, a Chinese parallel to the paradisaical Tree of Life elsewhere.

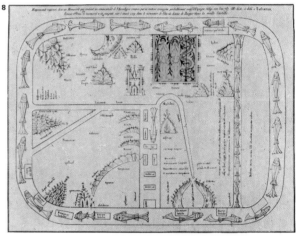

8 Myths are maps to guide and stimulate the imagination and orientate the individual in life. This | 8th-century map by a Spanish priest, Beatus, divided the world into three continents inherited by the sons | of Abraham. The surrounding ocean was not to be explored. Paradise was located in the east (top).

9 In "Resurrection" by the English painter Stanley Spencer (1891–1959), the Last Judgment has come for the good people of Cookham. Just like people | coming up from a cellar, folk are climbing out of their graves, quite uncorrupted and freshly dressed for the occasion. The picture is a literal and | earthly allegory of the sophisticated idea that souls can be reunited with their bodies in the building of a New Jerusalem. In many religions, emphasis | on heaven or hell is waning, as is the importance of past or future lives. It is conceded that corresponding states of mind can exist here and now.

The nature of religion

For an individual searching for the secret of how to live, the fundamental message of the religious traditions is that man does not know himself. He knows neither the extent of his weaknesses nor the possibilities of his greatness. Thus, at the heart of all the sacred traditions of the world there are ideas and disciplines that seem to acquaint man with both the "animal" and the "divinity" within him [Key]. The early Christian, for example, meditating in the desert of North Africa and practising the specifically Christian method of continuous inner prayer ("prayer of the heart"), directly experienced the extent to which his mind was distracted and filled with illusions about himself. Facing and accepting his weaknesses, he also discovered that he was the vehicle for the highest or most divine energies of the universe.

The common spiritual factor

When used wrongly, as a manipulative device to gratify egoistic aims, all the methods and practices of sacred traditions lose their real religious purpose. Thus, the prophets of Israel condemned even the most sacred rituals when they were performed externally without an inner recognition of personal helplessness and obligation to the source of life. Thus, too, the powerful meditative practices of Mahayana Buddhism, to take an example from Eastern traditions, are said to have a liberating effect only when used with the aim of benefiting all sentient beings.

In its most intensive form, religion offers man even more than the perception of his two opposing natures. When carried far enough, the practices of a great tradition are intended to bring about an actual transformation of human nature at the deepest level. The name given to this state of transformed being varies from tradition to tradition and also from one gradation or aspect of transformation to another. In the Western world it is spoken of as salvation, immortality, the attainment of the kingdom of God, among other terms. In the East it is nirvana, liberation, enlightenment or God-consciousness. Often the terms "wisdom" or "freedom" are used. But whatever words are used to describe the state, it is in the idea of transformation, that the common factor in all religions may be found.

The possibilities of human development envisaged by religious tradition are very great indeed. Man is understood as a potential microcosm [4], a being who contains in himself all the forces of creation and destruction that operate in the great universe. This concept of the microcosm forms the backbone of all ancient teachings, Eastern and Western, including so-called "primitive" religions.

Restoration of unity

The traditional teachings see the misery [5] and confusion of human life as rooted in man's failure to see, accept and live by the universal order of reality that is contained within himself. In man's "fallen" state, the divinity within himself is completely cut off from the animal. Thus divided within himself, man lives his life governed by impulses that were meant to be servants rather than masters. These "false masters" within man are desires, which are condemned not as such but only because man wrongly identifies himself with them and obeys them blindly and uncomprehendingly. To become a microcosm, that is, a mirror of the whole reach of

1 The phoenix is a perennial symbol of immortality. In legend, this beautiful bird lives in the wilderness of Arabia and is the only one of its kind. About every 500 years it burns itself on a funeral pyre, but rises from the flames reborn (as in this 13th-century English manuscript). The central message of many religious traditions is that it is only when man "dies" himself that a spiritual transformation, the birth of the Christian "new man", can take place. On a universal scale, the phoenix symbolizes the cosmic dance of birth, destruction and renewal in which man and all of nature are involved.

2 Kilimanjaro, the Tanzanian mountain sacred to several African peoples, is one of many peaks throughout the world that are regarded as places of communion with the spiritual world or as the king- doms of deities. Some of the most extensive sacred traditions surround Mount Meru in the Himalayas, the symbolic golden mountain of Hindu mythology. It stands at the centre of the uni- verse as the axis of the world and abode of the gods. Extending both upwards to the heavens and downwards to the nether regions, it is a bond both between earth and sky and between man and god.

3 Shintō – meaning "way to the gods" – is Japan's oldest religion. Like other anim- ists, Shintōists wor- ship many gods, or *kami*, which are the forces in mountains, rivers, trees and other parts of nature. Shinto emphasizes rituals and moral standards but does not stress life after death. One of the most compelling and unanalysable human experiences is that of identity – that "I am I". Many religions (so-called exotheic religi- ons) such as Shintōism turn upon a relation between Self and That – environment, fellow men and an external deity govern- ing the universe, with whom dialogue may be had and by whom duties are imposed. Together with such an approach, or in its place, may occur the feeling that the Self is not a citadel private to the person. In these so-called endotheic religions the Self is a microcosm and the That with which we experience dialogue is not outside but inside ourselves. It is in fact the Self of Selfhood; the believer senses that "I am That".

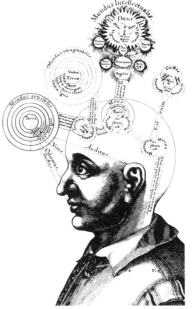

4 Man as a micro- cosm of the universe is depicted in this diagram by the Eng- lish alchemist Robert Fludd (1574– 1637), harmoniously integrating cosmic principles. All the great traditions teach of the exalted cosmic status of man, if only impli- citly. Hinduism has the idea of Primal Man, whose dispersal created the universe. The Judaeo-Christian tradition speaks of man being made in the "image of God". But these are con- ceptions of man in a state of perfection. Fallen man is neither microcosm, nor mir- ror of God, nor Primal Man. The work of spiritual discipline is to recover or recon- stitute the latent microcosmic nature of the human self.

the divine cosmic order, there must be forged within human nature a right relationship between the desires and the slumbering spiritual power with which all human beings are born.

The transformation of man (called the "second birth" in the Christian tradition) consists of the tangible establishment within the self of this right relationship, this extraordinary inner unity. Thus transformed, man may take his central place within the whole scheme of creation. He is then the Great King of the Chinese tradition, the Cosmic Man of Hinduism, the All-Containing Void of Buddhism and the image of God of Judaism and Christianity. He both reflects the whole of cosmic nature and becomes the conscious instrument of the creator within that universal order.

The cosmic pattern

In a general sense, the ideas, symbols and rituals of the traditions are meant to serve as instruments to help man experience what is taken to be his exalted cosmic destiny, both on an individual and on a social level. Thus, the structures of ancient society (called "theocracies") were designed to make human life conform to a cosmic pattern that is outside the range of modern scientific methods. Teachings about life after death [1], the "animistic" view of nature (a belief that all things are filled with life and consciousness), the role of shaman [7] and priest, the symbols of so-called "polytheism" and the function of "magic" may all be approached and studied from this point of view rather than from a conventional perspective which sees them as expressions of intellectually inferior cultures. It is especially revealing to study the rich and complex social orders that existed in ancient India and pharaonic Egypt from this standpoint.

Sacred tradition, whether Eastern or Western in origin, whether "primitive" [3] or "monotheistic", may therefore be defined as a means of transmitting ideas and ways of living that can guide individual men to pierce through the illusions that have become second nature to them and to realize in fact, and not just in fantasy, both the terrors of their present situation and the greatness of their possible inner evolution.

Janus, the Roman god of beginnings and custodian of the universe, was always depicted with two faces looking in opposite directions – an apt symbol of the duality of human nature. Many sacred traditions see man as being composed of two divergent elements: divinity and animal, spirit and body.

5 The vision of hell depicted in a panel of "The Garden of Earthly Delights" by the Flemish painter Hieronymus Bosch (c. 1460–1516) is full of medieval symbolism with a specific but now unknown meaning. But for many religious thinkers, and perhaps for Bosch himself, hell is a condition of the self rather than a literal external place in which lost souls suffer endless torments after death. In the subjective sense, hell is an infernal automatism and consists in continuing to tread the old paths, remaining a prisoner in the network of senseless illusions. Hell can thus be identified as a life that has not been transformed.

6 A North American Indian totem pole, like much of the symbolism of what used to be called "primitive" religions, expresses a vision of man in the universe that is as subtle and sophisticated as that of the "higher" religions. To grasp the meaning of these symbols it is necessary not only to have accurate information about the particular tradition in question; it is even more important to have an empathic approach to their form and content.

7 Shamans, as shown in this 15th-century Turkish painting, are ecstatic healers found in many cultures. They are believed to have links with the forces of the cosmos with whom they communicate during trances.

Judaism and Christianity

The fundamental message of Judaism is expressed by the prayer: "Hear, O Israel, the Lord our God, the Lord is One." This prayer is named by its first word – *Shema* ("listen", "hear", "understand", "obey"). It calls men to hear the truth that has been revealed, to take it to heart and to live by it in order to realize the unity of God in a relationship that demands of a man that he unify his own being.

The Jewish covenant and what it means
Judaism is the religion of a covenant between Yahweh (God) and the descendants of Abraham who was prepared, when tested, to sacrifice his own son to Him. From the Covenant radiates the mystery of an agreement between man and God. It is in the actions that make up the history of the chosen people that the teachings of the Jewish faith are set forth. The Lord appears in all the transcendence of His absolute power and at the same time in the immediacy of personal concern for His people. He brings Israel out of slavery in Egypt [4] and into the promised land. For the Jew, the "choosing" of his

people parallels the mystery of man's being created in God's image. As Israel is called to realize its covenant, so man is called to fulfil the promise of his being. The Jewish philosopher Martin Buber (1878–1965) writes: "Man must liberate himself because man is a microcosm and there is in him Pharaoh and Egypt; he is enslaving himself."

In Jewish mysticism, the symbolism of exile and return finds yet another level of interpretation, this time on a cosmic scale. Medieval Kabbalists, interpreters of the Torah – the law God gave to Moses [Key] – and its rabbinical commentaries, saw in the failure and exile of cosmic man – "Adam Kadmon" – the scattering of the sparks of the divine *Shekhinah*, the presence of God in the whole of creation. The redemption of man is thus intimately bound up with the redemption of creation.

This conception of man being responsible for the whole of creation had its greatest modern influence on Judaism in the communities of Hasidim, "the pious", which arose in Poland in the eighteenth century. In the Hasidic way of life, there is no separation

between sacred and profane. Everything that exists contains within it a divine spark waiting to be liberated. According to this teaching, there is in man a divine energy through which divine sparks that are present everywhere can be attracted and set free. All depends on intention, the condition of a man turned to God with his whole being. For the Hasid, everything he meets in the course of his day is holy and according to the Torah everything can be brought back to union with God.

Christianity and divine love
The whole of Christian religion is centred on the mystery of divine love. Man's task is to respond to that love. From the very source of Christianity come the words of Jesus addressed to the Jews in their own terms: "Thou shalt love the Lord thy God with all thy heart, and with all thy soul and with all thy mind. This is the first and great commandment. And the second is like unto it, Thou shalt love thy neighbour as thyself. On these two commandments hang all the law and the prophets" (Matt. xxii: 37–40).

Later the Christian faith came to include

CONNECTIONS

See also
216 The nature of religion
220 Islam, Hinduism and Buddhism
222 Methods of worship
224 Religion and the plight of modern man

In other volumes
58 History and Culture 1
110 History and Culture 1
304 History and Culture 2

1 The Western Wall (also called the Wailing Wall) in the Old City of Jerusalem is a place of prayer and pilgrimage that is sacred to the Jewish people. According to rabbinical belief the divine presence never departs from it. It is all that remains of the 2nd-century BC Second Temple destroyed by the Romans in AD 70.

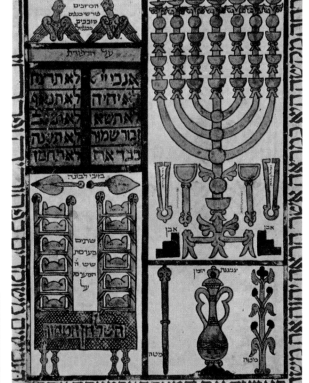

2 The original golden *menorah*, or ritual candelabrum, was shaped by Moses according to the pattern of the almond, Israel's most sacred tree. It is used during the eight-day festival of Chanukkah and its branches symbolize the seven days of creation. The middle cup signifies the sabbath.

3 Obedience to the Lord – the keeping of the Covenant – is fundamental to Judaism. The Ark of the Covenant, a gold-plated chest that housed the two tablets given by God to Moses, was constructed after Moses saw a fiery replica coming down from heaven. Originally kept at Shiloh and later at Jerusalem. After that it was seen only by the high priest on Yom Kippur, the Day of Atonement. Its eventual fate is completely unknown. brought out during battles (as depicted here in a 13th-century French manuscript), it was put by Solomon in the Holy of Holies in the Tabernacle at Jerusalem.

4 During the festival of Passover, Pesach, Jews commemorate their deliverance from Egypt, the prelude to the forging of their eternal covenant with God. The family gather for the Seder night service to retell the events from the *Haggadah* and answer children's ritual questions. Dishes symbolizing the hardships of slavery are prepared and only unleavened bread, *matza*, is eaten. Each person tries to feel as if he personally were saved by God and looks forward to a future redemption.

people who lacked the common basis of Judaism. In response to their need and in the face of the claims of then current systems of thought, Christianity began to take shape as an independent religion.

In its unique perspective, everything that exists was brought into being as an expression of divine love and is moved by that love to "be what it is", to fulfil its own nature as the plant reveals the secret contained in the seed. But in the order of creation, it is man, made in the image of God, who stands out as the element of uncertainty, the great risk freely undertaken by God so that His love might be freely returned.

In Adam's fall [5], the limited and separate existence of the natural world apart from its Creator asserted itself. Yet this failure is sometimes called "the happy fault" because it led on to the greatest act of divine love. The father's sacrifice of his son was fulfilled in a new covenant. The Son of God became man [6]. In the person of Christ, the way was reopened. The Incarnation mysteriously united in Christ the two natures – human and divine – and the passion and death of Jesus

demonstrated this unification in the perfect submission of human will to divine will. Finally the resurrection of Christ [7] promises the fruit of sacrifice, the "new man" in whom limited nature is transformed by divine life. "Unless the grain of wheat falling into the ground dies, itself alone remains; but if it dies it brings forth much fruit" (John xii: 24). And similarly: "He that findeth his life shall lose it; and he that loseth his life for my sake shall find it" (Matt. x: 39).

The search for God through contemplation
A deep and serious response to this call can be found in the Christian contemplative tradition. To the modern person the idea of contemplation may call forth associations of day-dreaming or sentimental ramblings. This is far from the contemplative's understanding of his work and its demand for a quality of awareness and impassioned searching that can bring him to the core of his being, there to discover his true need for God. Confused and alienated as a result of Adam's fall, man must struggle to discover within himself the central impulse of love, "Be what he is".

Moses, the great law-giver of the Jews, is depicted here by the French artist Gustave Doré (1833–73) with a gravity that is reflected in Mosaic law itself and which represents the absoluteness of God's word to man. According to Hebrew tradition, Moses led his people out of bondage in Egypt (probably between the fourteenth and twelfth centuries BC) and it was to him that Yahweh revealed the Ten Commandments in the Sinai wilderness. These laws, which the Jewish people carried with them to the promised land are the epitome of the demand for righteous action that characterizes the entire tradition of Judaism.

6 God's divine love, expressed in Botticelli's "Nativity" (1500), was translated by St Paul into ethics in the spiritual life. The result was *agapê* or *caritas* – the active concern for the well-being of others.

7 This resurrected figure of Christ by Piero della Francesca (1420–92) conveys the promise of rebirth after death. The suffering of Jesus on the Cross is an indictment of man's inherent drive to crucify or murder the Truth that can save. But the resurrection demonstrates God's loving forgiveness and the rebirth possible for men who can face their own corruption and accept the help of God.

5 The story of Adam and Eve, splendidly portrayed by the German painter Lucas Cranach (1472–1553), is an allegory of man's fall and his anguished sense of separation from cosmic unity. What was the sin of Adam and Eve in eating fruit of the Tree of the Knowledge of Good and Evil? The Trappist monk, Thomas Merton (1915–68) saw it as an act whereby man tried to appropriate for himself that which God would give out of His own love. The "original sin" is thus an act of pride stemming from lack of trust in the Goodness of God.

Islam, Hinduism and Buddhism

Islam, the youngest of the world religions, sounds again the message of God's unity – "There is no God but God". Recognition of this truth constitutes the act of submission by which a man becomes a Muslim – "one who submits". Conscious of his dependence, man acknowledges "I am not the Absolute". Yet one who is called to the inner path of Islam also comes to recognize, "I am nothing separate from or other than the Absolute". Unity is reflected everywhere, drawing itself out like a beautiful arabesque [1] that baffles the eye as it continually turns back on itself.

The basis of Mohammed's teaching
According to the Islamic perspective, man is in need of divine revelation to remind him of the One Reality, which is never directly manifested in the world. Judaism and Christianity are recognized as founded on authentic revelations and Islam is said to offer the third and final revelation.

Mohammed, the founder of Islam, was born in Mecca in *c.* AD 570 and began to fulfil his prophetic function by denouncing the prevailing Arab worship of many gods.

Confronted by powerful opposition, Mohammed and his followers became a social and political as well as a spiritual force. Following the teaching of the Koran, Islam developed both external and internal aspects of religion, providing laws for the guidance of a community as well as a way for the individual to unite with Allah (God).

Traditionally in Islam there is no separation between sacred and secular areas of life. There is no priesthood, no day reserved for worship. Instead, the law itself offers direction and an ideal of life that meets man's need. In the Islamic perspective, man sins not by wilful rejection of God, but by heedlessness or distraction. The required observances of Islam act as reminders of the relationship between that life and the Absolute.

The teachings and practice of Hinduism
For a Westerner the Hindu religion of India may be puzzling. In place of God and creation, he finds that Brahman (ultimate reality) is probably utterly impersonal and the phenomenal world is ultimately unreal. Even the idea of historical progression is over-

shadowed by the sense of a cyclic world drama of creation, preservation and destruction. Looking for clearly defined doctrines, the Westerner is instead plunged into a variety of methods and beliefs. For while Western monotheism seeks to protect the truth from distortion, in Hinduism the truth is left to protect itself [3].

The simplest – and therefore the most difficult – expression of the spirit of Hinduism is "Thou are That", which may be understood as a response to the deepest question men ask. Looking at the world around them, men saw in the sudden flash of lightning, in the invisible power of the wind, signs of energies beyond their control and asked: What is behind all this? Another form of the question is concerned with the mystery within man: Who am I? In a single moment of discovery comes the answer to both lines of questioning: The true Self (Atman) is the same as the ultimate ground of reality, Brahman – "Thou are That" [4].

The Hindu revelation is not the focus of an historical event such as the revelation given to Moses, and does not mark a unique

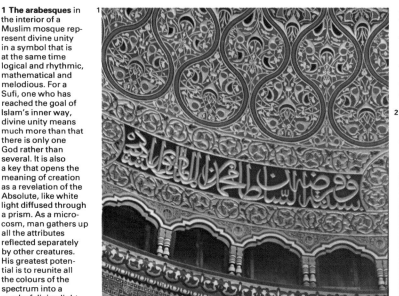

1 The arabesques in the interior of a Muslim mosque represent divine unity in a symbol that is at the same time logical and rhythmic, mathematical and melodious. For a Sufi, one who has reached the goal of Islam's inner way, divine unity means much more than that there is only one God rather than several. It is also a key that opens the meaning of creation as a revelation of the Absolute, like white light diffused through a prism. As a microcosm, man gathers up all the attributes reflected separately by other creatures. His greatest potential is to reunite all the colours of the spectrum into a spark of divine light.

2 Prayer at specified times, five times a day, is one of the fundamental practices prescribed by Mohammed and known as the Five Pillars of Islam. The other four are declaration of faith in Allah, almsgiving, fasting during the month of Ramadan from dawn until sunset and making a pilgrimage to Mecca in one's lifetime. Even fulfilling the law in an external way can help a Muslim understand his condition in life. A Muslim does not seek to go beyond the basic requirements by "doing more" in an external sense but to realize more deeply what he is already doing.

3 The many-armed Hindu god, Shiva, symbolizes the many modes of divine energy. The apparent polytheism that many Westerners see in Hinduism is in fact adaptability. Hinduism is monotheistic in a massive way, for in it all creation and experience are one. Judaeo-Christian monotheists would be asked by a Hindu: "Who are we to limit the forms in which Brahman may manifest itself?" A devotee may take hold of any of the forms, which are in fact generated only by our partial perception, to arrive at the One. He may use any method from asceticism to orgiastic abandon or from ritual poverty to industrious prosperity if it leads him to God-realization.

4 The AUM symbol (OM) is a ritual and sacred Hindu syllable, rendered in Sanskrit calligraphy, that is understood as the fundamental sound of the universe. It is chanted both for the effect of its vibration on the worshipper and as a tangible symbol of the one fundamental Reality: Brahman, the Absolute. One interpretation of its three sounds (A, U, M) is that they represent the trinity of Vishnu, Shiva and Brahma.

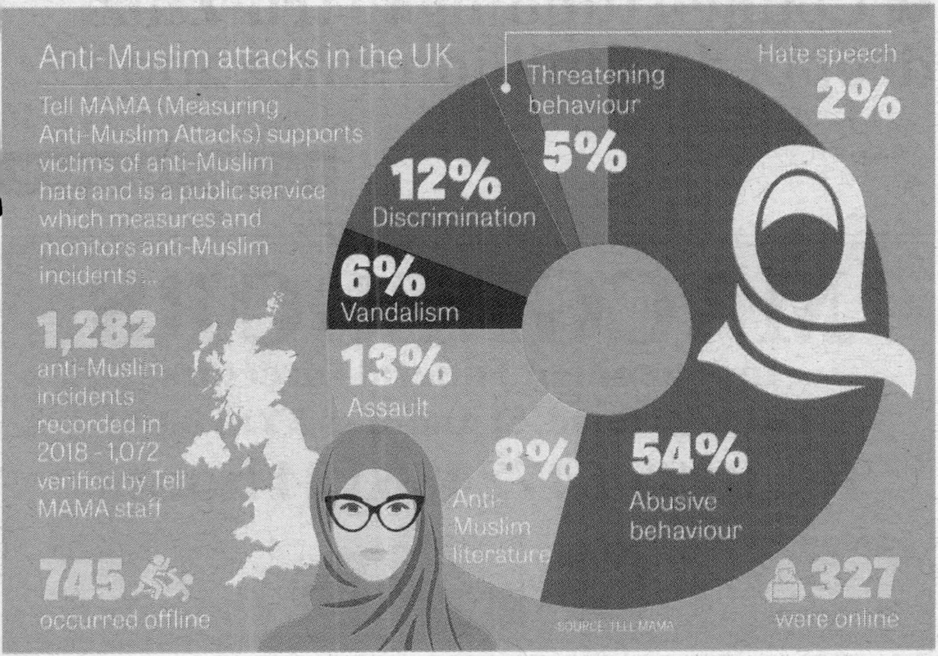

11 - 10 - 2019

Anti-Muslim attacks in the UK

Tell MAMA (Measuring Anti-Muslim Attacks) supports victims of anti-Muslim hate and is a public service which measures and monitors anti-Muslim incidents

1,282 anti-Muslim incidents recorded in 2018 - 1072 verified by Tell MAMA staff

745 occurred offline

12% Discrimination

5% Threatening behaviour

2% Hate speech

54% Abusive behaviour

8% Anti-Muslim literature

13% Assault

6% Vandalism

327 were online

SOURCE: TELL MAMA

Council funding to trust has dropped £4 million since 2008

FULL STORY ON PAGE 17

bridging of the gap between God and man such as that provided by the incarnation of Christ. It says that the Truth is in each person waiting to be realized. With this promise comes the warning "Neti, neti" ("Not this, not that"). One cannot identify either the Self or the Absolute with any particular thing. Belief in a separate self or ego is like an assumed identity that keeps us from realizing our true Self. In final wisdom the identity is laid aside and selfhood merged in an oceanic experience of That (Samadhi).

Fundamentals and precepts of Buddhism

Buddhism is more urgent and direct in its teaching than Hinduism, from which it grew. It sees ordinary existence as a nightmare that is not the less painful because it is unreal.

In the fifth century BC, Gautama Siddhartha, the son of an Indian king, woke from the nightmare. As the Buddha ("the awakened") [5] he was forever released from suffering and full of compassion for those who were still in darkness. The Buddhist believes that suffering is a universal fact of existence because of man's fundamental ignorance about himself and the world. The world is a process of continuous interaction of unstable compounds in which nothing lasts. Whatever a man may take to be himself – body, mind, feeling, perception – is an obstacle in the form of the assertion "This is mine; this am I; this is my ego", which makes him the centre of an imaginary drama of pleasure and pain, good and bad [7].

Some have seen in the Buddhist denial of the ego and emphasis on transience a pessimistic rejection of all values. What is negative in Buddhism, however, is not its truth but its way of presenting that truth. The goal is defined negatively (and practically) as release from the transitory evils of suffering, ignorance and selfishness.

Whatever has been shaped through the law of cause and effect can be reshaped by the same law. Codes of moral behaviour serve principally as a preparatory discipline, a method of purification for the most important task of cultivating "mindfulness" [8]. Direct insight into the workings of the causal law in oneself is what strikes at the root of all the illusions of the ego and its suffering.

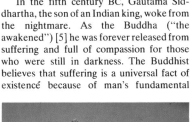

Hindu, Muslim and Buddhist symbols, reading from top to bottom, respectively, represent sacred aspects of their faith for a devoted following of nearly 1,300 million people. The vast majority of these are to be found in North African and Asian countries. The syllable AUM (*top*) is a mystic sound representing the Eternal Essence, and uttered by Hindus during the most solemn moments of worship. The arabesque (*centre*) is a rhythmically designed pattern of oneness in which, according to strict Muslim rules, no animate objects are represented. The Wheel of Life (*bottom*) means for Buddhists the continuing cycle of death and rebirth that traps mortals.

5 Gautama Buddha brought a teaching directed solely to one point: the extinction of suffering. He saw people everywhere making themselves miserable through deluded belief in the reality of the ego. Rejecting displays of miraculous powers and speculations about metaphysical questions, he urged his followers not to rely on the achievements or the understanding of others as this might simply be woven into their own fantasies. "Be ye a refuge unto yourselves. Betake yourselves to no external refuge. Hold fast to the Truth as a lamp. Look not for refuge to anyone besides yourselves."

6 Buddhism is split into two main streams – Mahayana and Theravada. Devotees of the latter revere the personality of the Buddha, his teachings and the order he founded. They hold that the ideal Buddhist is a faithful follower of the Eightfold Path. Mahayana Buddhists regard the Buddha as one of many who have appeared in many universes. They hold that the ideal Buddhist is a Bodhisattva or one vowed to become a Buddha through the six virtues – generosity, morality, patience, vigour, concentration and wisdom. The Tibetan Buddhists, here ready for the Tsam dance, hold to a mixture of Mahayana and Bönism – an indigenous worship of natural spirits through ritual.

7 The Wheel of Life, the great Buddhist symbol of *samsara*, is the endless round of birth and death in which all beings are trapped who have not pierced through the illusions of the ego and stilled its cravings. In other versions the hub of the wheel depicts three animals representing lust (dove or cock), hatred (snake) and delusion (pig). These impulses generate a universe of conditioned existence.

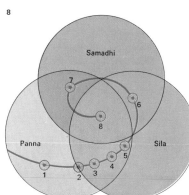

8 An Eightfold Path offers the Buddhist the only way to the blissful state of nirvana – release from the eternal cycle of rebirth. It is based on the fundamentals of Sila (morality), Samadhi (concentration) and Panna (wisdom) and its steps are right views [1], right intentions [2], right speech [3], right action [4], right livelihood [5], right effort [6], right mindfulness [7] and right concentration [8]. Several lifetimes are needed to reach nirvana.

9 Zen Buddhism first flourished in China in the 7th century AD and then spread to Japan in the 12th. There are two main branches – Sōtō and Rinzai. In the latter meditation on such paradoxes as the "sound of one hand clapping" is used to awaken insight into what transcends logical distinctions. In Sōtō adherents sit silently in gardens such as this and meditate on what illumination arises. Both believe in mind-to-mind instruction from master to disciple with the aim of awakening the Buddha-mind that lies within every individual. Sōtō concentrates on teaching the common people the good ways.

Methods of worship

Prayer might be defined as the method appropriate to belief in an external or transcendent God [1] and meditation as the method appropriate to a religion such as Hinduism or Buddhism directed towards realization of the divine principle within [3]. However, the Hindu prays before his chosen image of the Lord and the Christian contemplative engages in an activity that is no less properly called meditation than the "sitting" of a Buddhist monk.

In its essence, prayer is as little concerned with obtaining favours as is the practice of meditation. The experience of a nineteenth-century Hindu provides a vivid example. After several attempts to pray for the relief of his suffering family, he had to give up completely because each time he became aware of the deity he was so overwhelmed that he found it impossible to ask for anything at all.

The nature of prayer and meditation
In the Islamic tradition, prayer [Key] is the fundamental right and responsibility of man by virtue of his central place in the cosmic scheme. By his profession of faith – "There is no God but Allah" – the Muslim directly affirms the truth of which all creation is an indirect expression.

In the Judaeo-Christian tradition, prayer is the meeting of man and God. According to Hasidism, the Jewish mystical movement: "The people imagine that they pray before God. But this is not so, for prayer itself is the essence of divinity." Similarly, in the Christian contemplative tradition, prayer is an effort to find the place where a man is most himself – the ground of his being – which is, by the mystery of love, the place where he is most related to God.

Each of the religious traditions presents man with the startling claim that he is not really what he takes himself to be. For example, Christianity has the parable of a rich man's son who squanders his inheritance to live among swine but remains unchanged in the eyes of God. Hindu sages declare that the true Self is the infinite changeless Witness. Buddhism points to belief in a personal identity as the fundamental illusion that produces all suffering.

From the traditional point of view, meditation is the "laboratory work" in which a man can come to know himself as he is. A relaxed awareness is regarded as a condition of study, and the physical influences that contribute to this are taken into account.

The Bhagavad Gita, one of the greatest and most widely known texts of Hinduism, recommends a balanced posture for meditation – with "upright body, head and neck, which rest still and move not; with inner gaze which is not restless . . ." – and the support of a balanced way of life. "Yoga is a harmony. Not for him who eats too much, or for him who eats too little; nor for him who sleeps too little, or for him who sleeps too much."

The inner mystery of scripture
Jewish mystics studied the Torah [8], the five books of Moses, to discover the divine laws. Regarded in its mystical essence as the Name of God, the Torah was thought to serve as the instrument of creation. In Christianity it is Christ, the Word, who embodies the truth [9]. "Through Him everything came into being and without Him nothing that exists came into being" (John i:3).

1 An American Indian at prayer exposed himself to the power of the Great Spritit in "crying for a vision". For this, he needed courage and determination. Guided by a wise man through preparatory rituals, he faced his vigil almost naked and alone. All depended on his recognition of the depth of his need.

2 A monastery is like a laboratory where the conditions of life are arranged in such a way as to enable a man to face more directly the limitations of human nature and the need for supernatural help. Monasteries differ in their character and degree of asceticism, but in each, communal discipline is vital.

3 Hindu meditation (the lotus posture is shown here) depends on an approach as impartial as that of any scientist making an investigation – "master of his mind, hoping for nothing, desiring nothing". Working in this way a seeker may witness in himself the operation of cosmic laws that govern the play of nature and, by understanding its true forces, disentangle himself from their control.

4 The Egyptian Book of the Dead, depicting the weighing of a soul, is connected with the idea of an exact science of man. Much Egyptian sacred art attempted to express in a statue of a god or goddess an impersonal cosmic principle, guiding the observer past idolatry to an understanding of himself.

5 A Tibetan thanka, or sacred temple banner, represents deities in precise postures, expressions and colouring that will show the viewer a particular kind of awareness. In each painting there is an entire world, a subtle teaching and a benevolent influence for one who studies it. Yet in every case it is nothing other than the awareness of the viewer himself that is being discovered.

The link between existence and revelation is found also in the Hindu conception of the Vedas, which, as scripture, record what exceptional men have seen of the unchanging cosmic laws that govern all transformations of matter and energy. But the "eternal Vedas" are these universal laws themselves, to which the written record provides a key.

Yet if there is in scripture a mystery corresponding to the mystery that is in man, it is not on the surface. According to tradition, scripture responds to a person's preparation and the level at which he experiences his need for transformation. When the disciples asked Jesus why He spoke in parables, the answer was a paradox: "For whoever has will receive abundantly, but whoever has not will be deprived of whatever he has" (Matthew xiii: 12).

An indirect communication, scripture conceals and reveals its truth at the same time. Although its secrets demand preparation, its more accessible levels offer preparation. Commandments and special regulations guide a person in all conditions of life, directing his energies and reminding him of God. According to the great medieval Jewish philosopher Maimonides (1135–1204), even the apparent contradictions found in the Bible are intended to lead the reader to search for a deeper meaning.

The value of sacred symbols

In the Koran [10] which forms the heart and backbone of Islamic faith, the phonetic and symbolic qualities of the Arabic language itself guide the seeker. Like all ancient languages of revelation that are regarded as sacred, Arabic is inherently symbolic. A single word can convey several levels of meaning, from the name of an object to the subtle and elusive meaning of an abstract concept. For this reason the Koran is considered strictly untranslatable, since any rendering is necessarily limited to the translator's level of understanding.

Sacred art is a still more symbolic way of embodying truths that can make a deep impression on the inner man. Sometimes the artistic object is designed to embody a cosmic quality [4], sometimes to attract a spiritual influence [6] and sometimes simply to produce a particular kind of awareness [5].

The muezzin's call from the tower of a mosque reminds Muslims of the obedience they owe to Allah. Prayer involves a sequence of postures in which the individual stands, bows and prostrates himself, so that his body shares the act.

6 The sacred art of black Africa has unusual qualities of mystery and power. A characteristic example is this altarpiece from an ancestral shrine of the Yoruba people of Nigeria. In most African sculpture the proportions of a figure are carefully determined by the artist but apparently without any attempt at a naturalistic representation of the human body. The aim of the traditional African artist is not to make an accurate likeness but to capture a quality so accurately that the figure or carving can attract the corresponding cosmic influence or house the spirit of an ancestor.

7 Miraculous power was attributed by many Christians to religious objects such as this fifteenth-century icon of the Russian Orthodox Church, depicting Jesus and his mother, Mary. According to an Orthodox writer, "An icon or a cross does not exist simply to direct our imagination during our prayers. It is a material centre in which there reposes an energy, a divine force, which unites itself to human art". Unlike Western religious painting, which eventually developed individualistic forms of expression, icon art remained largely unchanging, its painters merging their identity into the sacred tradition.

8 The Torah, or book of Mosaic law, and the religious observances it prescribes underlie every aspect of Jewish life. The ancient scripture is regarded as sacred transmission from a higher source of a teaching that explains to man his covenant with God.

9 A portrait of St John from the illuminated manuscript the *Book of Kells* dates probably from tenth-century Ireland. The medieval artist approached the scriptures with reverence, regarding them in their essence as revelation – an expression of the same creative intelligence that brought forth man and the universe itself.

10 The Koran may appear to the casual reader a confusing collection of stories, religious and social regulations and enticing images of heaven. But the faithful Muslim can read in the historic struggles of his religion a symbolic account of an inner war against the forces of dispersal in his own being.

223

Religion and the plight of modern man

In recent years there has been evidence of renewed interest in the religious dimension of life – to such an extent that some observers speak of a twentieth-century "spiritual renaissance" in the West. Much of this rebirth is, however, taking place outside the structures of the historic religious institutions of the West. This new activity stems partly from efforts to bridge the gap between modern science and ancient spiritual world-views, and partly from an eruption – particularly in the United States – of "new religions", most of them influenced by the religions of the Orient.

Knowledge and belief

The conflict between religion [1] and Western science [2] is usually thought to date from the theories of Copernicus (1473–1543) and Galileo (1564–1642) concerning the movement of the Earth around the Sun [Key]. The popular view is that the Church regarded the Copernican-Galilean picture of the cosmos as a threat to the biblical conception of the Earth as the unmoving centre of the universe and that

from then on the quest for knowledge of nature was at loggerheads with the demands of faith. Eventually the explanatory power and pragmatic successes of science over-whelmed the teachings of the Church, and the scientific view prevailed.

According to this interpretation of events, the ideal of reason and knowledge triumphed over mere belief. In recent years, however, due in large measure to an influx of Oriental teachings, both the knowledge component of religion and the belief component of science have been more clearly realized.

In psychology, for example, it is now generally recognized that the great mystics of all spiritual traditions understood aspects of human nature that have eluded the vision of modern science. As a result, the whole idea of states of consciousness is becoming an increasingly important subject of research among Western psychologists [3]. The emphasis of inquiry is shifting away from pathological or hallucinatory states towards the study of states of consciousness characterized by increased general intelligence, moral power and freedom from egoistic emo-

tions. In the light of these studies, "normal" consciousness appears limited. Such a perspective is truly revolutionary, for a person's ability to perceive and explain is itself understood to be relative to his or her state. This challenges the orthodox scientific conception of reality much more decisively than any arguments from literal interpretations of the Bible. The point is that only by passionately embracing an inferior state of consciousness could mankind have arrived at its present dangerous predicament.

The key question

In addition to many psychologists and psychotherapists who are studying the mind in the light of traditional religious teachings, some physicists are turning to Oriental conceptions of cosmic order. There is also a significant movement among medical scientists to understand ancient systems of healing, such as Chinese acupuncture, that are rooted in a spiritual conception of human nature and a non-materialistic view of the universe. At the same time, many Westerners are actively practising methods of meditation

CONNECTIONS

See also
216 The nature of religion
218 Judaism and Christianity
220 Islam, Hinduism and Buddhism
222 Methods of worship
194 Meditation and consciousness
122 Non-medical healing
192 Body image and biofeedback
198 Parapsychology

In other volumes
304 History and Culture 2

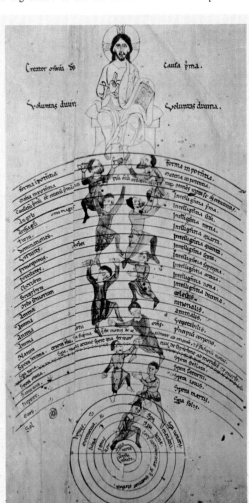

1 The ascent of the soul, as depicted in this 12th-century alchemical manuscript, shows a spiral progress towards God enthroned above the world. Reversing the original process of creation, man must pass through stages of gradual enlightenment as he travels outwards from the material world. The view of a universe with the world at its centre was not necessarily a naive astronomical theory. It was also, and mainly, a symbol of the idea that human life proceeds under the sway of many cosmic influences, both good and evil. Man must master all these influences to realize his divine destiny.

2 A universe of awesome dimensions is revealed through the telescopes of modern science, as in this view of the spiral galaxy in Ursa Major. The cosmic schemes of the ancient religious traditions were awesome too in their imaginative scope and grandeur. But one of the most fundamental differences between their visions and the rational universe presented by science is that the cosmos of the ancient religions exceeds man not only in size and physical power but also in consciousness and intelligence. Paradoxically, at the same time, religion suggests that man himself is an expression of infinity.

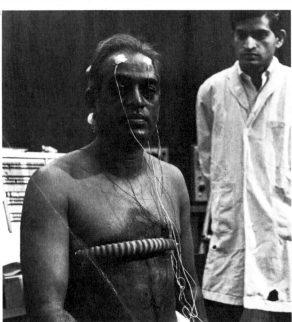

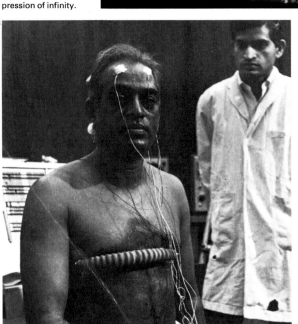

3 Experimental research is now being carried out to study the effects on the human organism of meditation techniques as practised by experienced yogis. Many Western scientists now practise meditation. Their interest is an acknowledgment that mysticism may contain a knowledge of the human psyche that has eluded Western science. Yoga is an aid to meditation with a view to acquiring enlightenment. It involves attainment of a resting state of mind by the help of manipulation carried out not on the anatomical body alone but also on the inner perception of it in the brain. Paranormal changes in self-perception occur as a result.

- some within the Buddhist framework, others within the framework of Hinduism. Transcendental meditation, for example, is a radical adaptation of certain aspects of the Hindu Vedanta system.

The question of critical importance is whether modern people can turn to meditation with the same intent as those who were helped by innumerable aspects of traditional culture, such as codes of morality that nourished their spiritual emotions. Will modern man make use of these fragments of ancient traditions in the same egoistic way that has characterized his use of the great modern scientific discoveries about the external world? Will he relate to his "inner environment" in the same way as he has related to his outer environment?

The "new religions"

The ambiguities of the current "religious renaissance" are strikingly apparent in the "new religions" that have taken root in the past decade, particularly in North America and Britain. Thousands of groups, small and large, throughout the Western world have formed around a teacher who has migrated from the East [5, 6]. At the same time there has been a revival of Fundamental Christianity that emphasizes emotional commitment to the person of Christ [4].

It is noticeable that the followers of the "new religions" tend to accept only those parts of ancient traditions that seem "relevant" or attractive. Can part of a tradition lead to a result that once required the complete tradition?

In the spiritual history of mankind, the tendency of the mind to select from a teaching only those aspects that it likes, thereby creating a subjective religion out of a carefully interconnected totality, has always been a problem. It is one of the most fundamental meanings of the term "idolatry" in the Judaeo-Christian teachings: man must not create his own god. Many of the teachers who have moved to the West from Asia are wrestling with this question now. No one yet can say whether they will succeed in transmitting to modern people the workable essence of religion, while adapting the outer aspects to the modern temperament.

The battle between the Church and Galileo ended with the retraction of his belief that the Sun was at the centre of the universe. The Church's innate conservatism, revealed by its reaction to an issue that was largely irrelevant to the main issues of religion, has contributed to the decline of organized religion.

4 Billy Graham (1918–) is typical of modern Fundamentalist Christian preachers, with his mixture of modern media methods and a simple Gospel message of faith in Christ. He offers his followers a form of intense religious experience that until recently was limited geographically and socially to advanced Western societies.

5 The Hare Krishna sect is one of the better-known examples of "new religions" that have captured many young Western people. Most US cities also have centres for the practical pursuit of Zen Buddhism. Lamas from Tibet have their followers and so has Islamic mysticism, or Sufism.

6 The Divine Light Mission, led by the Guru Maharaj Ji, is one of the many "new religions" that are inspired by the ancient teachings of the Orient. The translation of an Eastern-style, endotheic religion, with all its strong cultural traditions, into the context and social environment of the Western industrialized world poses problems. For instance, are the forms by which truth was once transmitted applicable to the conditions of modern life?

7 Rejecting the established structure of organized religion and returning to the essentially personal and basic Christian message of love and brotherhood has become one of the growth points of religion in the West.

What is philosophy?

The word philosophy has almost as many meanings as there have been philosophers. In its broadest sense, every man or woman who exists is a philosopher and has evolved a philosophy: a point of view, an opinion of the world and of how life should be lived. By extension the word philosophy has come to include the personal qualities – calm, balance, the capacity for reflection or detachment – expected of a philosopher or sage.

However, in the strict meaning of the word philosophy is a technical study of human thought and knowledge. The word has its roots in a Greek expression meaning "love of knowledge" and students who wish to become "lovers of knowledge" are required to embrace a discipline as rigorous as its subject is potentially vast.

Various fields of philosophy
Because philosophy is concerned, directly or indirectly, with almost everything in the known universe, its practitioners have found it necessary to break it down into a number of smaller, although often overlapping studies: epistemology, or the study of the origins, nature and limitations of knowledge; metaphysics, the search for reality beyond what we know from our senses; ethics, the study of how men should behave towards each other; and logic, the study of the rules and methods of correct reasoning.

Thus philosophy stands between the sciences and religion. Like science it appeals to reason rather than authority, whether traditional or revealed, but is not solely concerned with a knowledge of the facts. Like theology, it deals with matters about which definite knowledge is not, so far at least, possible.

Philosophical method
In the primary, technical sense, philosophy is essentially argumentative and it is concerned largely with abstract questions that may often seem tiresomely hair splitting and of little practical value. For example, is the world divided into mind and matter and, if so, what are they and how do they differ? Is the universe haphazard or planned? What is "good" and must it be eternal to be valued? Are there laws of nature or do we invent them to satisfy our innate sense of order? Concern with questions of this kind is what preoccupied those philosophers who are by common consent considered important in the European tradition such as Plato (*c.* 427–347 BC), Aristotle (384–322 BC), Thomas Aquinas (*c.* 1225–74), René Descartes (1596–1650), Gottfried Leibniz (1646–1716), David Hume (1711–76), Immanuel Kant (1724–1804), Georg Hegel (1770–1831) and, in our own day, Bertrand Russell (1872–1970) and Ludwig Wittgenstein (1889–1951). The same concern is shared by the Persian Avicenna, the Arab Averrhöes [7] and the Indian Sankara (780–820).

Socrates (*c.* 469–399 BC) [Key] in Europe and Confucius (*c.* 571–479 BC) in China were both outstanding teachers of a philosophical way of life. But Socrates was also a philosopher in the technical sense. He spent his life arguing and teaching in the market-place of Athens. Characteristically he raised and pressed questions in his dialogues about virtues or values, like "What is justice?" He introduced the idea of universal definitions which could be arrived at by arguing from particular facts. This method

1 Plato, in the *Republic,* compares those who live by conventional wisdom and who are dominated by their senses and appetites to prisoners living in an underground cave, chained since childhood, so that they can see only what is directly in front of them. A fire behind them throws the shadows of objects on to the wall. Our knowledge of real subjects is as incomplete as those prisoners' would be. It is only through philosophy and intelligent reflection that we can escape from this world of shadows of puppets and see true realities. If anyone returned to the cave to point out the illusion, he would be ridiculed.

3 Plato's One Over Many Argument applies to every general word, whether material such as tree or abstract such as piety. "Tree" may describe a number of different examples. These change and none of them is perfect, but still we recognize the general class. The Form or Idea is eternal, unchanging, incorruptible and immaterial. It is somehow simultaneously both more real and more ideal than any particular manifestation. These perfect Forms can never be attained in the everyday world and can only be known by the intellect, not the senses. True knowledge is therefore the knowledge of Forms. Plato is not clear as to whether particular examples are caused by the Forms or are merely a shadowy likeness.

2 Heraclitus flourished about 500 BC in Ephesus. As with all other Greek pre-Socratic philosophers, any knowledge of his ideas has to be derived from a few surviving fragments and the often malicious gossip of rivals. His theory that all things are in a state of flux or change stimulated Plato into producing his theory of unchanging Ideas that provided stable standards for conduct.

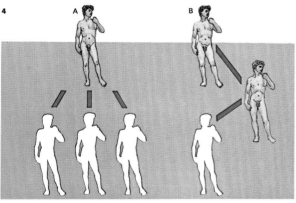

4 Aristotle advanced several common-sense criticisms of the Theory of Forms. Plato had given the fact that things have common characteristics much too great significance. It was unnecessary to postulate a separate mystical realm where pure being exists and which is never experienced. Forms, he contended, are no more than those qualities that are experienced as similar within things. Aristotle also put forward the well-known Third Man argument. Wherever two entities are discovered with a common quality, Plato postulates a Form [A]. But the original entity and the Form now also share a common quality. Therefore we have to postulate yet another Form in which they both share [B]. This process can be extended infinitely.

is called inductive reasoning. Often there are no definite conclusions from his dialogues but the aim remains the same – to arrive at answers about meaning through argument.

The gap between appearance and reality

The search for universal definitions is triggered by the need to discover what things, if any, we can know with absolute certainty. Generally people are aware, at least from time to time, that there is a gap between appearance and reality even if it is only that the person we see in the mirror parts his hair on the right whereas in reality he parts it on the left. In a world where one of the most salient characteristics is change or flux, how do we gain knowledge of the "real"?

Plato, a pupil of Socrates, identified flux with appearances or what we "know" through our senses, appetites and emotions. He argued that reality is something other than that which is perceived by our senses – the so-called Theory of Forms or One Over Many Argument.

Plato believed that what is common to things described by the same name, such as

trees, is their Form or Idea [3]. The Form represents being, the particular examples of that form represent becoming, and these two realms are separate. Plato's arguments are illustrated by curiously haunting imagery. His four ultimate categories of existence (and therefore of knowledge) constitute a hierarchy. These levels correspond to the four sections of what Plato called The Line, which he divided first into two unequal parts with each then subdivided in the same proportions. At the top are Ideas and the knowledge of them; immediately underneath, the purest of pure mathematics. For Plato this ideal world is alone truly real. A long way below falls our everyday world: physical objects on top and, under these, shadows and reflections. To move from lower to higher is to pass from shadow to substance.

Yet any such move, up or down, is disturbing. To be confined to the nether region is to be like a prisoner in a cave [1], seeing nothing but shadows cast by artificial light. For the released prisoner it hurts at first to look at things in the sunlight, and still more to confront the sun itself.

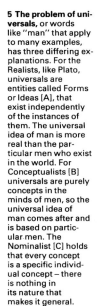

KEY

Socrates frequented the market-place in Athens, spending his time, to the annoyance of his wife, arguing and challenging the conventional wisdom of the day rather than earning a living. Few could stand up to his style of cross-examination and he made many enemies. Eventually he was tried on charges of corrupting youth and being irreligious and was sentenced to death. He could have escaped but argued that it would be inconsistent with all that he had taught. He drank hemlock and died comforting his friends.

5 The problem of universals, or words like "man" that apply to many examples, has three differing explanations. For the Realists, like Plato, universals are entities called Forms or Ideas [A], that exist independently of the instances of them. The universal idea of man is more real than the particular men who exist in the world. For Conceptualists [B] universals are purely concepts in the minds of men, so the universal idea of man comes after and is based on particular men. The Nominalist [C] holds that every concept is a specific individual concept – there is nothing in its nature that makes it general.

6 Zeno of Elea (*fl.* 450 BC) questioned our notions of time and change through a paradox. Achilles can run ten times as fast as the tortoise, who has a ten-unit start. When Achilles has run his first ten the tortoise will be one ahead. When Achilles has run that one the tortoise will be one-tenth ahead. And so on. Logically, Achilles cannot win, because Zeno defines the race mathematically, but our senses tell us that Achilles does win.

8 Philosophers were described by Plato as being split between two great armies: those of the Gods and the Giants. The Gods, fighting from the heaven of the ideal world, maintain that ideas are fundamental and are all that has any existence or reality; they are Idealists. The Giants, by contrast, struggle to pull everything down to earth and maintain that matter is primary, or even that it is all there is; they are Materialists. In his dialogue the *Sophist*, Plato goes on to say that in the great battle between Materialists and Idealists neither side can defend itself. If, as the Materialists say, reality is what we can grasp with our hands we deny "justice" or "wisdom". If we say only ideas are real, we deny living things.

7 Avicenna (980–1037) [A] and Averrhöes (1126–98) [B] were Islamic philosophers. Two centuries before Aquinas, Avicenna attempted an Islamic scholasticism, a synthesis of the best of ancient Greek philosophy with the teachings of Mohammed. Some of his work was attacked as heretical, but later both the Jew Maimonides and the Christian Aquinas adopted Avicenna's suggestion that in God essence and existence are one. Averrhöes, born in Moorish Spain, became best known in Christendom for commentaries on Aristotle. Translations of these were at one time regularly bound up with Latin versions of Aristotle's *Works*. Averrhöes founded a Muslim philosophy of religion.

227

Logic and the tools of philosophy

Logic as a discipline was invented by Aristotle [Key]. No doubt people had been reasoning in consistent and logical ways long before Aristotle; but he seems to have been the first to attempt to spell out and formalize the rules of valid inferences [1].

Logic is concerned less with truth or falsity as such than with the transmission of truth or falsity from one set of statements (the premises) to another (the conclusion). Its central concepts are those of logical consequence and of valid inference. If some statement q is a logical consequence of a statement p, then if p is true, so is q; if q is false, so is p. An inference is valid if the conclusion is a logical consequence of the premises from which it was inferred.

Invalid inferences
In the classical form of the syllogism, with two premises and a conclusion, one example of an invalid inference is known as "the fallacy of undistributed middle": All cows are animals, all herbivores are animals; therefore all cows are herbivores. Here the premises are true, and so is the conclusion, but only accidentally, not by logical necessity. That the inference is invalid can be shown by choosing replacements for each of the descriptive terms in the argument in such a way that although the premises remain true, the conclusion is false. Thus the same reasoning from the true premises: All men are mortal, all gorillas are mortal; would give the false conclusion: All men are gorillas.

Some famous proofs in the early development of mathematics were *reductio ad absurdum* proofs, a proposition proved by showing that its denial, combined with other true propositions, would lead to an absurd conclusion.

In the nineteenth and twentieth centuries there were many important new developments in mathematical logic.

Aristotelian logic could handle only very limited kinds of deductive reasoning. For instance, Euclidean geometry had long been regarded as a superb example of deductive reasoning; yet Aristotelian logic could say almost nothing about the validity or otherwise of Euclid's inferences.

Kant had endowed mathematical knowledge with a special status essentially different from that of both physics and of logic. Since Kant's view implied that no alternative to Euclidean geometry was conceivable, it became untenable when non-Euclidean geometries were developed. John Stuart Mill (1806–73) tried the alternative of interpreting mathematics as a part of empirical science; but there were overwhelming objections to this interpretation. A remaining alternative was to interpret it as a branch of logic. It was Gottlob Frege (1848–1925) who first undertook the task of showing that all pure mathematics is deducible from premises that contain only logical terms and are logically true. (This programme is known as logicism.) Just as he seemed to have succeeded, Bertrand Russell [2] discovered that the logical foundations of mathematics itself contained fundamental paradoxes.

Paradox and truth
Logic requires an adequate concept of truth [6] since it deals with the transmission of truth. But the traditional theory of truth was also beset by paradoxes. One of these, the

1 **In Aristotelian logic**, propositions containing two terms are classified into four kinds (A, E, I and O) and displayed in a Square of Opposition. Propositions labelled A are universal and affirmative (All men are brave); E are universal and negative (No men are brave); I are particular and affirmative (Some, meaning one at least, men are brave); O are particular and negative (Some men are not brave). Specific relationships of truth and falsehood follow from the positions in the square of statements referring to the same two entities. Thus, if proposition A (All men are brave) is false, then its contradictory, O (Some men are not brave), must be true; while A's contrary, E (No men are brave), could be either true or false and is undetermined.

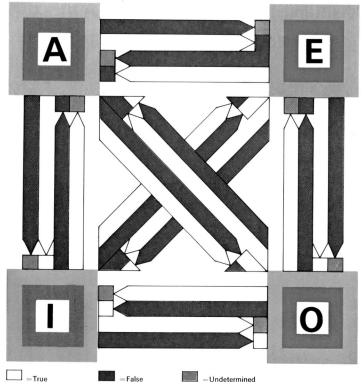

☐ = True   ■ = False   ▨ = Undetermined

A ▷ E = When A is true, E is false

2 **Bertrand Russell** (1872–1970) was joint author with A. N. Whitehead (1861–1947) of *Principia Mathematica*. Their main philosophical aim was to show that mathematics was ultimately reducible to logic. That is, every true mathematical statement could be shown to be deducible from premises containing only logical concepts and which are logically true. Their logical system dissolved the traditional Aristotelian distinction between subject and predicate. Russell's theory of descriptions translates the subject-predicate sentence "The King of France is bald" into one that is no longer of the subject-predicate form: "There is an x such that x is King of France and for all y, if y is King of France then y is x; x is also bald."

3 **"Category"** is a key word in philosophy. Aristotle introduced it first, as a label for ten items he took to be fundamental and irreducible in human discourse. These items are, first, substance (what any statement is about) and next, the nine kinds of statement that can be made about it. For example, Socrates, the central substance in the diagram, can be discussed in terms of quality (he is wise), quantity (he is tall), relations (he is the teacher of his students), place (in the Agora), time (at midday), position (he is standing up), state (he is poor), action (he is arguing) and lastly, affection (he is being verbally abused by some of the students). The study of language was an important philosophical pursuit for Aristotle and his attempt to define its component parts in this way has been followed by many other logicians concerned with the nature and relationships of substances.

Liar, had been known in antiquity. The statement: "This statement is false"; if true, is false, and if false, is true. A similar paradox arises as follows. Some adjectives (for example "polysyllabic" and "short") possess the property they denote. These are called "homological". Others (for example, "monosyllabic" and "long") do not. These are called "heterological". Is the adjective "heterological" itself heterological? If it is, then it is not; if it is not, then it is.

Alfred Tarski (1902–) eliminated such paradoxes with his semantic theory of truth, which involved a sharp separation between an object-language (the language spoken about) and a meta-language (the language in which the object-language is spoken about).

But when such difficulties can be found in the formulae of mathematics and the language of logic, the problems of establishing a coherent system of thought and then using it to establish scientific truths become obvious. For argument from experience is very different from valid inference, in which the truth of a conclusion can be proved to be logically necessary because denial would involve a

contradiction. David Hume (1711–76) pointed out that since the conclusion of a valid inference can contain no information not found in the premises, there can be no valid inference from observed to unobserved instances. Thus we cannot logically infer that all A's are B, or even that the next A will be B, from the premise that we have observed billions of instances of A's that are B and no instance of an A that is not B. Thus all the laws of science, and nearly all common-sense beliefs, are logically unjustified, this is known as the problem of induction.

Testing scientific hypotheses

One attempted solution, associated with Karl Popper (1902–) is to abandon any sort of justifying inference from evidence and to ask of scientific hypotheses that they be subjected to searching attempts to falsify predictions derived from them. If such attempts are successful, the hypothesis has to be rejected. If the hypothesis withstands testing we may not conclude that it is true (the fallacy of affirming the consequent) but we may retain it until we hit upon a less falsifiable one.

Aristotle (384–322 BC), one of the greatest philosophers, founded logic as an academic discipline. For a time, he tutored the boy who was soon to conquer the known world, Alexander the Great. Aristotle directed the first programme of research in comparative political science. He both systematized and advanced biological studies. He founded and led the Lyceum (the second university) after withdrawing from the first – Plato's Academy. The Middle Ages spoke of Aristotle as "the master of those that know". So in the 1500s and 1600s his name became for all the pathfinders of the new science the epitome of traditional conservative thinking.

4 Venn diagrams are devices for the visual representation of logical relations. If all swans [S] are white [W] and not all white things are swans, the S circle must be wholly within the W circle. If some swans [S] are black [B] and not all black things are swans, the S circle must overlap the B. If all unicorns [U] are one-horned [O] and all one-horned creatures are unicorns, then the U and O circles must coincide exactly.

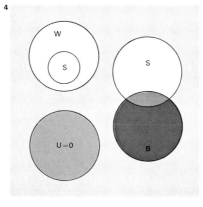

5 A doctrine of Four Causes was proposed by Aristotle as a means of framing fundamental questions about material and form, means and end. He labelled these four Material, Formal, Efficient and Final. Of a statue [E], for example, the Material cause is stone [A], its Formal pattern is that of a man [B], its Efficient cause is a sculptor [C] and its Final end is as part of a decorative frieze [D]. In Aristotle's terminology, the Final Cause is not necessarily someone's conscious intention. The Final Cause of a fetus, for instance, is to achieve its intrinsic end by developing into an adult organism.

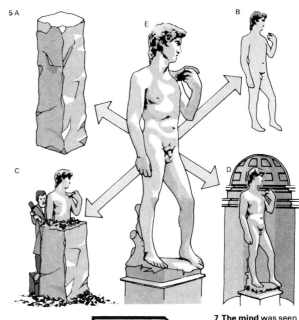

7 The mind was seen by empirical philosophers such as John Locke (1632–1704) as rather like an empty box [A] "void of all characters, without any ideas". According to this view, human knowledge has to be derived from whatever experience comes flowing in, and the box develops an internal structure only gradually. Against this picture of open-mindedness, the German philosopher Immanuel Kant (1724–1804) believed that the mind brings a prior system of categories to organize and interpret data from the senses. Kant's box [B] therefore has its own structure, imposing order on the intrinsically unknown or disorderly materials of experience (although influenced by it).

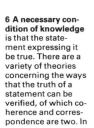

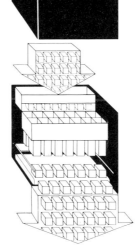

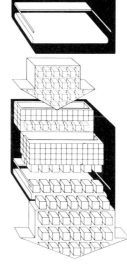

6 A necessary condition of knowledge is that the statement expressing it be true. There are a variety of theories concerning the ways that the truth of a statement can be verified, of which coherence and correspondence are two. In the coherence theory [A] statements are judged to be true if they form a coherent system in conjunction with other propositions – like the pieces of a jigsaw which one knows to be "right" because they interlock to form a whole. In the correspondence theory, [B], a proposition is judged to be true only if there is a fact such as the proposition asserts. Hence the meaning of truth is correspondence with fact that each part of the jigsaw matches part of the known whole.

Philosophy and religion

Religion and philosophy are not the same thing. Nor are they rivals fighting to occupy the same ground, although philosophical conclusions may sometimes support or deny religious claims and philosophical questions can arise out of and about religious beliefs.

The impact of St Thomas Aquinas

The philosopher and theologian St Thomas Aquinas (c. 1225–74) [Key] claimed that the existence of God could be proved in five ways. He was responding to the challenge of newly revived Aristotelian studies, with their underlying question: "Is there any need to go beyond whatever may be the fundamental laws of nature?" Aquinas started from broad and uncontentious premises: that in our universe we find causation, motion, order and so on. He argued that these presuppose that there must be a "first cause", a "great orderer" "which all men call God".

His "first cause" was not defined as that which started things off "in the beginning". Rather it was the ultimate sustaining cause, operating now and for as long as causation continues [2]. Aquinas maintained, despite accusations of heresy, that it was not possible to prove by philosophical argument that the universe did in fact have a beginning. That was something to be accepted simply on faith, as being taught by the Holy Catholic Church. He argued that philosophy is based on reason and their is no conflict between faith and reason as two sources of knowledge.

Aquinas's argument that God's existence is demonstrated by familiar general facts took no account of distinctions that were set out later by David Hume (1711-76) [7]. Hume attacked the whole idea of natural (as opposed to revealed) theology: that is, the attempt to reach positive conclusions about God and the soul by philosophical argument. He argued that any attempt to deduce from general facts about the universe that it must have a cause was unsound: "Any thing may be the cause of anything."

A popular argument, based on human experience, holds that there cannot be order without design. But Hume argued that even if this were true to the facts of, say, biology, the universe as a whole was by definition unique; and so man could have no experience of the origin of the universe. So why not say that its observed order is the order of the universe itself, and not one imposed upon it by an outside Power? (Hume unlike Aquinas was an agnostic and he did not believe the existence of God could be proved.)

Descartes and the Cartesian philosophy

Where Aquinas held the existence of God to be implied by general facts about the world, René Descartes (1596–1650) [3] inverted the argument. He believed these facts could not be known without a knowledge of God. In order to find unshakeable foundations for knowledge, Descartes began by systematically doubting everything he could doubt. This left him certain only of his own existence as a being incapable of doubt: "I think, therefore I am" ("*Cogito ergo sum*"). Accepting that the idea of God was so perfect that only a perfect God could have caused it to arise, he concluded that his own God-given senses would not deceive him provided they were properly employed. This is the Cartesian solution (Descartes' philosophy is known as Cartesian) to how we can have certain

1 Scholastic philosophy was the preserve of monks and priests in medieval times. This 12th-century manuscript shows Bede at work. The teaching was subject to the authority of Christian theology.

2 To explain why the light is burning, something needs to be said both about switching it on and about the continuing flow of current. The former is the initiating, and the latter the sustaining, cause.

The five ways of Aquinas were intended to prove that the universe has a Creator, in the sense of a sustaining cause: without His support all creation must collapse into non-existence.

3 René Descartes is considered the philosopher who ended the medieval period and initiated the modern period of philosophy. He created analytic geometry and did innovative work in physics. Descartes broke the hold that theology had on philosophy by beginning systematically to doubt everything he had been taught in school. He then resolved to believe nothing, unless it was logically necessary, regardless of the prestige of any belief. He noted that it was not always possible to perceive things clearly and that he must never rely on empirical knowledge. Yet he was certain of his existence and so his philosophy is based on knowledge of the self.

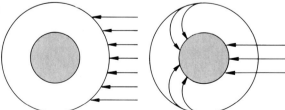

4 From his own immediate consciousness Descartes then proceeded to contend that he was just an immaterial object of consciousness and to discover how such a being knew what was going on around him. As all other animals were to be regarded as machines, he had to ask how it was proved that human machines were inhabited. His answer was their ability to think [A]. From this Descartes argued that the external world [arrows] included not only that which lay outside the body, but also the body itself [B]. Man was, therefore, essentially not the body but the spirit inhabiting it [shaded]. The English philosopher Gilbert Ryle (1900–) described man as "the ghost in the machine".

knowledge that an external world exists.

For one of his proofs of God's existence, Descartes argued from the definition of the word "God": since God is defined as embracing all perfections and existence is a perfection, therefore God must exist.

If it were valid, this argument (called the ontological argument) would provide a certain foundation for rationalist thought. The three classical rationalist thinkers – Descartes, Leibniz and Spinoza – made the most of it. Rationalism in this sense is contrasted with Empiricism [5]. The Rationalist hopes to produce a deductive system consisting, like pure geometry, of logically necessary truths – but truths that, unlike those of pure geometry, tell us about the universe and ourselves. The Empiricist believes this is a will-of-the-wisp, insisting that there can be no knowledge that does not refer to actual experience of the world.

Philosophy of religion

The philosophy of religion deals mainly with questions of how religious beliefs can be both coherent and significant. Thus Leibniz in his *Theodicy* (1710) tried to solve the problem of evil by showing that it is not inconsistent to say that evil exists and that God is perfect. His key idea was that some virtues logically presuppose some evils [6]. For example, it is impossible, both as a matter of fact and of logic, to forgive unless there is an injury to be forgiven. And since there is a perfect God (assuming the validity of the ontological argument) actual evils ultimately must have a justification. It therefore follows logically that this be the best of all possible worlds.

Some religious philosophers have urged that the idea of personal survival and immortality is senseless. Followers of Plato and Descartes believe that men are composed of two elements, body and soul, of which the immaterial soul is truly the person. But followers of Aristotle insist that words like "mind" and "personality" refer to the qualities and capacities of a unitary organism. To suggest therefore that the mind or the personality might survive the death and dissolution of the organism is as absurd as Lewis Carroll's suggestion that the grin of the Cheshire Cat could outlast its face.

KEY

Thomas Aquinas was born in Italy. Against the objections of his family he entered the Dominican Order, in which his whole life was devoted to study and teaching. Canonized in 1323, he was proclaimed a Doctor of the Church in 1567. By the papal bull *Aeterni Patris* (1879) of of Pope Leo XIII his works were given special status in the training of priests. Of these works, the most important are the *Summa contra gentiles* and the *Summa theologica*. Aquinas was always concerned with the relations between faith and reason, and with assimilating into a new Christian synthesis the then recently rediscovered works of Aristotle.

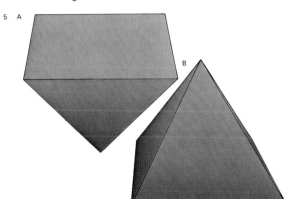

5 **Rationalists**, in the technical sense of the word, see all sound knowledge as an inverted pyramid [A]; everything else depends on and is to be deduced from a few fundamental, self-evident and necessarily true principles. Their Empiricist rivals favour instead a structure built up upon a broad base of observations of how things actually are; each truth rests on a wide foundation [B].

6 **Leibniz** tried to explain in his *Theodicy* how evil can exist in a world created by a perfect God – this problem was also explored in the story of the temptations of Job in the Bible.

7 **David Hume** was most famous in his own time for what was then the best-selling *History of England*. His *Treatise of Human Nature* was the first comprehensive radical work of philosophy written in the English language. However it was the later *Inquiry concerning Human Understanding* and the posthumous *Dialogues concerning Natural Religion* that so strongly influenced Kant.

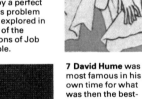

8 **Two of the most influential Existentialists** are Sören Kierkegaard (1813–55) [A] and Jean-Paul Sartre (1905–) [B]. A main feature of Existentialism is that it sees our essence, or what we are, as being determined by our existence, or what we do. Implicit in this attitude is the notion that existence is always prior to essence and therefore that of our lives are always completely determined by our choices. The existentialism of Sartre has been consistently atheist.

231

Fields of philosophy

Philosophy is classically divided into the major fields of logic, the theory of knowledge (epistemology), metaphysics, ethics and aesthetics. Each studies a set of concepts that constitute its subject matter [Key].

Elements of metaphysics

Some important concepts discussed in metaphysics are existence, essence, space, time, self, God, cause, event, change, permanence, determinism and free will.

The question of free will and determinism is a battleground shared by the philosophies of many disciplines. Because the issues are so easily clouded, this is an area where the philosopher is needed and can make a significant contribution. The genuine issues are, as always, issues of meaning, presupposition, implication, compatibility and incompatibility. How far are the apparently deterministic presuppositions and implications of work in this or that specialist field compatible or incompatible with everyday commonsense assumptions about action and choice?

Ordinarily we do not think the world changes randomly. We think that some kinds of events are regularly followed by other kinds of events. One we call the cause, the other the effect. A scientific law states a causal relation between events; for example, that mercury expands when heated.

Determinists, such as Baruch Spinoza, say that every event has a cause, including choices, decisions and human actions. It follows therefore, that man does not have free will. However Libertarians, such as Henri Bergson, argue that moral responsibility is not possible in a wholly determined universe. We cannot blame or praise a person for an act if they could not have done otherwise. Libertarians must, therefore, try to show that the will is not invariably predetermined and that there is an element of self-determination in our actions, choices and decisions.

The views of David Hume are particularly relevant to the problem of causation and of free will. His criticism of the idea of necessary causal connections is one of the landmarks of the history of philosophy. Hume's thinking can be illustrated by the example of one billiard ball striking another. Hume denies that there is a power in the first ball that causes the second ball to move, and says that our idea of such a causal relation is only a feeling produced in us by experience of the repeated conjunction of events. Objectively, a causal law is derived from merely a regular succession of two or more events.

The theory of knowledge

Concepts central to discussions in epistemology are knowledge, truth, theory, method and evidence. For example, is the concept of "know" the same in the expression "I know the sum of two numbers" as it is in "I know my own name"?

Knowledge and truth (and falsity) are two concepts intimately interconnected. We can be said to know something only if what we believe is true. I cannot say that I know 5+7=11 because that equation is false and because the idea of false knowledge is self-contradictory. It would seem, then, that I cannot say I know anything unless I can also say that I know that what I say is true.

What do I have to know in order to know that some proposition is true? Three major theories of truth are correspondence, coher-

CONNECTIONS

See also
226 What is philosophy?
228 Logic and the tools of philosophy
230 Philosophy and religion
234 Philosophy and ethics

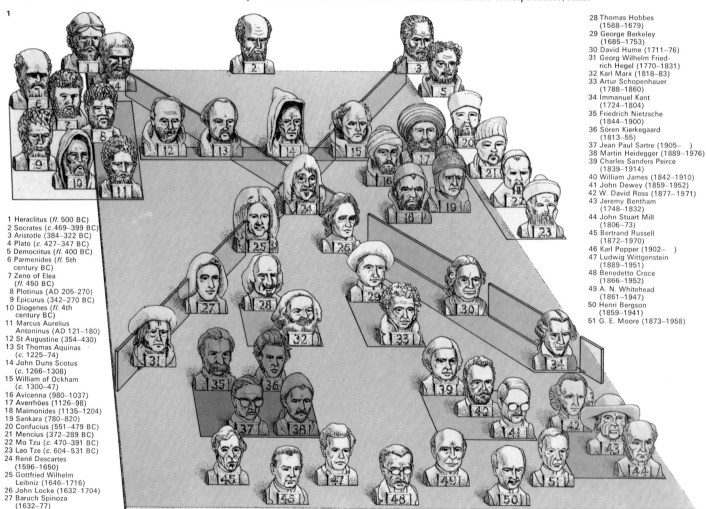

1 Heraclitus (*fl.* 500 BC)
2 Socrates (*c.* 469–399 BC)
3 Aristotle (384–322 BC)
4 Plato (*c.* 427–347 BC)
5 Democritus (*fl.* 400 BC)
6 Parmenides (*fl.* 5th century BC)
7 Zeno of Elea (*fl.* 450 BC)
8 Plotinus (AD 205–270)
9 Epicurus (342–270 BC)
10 Diogenes (*fl.* 4th century BC)
11 Marcus Aurelius Antoninus (AD 121–180)
12 St Augustine (354–430)
13 St Thomas Aquinas (*c.* 1225–74)
14 John Duns Scotus (*c.* 1266–1308)
15 William of Ockham (*c.* 1300–47)
16 Avicenna (980–1037)
17 Averrhöes (1126–98)
18 Maimonides (1135–1204)
19 Sankara (780–820)
20 Confucius (551–479 BC)
21 Mencius (372–289 BC)
22 Mo Tzu (*c.* 470–391 BC)
23 Lao Tze (*c.* 604–531 BC)
24 René Descartes (1596–1650)
25 Gottfried Wilhelm Leibniz (1646–1716)
26 John Locke (1632–1704)
27 Baruch Spinoza (1632–77)

28 Thomas Hobbes (1588–1679)
29 George Berkeley (1685–1753)
30 David Hume (1711–76)
31 Georg Wilhelm Friedrich Hegel (1770–1831)
32 Karl Marx (1818–83)
33 Artur Schopenhauer (1788–1860)
34 Immanuel Kant (1724–1804)
35 Friedrich Nietzsche (1844–1900)
36 Sören Kierkegaard (1813–55)
37 Jean Paul Sartre (1905–)
38 Martin Heidegger (1889–1976)
39 Charles Sanders Peirce (1839–1914)
40 William James (1842–1910)
41 John Dewey (1859–1952)
42 W. David Ross (1877–1971)
43 Jeremy Bentham (1748–1832)
44 John Stuart Mill (1806–73)
45 Bertrand Russell (1872–1970)
46 Karl Popper (1902–)
47 Ludwig Wittgenstein (1889–1951)
48 Benedetto Croce (1866–1952)
49 A. N. Whitehead (1861–1947)
50 Henri Bergson (1859–1941)
51 G. E. Moore (1873–1958)

1 Classifying philosophers is not an easy task, but this chart does bring out a few major similarities and dissimilarities, as well as a few major lines of development. The Chinese [20–23] are shown as peripheral if only because their work is not, in the strict sense, philosophical. The Greeks are placed first and at the top because they created the Western philosophical tradition. Aristotle [3] and Plato [4] stand out from the rest in both influence and achievement. After the collapse of the old pagan world, thinkers laboured for centuries to reconcile its philosophical achievements with, and to put them at the service of, religion: Christians such as Augustine [12], Aquinas [13], Duns Scotus [14] and Ockham [15]; and also Maimonides [18] in Judaism and Avicenna [16] and Averrhöes [17] in Islam. Descartes [24] is at the centre of the picture as he laid the foundations of modern philosophy. By persuading succeeding generations that private consciousness is the only sure starting point for knowledge, he made the epistemological problem fundamental: Do we know, and if so how, anything of the universe? One line of development then passes through Locke [26], Hume [30] and Kant [34] with the emphasis on discovering the nature and limitations of our learning apparatus. Another line, by way of Leibniz [25], Spinoza [27] and Hegel [31], takes up Descartes's rationalism. The existentialists [35–38] can be seen as reacting against a rationalist picture of the world, as can the pragmatists [39–41] who were concerned with practical bearings. Of the 20th-century group Wittgenstein's [47] works are most discussed by philosophers. And although Russell [45] was most in the public eye, it is Popper [46] whose ideas are having most influence.

ence and pragmatic certainty.

One view is that what I am thinking, say that the apple before me is red, is true because it corresponds with the fact of the apple's being red. Two difficulties with this view are urged. The first is that we only experience the sensory effect that the supposed red apple has on me, never the red apple out there itself. To say, then, that "the apple is red" is true because it corresponds to the fact, is like claiming that I know a portrait corresponds to a face even though I have never seen and can never see the face.

Since there are no observable facts to correspond with $7+5=12$, this equation's truth has to be explained in another way. The coherence theory would do so by calling attention to the logical relation it has to other equations. Thus, since $5=3+2$, and $7=3+2+2$, and $(3+2)+(3+2+2)=12$ all cohere, and because $5+7=12$ coheres with them, we may take $5+7=12$ to be true.

The American Pragmatists Charles Sanders Peirce and John Dewey held that our knowledge of truth is acquired through a process of verification. Our thought about the

redness of the apple enables us to anticipate its ripeness, and so, further, to anticipate its sweetness. The truth of this is verified when future experience of the apple is in accord with what we predicted.

Other areas of philosophy
Other sets of concepts that philosophers study can be classed under the heading of "philosophy of" such as the philosophies of art, language, politics, history, science (natural and social), law and mathematics.

Philosophy has links with both arts and sciences, influencing them and being influenced by them. Pure mathematics, for example, inspired the philosophical thinking of the Rationalists. Of these René Descartes and Gottfried Leibniz were themselves also major creative mathematicians; while Plato's thought has stimulated generations of philosophers to think of mathematical entities, such as numbers and geometrical relationships, as timelessly existing. Both the theories of Newtonian physics and Darwinian evolution have had a major impact on the development of philosophical thinking.

Logic
argument
validity, proof
definition, consistency

Epistemology
knowledge, truth
theory, method
evidence, analysis

Metaphysics
existence, essence
space, time, self
God, cause

Ethics
good, right, duty
responsibility
utility

Aesthetics
beauty, art, taste
standard, judgment
criticism

The traditional labels for branches of philosophy are all more or less unsatisfactory because what earns its place under one heading will often serve under another. Metaphysics is defined as the search for fundamental categories. Ultimately what sorts of things are there? Epistemology asks whether we know, what we know and how we know. Answers to general epistemological questions are thus in one aspect metaphysical or have metaphysical presuppositions and implications. Logic is the study to valid and invalid forms of argument. Philosophical ethics investigates logical characteristics of moral discourse. Aesthetics deals with appreciation of beauty. Shown here are some of the concepts that are examined in the various branches.

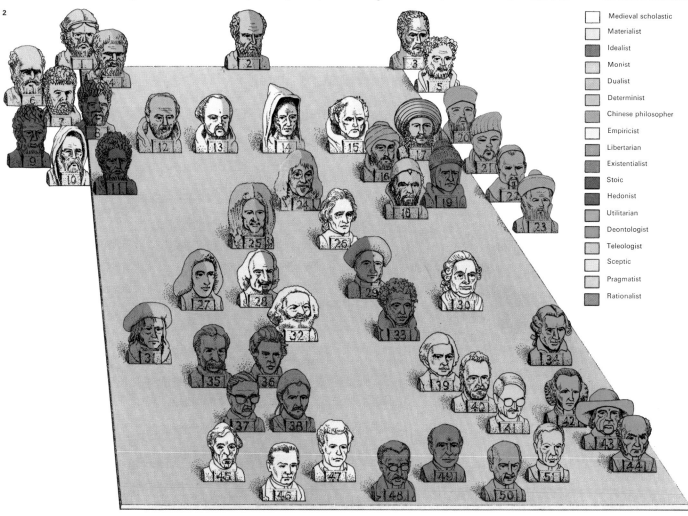

Medieval scholastic
Materialist
Idealist
Monist
Dualist
Determinist
Chinese philosopher
Empiricist
Libertarian
Existentialist
Stoic
Hedonist
Utilitarian
Deontologist
Teleologist
Sceptic
Pragmatist
Rationalist

2 Philosophers can also be divided according to where their strongest interest lies. Within the field of metaphysics, materialists believe that matter is the only kind of entity that exists in contrast to idealists who hold that matter is an illusion. Monists believe that only one kind of ultimate stuff exists, while dualists maintain there are two kinds – mind and matter. Determinists hold that events are caused by other events and are predictable according to laws; libertarians maintain that there are uncaused events – human free will. In epistemology, the study of knowledge, the empiricists trace the truth of propositions to observations and experience; positivists are extreme empiricists claiming that anything that is neither a part of logic and mathematics nor of empirical science is meaningless. Rationalists claim humans have innate ideas that are prior to experience and necessarily true. Pragmatists claim that knowledge comes from practical action. Sceptics deny that any knowledge is possible because our senses and reason are so misleading. In ethics a teleologist maintains that the concept of good is more basic than right (right action is determined by its consequences). The deontologist holds the opposite – an action is right or wrong regardless of the value of the consequences. Utilitarians measure the goodness of an act by its utility. Hedonists maintain the only thing good in its own right is the experience of pleasure or the absence of pain. Stoics emphasize the practical aspect of philosophy as a guide to living. Reason, not our desires, should be our guide to action. Existentialists maintain that man's existence precedes his essential nature which is not given to him but is made by him in the choices he makes.

Philosophy and ethics

Philosophical ethics is concerned with how men should behave and involves moral discourse about such concepts as good, right, duty, responsibility and punishment.

The concepts of good and right

There are many definitions of what constitutes "good". Naturalists identify the concept of good with the concept of some natural, psychological feature. Hedonists say this feature is pleasure, others that it is the object of desire and still others that it is the satisfaction of a need. Non-naturalists dispute these definitions. Plato (c. 427–347 BC) pointed out that there are morally bad pleasures and that if pleasures and good were identical we would then have the self-contradictory notion of a bad good.

In discussing "right" some philosophers assert that the concept of good is more fundamental than the concept of right. In other words, to say of an act that it is right is to say that it is productive of a greater balance of good over evil than any other act open to the agent. Typical of those philosophers who hold this view are the Utilitarians, such as Jeremy Bentham and John Stuart Mill [4].

Opposed to this view are those who say that right cannot be defined in terms of good. Otherwise it would be possible to judge the intimidation of an innocent person, with the aim of deterring crime, as right because it produced a balance of good over evil.

The characteristics of moral discourse

To be genuinely moral, discourse must have several characteristics. First, it is prescriptive as opposed to descriptive [1]. That is, it contains statements about what ought to be rather than what is. From the confusion of these two comes the naturalistic fallacy of invalidly deducing an "ought" statement from an "is" statement. One popular form of this fallacy is the move from saying that something is natural (in the sense that it happens or tends to happen) to the conclusion that anything else would be unnatural and wrong – not in the sense that it *will* not happen, but that it *ought* not to happen.

This seems obvious once clearly stated. But it is quite another thing to see all the implications of this ought/is dichotomy.

Obviousness is essentially relative to time, place and person. It was not obvious to Aquinas (c. 1225–74) when, if only temporarily, he overlooked the crucial difference between those Laws of God that are the scientists' laws of nature and cannot be "disobeyed", and those Laws of God that are prescriptive laws that rule human conduct but are notoriously ignored or breached.

Not all prescriptive utterance is moral. Immanuel Kant [Key] distinguishes the hypothetical from the categorical imperative. The former may suggest a course of action in certain cases but is never absolutely categorical, in contrast to a moral imperative such as "Thou shalt not kill".

Kant further suggested that there are two other conditions that distinguish the authentically moral. The first of these two conditions is universality. If anything is to count, not necessarily as a correct moral principle but as genuinely moral, then it has to be a principle applied universally and impartially. If you claim that the use of chemical weapons or torture is immoral, then this claim must be universally applied; to protest

1 A prescriptive law states that some action ought or ought not to be done. Transgression of such a law, for which the woman taken in adultery was condemned to be stoned [A], does not prove that the law itself is invalid. A law of nature, by contrast, claims to be a description of a state of nature and stands or falls by whether it is "transgressed" or not. The stargazer [B] who observes something that confounds a law of nature must record that the law does not, after all, hold true. He is the spectator, speaking a language of non-participatory description, whom Kant, in his examination of practical reason, contrasts with man as active agent. He says that when we abandon the roll of spectator then our language ceases to be neutral and we begin to ascribe values to what we see and to make moral judgments.

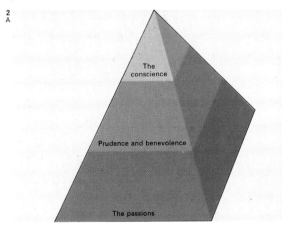

The conscience

Prudence and benevolence

The passions

2 Joseph Butler (1692-1752) appealed to the notion of a self-evidently authoritative hierarchy in the principles of our nature [A]. Just as the passions are subordinate to prudence and benevolence, so these in their turn must be subordinate to the rational and reflective control of conscience. By comparison Plato's *Republic* represents the Man Reason [B], controlling with assistance from the Lion of Self-assertion the Many-headed Monster of Passion.

against offences by those regimes to which you are politically opposed, while remaining silent about those regimes which you favour politically, is not to voice an authentic and sincere moral protest.

Kant characterizes the second of his two formal conditions for rating discourse as moral rather awkwardly. He maintains that moral discourse has to be autonomous as opposed to heteronomous. The idea is that each of us must somehow impose his own moral principles on himself, in contrast to the laws of the land which are imposed on us from without. (However, it is often rightly remarked that all such legislation is also in principle subject to the moral assessment of the individual: "I know that this is the law, but is it right, ought it to be the law?").

The problem of subjectivism

The problem is to retain this notion of autonomy without succumbing to some form of subjectivism. In the strictest sense, a moral subjectivist holds that moral words do no more than express the reactions of the person who is using them. But the social "subjec-

tivist" may also, by extension, be one who defines moral words in terms of the class or tribe to which the speaker belongs.

Clearly, any moral argument between two strict subjectivists is impracticable, just as it is between two people, one of whom says "I like chocolate" whereas the other says "I don't like chocolate". The social subjectivist, can at least argue that a group dispute may be settled by a simple vote. Both the subjectivist and the objectivist see the ought/is problem as fallacious, since one believes that morality is based on personal opinion, and the other, in contrast, believes that morality can be determined by the facts alone.

These conflicting approaches, if rigidly held, would make effective moral argument impossible. Any universal moral principle, however, unless susceptible of modification by the proposer, will have consequences unforeseen by him and it is this fact that makes moral argument possible. One of the major purposes of a critical moral discussion is to expose such unacceptable consequences and thus persuade the proposer to amend or abandon the offending principle.

Immanuel Kant (1724–1804) lived in the East Prussian city of Königsberg, (Kaliningrad). He was raised in the traditions of Leibnizian ration-alism, and his own distinctive "critical philosophy" was a response to the chal-lenge of Hume's radic-ally sceptical empir-icism. Kant's three main works (*Critique of Pure Reason, Critique of Practical Reason* and *Critique of Judgment*) set out to find the nat-ure and limitations of our learning cap-acities. Kant empha-sized the contribution of the knower to knowledge, whereas Locke and Hume had been inclined to speak of materials from the external world im-pressing themselves upon an originally blank and inert mind.

3 Friedrich Niet-zsche (1844–1900), seen with his mother, radically broke with the established concep-tion of ancient Greek culture in his first book *The Birth of Tragedy. Thus Spake Zara-thustra*, his most famous work, argues that the "will to power" is the primary human drive. Anti-Christian polemics become pro-gressively more cen-tral in his later books, in which critics dis-cern signs of the mad-ness that overtook him.

4 John Stuart Mill (1806–73) was an intel-lectual leader of the Philosophical Radi-cals. Active in all liberal causes, he pro-duced standard works on economics and on logic and scientific method, as well as the libertarian classic *On Liberty*. Also, follow-ing Jeremy Bentham (1748–1832), he argued for ethical utilitarian-ism – a system of morality governed by the idea of the "great-est happiness of the greatest number".

5 Jonathan Edwards (1703–58) was the first major philosopher born in North Ameri-ca. Just as Aquinas had seen his mission as the incorporation of the best of ancient Greek philosophy in-to a new Catholic syn-thesis, so Edwards laboured to incor-porate Newton and Locke into a revived Calvinism. In phil-osophical psychology his *Religious Affec-tions* anticipated much of William James's *Varieties of Re-ligious Experience*.

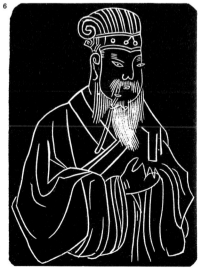

6 Mencius (372-289 BC) was both in his life and thought extraordinarily simi-lar to Confucius. Born in the same province, both lived as professional moral teachers. Both shared concerns for filial piety and for es-tablished rites and both respected the Sage Kings. Both agreed that "Once the ruler is rectified, the whole kingdom will be at peace", a statement from the *Book of Mencius*. However, it was Mencius alone who made explicit and defended his con-viction that human nature is essentially good. Apart from his philosophical teaching, Mencius urged many practical reforms.

7 Confucius (551–479 BC) was born within modern Shantung. His ideas became the greatest single force shaping traditional Chinese civilization. However, his thought was not in our sense philosophical. When, for example, Tzu Kung asked him whether the True Way could be epitom-ized in a word, Con-fucius replied: "Is not reciprocity the word? Do not to others what you do not want them to do unto you." But he never made an attempt to show, as centuries later Kant was to try to show, that this is an es-sential element distinguishing some categorical im-peratives as moral.

Talking without words

People communicate with each other in many ways other than by speaking or writing. Animals build quite complex societies by means of non-verbal communication and humans reveal a surprising amount about themselves by the use of gesture, posture, facial expression and other forms of behaviour. Several thousand pieces of information can be exchanged within seconds of a meeting.

Ways of expressing emotion
In addition to outward facial and bodily signals, including the use of costume and adornment, there are less obvious means of communication – the use of time and space. Touching [5] and distance or nearness of approach can reveal much about a relationship. Punctuality can convey eagerness. On the other hand, a person who keeps someone else waiting can convey an impression of busyness that may be either genuine or false.

In evolutionary terms non-verbal communication preceded speech, and the existence of a system for expressing emotions was particularly significant to man's development. More than a century ago

Charles Darwin (1809–82) suggested that emotions help a species to survive and that feelings such as happiness, sadness, fear, anger, surprise and disgust each have unique forms of display.

While the expression of these emotions looks similar in all humans [2], cultures differ in the degree to which emotional displays are encouraged or discouraged. In many Western societies, for example, men are not supposed to cry and women are discouraged from showing extreme anger.

Accurate reading of emotion from the face is complicated by man's ability to control his expressions by masking some feelings and fabricating the appearance of others [1]. In addition, many facial displays last only a fraction of an instant. Displays may also be partial – occurring on only one area of the face [3] – or they may be a blend, combining two or more emotions in one expression [4]. People differ in their ability to decode expressions, particularly fleeting ones.

Specific emotions are even more difficult to read from body posture. But overall attitudes, positive or negative, are easily

recognizable [Key, 6]. Interest or disinterest is shown by whether the body is erect or slouching, leaning backwards or forwards. Attraction or dislike is revealed by approach or avoidance. Status is indicated by the assumption of a higher or more dominant position in a group.

The significance of gestures
Gestures can be classified in a number of ways. Adaptors are movements that, at least originally, helped man to adapt to his environment. Scratching, wiping and fondling are examples. Regulators are movements that guide the flow of speech or contact between people; nods and eye signals are used, for example, to encourage a speaker to continue or to indicate a wish to interrupt. Conversation without regulator gestures (as when someone does not react at all to what is said to them) can be disconcerting. Illustrators are movements that help to elaborate, punctuate and clarify speech. Emblems have specific word-like meanings and often replace words and phrases.

Adaptors are often used unconsciously

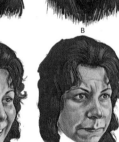

1 Facial expressions in both humans and primates show many similarities. The smile is usually a sign of happiness. But it is also one of the easiest expressions to "put on" and may be used to conceal real feelings of fear, anxiety or dislike. The tense grin adopted by a chimpanzee when approaching a more dominant male parallels the nervous smile a woman may wear in a stressful social situation [A]. In both the expression is meant to demonstrate lack of hostility. When aggressive or angry [B] both primate and human display a stare and, at least initially, a firmly closed mouth.

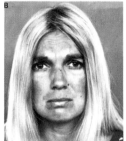
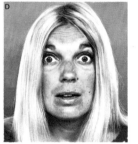

2 Each of the primary emotions has a unique facial display that can be recognized and correctly identified all over the world. Interpretation of photographs shown to people as diverse as New Guinea hunters and New England socialites is remarkably uniform. Happiness is associated with the characteristic upturned mouth [A]; sadness with downturned mouth and slightly up-turned brow [B]; anger with down-turned mouth and in this case an aggressive compression of the lips [C]; surprise with raised brow and open eyes and mouth [D]; fear with tension in the central forehead, the lower eyelids and the corners of the mouth [E]; disgust with a distinctive wrinkling of the nose and shape of the mouth [F]. Happiness is easiest to recognize. It is harder to distinguish fear from surprise or anger from disgust.

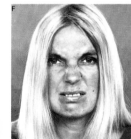

4 A blend expression, in which more than one emotion is shown on the face at once, can often be seen as a person's response changes, producing startling effects such as an angry brow and a laughing mouth.

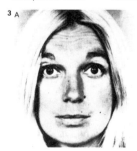

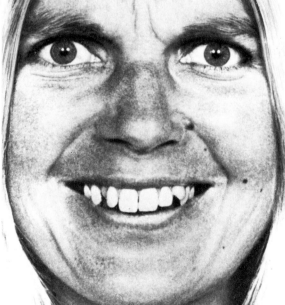

3 Partial expressions are more difficult to interpret than full-face displays of emotion. The human face is tremendously flexible and is under considerable voluntary control. An emotion such as surprise may therefore appear in only one part of the face and then only for a very short time. The main areas in which such fleeting expressions of emotion may be detected are the brow [A], the eyes [B] and the mouth [C]. To make it easier to separate the characteristic signs of surprise, each area is shown superimposed on a neutral face.

with no intention of communicating anything, although an observer may find them informative. A psychotherapist, for instance, may notice that a person fiddles with his hands when a troublesome topic is broached.

Regulators, too, are used with little awareness. In many cultures gestures of greeting serve as more conscious regulators. The appropriate bow, handshake, hug or kiss must be delivered before conversation can take place.

Illustrators are classified into sub-categories such as pointers, indicating which object is being discussed; spatials, indicating size or space relationships; batons, used for punctuation or emphasis; pictographs, outlining or portraying an object; ideographs, tracing the flow of an idea; and kinetographs, re-enacting some bodily movement. In Germany, Nazi theorists once argued that illustrative gestures were innate – that, for example, Jewish people had distinctive, innate gestures. A pioneering work by David Efron in New York showed that gestures did differ between eastern European Jews and southern Italians. But it showed also that

these gestures changed in second-generation immigrant groups. Thus illustrators are learned, just as language is also learned.

Emblems in communication
Emblems are usually employed consciously to communicate. Examples include the direction-indicating thumb of the hitch-hiker and the two-finger "V" for victory (or peace) signal. The significance of emblems differs sharply from one language to another [7, 8]. The American "A–OK" gesture, with forefinger and thumb forming a circle, for instance, has vulgar, derogatory connotations in many parts of the world.

Advances in modern technology have made non-verbal messages become increasingly important, especially in visual media such as television and the cinema. There has been a corresponding increase in research into the ways in which these communication patterns have evolved and are used in different cultures today. The term kinesics has been applied to the study of body movement and efforts have been made to analyse it in the same way as linguists analyse language.

KEY
A

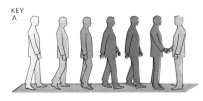
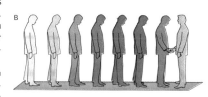

When people meet, their posture and movements can reveal a great deal about their emotional state. Normally, when a man walks across a room to shake hands, he holds himself erect, swings his arms by his sides and remains balanced and in control [A]. If he is depressed, on the other hand, he tends to bend his head and shoulders forwards, taking small, shuffling steps and holding his arms by his sides [B]. When a man is elated and excited his body may lose its natural balance and erectness, his arms swing widely and his gait becomes erratic [C]. These movements are often seen when old friends meet. They can also be associated with a manic state of mind.

5 The extent to which people touch each other varies widely according to culture and relationship. A 1966 American study showed where male and female students were touched most often by parents and friends.

5

Percentage contact
☐ 0–25
☐ 26–50
☐ 51–75
■ 76–100

Mother   Father   Same sex friend (female)   Same sex friend (male)   Opposite sex friends

6 Posture and gesture make up a fundamental part of the repertoire of non-verbal communication. One look at a collection of people reveals much about their characters and moods. A man with head in hands and drooping eyes 1 is shut off from the group and does not hide his boredom or sadness. Another 2 shows self-confidence or smugness by forming a steeple of his hands while the open hands of 3 suggest sincerity and warmth. Sorrow or shame are typified by the way that 4 hides his face. The erect posture with hands on hips of 5 conveys assertiveness. An attentive posture is taken by 6; he sits on the edge of his seat leaning for-

wards with hands on mid-thighs. The fact that 7 has chosen the highest place may show that he is the most dominant of the group or simply that he is aloof. Touching

or rubbing the nose 8 is associated with doubt while the arm-gripping, defensive posture of 9

suggests that he is nervous (or perhaps sitting in a draught). A nervous person may wrap his fingers

around his biceps so hard that the

knuckles show white. Finally 10 shows lack of interest by turning away from the group entirely.

7 Gestures can convey a common emotion such as approval in many ways.

8 An insult to one person may be a compliment to another. Tucking a thumb beneath the forefinger [A] is a good luck charm in Brazil but a jeer or obscenity elsewhere. A raised forefinger [B], vulgar in Anglo-Saxon countries, means "Wait, I have an idea" in Italy, Austria and among Jews but "God is my witness" to gypsies. It may also be an auction bid.

7

"Excellent" USA, Europe, Levant, Iran

"Excellent" Sicily

"Excellent" Brazil

"Beautiful" Italy

"Beautiful" Latin countries

"What a smasher" Brazil

"Hello beautiful" Spain, Portugal

"Excellent" Anglo-Saxon

8

A

B

Communication through speech: 1

Any system used by a social group to communicate information, whether drum beats, smoke signals or finger movements, may loosely be called a language. But of all language forms that exist by far the most flexible and expressive is human speech.

Origins of speech

Language and thought are so intertwined that it is sometimes forgotten that men originally had to learn to talk to each other by inventing arbitrary vocal symbols for a whole world of nameless objects, actions and emotions. The idea of naming things is assumed to have evolved from more simple forms of communication such as gestures, facial and bodily movements and the kind of cries, grunts, snorts, whistles or clicks uttered by animals or birds [1]. Experiments with animals ranging from apes to dolphins have shown that some are capable of imitating human speech or of responding to a limited number of sounds [2, 3]. The crucial difference between speech and animal communication is man's ability to cope with complicated ideas, particularly involving time and space.

It seems certain that language and thought evolved together, one quickening the other. Man's ability to perpetuate and extend his knowledge gave him the power of swift cultural development, increasing his dominance over other animals. The idea that language was originally a divine gift appears in many mythologies.

Although the different shape of man's vocal organs allows him to make more varied sounds [5] than the hominoid ape, his linguistic superiority is primarily an intellectual one. Every normal child appears to be born with the capacity to learn a language simply by watching and listening to people around him. Since few sentences are ever repeated in precisely the same way, this remarkable achievement implies an innate faculty not only to learn words and their meanings but also to grasp grammatical structure at a deep level.

The question of whether all languages descend from one common source language or rather evolved among separate groups independently in different parts of the world is impossible to determine. The length of

man's history, the transitory nature of early cultures which must have possessed language, and the superimposition of later tongues through trade or invasion, have all weakened the scent.

Linguistic studies, on the other hand, have shown that one historic language, Indo-European, has been the parent of existing languages spoken by about half the world's population. In western Europe only one regional tongue, Basque, is not descended from it. Further investigation of a small group of words for plants and animals that had a common origin [6] narrowed down the geographical origin of the proto-language to a small area of northern Europe from which chariot-riding marauders began to spread out in about 2000 BC, mingling their language with those of the lands they reached.

Limitations and resources

Each particular language conditions a speaker's way of seeing the world and of feeling and acting in it. The limitations of language have preoccupied many twentieth-century writers and thinkers. At the same

1 Wolves, like humans, form integrated social groups and have a flexible communication system to express both emotions and status within the hierarchy of the family pack. Apart from vocal noises and facial or body positions, a wolf can use its tail to indicate confidence [A], confident threat [B], relaxation [C], uncertain threat [D], relaxed feeding [E], depression [F], defensive threat [G], active submission, with wagging [H] and abject submission [I].

2 Dolphins communicate with each other by a system of distinct sounds ranging from clicks to whistles. These highly developed marine mammals can also locate small objects from relatively long distances by sending and receiving echo-locating pulses. To test the ability of dolphins to communicate information through sound codes, scientists noted the time a dolphin took to learn a trick [A]. The skilled dolphin was then placed in a tank alongside an unskilled one which could hear but not see it [B]. In all tests the second dolphin learned to perform the trick much faster, apparently because it was being prompted by sound messages.

3 Chimpanzees have been taught a rudimentary vocabulary as a result of experiments conducted in the USA since the early 1950s. In attempting to find out if animals could be taught to com—municate directly with man, scientists chose the chimpanzee both because its physical and mental capacities are closest to those of humans and because it is able to form strong emotional ties with them. In early experiments efforts were made to teach spoken words to a chimpanzee reared from birth with a family. But it managed to learn only four words. More recent experiments have used systems of geometric shapes or hand signs. Having noticed the use of gestures by chimpanzees in the wild, one researcher adopted a year-old animal into his family and began teaching it American sign language, a system used widely by the deaf. By instruction through play the chimpanzee steadily amassed a vocabulary of signs. Seen here are the signs for listen [A], eat [B] and tooth-brush [C]. After four years of study it knew about 150 signs and began to use them in combinations to form meanings. Signs for bird and water were used to represent a duck, for instance. The sign for dirty, which previously had been used only about soiled objects, was combined with monkey when the chimpanzee described a macaque that threatened it. Researchers are now investigating whether chimpanzees will use learned sign language between themselves.

238

time, the film image has become, through cinema and television, an important supplement to language as a cultural tool. Yet the influence of spoken language on man's development remains fundamental. Only through language can people communicate diversities of meaning with sufficient precision to allow complex social, economic and cultural systems to work.

Vocabulary itself is only one of an armoury of speech weapons. Variations in word selection and sentence structure, intonation and emphasis, can convey infinite nuances. When supplemented by facial expression and gesture, language becomes more expressive still. The Russian actor-manager Konstantin Stanislavsky (1865–1938) used to ask his students to say the word "tonight" in 50 different ways, ranging from inquiry, surprise and doubt to rage, fear, relief and excitement.

When we speak we do not utter words but what may be termed "breath groups" – an uninterrupted flow of breath that is often less than a sentence (depending sometimes on lung capacity). Listeners understand three things: logical value as indicated by choice,

order, emphasis; intonation; and something beyond that which word symbols convey.

The study of meaning at its deepest level is the most complicated and difficult aspect of language study, not only because of the diversity of human thought and experience, but also because the vocabulary of a living language may change rapidly while grammatical and phonetic changes evolve slowly.

The analysis of language
Grammar is the description of the structure of a language and has two necessary subjects: the sounds and the meaningful combinations of sounds. The sounds are studied in phonetics. The description of the meaningful terms is the morphology; a minimal meaningful unit is the morpheme. For example, "man" is a basic morpheme from which "manly" and "manliness" are derived. In English, a free entity of one or more morphemes bound together is a word. The study of the history, development and origin of words is called etymology; the arrangements of words is called syntax. Semantics is the study of the meaning of language.

International organizations are the modern Towers of Babel. At the UN every speech, no matter what its original language, is simultaneously translated into the five official languages – Chinese, English, French, Russian and Spanish.

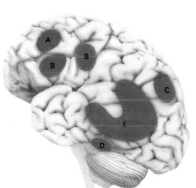

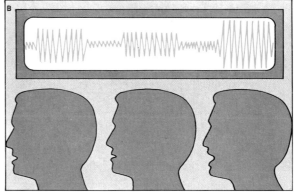

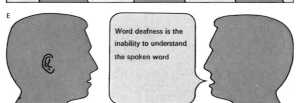

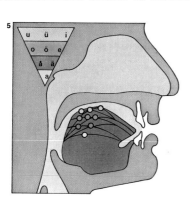

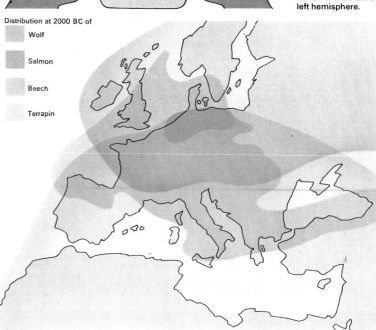

4 Language control areas in the brain are vital for several different functions. Writing ability is located in the frontal lobe [A], together with two areas that control voice production [B]. Alexia, a form of word blindness in which letters are confused, can result from damage to an area of the parietal lobe [C], while deafness can result from damage to the cortex of the parietal and temporal lobes where ability to comprehend the spoken word is located [E]. Another area of the temporal lobe apparently controls the ability to name things correctly [D]. In most people, the processes are controlled in the brain's left hemisphere.

5 Vowel sounds are controlled largely by the position of the tongue in the mouth. A triangle diagram named after the scientist who originally formulated it, Hellwag, represents the limits of the tongue positions for the major vowels, although these vary slightly for different languages. Depending on their vertical position, vowels are classified as front, central or back, while the horizontal classification is closed, half-closed, half-open and open.

6 Language detective work by scholars and scientists has led to the theory that an area of northern Europe was the place of origin for all the languages of the vast Indo-European group. The names of a small group of plants and animals are similar in all these tongues. But fossil-dating techniques indicate that before 2000 BC only one area, close to the shores of the Baltic, provided a habitat for all four species – wolf, salmon, terrapin and beech tree. The theory is that a race of people who spread out from this area imposed their own names for these species on other tongues, but adopted various local names for species with which they were unfamiliar.

Distribution at 2000 BC of
Wolf
Salmon
Beech
Terrapin

Communication through speech: 2

More than 4,000 languages, living or dead, have been identified. The migration of early nomadic peoples produced an astonishing diversity of speech forms as each group experimented with language, discovering new things to be named, borrowing words from other tribes and slowly changing the sound and the grammatical construction of its own tongue in so doing.

How languages are classified

Nearly half the population of the world speaks one of the Indo-European group of languages [Key], all of which derive from a common tongue spoken in northern Europe about 5,000 years ago. But the language divided into eight major branches, five of which split and resplit as words were shortened, lengthened, coined or swopped, as syllables were added or dropped and as vowels and consonants changed. Because Eastern peoples, for example, were not accustomed to pronouncing the Aryan sound "k", they altered it to "s" or "sh" when invaders from the West settled amongst them. As grammatical changes were made, the original links became obscured.

A few languages, called isolates, seem unrelated to others; Basque is an example. But some features of grammatical structure are common to all forms of speech and many languages are historically related. Classification of the difference between them is of two kinds: genetic and typological. Genetic classification is based on word derivation, common history and literary traditions and on socio-cultural factors. Within the Indo-European family of languages genetic classification identifies such sub-families as the German, English, Dutch, Swedish, Norwegian and Danish group.

Typological classification groups languages according to their structure as isolating, agglutinating and inflecting types. An isolating language is one that indicates grammar mainly by word order with each word being a single grammatical unit called a morph. Vietnamese is an example. An agglutinating language is one in which individual words can be composed of several morphs glued together, as in Turkish. An inflecting language is one in which there is no specific correspondence between particular segments of a word and particular grammatical functions. An example is the English word "mice" in which plurality is indicated by a syllable change instead of an added morph. English, like many languages, combines all three typological features.

Differences within languages

Adding to the enormous diversity of language are the subtleties of accent and tonal change that can be produced by the speech organs [2]. Even within a single language there are subdivisions, or dialects, based on regional, social or occupational differences. There are American, Australian and Scottish dialects of English, for example. They contrast with Gaelic, which is quite a different language from Scottish English, although the two may be spoken within the same Highland village. The point at which dialects become separate languages is not always clear. The Dutch-German speech community, for example, spans a continuous area of intelligibility from Flanders (Dutch) to Styria (German), but speakers of Flemish and Sty-

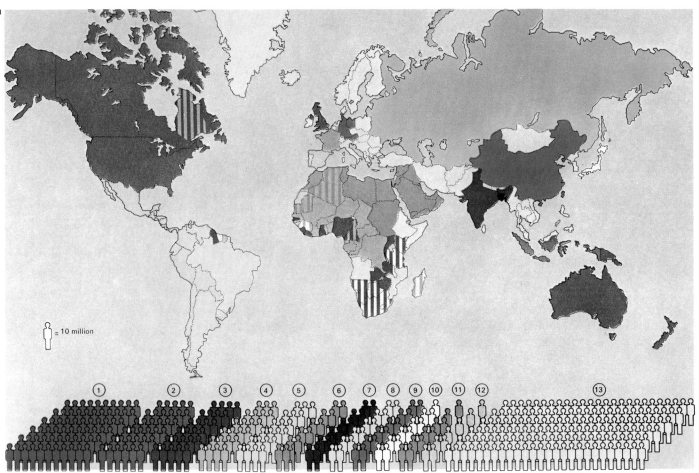

1 Each of the world's 4,000 million people belongs to a speech community, a group speaking the same language. About 1,500 different languages are spoken, the largest block being Mandarin Chinese, spoken by more than 650 million, but this does not include those speaking Wu (70 million), Cantonese (47 million) and other Chinese tongues. The map shows the principal official languages throughout the world. These twelve take in almost 2,500 million people. The rest of the world is split into smaller language blocks such as Italian (60 million), Tamil (55 million), Korean (50 million), Punjabi (50 million), Dutch and Flemish (19 million) and Greek (10 million). About 100 languages have more than one million speakers. A multiplicity of other languages are each spoken by a smaller number of people than this. Superimposed on the thousands of local languages and dialects around the world are the six major international languages of the historic colonialist powers: English, Spanish, Russian, Arabic, Portuguese and French. Easy communications may increase the dominance of some of these languages, and of Japanese and Chinese, as mediums of commerce, education and cultural exchange. The Indo-European group of languages is the most widespread. More people now speak these languages outside Europe than inside it. North America has the biggest English-speaking population and South America has most speakers of Spanish. A growing number of people throughout the world can speak one of the major international languages as well as their native tongue. At the same time, the number of speakers of minority tongues is steadily dwindling. Efforts to establish an artificially created language such as Esperanto have been limited by the fact that such languages lack a cultural base.

1 Mandarin Chinese
2 English
3 Hindi and Urdu
4 Russian
5 Spanish
6 Arabic
7 Bengali
8 Portuguese
9 German
10 Japanese
11 Indonesian
12 French
13 Other languages

rian dialects cannot understand each other.

Social groups may use speech differences to heighten group identity, to exclude others speaking the same language or to underline social divisions by the snobbery of a preferred accent. When extended to political or international rivalry the divisive aspects of language can be dangerous. In India, for example, the end of British rule led to rioting over whether Hindi (an Indo-European tongue) or one of the earlier Dravidian tongues, such as Tamil, should become the country's official language. Similarly, language rivalry in Belgium and Canada has also caused political tension.

The potential of language as a unifying force was recognized by the Romans who used Latin in the west and Greek in the east of their empire to weld together the peoples they controlled. Latin provided scholars throughout medieval Europe with a *lingua franca* (common language) and survives today both in professions such as law and medicine and as a source for new scientific words. The prestige of France and the clarity of its language made French the language of European diplomacy for many years, while in the twentieth century the dominance of English-speaking peoples in technology and commerce has led to increasing interest in the acquisition of English as a second language.

Improving communication
A simplified language form called pidgin was used by several European nations to facilitate trade or to communicate with peoples they colonized. At a more sophisticated level there have been at least 300 attempts to invent a universal language of which Esperanto alone (with more than 750,000 speakers) has made some headway.

A whole range of "languages" has also been developed for computers. These codes are made up of unambiguous words and symbols. The message is first translated into the special language. A typical example in COBOL (Common Business Oriented Language) might read "Multiply hours-worked by rate-for-job giving wage-payable". This is fed into the machine, which translates it into basic instructions that trigger further commands to the machine and produce the result.

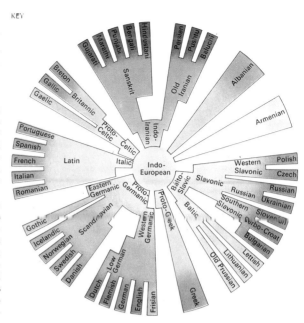

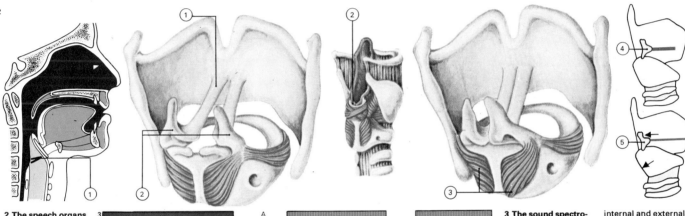

2 The speech organs consist of the lungs, larynx, nasal cavity, tongue, mouth and lips. Speech begins with voice production as the lungs push air through the larynx. During normal breathing the vocal cords [1] in the larynx are held apart forming a triangular opening [2]. When speech begins, the articulating cartilages of the larynx are drawn together by muscular action [3] and a linear chink is left between the vocal cords. The faster air is forced through this chink, the louder the voice. The pitch of the voice depends on the tension of the vocal cords and the parts of the cords that are made to vibrate. When the articulating cartilages [4] are tilted [5] changes are caused in cord tension. The tighter the cords, the higher the note produced in the larynx. As air passes through the mouth the voice is modulated and broken up by changes in the shape and position of the other organs.

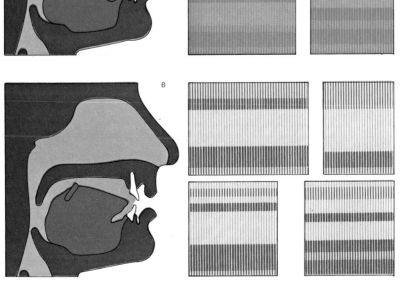

3 The sound spectrograph analyses the sound waves made by speech into component frequencies and intensities. A spectrogram is produced on electrically sensitized paper on which a registering stylus burns dark traces to show the concentration of energy in the appropriate frequency areas. The tongue position that produces spectrograms of front vowels [A] is slightly different from that which produces back vowels [B]. No two vowel patterns are exactly alike; the significant variations are seen in the position of the resonance bars – the darkest horizontal bands. Each person produces a distinctive spectrogram because the number of permutations in voice production are infinite, involving about 100 paired and single muscles and dependent on very delicate adjustments in the air currents escaping from the lungs. All this is associated with contraction of internal and external muscles of the larynx to maintain the margins of the glottis at the right degree of elasticity. The complex sound wave produced by regular vibrations of the glottal folds consists of a basic vibration (called the fundamental, related to our perception of pitch) and a series of overtones. The relative pitches and ranges of these overtones determine the quality and the character of the voice produced. In speech and song, sounds (usually initiated by lung air, with or without glottal vibration) are modified by the shape of the resonators provided especially by the pharynx, the nasal cavities and the mouth, which is particularly flexible and important in "moulding" the emerging stream of air. An individual spectrogram is unique, as is a fingerprint, although perhaps more variable, opening the way for the use of "voiceprints" as a means of identification.

Codes, cyphers and secret messages

There are many ways of communicating besides speech and writing. Some are secret, others open. Examples of the former are military and diplomatic codes and cyphers; of the latter, commercial codes, Morse code, semaphore, flag codes, Indian smoke signals, jungle drums and braille used by the blind.

Concealment and disguise

Many ways of concealing messages exist, but basically the plain-text can be disguised in two ways. The first is by steganography, which conceals the message's very existence by writing it in invisible ink [1], or reducing it either to a microdot about the size of a full stop (by photographing it through the wrong end of microscope) or to a micropulse (by recording its transmission at higher and higher speeds). A message can also be arranged in such a way that the first or last letter of each word in an apparently innocuous text spells out a real message.

The second way to disguise plain-text is to transform it in one of two fundamental ways. The first is by transposing the letters of the message in a prearranged order to make a cypher [9]; the second is to substitute the letters of the plain-text by other letters, numbers or symbols to form a code.

Forms of codes

A code, as distinct from a cypher, involves the substitution of prearranged code words, code numbers or code groups for words, phrases or syllables. The trilingual international commercial code in English, French and Spanish invented by Marconi, for example, has five-letter groups representing common words in each language (UVYDU = railway sleeper, traverse, traviesa). The groups are recorded in a dictionary and are used to save telegram costs. This type of code can be secret or open, but military or diplomatic codes based on this principle have to be further disguised to make them secure. An example is the military one-time pad code used by the British during World War II.

In this, to send the message "To the C-in-C: Tanks will move tomorrow at dawn", the sender looks up the basic code which contains a list of words and phrases commonly used in military operations, thus: "To the C-in-C = 5475; Tanks will move = 9835; Tomorrow = 4439; At dawn = 7463". Since code books are liable to loss or capture, it would not be safe to send this message as 5475:9835:4439:7463. Frequent repetition of such phrases as "to the C-in-C" would soon be identified and the code cracked. It is therefore necessary to disguise these four-figure groups. The safest way of doing this is for both sender and recipient to have what is known as a one-time pad on each page of which are columns of four-digit groups printed at random. The sender indicates the page, column and line where the message is to start, thus: 1549, which would mean page 15, column 4, line 9. If the next four groups beginning on page 15, column 4, line 9 are 4431:7628:5016:4881, the code text to be sent (5475:9835:4439:7463) is written down and to each group are added the figures taken from the one-time pad, ie:

$$4431:7628:5016:4881$$
$$5475:9835:4439:7463$$

$$1549:9906:7463:9455:2344$$

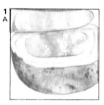

1 Secret inks can be made from many substances ranging from potato juice to the most sophisticated chemical compounds. To make a potato ink-well, scoop out the centre of a potato and scrape juice into the hole [A]. A message written with this juice [B] will be invisible, but can be read when the paper is heated to 120°C (248°F) [C]. If paper is waxed by rubbing it with a candle and laid waxed face down on plain paper, then a message written on the plain paper will be impressed as a waxen line and become legible after powdered coffee, soot or some other dark dust has been scattered on to it.

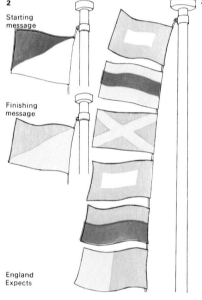

2 Single coloured flags are used by ships to send specified messages. A yellow flag means "I have a quarantine case on board". Navies also use the 26 different designs that relate to the letters of the alphabet in order to send messages according to prearranged codes. Shown is the start of Nelson's famous "England expects" message sent before the start of the battle of Trafalgar (1805).

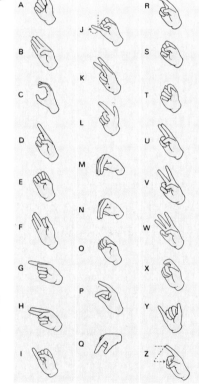

3 Deaf and dumb sign language is a system in which the fingers and hands are used to illustrate the letters of the alphabet. The one-handed system is most common, although a two-handed version is used in the UK. The speed of both systems is similar to a slow, ponderous conversation. Normally, however, the deaf and dumb use a mixture of signs, gestures and expressions, which is as fast as normal speech, reserving the formal hand signals for difficulties.

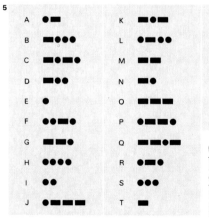

4 Semaphore makes use of two flags or lights to send messages and is based on the circular movement of the two hands of a clock. The angle of the flags, which should be held with straight arms, stands for a letter of the alphabet. B = 6.45, D = 6.00, F = 6.15 and so on. Semaphore was originally designed to communicate over a short distance where the sender and recipient could see each other but were out of earshot. It was used extensively during the Boer War (1899–1902) before wireless made it largely redundant. It is still used to a limited extent at sea and where radio silence is imperative. Rail signals have semaphore markers as part of the points mechanism, to enable drivers to tell at a glance the way the points are set.

5 In Morse code letters are represented by combinations of dots and dashes. Invented by Samuel Morse (1791–1872) in about 1837, it is signalled by a light or buzzer.

To decode, the recipient subtracts the one-time pad groups from the groups received; then refers to a code "dictionary".

Cyphers and secrecy

Simple substitution cyphers afford little protection since substituting one letter or symbol for one letter of plain-text reflects the frequency of letters occuring in the language.

To encypher the message "Tanks will move tomorrow at dawn" by a simple letter-for-letter substitution cypher based on the following key:

Plain-text abcdefghijklmnopqrstuvwxyz
Cypher-text monseratbcdfghijklpqzyxwvu

the result is as follows:

Tanks will move tomorrow at dawn
qmhdp xbff giye qigillix mq smxh

A further elaboration that helps disguise the letter frequencies is the bigrammatic Playfair system in which the message is first split up into groups of two letters regardless of word breaks and then changed according to the following rules. Letters of the alphabet are arranged in five columns of five letters each (I and J occupy one space), ie

```
M O N S E
R A T B C
D F G H IJ
K L P Q U
V W X Y Z
```

If the two letters of the groups form the diagonals of an imaginary square within the five columns, the letters at the opposite sides of the square are chosen (plain-text NK = cypher mp). If they are on the same line, then move one letter to the right (TA = bt). If they occur in the same column, then the letters immediately below are chosen (WA = of). To encypher "Tanks will move tomorrow at dawn" the letters are written out: TA NK SW IL LM OV ET OM OR RO WA TD AW NX and the encyphered text will read: bt mp oy fu ko mw nc no ma am of rg fo tn.

An efficient way of preventing wire-tapping is to shift the frequency of the sounds of speech, or scramble. Time Division Scrambling (TDS) cuts the stream of speech into split-second portions and shuffles them by first tape recording the conversation and then picking off segments in jumbled order.

KEY

Giovanni Battista Porta (1535–1615), an Italian physicist-inventor, produced one of the earliest substitution cyphers. He used symbols to replace letters and figures. The symbols were chosen by turning the pointer in the centre to the necessary letter or number in the two outer rings, then reading off the corresponding symbol in the inner ring. The cypher could be changed at any time by rotating the discs themselves relative to each other, according to a prearranged schedule. The lines of symbols below the disc spell out the inventor's name. A somewhat simpler version was invented by Leon Battista Alberti (1404–72) regarded as the "father of cryptography".

6 Thomas Jefferson (1743–1826) devised a wheel cypher consisting of 36 wheels that can be assembled in any order. Upon each of them the letters of the alphabet are written in jumbled order and the number of different combinations this gives amounts to many millions. It was by far the most advanced cypher of its day and was adopted by the American army almost at once. It is the basis of some modern machines.

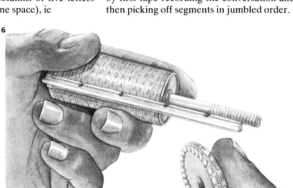

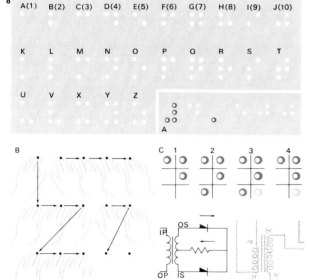

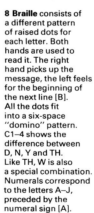

Transformer printed normally / and in Braille

7 The Enigma cypher machine, invented in the 1920s by Arthur Scherbius, handled the Germans' most secret information during World War II. It had a keyboard like a typewriter's but the keys were interconnected electrically so that by pressing a b, say, it printed an x. This coupling was changed constantly so there was never any consistency of relationship between key hit and letter produced. First the message was typed and scrambled by Enigma and then sent by Morse. The recipient fed the garbled words back through a similar machine to obtain the correct message out – provided he knew the original links between letters.

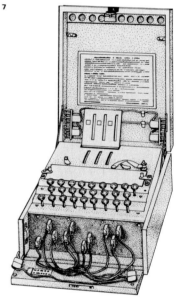

8 Braille consists of a different pattern of raised dots for each letter. Both hands are used to read it. The right hand picks up the message, the left feels for the beginning of the next line [B]. All the dots fit into a six-space "domino" pattern. C1–4 shows the difference between D, N, Y and TH. Like TH, W is also a special combination. Numerals correspond to the letters A–J, preceded by the numeral sign [A].

9 Simple transposition of the letters of a plain-text message [A] is achieved by writing it in rows [C] beneath a code word, in this case "SPEC–TRUM" [B]. The letters of the code word are numbered according to their position in the alphabet (for SPECTRUM it is CEMPRSTU) [D]. Then the message is written in the order of the numbered columns [E]. To decipher, you need only know the code word.

10 Pencils or cigarettes can be used to make a simple transposition cypher. Write a message on a narrow strip of paper wound in a spiral round one and send it off. The receiver needs a tube of the same diameter to make sense of it.

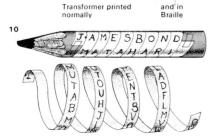

LE ROI EST MORT VIVE LE ROI

S P E C T R U M

L E R O I E S T
M O R T V I V E
L E R O I X X X

6 4 2 1 7 5 8 3

OTO RRR TEX EOE EIX LML IVI SVX

243

Communication through writing: 1

Writing is an essential ingredient in all but the most primitive civilizations. It is not only one of the most important means of communication, but until recently it was the only one able to defy time; the presence of written records is usually held to mark the watershed between history and prehistory.

The earliest known scripts
Palaeolithic man's cave paintings may have been an attempt to convey his ideas; communication by means of pictures is a recurrent feature in primitive societies [1]. From this it is only a short step to picture-writing, in which a circle, for example, may represent the sun. Ideograms are a natural progress from picture-writing with the circle of the "sun" representing related ideas such as "heat", "light" and, eventually, "day".

Ideograms form a recognizable feature of such diverse writing as Mayan "glyphs" [2] and the mysterious script of Easter Island in the Pacific Ocean [3]. They can still be recognized in much more sophisticated writing systems such as Egyptian hieroglyphics [4]. Of its 600 signs some are words ("man",

"look"), some determinatives that indicate how the preceding signs are to be understood (for example, as an abstract idea), whereas still others are used for syllables and single sounds, thus foreshadowing a true alphabet.

The earliest known script that can properly be called writing was devised by the Sumerians in the fourth millennium BC. The cuneiform (wedge-shaped) script derives ultimately from ideograms, but soon became a system of conventionalized characters, each having a distinct phonetic (sound) value [5].

Most important of the non-alphabetic scripts is Chinese. Its characters, derived from ideograms, each represent a complete word, with the result that the Chinese scholar must be conversant with about 9,000 different symbols whereas everyday writing uses as many as 1,500 to 2,000 characters [6]. Most characters are compounded from a radical element, giving the meaning, and a phonetic element indicating the pronunciation. The phonetic "fang" when compounded with the radical "tree" gives the character for "timber"; whereas when compounded with the radical "word" it produces

the character for "to enquire". Traditionally Chinese was written right to left in columns; today it is usually in lines from left to right.

The development of the alphabet
Picture-writing and ideographic scripts arose independently in different parts of the world. In contrast the alphabet was probably invented only once, by the Semitic people inhabiting the Syria–Palestine area in the latter part of the second millennium BC. This first alphabet (North Semitic) consisted of 22 characters, each representing a consonant, written right to left [7]. It had two main branches, one of which led directly to classical Hebrew and Aramaic and, indirectly, to Arabic. In its present form Arabic consists of 28 letters and is the most widely used alphabet in the world apart from the Latin alphabet [9]. All these alphabets are consonantal and written right to left.

In the spread of the alphabet eastwards, the role of Aramaic, the *lingua franca* of the Middle East for several centuries from the sixth century BC, was crucial. It is from this, most scholars now believe, that all the vari-

1 The wampum belts used by North American Indians conveyed messages by means of pictures. Here the two figures represent the Moqui and the president of the USA; the Moqui are offering friendship and negotiation. These belts are better classed as "mnemonics", that is aids to the speaker, than as writing in the form of pictures.

2 The Maya had a highly developed writing system, but their manuscripts were almost totally destroyed by the Spanish invaders.

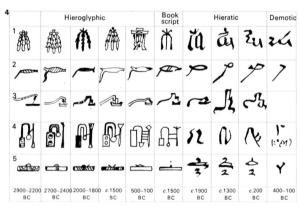

3 The script from Easter Island (as yet undeciphered) has carved characters incised on wood with a shark's tooth.

	Hieroglyphic				Book script	Hieratic				Demotic
1										
2										
3										
4										
5										
	2900–2200 BC	2700–2400 BC	2000–1800 BC	c.1500 BC	500–100 BC	c.1500 BC	c.1900 BC	c.1300 BC	c.200 BC	400–100 BC

4 Hieroglyphics, "sacred carvings" familiar from temples and pyramids, is the earliest form of Egyptian writing, first used in 3000 BC. It underwent no substantial modification in the three thousand years of its history, but a more cursive (rapidly written) form was soon developed for writing on papyrus – known as hieratic. A highly cursive form of the same language, known from 600 BC, is demotic (of the people) although the derivation of most of its signs cannot be deduced. Shown here are the changes in five hieroglyphs and their equivalents in book script, hieratic and demotic. The first two are just phonetic, *ms* and *mh* respectively; 3 means "to write"; 4 means "to choose"; and 5 indicates "abstract idea".

5 The wedge-shaped writing known as cuneiform was developed because of the ease with which such strokes could be imprinted on wet clay, the normal writing material used for this script. This example is of a relatively late cuneiform script developed in the ancient Persian Empire in 500 BC. The external form of the script is the sole connection with the earlier ideographic cuneiform writing of Mesopotamia. The 42 signs shown were created from the Neo-Babylonian model: 36 of these have the value of either a vowel or a vowel plus a consonant; four signs are true ideograms and two are word dividers. The script was written left to right.

Phonetic	a	b(a)	č(a)	d(a)	d(i)	d(u)	f(a)
Ancient Persian							
Phonetic	g(a)	g(u)	h(a)	h(a)	i	j(a)	j(i)
Ancient Persian							
Phonetic	k(a)	k(u)	l(a)	m(a)	m(i)	m(u)	n(a)
Ancient Persian							
Phonetic	n(u)	p(a)	r(a)	r(u)	š(a)	s(a)	t(a)
Ancient Persian							
Phonetic	t(u)	t(a)	u	v(a)	v(i)	y(a)	z(a)
Ancient Persian							
Phonetic	tr(a)	Word divider	King	Country	two forms	Divine name	Earth
Ancient Persian							

ous alphabets used in India and South-East Asia ultimately derive, both the angular forms such as Devanagari and Sanskrit and the round forms such as Burmese [8].

The other main branch of the alphabet was brought to the West by the Phoenicians, reaching the Greeks by about the ninth century BC. It is certain that the Greeks derived their alphabet from North Semitic; for example, in the Semitic alphabet each letter was given a name: aleph "ox", beth "house", and so on. These were taken over by the Greeks as alpha, beta, etc (thus giving our word "alphabet"), although they have no meaning in Greek. The Greeks made one fundamental innovation: the introduction of vowels. Their alphabet came to consist of 17 consonants and seven vowels.

The classical Latin alphabet

The Greek alphabet reached the Romans through the Etruscans and resulted in the classical Latin alphabet of 23 letters, the parent of the present-day Western alphabet. Subsequent changes have been limited to the introduction of w and of separate forms for i and j, and for u and v. The important part played in the dissemination of this alphabet by the Roman Catholic Church is clearly shown by the scripts of the Slavonic peoples; those converted to Roman Catholicism (Poland) use the Latin alphabet whereas those converted to the Greek Orthodox religion (Russia) use the Cyrillic alphabet. Cyrillic, invented in the ninth century and traditionally ascribed to St Cyril of Byzantium, consists of 24 letters taken over from Greek uncial (a form of writing in rounded letters) to which 19 new characters have been added.

An ideal alphabet should consist of symbols that are distinct and easy to remember, but at the same time capable of representing all significant speech sounds. In our alphabet the first requirement is admirably fulfilled, but the second much less so. In English, for example, five vowel symbols must represent 12 vowel sounds and eight diphthongs. Some modern languages compensate for this limitation by introducing additional, distinguishing marks, such as the umlaut that is used in Germanic languages.

KEY

As seen in this diagram of the history of the alphabet, North Semitic is the precursor of all alphabetic writing systems, but no link has yet been established between it and Egyptian hieroglyphics or cuneiform writing.

6 It is generally accepted that Chinese characters are composed of eight basic strokes, all of which are to be found in the character "yung", meaning eternity. The diagram shows the sequence of strokes by which the character evolves (with the arrows giving the direction of the brush-strokes). Chinese writing is painted with a brush, often on silk or some similar material. It is said that Wang Hsi-chih, the fourth-century calligrapher, took 15 years to perfect this character.

7 The development of the modern alphabet from North Semitic, to Greek, Etruscan and Latin is illustrated here.

North-Semitic		Greek		Etruscan	Latin		Modern caps
Early Phoenician	Phoenician	Early	Classical	Classical	Early	Classical	Roman
							A B C D E F G H I J K L M N O P Q R S T U V W

8 Six different scripts used today are: Russian, the most important alphabet of the Cyrillic family; Burmese, perhaps the most beautiful of the rounded scripts used in SE Asia; modern Greek print, essentially a derivative of Byzantine minuscule; modern Hebrew, which is written right to left; Sanskrit, ultimately derived, like Burmese, from Aramaic (Devanagari, widely used to write Hindi, is similar in form to Sanskrit); and classical Chinese characters.

8 Russian
Идея использования квантовых систем для ге радиоволн оказалась весьма плодотворной и недостижимые для обычной радиотехники резу.

Burmese

Greek
'Ο 'Οδυσσεὺς καὶ οἱ σύντροφοι αὐτοῦ πλοῖα, τὰ ὁποῖα ἦσαν πλήρη λαφύρων, ἀ Τρωάδος, ἐπιθυμοῦντες νὰ φθάσωσιν ὅσο

Hebrew

Sanskrit

9 Classical Arabic is a consonantal script that is written left to right.

10 Runic writing, found on artefacts in Britain and Scandinavia, dates from the 3rd to the 16th centuries. Its origins are obscure, but it can be deciphered because of the existence of manuscripts giving the phonetic value of its 24 characters. Some are regularly found in English medieval writing – in "ye olde Englishe" the letter y is really Runic "thorn", pronounced th.

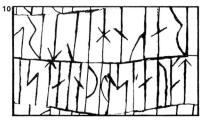

245

Communication through writing: 2

Over the centuries man has employed a tremendous variety of materials to write on. Hard materials, mainly stone but also metals, were incised with a sharp instrument. Soft materials might also be incised, as were clay tablets in Mesopotamia [1] and the wax-covered tablets of the Greeks and Romans [4], but were more often written on with pen and ink or painted on with a brush. Leaves, bark and textiles have also been used, although the principal materials were papyrus, parchment and paper.

The earliest materials
The Egyptians were writing on papyrus as early as 3000 BC and it remained in use as a writing material for 4,000 years. Papyrus was made out of strips cut from the stem of the papyrus plant, one set being laid with the fibres running vertically and a second set placed on top with the fibres running horizontally. Pressure was applied to bond the two sets together to form a sheet and a number of these sheets were then glued to one another to form a long strip that could be wound into a roll.

During the first four centuries AD the roll was gradually superseded by the codex. In this format several leaves of writing material ("folios") were folded down the middle and then sewn to other sets or gatherings, in much the same way as most modern books are produced. The new form began to predominate at the same time as parchment began to replace papyrus as the most common writing material. Animal skins could be prepared for writing by tanning, which produced leather [3], or by tawing (soaking in a solution of alum and salt), which produced parchment.

Paper is traditionally supposed to have been invented by the Chinese in the first century AD. The earliest paper was made of rag (paper made from wood dates from the mid-nineteenth century). The pens used in early times were normally made from reeds and the ink was carbon based using, for example, lamp-black. Later, metallic ink was used and, by the sixth century, quill pens.

Deciphering early scripts
The decipherment of ancient scripts is obviously most difficult when the language is unknown and it generally depends on the discovery of a bilingual document with one text in a known language. Thus the Rosetta Stone [Key], inscribed in Egyptian and Greek, enabled Jean Champollion (1790–1832) to penetrate the mysteries of Egyptian hieroglyphics and restore knowledge of a language that had been lost for 1,500 years. Similar bilingual texts made it possible to decipher Sumerian cuneiform. Even when such convenient documents do not exist, ingenuity can sometimes triumph. In the 1950s, Michael Ventris (1922–56), using techniques learned while code-breaking during World War II, deciphered the Minoan syllabic script known as Linear B [5].

Even when the language is known, considerable problems of decipherment remain, especially with examples of writing on soft materials extending over several hundred years. Styles of writing naturally tend to change, sometimes radically. (An exception to the rule is Chinese, the calligraphy of which has remained essentially unaltered for some 2,000 years.) In Greek, for example, the development from monumental capitals

1 This seventh-century Babylonian tablet is a calendar listing the lucky and unlucky days of the year. Cuneiform writing was originally inscribed on clay with a specially cut reed stylus.

2 The papyrus roll was the usual writing material in the eastern Mediterranean area for 4,000 years. This letter on papyrus (third century BC) complains that a previous letter has been eaten by mice. Many works of Greek literature are extant only on papyrus and it is the material on which the earliest manuscripts of the New Testament survived. It is found only in Egypt and in the Near East.

3 The finding of the Dead Sea Scrolls in the Qumran area since the 1940s has aroused considerable interest, principally because of the close relationship between the language and the ideas to be found in the scrolls and those in the New Testament. It is now accepted that the scrolls antedate the Christian era and that they belonged to a sect of Essenes. Many of them contain books of the Old Testament, others regulate the sectarian discipline of Essenes. A few are written on papyrus but most are written on leather and nearly always in the Hebrew language.

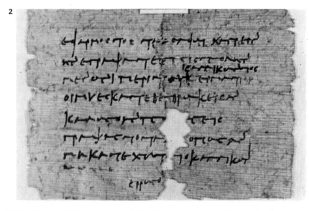

4 This wall painting from Pompeii (first century AD) represents a lady holding wooden tablets and a stylus. The tablets were covered with wax, then incised with a sharp-pointed stylus and finally all bound together. The codex form, which at about this time began to replace the roll for works written on papyrus and parchment and which resembles the pages of a modern book, may have developed from this.

5 Linear B, the archaic Minoan script, was deciphered by Michael Ventris using a grid [A] on which, according to known or supposed phonetic values, the different symbols used in the Minoan clay tablets were plotted in squares, each symbol being assumed to represent a syllable, consonant plus vowel. After many trials, the language turned out to be Greek. This fragment of one of the inscribed tablets (c. 1400 BC) [B] includes vertical strokes denoting numerals.

6 *Codex Sinaiticus,* a fourth-century manuscript of the Bible, is written in Greek uncial or biblical majuscule script which is a rounded form of the capital script used in inscriptions and the source of the Cyrillic alphabet. The manuscript was found in the last century in a monastery on Mt Sinai by Konstantin Tischendorf (1815-74) and was eventually purchased by the British Museum. It is written on parchment of a fine quality.

to the uncial (rounded characters) of the early biblical manuscripts [6] is readily detectable. But it is difficult to see any connection between these scripts and the cursive script with its typical flowing characters employed in a papyrus of the third century BC [2], or between the latter and the minuscule, or small letter script, used in a Byzantine manuscript 700 years later [7].

Development of writing styles
Development of writing styles is to be seen at its most diverse in Latin script [10]. In majuscule (capital letter script), the lettering of inscriptions of the classical period shows a clear connection with modern capital letters, a connection still easily seen in the rounded version (uncials) used in parchment manuscripts from the fourth century [9]. However, the Latin script on papyrus and wood from the first two centuries AD [8] bears little resemblance to our own alphabet, or even to the minuscule used for everyday writing from about AD 300. A direct descendant of this minuscule is the script known as Carolingian minuscule. This simple and clear script soon

became the accepted form for writing books throughout Latin-speaking Europe. At the same time it came to be written with increasing angularity, so that by the thirteenth century it had developed into Gothic or black letter script.

When printing was invented in the West, Gothic was the script employed for the first books. In Italy, however, fifteenth-century humanists reverted – in conscious reaction to Gothic – to a script based on ninth-century Carolingian and this was ultimately to become the universal script of western Europe (although in Germany this did not occur until the present century). A Carolingian hand, therefore, is immediately recognizable as being in essence the script of modern print.

With the invention of printing the history of manuscript books ends, but styles of writing used in documents and letters continue to develop. The increasing literacy throughout the world has meant that today there are not just one or two recognizable styles of writing but countless individual and idiosyncratic variations.

The Rosetta Stone was found by Napoleon's Egyptian expedition in 1801 and is now in the British Museum. The stone contains the text of a priestly decree (c. 197 BC). The stone provided the key to the decipherment of Egyptian hieroglyphics and opened the way to the development of Egyptology as a scientific study. The top part is written in hieroglyphics, the middle in demotic and the lower in Greek. Comparison of the Greek with the hieroglyphic eventually enabled the latter to be deciphered. A major step was the recognition that characters inside circles, known as "cartouches", represented the names of the rulers of Egypt.

7

8

9

7 This Byzantine manuscript, dating from the ninth century AD, is written in Greek minuscule script, which ultimately derives from the cursive writing on the third-century BC papyrus.

8 A Latin writing tablet from Vindolanda, a Roman fort in the north of England, records a list of military stores. It is part of a remarkable archaeological discovery of wooden tablets written in ink, the first of which were unearthed in 1973. The script is the normal everyday Latin writing of the second century AD, descended from monumental capital but written much more fluently and rapidly.

9 This manuscript is an example of Latin uncial writing and dates from the sixth century AD. The script is a rounded form of the capitals found in inscriptions; it was also adapted

for writing on parchment manuscripts. It was in regular use from the fourth to the eighth centuries, but remained generally popular for headings and titles for considerably longer than this.

10 Monumental capital [A] is the type normally found in inscriptions and resembles modern capital letters. It is from minuscule cursive [B], of the fourth century AD, that small letters originally derive. Carolingian minuscule [C] was a deliberate innovation of the period of Charlemagne (c. AD 800). Its clarity and simplicity won it acceptance as the script in regular use throughout western Europe. Twelfth-century Gothic minuscule [D] is sometimes known as black letter. The letters are in basically the same form as in Carolingian minuscule but have become more angular over the centuries with a considerable loss of clarity.

11 Chinese is not an alphabetic or syllabic script; it employs a different character for each word and therefore a special typewriter has to be used. This consists of a tray, which contains the 2,000 characters in regular use, and a

normal typewriter roller. Both parts are movable, enabling the roller to be positioned over the character required, which can then be typed. It is necessary to move the roller along the rows and columns to the next character

required and so on. Additional characters are contained in separate trays and must be picked up individually and positioned beneath the roller as required. The apparatus resembles a primitive printing machine.

The study of man

The "discovery" of primitive or tribal peoples has occurred repeatedly. Ancient societies, such as those of the Greeks, Romans and Chinese [1], "discovered" them in conquering the less developed societies on their frontiers. But tribal customs and institutions were of less interest to them than the military or administrative problems and any judgments were ethnocentric, that is, made in the light of their own cultural values.

For the emergent group of anthropologists of the nineteenth century the ways of life of people whose appearance was strange and who practised unfamiliar customs became subjects of scientific investigation. Initially, the objective was to use facts concerning simple cultures to determine the broad outlines of human history. This gave way to the goal of modern anthropology – the description and explanation of differences and similarities in human culture.

European expansion
The geographical explorations of the Portuguese, Spanish, English, French, Dutch and others, begun in the late fifteenth century, generated a flood of knowledge concerning the diverse ways of life of non-Western and "uncivilized" societies. Navigators, explorers [Key, 3, 6], traders, fur trappers, soldiers and missionaries all contributed their observations. Much of the information that accumulated was incidental to other purposes: the pursuit of trade routes, of possible missionary settlements, of lands for conquest and colonization, not to mention El Dorados, fountains of youth and the hypothetical Southern Continent. Particularly important in the pre-anthropological era were the observations of trained naturalists and scientists [7] who, while they seldom penetrated the inner workings of native life, offered accounts that were far richer in detail and more objective in perspective than the typically superficial travellers' tales [2].

In the latter half of the nineteenth century the growing knowledge of primitive societies in Africa, the Pacific and Asia paralleled the course of European imperialism as much of the non-Western world was partitioned into the colonial domains of a few European countries. Knowledge of the life-styles of subject peoples was sought and applied by the administrators of the newly founded colonies in varying degrees. Some felt they could proceed just as effectively in ignorance of the complexities of native custom. To them, natives were natives, all destined for conversion to "civilization" and Christianity, and a cheap labour supply for the mines and plantations owned by Europeans.

The science of anthropology
By the 1860s knowledge of mankind's diverse cultural manifestations was sufficiently detailed to call forth an organized scientific enterprise – anthropology – to bring some order to the mass of detail. What meaning could be assigned to such strange customs as a man's taking to his hammock during his wife's pregnancy, uttering an oath when a companion sneezed, or referring to his wife's sister as "my wife"? If tribes in different parts of the world practised the same custom did this prove there was an historical link or was it an independent development? These were among the questions posed by the first anthropologists. Some

1 The Chinese of Han times (206 BC–AD 220) knew of many less developed societies on their borders. Their view of these "barbarians" was expressed in the way they rendered their names. People held in high esteem were honoured by having their names written in combination with the radical form of the character *jen*, meaning human being [A]. The names of people on poor terms with the emperor, or held in low esteem, were combined with *ch'uan* meaning dog [B]. Peoples whose cultures differed greatly and whose customs were repellent had affixed to their names a form of the character *ch'ung* or insect [C].

2 The ancient Scythian nomads living north of the Black Sea had much in common with the central Asian nomadic peoples observed by Europeans 2,000 years later. Herodotus (*c.* 485–425 BC) left an extensive account of the customs and institutions of the Scythians. Such writings were rare among the ancients and this one is valuable although partly based on second-hand information obtained from Greeks living in the Black Sea town of Olbia. The Scythians appear to have been cultured people who traded with the Greeks. The decorations on the jewellery seen here showed Greek influences.

Typical of the ancients, however, were the tales Herodotus reported of races of men beyond Scythia: some had only one eye, others hibernated for half a year, still others had the feet of goats.

3 Captain James Cook (1728–79) commanded three voyages of exploration in the Pacific Ocean that resulted in the discovery of many islands and their societies. Among the most politically developed and warlike of these societies was Hawaii, where Cook was killed in the course of his third voyage, during which he proved that there was no direct sea route from the Pacific Ocean to Hudson Bay. There was little Pacific pioneering by navigation to be done after Cook. The various records of his voyages were, for later anthropologists, an invaluable source of information on island cultures, many of which Cook and his men were the first to observe.

4 The English philosopher Thomas Hobbes (1588–1679) characterized primitive life as "nasty, brutish and short". Although extreme, the Hobbesian view could have helped to correct the equally one-sided conception of the "noble savage".

Hobbes perceived that if all men were equal by virtue of the conditions of life, then every man was a law unto himself. With no common or sovereign power to keep him in awe, life would be a constant state of war – real or potential.

of them, such as Edward B. Tylor (1832–1917), were largely content to sift through the recorded observations of explorers, missionaries and travellers. Others examined primitive life at first hand. Lewis H. Morgan (1818–81), for example, produced the first systematic account of a primitive culture, namely the Iroquois Indians.

In posing questions concerning the meaning and interconnections of primitive institutions and in devising methods and theories to answer their questions Morgan, Tylor and others opened the way for the genuine discovery of primitive peoples.

Anthropologists had clearly gained an idea of man as a species, with differing cultures, and they offered various schemes outlining a universal history of mankind. These showed that throughout most of his history man had lived in small, kin-based or primitive societies of a kind that can still be observed in the contemporary world. What was lacking as a firm basis for generalization, however, was sufficient "ethnography", that is, detailed accounts of the culture of particular primitive societies. Morgan's study of

the Iroquois was ethnographic, but most other amassed data on primitive life consisted of scattered observations coloured by a variety of European viewpoints. While careful comparison could extract meaning from such facts it was clear that anthropologists must henceforth collect their own data if understanding of primitive cultures were to advance further.

Ethnography, the long-term analysis
The method of ethnography is as simple to describe as it is exacting to apply. It calls for intensive, long-term study of native cultures through the medium of the native language and through the participation, as far as possible, in native life in order to gain an understanding of the culture from the point of view of its own members. This method of intensive ethnography was pioneered by the Polish anthropologist Bronislaw Malinowski (1884–1942) in his study of the Trobriand Islanders in the early 1920s. Our present understanding of primitive man rests upon the many hundreds of ethnographies carried out since then in all parts of the world.

KEY

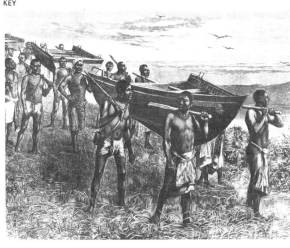

Henry Morton Stanley (1841-1904) was one of the most successful of the great 19th-century explorers. He travelled partly in boats that could be dismantled for easy portage and partly on foot. During his first journey, on assignment to the *New York Herald* the second (1857–8) they found Lake Tanganyika. For most explorers survival and public acclaim for some important "first", such as finding the source of the Zambezi and opening up territory for (1871-2), he found David Livingstone (1813-73) living by Lake Tanganyika. His later trips included a trans-Africa expedition in 1874–7. colonization, was most important. By the time the trained anthropologists arrived many primitive cultures had already been destroyed through the shattering impact of European civilization.

5 The "noble savage", an idealized image exemplified in this romantic portrait of King Kamehameha of Hawaii, was part of the European reaction to the new knowledge concerning tribal societies. Jean Jacques Rousseau (1712–78), influenced partly by travellers' accounts, thought of primitive man as uncorrupted by civilization and conceived of an original "state of nature" in which all men were equal (inequality being the prime source of evil). In 18th-century Europe the noble savage entered into philosophical and political debates about the relationship of the individual to society and to government.

6 Observing and recording the tribes and cultures they discovered were not important aims for the 19th-century explorers of Africa. A notable exception was Richard Burton (1821–90) who published 43 volumes about his travels in Africa, India, the Near East and the Americas. With John Speke (1827–64) he led two unsuccessful expeditions to discover the source of the White Nile. On

7 Many South American tribes, with the same life-styles as these Kamairu Indians on a tributary of the River Xingu in Brazil, were observed by the German scientist and explorer Alexander von Humboldt (1769–1859), whose work is an example of the contributions of trained observers during the pre-anthropological era. Between 1799 and 1804, with his French colleague, the botanist Aimé Bonpland (1773–1858), Humboldt travelled thousands of kilometres on foot, on horseback and by canoe in Central and South America. He collected scientific data on previously unknown tribes.

[Map of Africa showing explorers' routes]

Tangier, Oran, Tripoli, Benghazi, Alexandria, Cairo, Timbuktu, Bathurst, Suakin, Khartoum, Mesewa, Lake Chad, Kukawa, Niger, White Nile, Blue Nile, Freetown, Lagos, Masindi, Lake Victoria, Congo, Ujiji, Banana, Luanda, Lake Tanganyika, Bagamoyo, Mikindani, Benguela, Zambezi, Chinde, Tananarive, Pretoria, Port Nolloth, Durban, Cape Town, Port Elizabeth

0   2,000km

British
••••• Baker 1862-65
= = = Bruce 1769-72
━━━ Burton, Grant, Speke 1851-59, 1860-63
━ ━ Clapperton, Denham, Oudney 1822-25
•━•━ Laing 1825-26
•••• Livingstone 1840-73
•━•━ Mungo Park 1795-97, 1805-06

French
──── Caillié 1827-28
─ ─ ─ Grandidier 1868-70

German
──── Barth 1850-55
─ ─ ─ Junker 1875-8 '79-86
•••••• Nachtigal 1869-74
•••••• Rohlfs 1861-9 '73-80
── ── Wissmann 1880-87

Portuguese
──── Porto 1852-53
= = = Serpa Pinto 1877-79

Swedish
──── Andersson 1851-53

American
──── Stanley 1871, 1874-77, 1887-89

Area little explored before the beginning of the nineteenth century

Origins of human society

Human society is at most a few million years old and at the least tens of thousands of years old. But whereas our knowledge of the earliest societies remains fragmentary, the general character is no longer a mystery.

Changing theories

The origins of human society became a subject of scientific scrutiny with the speculations of the eighteenth- and nineteenth-century social evolutionists, among them Adam Ferguson (1723–1816), Lewis H. Morgan (1818–81), J. J. Bachofen (1815–87), Edward B. Tylor (1832–1917), Andrew Lang (1844–1912) and J. F. McLennan (1827–81). These scholars focused on fundamental and universal institutions such as the family, the incest taboo and kinship; and they constructed models of early human society.

Of all the classical evolutionists, Morgan was perhaps the most influential. His *Ancient Society* (1877) contributed to the development of Marxist theory, and his discovery of classificatory kinship (calling by the same name people of the immediate family and more distant relatives), which is the principal basis of organization in tribal society, greatly affected the anthropology of his successors.

Tylor also made important contributions. Seeking an explanation for the universal practice of exogamy, a ban on marriage within the immediate group, Tylor argued that marrying out created wider circles of co-operation and mutual aid. Rather than relying upon its own numbers and resources, a local group or band that was tied to others through marriage and kinship could call on them for aid. It was probably a question, said Tylor, of "marrying out or being killed out".

Modern theories have rejected some of the assumptions of earlier anthropologists. For example, it was thought that social evolution could be explained as an intellectual process in which early man "reasoned out" social rules and customs and that the family evolved from lower to higher, or from simple to complex forms (the Victorian monogamous family being regarded as the "highest" form). Actually, the family does not so evolve, and may even become simpler as other institutions take over functions that used to be performed by the family. There has also been an enormous increase in factual knowledge, through ethnographic reports on the simplest human societies of hunters and gatherers; field studies of the social life of various species of monkeys and apes in their natural habitats; and the archaeological and fossil records, which can tell us about early man, what tools he used and what foods he ate.

Basic human institutions

The result of this study has been to show, initially, that even the simplest human society, as typified by the hunting and gathering cultures of, for example, the Bushmen of the Kalahari Desert or the Australian Aborigines [2], is a complex system of universal and perhaps primeval human institutions. These are the incest taboo, the prohibition of marriage or sexual relations among immediate family; exogamy, rules ensuring marriage outside of a certain group, usually larger than the primary group; kinship, the recognition of various categories of kin who behave towards one another in prescribed

CONNECTIONS

See also
258 Origins of civilization
260 Ancient states and empires

In other volumes
172 The Natural World
170 The Natural World
20 Man and Machines
24 Man and Machines
26 Man and Machines
28 Man and Machines

1 **Like early man**, baboons are social animals living in groups. They eat a wide variety of foods and are distributed over a large part of the Old World to which they adapted by varying their behaviour rather than their biological structure. But unlike early man's way of life, the baboon's is not dependent on tools and its social life seems to rely on the constant use of aggression within the band.

2 **Tools** of the kind used by such hunting and gathering peoples as the Aborigines (as here) and Bushmen appeared within the last 50,000 years. The rapid development of technology has led some anthropologists to propose that human society emerged, together with new tool traditions and anatomically modern man, 50,000 years ago. This is contrary to the view that human origins date from a million or more years ago.

3 **Since marriage is a universal** human institution, all humans would appear to be the "marrying kind" (seen here is a French marriage). Although varying in form, the functions of marriage seem to vary little, even if the marriage practices of complex human societies are included. In all known societies, the incest taboo decrees that of the four cross-sex relationships inherent in a nuclear family – husband-wife, brother-sister, mother-son, father-daughter – only the first is allowed to involve sex. The universality of these proscriptions leads to the assumption that they are rules of fundamental importance because they remove the inevitable disruption of sexual jealousy. Since early human societies were composed of family groups interrelated by kinship (that is both biological and social relationships) then marriage, the incest taboo and marriage rules as a whole, can be viewed as foundations of society.

ways; marriage [3], which universally (in known societies) legitimizes offspring and creates affinal (in-law) relationships; the family, which is the basic economic unit; a division of labour based on sex and age; reciprocity, the sharing of food and other commodities; and the notion of territory, including concepts of property.

The classical evolutionists thought of early man as mating promiscuously, only later evolving rules governing marriage. If we examine mating patterns among monkeys and apes, it appears that our human precursors might have lived in "promiscuous hordes". Whatever the mating behaviour of subhuman primates, it does not correspond to the human pattern in which all or many of the most accessible females are off limit because they are thought of as "mothers", "daughters" or "sisters". A male gibbon, threatened by the sexual competition of his male offspring, drives the latter away. But for a human father to do the same would spell disaster: an adolescent human male would be unlikely to survive on his own. Sexual competition is a highly disruptive force

which, if permitted free expression among family members, would destroy the family as a unit for survival. The threat of conflict over both women and resources among different groups was equally serious. Marital alliance through exogamy could at least mitigate it.

The simplest form of alliance is that of groups of men exchanging sisters generation after generation. In more complicated systems, one group does not give wives to the same group as it receives wives from; a larger number of groups is required for exchange – thus a larger alliance system is upheld [4].

Social origins
Human social origins are thus found in the origins of these basic and interrelated institutions. Many forces helped to account for their emergency including the prolonged dependency period of human children; developments such as tool-using, which long antedated the first human societies; and basic needs of social life such as sharing food, providing for offence and defence, and transmitting technical and social knowledge from one generation to the next.

Speculating about which of the great ape species man's ancestors might have resembled is an intriguing guessing game. This picture shows a child playing with an orang-utan.

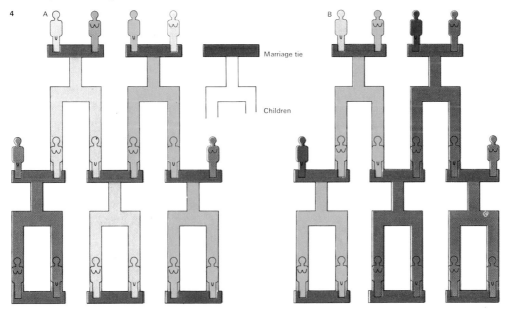

4 Rules of descent were probably early social inventions. Descent traced from a common ancestor through the males (called patrilineal) [A] or through females (matrilineal) [B] places each person into one of a number of groups, membership of which gives the individual many of his or her most important rights and responsibilities. People do not marry anyone belonging to the same descent group. In the patrilineal system, inheritance and succession pass through the man. A husband has sexual and economic rights over a woman and her children. Here a blue lineage man marries a green lineage woman. Their children belong

Marriage tie

Children

to the blue lineage. In a matrilineal descent system, membership of the group is inherited through the mother, but this does not imply a matriarchy or rule by women. The maternal uncle is usually the key figure. J. J. Bachofen believed that matrilineal societies represented a simpler stage of social development that predated the emergence of patrilineal societies, because, he reasoned, only maternity could be determined with any certainty. Later anthropologists rejected this. Patrilineal and matrilineal descent are alternative organizing principles. Patrilineality is common, but both are found in non-industrial societies.

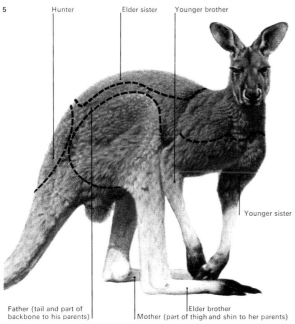

Hunter   Elder sister   Younger brother

Younger sister

Father (tail and part of backbone to his parents)   Elder brother   Mother (part of thigh and shin to her parents)

5 Australian Aboriginal customs show that real rights and duties are associated with kin statuses. A hunter who kills a kangaroo is obliged to give particular portions of the animal to specifed relatives. It is presumed that rules governing the sharing of food were among man's earliest cultural inventions.

6 Diffusion, through borrowing or by cultures expanding into new areas, is a pervasive influence in social evolution. The spread of iron working from the ancient Near East made better tools and weapons available. These gave the peoples who possessed them advantages over their neighbours. The resulting invasions brought many social changes.

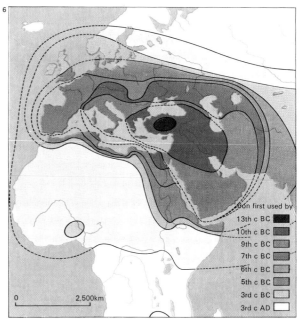

Iron first used by
13th c BC
10th c BC
9th c BC
7th c BC
6th c BC
5th c BC
3rd c BC
3rd c AD

0   2,500km

Stages of social evolution

History is a record of past events or our interpretation of those events. But history alone will not enable us to understand the past. We need also to study social evolution – how societies change their organization.

Types of human society

The historian's arrangement of events into periods or ages does not necessarily coincide with a classfication of societies into stages of social evolution. A division of European history into ancient, medieval and modern periods, for instance, would not correspond to the stages of its social evolution. While ancient Europe, or that part of it under Roman rule, exemplified the stage of archaic state-organized society, the succeeding medieval societies were merely smaller and more rudimentary versions of the same stage. Modern European societies, however, beginning with the emergence of industrial capitalism, exemplify an emergent general type, that of the industrial nation state. Different stages of social evolution are really based on distinctive features of organization rather than on time periods, regions or

specific events. Hence, evolutionary stages are not marked off by, nor identified with, specific technologies. Pastoral nomads, for example, may belong to small tribes or to centrally organized chiefdoms [2]. Nor do societies follow an invariable, straight line through a set pattern of stages [7].

The evolutionary process does not follow any preordained path; rather it is "opportunistic", with societies adapting in response to the challenges of their environments, including both their physical settings and their relationships to neighbouring societies [5]. Occasionally, however, adaptation to an environment has led to a "higher" form of organization, one in which a society is organized in a more complex fashion than its predecessor. This happened, for example, in northern Europe when a new kind of society formed as a result of the Industrial Revolution, or in the ancient Near East when agriculture first appeared. Thus, while evolution is, in the main, a process of diversification as societies adapt to different environments – producing variations on the same organizational themes – there have been

changes involving new organization.

The purpose of distinguishing between stages or types of societies is to locate these major changes in the course of social evolution [Key] and to isolate their causes. It is precisely here, however, that anthropologists have often disagreed. One such disagreement concerns the Neolithic period. Since the researches of the British archaeologist V. Gordon Childe (1892–1957), it has been held that the "Neolithic Revolution" constituted an evolutionary social transformation.

The advent of the Neolithic period when agriculture began did involve major changes. Through domestication of various plants and animals (the technical breakthrough underlying agriculture) man's relationship to his environment was radically altered. In addition agriculture permitted an increase in human population, perhaps a 20-fold increase over average population levels of the cultures of foragers and hunters. Yet Neolithic societies, though often larger and more sedentary than their predecessors, did not necessarily display any new means of organization. They remained egalitarian,

1 Hunting-gathering or fishing peoples, such as the Bushmen [A] of the Kalahari Desert in southern Africa or the Eskimos [B] of Greenland, are among the last remnants of the simplest and earliest form of social organization. The harsh environments that they live in are unsuitable for the growth of more complex tribes or chiefdoms because they are unable to support a large or settled population.

2 Nomadic pastoralists, such as these in Afghanistan, once inhabited the great African-Asian belt of desert and steppe, subsisting on their herds but also raiding oases and agricultural communities. Many herding societies, such as the Tuareg of North Africa and the Mongols of Asia, were organized as chiefdoms whose mobility could make them effective militarily. The Mongols, for instance, conquered a vast empire.

3 Shifting horticulture, in which gardens are annually or frequently moved to newly cleared plots, was probably man's earliest system of cultivation. It survives in tropical areas of Africa, Asia, the Pacific and the Americas. While sometimes regarded as wasteful of land, it has served as the productive basis of societies ranging in complexity from the tribal Jivaro of South America to the ancient Maya civilization of Yucatán and Guatemala.

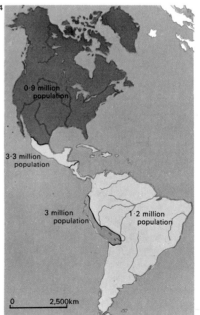

0·9 million population

3·3 million population

3 million population

1·2 million population

0   2,500km

4 Relative densities of population in the New World at the time of the Spanish conquest give a clear indication of the significant role played in population growth by the existence of a state-organized form of society. In the areas that were dominated by the Aztec and Inca empires in Central and South America, the population density was much greater than in the bigger areas of North America and east of the Andes in South America. This difference reflects the superior economic system made possible by state organization. The figures are estimates of total populations of each area at the time of the European discovery of the Americas in 1500.

kin-based societies, indistinguishable in any basic way from pre-agricultural societies of hunting-gathering and fishing peoples.

Chiefdom societies

In time, however, and in certain localities, Neolithic technology and economy did give rise to more advanced or complex societies based on a chiefdom or hierarchy. What distinguished chiefdoms from the preceding tribal, egalitarian societies, was a measure of institutional centralization. The critical development was the emergence of hereditary chieftainship in place of the rather ephemeral leadership of tribal societies (indeed, chieftainship has been compared in its importance with the development of the central nervous system in biological evolution). Chiefdoms were larger, more complex and more firmly integrated than tribal societies. For the first time, persons, families and larger subgroupings became differentiated in political power, command over resources and role in economic life.

Chiefdom societies had the potential for expansion and increasing centralization.

Presumably, such societies were the immediate precursors of the first state-organized societies in ancient Mesopotamia and other areas. Such states continued the line of evolutionary advance initiated by chiefdoms by substituting for ruling families a government or ruling group with a monopoly of coercive force. This development of a powerful means of integrating diverse groups, allowing population growth [4] and centralization, initiated a new stage and led to the disappearance of many simpler societies.

The nation state

In terms of evolution, therefore, during most of human history societies were of one type: small-scale egalitarian societies integrated by ties of kinship and marriage and by tribal leaders who, in most societies, were scarcely distinguished from their fellow tribesmen. In favoured environments, and particularly after the invention of agriculture, chiefdom societies developed, some of them evolving further into true states. Later, nation states emerged in western Europe to become the universal and dominant social organization.

KEY

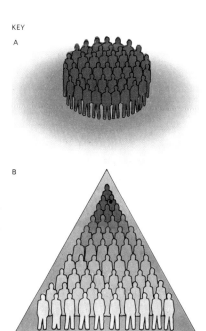

The change from egalitarian, kin-based societies [A] to hierarchical chiefdoms [B] was an advance in social evolution of the greatest significance. By improving and so enormously extending social cohesion, it encouraged population expansion by making possible its effective organization. The social mechanisms of egalitarian societies are not strong enough to hold in check the conflicting needs of large numbers of people. Egalitarian societies are therefore limited to areas of kinship influence, whereas chiefdoms and state societies have definite physical boundaries and a sovereign government set above the population.

5 The cultural evolution of any particular society is usually the result of outside invasions by societies with higher stages of development. Consistent growth within the society is seldom as important. By at least 1,000 years ago in the southeastern United States, the region south of the Ohio River and east of the Mississippi River, tribal Indian societies [A] gave way to more complex chiefdoms owing, in part, to cultural influences from Mexico [B]. European colonists then erected the slave-based agrarian society of the old South [C]. Beaten by the industrial North during the Civil War, the South was not transformed into an industrial urban society until after World War II [D]. In the 1960s the South provided sites for launching America into the space age [E] and so became a part of what may develop into a post-industrial electronic society.

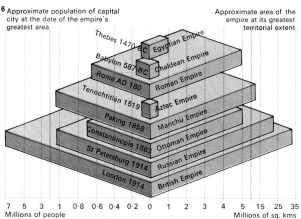

6 Approximate population of capital city at the date of the empire's greatest area

Approximate area of the empire at its greatest territorial extent

Thebes 1470 BC — Egyptian Empire
Babylon 587 BC — Chaldean Empire
Rome AD 180 — Roman Empire
Tenochtitlán 1519 — Aztec Empire
Peking 1858 — Manchu Empire
Constantinople 1863 — Ottoman Empire
St Petersburg 1914 — Russian Empire
London 1914 — British Empire

7 5 3 1 0·8 0·6 0·4 0·2 0 1 2 3 4 5 15 25 35
Millions of people   Millions of sq. kms

6 Effective central control plays an important role in controlling the size of an empire. This, in turn, is reflected in the size of the imperial capital.

7 "Ethnical periods" or stages listed by Lewis H. Morgan in *Ancient Society* (1877) were based on technological criteria, implying that all societies progressed through each stage. This idea of unilinear evolution is no longer accepted and stages are now based on social criteria.

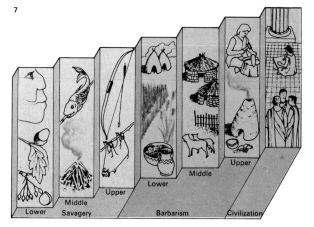

Lower Middle Upper Lower Middle Upper
Savagery   Barbarism   Civilization

253

Simplest human societies

Much of human history is the history of primitive cultures. As recently as five centuries ago, on the eve of European expansion, such cultures were spread over a large part of the world – in Africa, northern Eurasia, Australasia, the Pacific Islands and most of the Americas [1]. Our knowledge of man's way of life during most of his existence is derived principally from the "modern" representatives of primitive cultures.

If anthropologists refer to the simplest cultural adaptations as primitive, this is not an attempt to disparage them. Rather "primitive" refers to cultures that are simple technologically, small in population, based on kinship ties, egalitarian and still remaining unspecialized institutionally.

Family units in early cultures
Primitive technologies are directed primarily to securing food by collecting wild foods, fishing, horticulture or herding. Men and women normally perform different tasks, but each family has the required skills, tools and resources although some tasks involve the co-operation of people from many families.

Primitive cultures are small, averaging only a few hundred members. A lower limit of two to three dozen people is set by the requirements or advantages of co-operation in economic activities, sharing of food and caring for the sick. An upper limit is determined by food supply but also by the fact that this way of organizing people cannot cope with more than several hundred people without great stress and strain. Primitive communities that grow too large to function in a peaceful way often split in two.

As is often observed, primitive cultures are family cultures [Key]. The so-called nuclear family of parents and children is often part of a larger unit, the extended or three-generation family. Beyond the family, social relationships are still centred on kinship. Indeed, the society as a whole is often conceived as a body of kin. Prominent as a focus of ceremonial life are individual "life crises" or rites of passage: birth, puberty, marriage and death.

Primitive cultures are not lacking in social distinctions: men have more favoured social positions than women; older people have an advantage over their juniors; the successful hunter or industrious cultivator is accorded prestige. But primitive societies are egalitarian in that there are as many positions of prestige as there are people with the characteristics needed to fill them.

Multiple roles within kinship groups
The family-based character of primitive societies is the key to the simple way they are organized. For the family is not only at the centre of economic life, but is also a member of larger groups of kin. It has a social, political, economic and ritual role to play.

While it is correct to think of primitive cultures as institutions lacking special purpose, all cultures are more or less specialized in the way they adapt to the environment in order to survive. Specialization of some sort is the invariable outcome of adaptation to specific challenges or opportunities. Primitive cultures are distinguished by their dependence on a narrow range of resources, or even upon a single resource. The acorn, for example, was vital to Californian Indians, as were the ubiquitous reindeer to Siberian

1 **Tribal cultures reigned supreme** in the world of 5,000 years ago. They were challenged by the emergence and spread of the ancient civilizations, which converted tribal people into ethnic groups within a more complex social organization. This process took place when the Romans conquered and Romanized the tribal Britons, for example. The contraction of the tribal world was greatly accelerated by Western exploration and colonization, particularly during the past two centuries. Tribal peoples now comprise much less than 1% of world population. The map compares their settlement today with areas of tribal culture in 1500.

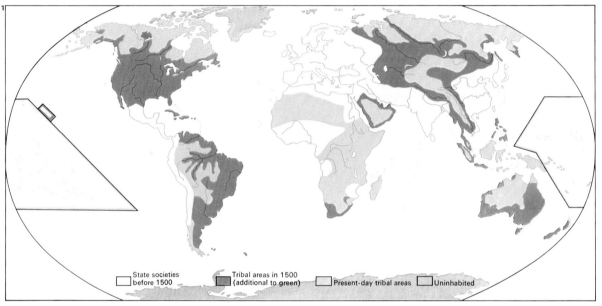

State societies before 1500 | Tribal areas in 1500 (additional to green) | Present-day tribal areas | Uninhabited

2 A **"snowmobile" revolution**, beginning in the 1960s, has already had a significant impact on the way of life of Arctic peoples, including both North American Eskimos and the Laplanders. The range and speed of the snowmobile allow the Eskimos to reach their hunting grounds more quickly and so permit compact settlements and more frequent visiting among people still living in widely scattered groups. Under the impact of Western materialism, the distinction between "rich" (snowmobile owners) and "poor" (dog-sled people) has become important. Such a distinction between rich and poor is seldom made and of little significance in tribal cultures.

3 **The Pueblo Indians** of the American Southwest practised flood-water farming, and growing crops of the Amerindian complex – maize, beans and squash. They developed an elaborate ceremonial life in which the needs of agriculture and the harshness of their desert habitat were central themes. Although long a part of the Spanish Empire, the Pueblos have retained much of their traditional culture in modern times. Seen here is a Taos Indian from New Mexico in the United States. In spite of the need for a degree of centralized organization to make their type of farming successful, the Pueblos never developed a hierarchical structure or chiefdom.

hunters and pastoralists.

Specialization may prove to be limiting when environmental change decrees the need for social change. A frequent cause of change is the advent of new relationships with neighbouring societies. Change of this kind occurred on a wide scale following the development of agriculture. Agricultural societies, more populous and powerful than most hunting-and-gathering societies, became dominant over a large part of the world because their new skills permitted more effective use of many environments. As a result hunting-and-gathering peoples were confined to "marginal" areas such as deserts, tundra, the arctic wastes and other habitats unsuited for or unwanted by primitive agriculturists. For most of human history, of course, wild food was man's only means of subsistence and allowed him to survive, if not always thrive, in almost every known habitat.

The impact of the Europeans

The rapid expansion of Western culture in the sixteenth century posed a challenge that primitive cultures could not meet. If their ter-
ritories were wanted for settlement by Europeans they were pushed aside – exterminated by a combination of introduced diseases and force of arms. Surviving remnants were finally confined to tribal reserves on marginal lands. Exploitation of other foreign territories was achieved partly through native populations recruited, initially by forcible means, to supply labour for European plantations, farms and mines. Primitive societies harnessed to European economic enterprises were doubly penalized by losing territories and resources as well as many of the active men who were drawn off as wage earners [9].

European trade had far-reaching effects even upon societies that had no contact with Europeans. For example, the fur trade in North America led to a significant increase in Indian warfare as some societies formed tribal confederacies to strengthen their position in the trade. Modification of primitive cultures elsewhere has been rapid during the past century and although many new nations have recent experience of them, these cultures have virtually passed out of existence.

Social organization in the simplest cultures is centred on the family, such as this Zulu group. Tribal societies, never more than a few hundred strong, are closely bound together by marriage. Leadership is not in the hands of any one person for all activities, but is taken by the most talented in each field.

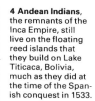

4 Andean Indians, the remnants of the Inca Empire, still live on the floating reed islands that they build on Lake Titicaca, Bolivia, much as they did at the time of the Spanish conquest in 1533.

5 Argentinian gauchos are descended from nomadic Indians who began hunting on horseback after the Spaniards introduced horses. Their way of life resembled that of the North American Plains Indians.

6 Australia was colonized by hunter-and-gatherer people at least 30,000 years ago. Bands of a few dozen people had contact with others during their annual wandering across the arid continent. Seasonal abundances of food in certain localities enabled groups numbering in the hundreds to assemble for ceremonial activities such as the dance for which these Aborigines have liberally painted their bodies.

7 Aborigines today frequently live in poverty on the fringes of Australian society, both in urban areas and in the outback, in spite of an official policy of assimilation. Aborigines numbered about 300,000 when the white settlement of Australia began during the eighteenth century but only a few thousand, mainly in dry inland areas, still carry on a semi-traditional way of life based on the tribal culture of their ancestors.

8 The Bantu peoples of Africa are descended from tribal groups that began to expand from a probable base in the Cameroons a few centuries before the birth of Christ. They spread their language, agriculture and iron tools over much of Africa south of the Sahara. Some Bantu developed chiefdoms that in turn evolved into primitive states under the pressures of the slave trade and European colonization. Prominent groups are the Zulu, Swahili and Kikuyu.

9 A Zulu working as a rickshaw boy in Durban, South Africa, illustrates one of the sadder aspects of the way in which African villagers have been drawn into towns, mines, plantations and farms to work for wages or to prostitute their original culture for the amusement of tourists, particularly as white settlement has pushed their tribes into poorer, more marginal land. Bantu working for whites during the early phase of European settlement kept strong tribal ties.

Prestate societies

Primitive man has been pictured as both an individualist and a rigid conformist – a "slave to custom". These characterizations, however, are not completely opposed. In a society lacking a formal legal system backed by state power, people must adhere strictly to rules of custom lest the conflict of individual interests leads to violence and anarchy.

In modern society, the state can be seen either as simply providing security for people and their possessions or as an instrument for constraining individual freedoms. Whatever the viewpoint, it is difficult to conceive of an orderly social life in the absence of a supreme political authority, yet primitive or prestate societies lack such an authority. To understand their organization is to appreciate what the British anthropologist Mary Douglas has called the "miracle of social order in the absence of radio police cars".

The egalitarian or tribal society

Every organization is a system of integrated parts. In these terms, prestate societies exhibit two kinds of organization, which differ in size, the nature of their parts and in the means of bringing them together.

The first type, the tribal egalitarian society, is built on a segmental plan. The segments – such as families, lineages [1] and clans [2] – are unspecialized, basically equal and linked together by relationships of kinship, marriage and descent [Key, 3]. Within the segments the head of a family household wields authority over women and children in the domestic sphere, lineage elders have a greater say than junior members in family affairs, and yet no leader or group coordinates the activities or relationships of one segment with another.

Men and women in their respective spheres of work engage in the same pursuits and hence each family is involved in the same round of economic activities [4]. People have a common style of housing, dress and personal adornment. They use the same kind of tools, eat the same foods, observe similar rituals and worship the same gods. The cultural sameness of a number of such social segments helps to produce a social unity based on likeness or what the French sociologist and anthropologist Emile Durkheim (1858–1917) termed "mechanical solidarity". The weakness of this system, however, is that it is not really integrated because the social segments are so self-sufficient. The loss of one or more segments does not impair or destroy the society.

Segmental societies are not wholly without leaders, but the tribal leader generally lacks the ability to give commands or, if he has this ability, it is limited by context and duration; he may, for example, exercise it only during a hunt, a war party or a ceremony. Because the tribal leader's influence does not extend beyond his own social segment – perhaps a kin group or hamlet – leaders of other segments could be his rivals.

Hierarchical society and the family

The second type of prestate organization, the chiefdom or hierarchical society, achieves a measure of "organic solidarity", because specialized parts depend on each other. The same kinds of segments are present, but they differ in rank or status and in their political function and economic role. Some families and groups rank as chiefs and others as com-

CONNECTIONS

See also
258 Origins of civilization
248 The study of man
250 Origins of human society
252 Stages of social evolution
254 Simplest human societies
260 Ancient states and empires
264 Conflict, power and social inequality
272 Fundamental political ideas

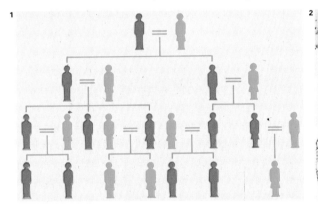

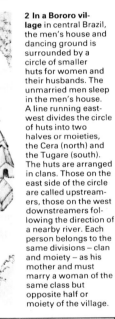

1 Lineages are descent groups. Membership is determined by descent from a common ancestor through lines of either males (patriliny) or females (matriliny). They are a common type of social group in prestate societies. Since lineages are exogamous – mates must be acquired from other groups – a society must be composed of more than one lineage. Some prestate societies consist of a number of inter-related lineages. Illustrated here are the interconnections between two lineages, brown and green, in a patrilineal society where the children of a marriage belong automatically to the father's lineage.

2 In a Bororo village in central Brazil, the men's house and dancing ground is surrounded by a circle of smaller huts for women and their husbands. The unmarried men sleep in the men's house. A line running east-west divides the circle of huts into two halves or moieties, the Cera (north) and the Tugare (south). The huts are arranged in clans. Those on the east side of the circle are called upstreamers, those on the west downstreamers following the direction of a nearby river. Each person belongs to the same divisions – clan and moiety – as his mother and must marry a woman of the same class but opposite half or moiety of the village.

3 Village divided into 2 moieties

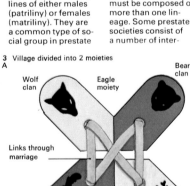

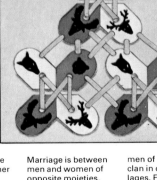

3 Tribal societies consist of various kinds of social segments interrelated through kinship, marriage, traditions of common descent, exchange and perhaps a division of labour in ceremony. In this example, the tribal village contains local segments or lineages of four clans, which are arranged in two intermarrying halves or moieties, the Eagle and the Crow [A]. Lineage is a system of actual blood ties. A clan is more of a cultural grouping with a common interest, such as war or religion, that cuts across lineage providing other than family ties for its members. The Wolf and Bear clan people have a tradition of common descent and since they regard themselves as too closely related to marry, they must obtain their spouses from either of the paired clans (Elk and Deer) of the Crow moiety. As local groups of clanspeople increase in population, they break up and some of their members go off to colonize a new village [B]. In time they are joined here by segments of other split lineages. The clan and moiety organization is thus duplicated in the new settlement [C]. Through growth and break up, members of the four clans are distributed over the tribal territory. Marriage is between men and women of opposite moieties, whether of the same or different villages [D]. Inter-moiety marriage and cross-cutting clanship are the ties that unify the tribe. A Wolf clan member has fellow clansmen and men of a brother clan in other villages. From these people he can expect hospitality, protection and assistance in time of need. In addition he is a relative of Elk and Deer clan people, who have married his clan sisters.

256

moners. In addition, individuals, families and village communities begin to specialize their pursuits, due in large part to the administrative role of chiefs [5].

In prestate societies social activity takes place predominantly or exclusively between people who are kin. All the men of the elder generation of one's own band or lineage are "father", and all their wives are "mother"; all the clan members of one's own generation are "brothers" or "sisters"; the women of one's wife's group are all "wives" and their brothers are "brothers-in-law", and so on. Relationships are direct or face-to-face among people whose behaviour is governed by rules of kin etiquette. Breaches of etiquette – for example, the failure to treat a "father" with proper respect, an allusion to sexual matters in the presence of a "sister", a refusal to share food with a clan "brother" – are immediately seized on and punished by ridicule, withdrawal of support or ostracism.

Dealing with wrongdoing

Prestate societies, although lacking a formal law-enforcing body, do have means of dealing with wrongdoing. An obvious approach is for the injured party to take it upon himself to recover stolen property, punish an adulterer or exact restitution. This is termed "self-help" and is good evidence for primitive individualism. Often, however, people turn to their kin group for assistance.

Feuding of some kind is universal in prestate societies, whereas neither the feud nor most forms of individual self-help are condoned in state societies. Feuding can escalate as revenge is met by counter-revenge, and this poses a serious threat to the social order. In segmental societies, social ties between different groups and the efforts of those who are outside the fray but who risk injury themselves if the conflict widens, exert pressure on the parties to settle their differences.

In chiefdom societies there is also the role of the chief in curtailing self-help. Although chiefs are not unbiased judges or mediators – because they may be more closely related to one party than another – their villages often serve as places of refuge where a wrongdoer can pay a fine to the chief rather than expose himself to attack by an avenging group.

A tribal village, like this one in northern India, although outwardly one unit, is in fact divided into different clans and lineages, whose formal interrela- tionships are essential to the smooth functioning of the whole society.

4 A

B

4 The division of labour in simple societies [A] is based on age and sex but since each family household contains this division of lab- our, families are structurally iden- tical and largely self- sufficient. Larger social units, such as lineages and clans, also have identical functions. The in- stitutions of complex societies [B] are specialized and thus functionally interdependent, like assembly line tasks.

5 A chief's power differs from that of a tribal leader in several ways. First, chiefs act as economic administrators [A]. This includes the distribution of land to ensure that the people have all that they need for house building, cul- tivation and pasture; enforcing seasonal taboos on the use of certain resources, and organizing col- lective hunts. They can also subsidize the work of special- ist craftsmen [B] by redistributing the taxes and so on. They also have religious power [C], since their ancestral spirits are elevated to the status of gods. This super- natural power was re- flected in the custom of carrying the para- mount chiefs of Hawaii so that con- tact with their bodies would not make the land taboo to its owner. Finally, they play an important jud- icial role in mediat- ing and adjudicating between disputants [D], although they do not have an effective monopoly of coercive power.

Origins of civilization

Civilization is often viewed as a complex of elements that include writing, government, law, cities, monumental architecture and art, metallurgy, science, craft specialization, commerce and large-scale warfare. The problem of the origin of civilization would therefore appear to demand the study of the origins of these various elements.

Examining the evidence

Scholars have tended to focus on just one or another of these as being critical. Examples are metallurgy by which the earliest Old World civilizations are classified as "Bronze Age"; writing, as suggested by the distinction between literate (civilized) and nonliterate (primitive) societies; and cities, as indicated by the fact that the appearance of civilization has been described by some authorities as the "urban revolution".

From a comparative viewpoint, however, neither metallurgy nor writing nor urbanism can serve as a defining criterion of civilization. Metalworking preceded the development of civilized societies in some areas while in others, such as the Americas, it never attained great technological importance. Writing was a feature of some early states such as Sumer but ancient Peruvian civilization lacked a writing system and the role of writing in Central America and the Indus valley has not been fully determined. Nor was urbanism an inevitable part of the earliest civilizations. In early Egypt and Central America ceremonial centres appeared rather than cities and they were probably characteristic of the first phase of development in Mesopotamia and India as well.

There is growing agreement that political organization was the active element in the formation of civilization. This organization – the state – was a powerful new means of regulating the affairs of large and complex societies. It was the state that built cities, pyramids [7], temples and irrigation systems; organized commerce; carried on wars of conquest; subsidized craftsmen; provided for the discovery, extraction and smelting of metals; and made use of scribes to keep records.

Ancient or modern, the political state involves a governing group that monopolizes the legitimate right to use force. The relationship of the state to society, however, has long been debated. The issue is probably best resolved by examining the original state-organized societies. States did not spring directly from a tribal milieu. Rather, they resulted from the evolution of hierarchical societies that already had hereditary chiefs, class divisions and central direction of economic life. In such societies chiefs did not possess a monopoly of the right to use force – they were not yet kings.

Theories of states

A number of theories exist that seek to isolate the key factors in the emergence of the state. The conquest theory, one of the earliest, attributes state formation to the conquest of one society by another, the conquering group forming the governing class [8]. Historically, however, either the conquering group or both the conquerors and the conquered were already complex societies. In other words, this explains the formation of secondary states rather than original ones.

The cross-roads theory takes trade as the principal factor in the development of the

1 The transition from a simple, egalitarian society [E] to a complex hierarchical one [H] is the key to the origin of the state. In theory, the type of environment favouring this transition is a circumscribed area with agricultural potential. With egalitarian societies as a starting-point, the numbered areas represent various intermediate or final phases of development suggested by the history of societies. Most egalitarian societies reached a dead end in areas 1, 3, 4 or 7. The blue pathway shows the development of tribal societies of Highland New Guinea, which reached area 4 via population growth, intensified production and warfare. In different circumstances egalitarian societies reaching areas 2, 5 or 6, might emerge as hierarchical societies. This is shown by the red path, which relates to certain island chiefdoms of Polynesia.

1 Potential for agricultural development not realized.
2 Increased agricultural production leads to population growth.
3 Surplus production used to sustain social, ceremonial and artistic activity.
4 Population stabilizes at higher level.
5 Resources scarce as population increases.
6 Agricultural production intensified to meet needs. New population increase.
7 Agriculture intensified by new methods, eg irrigation, requiring formation of hierarchy.

2 Karl Wittfogel, German-born historian and anthropologist, has presented the most systematic version of the theory relating the origins of civilization to irrigation. In his view, the first civilizations were bureaucratic and despotic in response to the requirements of constructing and maintaining large-scale hydraulic works. Criticisms of this theory include the fact that the extent of despotism in ancient societies is exaggerated.

3 Ancient Chinese civilization, in common with other early civilized societies, was irrigation-based. Irrigation agriculture, by permitting two or perhaps three crops a year, produced the surplus food needed to support the non-agricultural classes. Moreover, it has been argued that the first centralized state bureaucracies developed to mobilize and co-ordinate the masses of labourers required to build and maintain large-scale irrigation and flood-control systems. But, as with the development of Mesoamerica and Mesopotamia, archaeologists have found no clear evidence of large-scale hydraulic works associated with the early civilizations.

state. At the intersection of major trade routes, communities of traders developed who were drawn from diverse societies. Because they lacked a common culture and common social institutions these internally different communities needed an authority – the state – to regulate their affairs. But since long-distance trade was carried on only by established state societies, the theory presupposes the existence of the very institution it is trying to explain.

The hydraulic theory [2] of origins is based on a correlation between certain forms of government and arid or semi-arid river valleys. In the valleys of the Yellow River of China, the Indus, the Tigris-Euphrates, the Nile and coastal Peru, irrigation and flood control had to be carried out on a large scale. Thus an organization was required to supervise and supply mass labour. This organization was the state.

While there is no doubt about the importance of irrigation in the early civilizations, hydraulic works may have depended upon, rather than caused, state formation, for it appears that large irrigation systems could

have resulted from the linking together of smaller systems whose construction was well within the capabilities of tribal communities.

The circumscription theory

The circumscription theory [1] combines a number of factors – agricultural development, growth of population and conflict – to account for the origin of the state. The distinctive feature of the theory is its insistence that areas bounded by geographical barriers such as deserts, mountains or seas, or areas in which resources were concentrated, were favourable for political evolution. In such areas agricultural production permitted an increase in population [4], which in turn led to intensification of production and a further increase in numbers. Eventually population pressure led to competition for resources in the form of economic warfare. The vanquished chose to submit to the victors and thus the major evolutionary step was taken: the formation of multi-village organizations or chiefdoms. Further warfare led to the formation of states and their ultimate expansion into empires.

The power to organize and direct the work of society is an important advance towards the establishment of a state society. These village chiefs in Samoa wield much more power than the leaders of egalitarian tribes whose power is not institutionalized.

4 The British archaeologist V. Gordon Childe (1892-1957) viewed the origin of civilized society as the automatic outcome of agriculture, which permitted the production of a surplus and hence a division of society into food producers and non-food producers. His "surplus theory" did not explain why tribal agricultural societies did not develop further or what made the cultivators give up their surpluses.

5 The energy theory of cultural development held by American anthropologist Leslie White (1900-) states that societies evolve as the energy harnessed and put to work increases. At the bottom of the scale is human energy, then that of domesticated animals, then wind and water power, then fossil fuels such as oil and coal, and then atomic energy. State-organized societies can realize this potential.

6 Julian Steward, (1902–72) an American anthropologist, drew attention to the parallels in development of the early civilizations of Mexico, Peru, Mesopotamia, Egypt, China and the Indus valley. Such parallel patterns of evolution suggest that similar factors, primarily environmental, were at work and that comparison between the six areas would reveal uniform causes of the origins of civilization.

7 The sizes of ancient monuments do not indicate the kind of organization required to build them, since their construction could have been achieved with different combinations of time and labour – small labour forces working over two or three generations or large labour forces working over a shorter period. The Egyptian pyramids were built by large numbers of men and their construction periods overlapped one another. This, and the fact that the large pyramids were built during the early phase of Egyptian civilization, suggests that pyramid building was an important factor in consolidating the new state's control over labour and revenue.

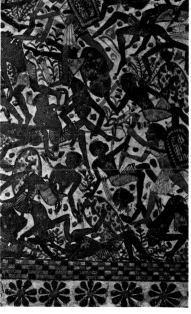

8 War of conquest, as depicted in this scene from the tomb of the Egyptian pharaoh Tutankhamen (r. 1361–1352 BC), aimed at capturing territory, population and resources. Some theorists have argued that conquest was a primary factor in the formation of societies with formal government and social classes. However, the earliest civilized societies had small populations and territories and were apparently peaceful, so it is doubtful they were born in warfare.

Ancient states and empires

Civilizations are often judged by their technological inventiveness and in that light ancient civilizations are apt to appear uninventive, even stagnant. However, the civilizations of the past were based on the greatest innovation of all – the organization of a political state. The political ideologies, social institutions and means of organization evolved by these civilizations include democratic and totalitarian forms of government; bureaucracy; class and caste systems; cities [7] and municipal government; professional armies [4]; the census [6]; taxation; writing; science [2]; codified laws [5] and law courts; police; money [8]; "world" religions and priesthoods. Indeed, modern technological civilization has added little radically new to the inventory of the socio-political institutions produced by ancient states.

The development of early states and empires

The first ancient states, some of them taking the form of city states, such as Sumer and Akkad, comprised relatively small populations and territories. They were theocratic in character, the business of government being largely controlled by a powerful priesthood [3]. The gradual emergence of civil rulers whose power was based on the leadership of armies and the command of resources to feed, arm and equip them, led to military expansion. The first empires, such as the Assyrian, Hittite, Egyptian and Persian, were individual states that were joined together by conquest.

Ancient empires tended, however, to be weakly integrated. Conquest, imperial administration and a thin overlay of imperial culture did not erase deeply ingrained ethnic and regional divisions. Furthermore, the political organization of these empires tended to reinforce this inbuilt bias against centralized government. Beyond a core area surrounding the seat of central government were subject territories ruled by provincial governors who, in command of local revenue systems and armies, might even be able to defy the authority of the central ruler. Outside the provinces, imperial authority trailed off into uncertain border areas of unconquered territories that were often inhabited by tribal nomads.

These forces of disintegration eventually led to parts of an empire either breaking away, challenging the centre, or simply fragmenting to the point where the whole empire was vulnerable to conquest by neighbouring states or tribal nomads.

In addition to developing a central bureaucracy, ancient rulers devised a number of partial solutions to the problem of "creeping decentralization". These included rotation of provincial officials; alliances through marriage between the royal family and families of provincial governors; the keeping of hostages as pledges for an official's loyalty; systems of inspection, auditing and spying; concessions to merchants, cities, the priesthood or other special groups in return for political support; and the development of roads, canals, messenger systems and shipping to facilitate administrative communications [1].

The way of life of the peasants

Written histories of ancient civilizations tend to focus on the activities of kings and emperors, priests and philosophers, artisans

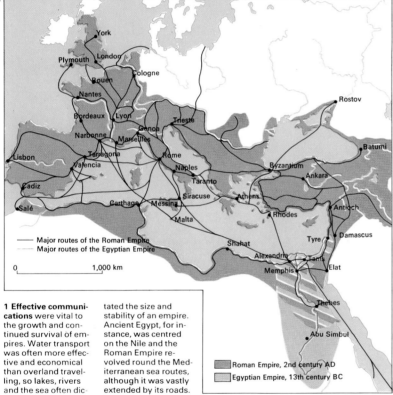

1 **Effective communications** were vital to the growth and continued survival of empires. Water transport was often more effective and economical than overland travelling, so lakes, rivers and the sea often dictated the size and stability of an empire. Ancient Egypt, for instance, was centred on the Nile and the Roman Empire revolved round the Mediterranean sea routes, although it was vastly extended by its roads.

— Major routes of the Roman Empire
--- Major routes of the Egyptian Empire

0 ____ 1,000 km

Roman Empire, 2nd century AD
Egyptian Empire, 13th century BC

2 **Early scientists**, such as the astronomer and geographer Ptolemy (c. AD 90–168), were encouraged by the state because much of their work was useful to the central government in planning and administration. The ancient Egyptians' discovery of the solar year's 365-day cycle resulted from their observations of Sirius and the time intervals between the annual floods of the Nile. The Egyptians' primary concern was to measure the floodwater level annually. From this, land inundation could be estimated and thus grain production and government revenue predicted.

3 **The role of the priest** was important in ancient civilizations, with priest-kings acting as mediators between man and gods. Such theocracies were very efficient; the populace was bound with a minimum of coercion to a round of religious duties and rituals that benefited both individual and society. Civil obedience and service to the state were thus assured. Here Aztec priests conduct a ritual sacrifice. Captives were often victims.

4 **The cultural diversity** and territorial extent of ancient empires meant that professional armies were as necessary for maintaining internal order as they were for the purposes of defence and external conquest. The Roman Empire was particularly successful at integrating the subjects of its diverse territories and client kingdoms. As well as instilling a sense of collective identity in their soldiers, the Romans were also adept at using their armies for the control of strategic resources and the improvement of communications. Cohesiveness and mobility were the results of this policy and made the Roman army a formidable policing and fighting instrument throughout the Roman Empire.

and city dwellers. However, most of the people of past civilizations were peasant farmers whose lives might be only slightly affected by long-term political developments but whose surplus production was the basis of the achievements of the non-agricultural specialists. In return for such benefits as state-organized irrigation and flood-control systems, military defence and perhaps government relief in time of famine, the peasant committed a large part of his land and labour to the support of others [9]. In addition, he or his sons might be conscripted into the army.

The peasant's way of life, as far as material standards and much of his daily routine were concerned, did not differ greatly from that of the agricultural tribesman who preceded him. Yet the peasant was a creation of the ancient state. With permanent fields, irrigation and animal-drawn ploughs, he worked in a more advanced agricultural system requiring more disciplined labour. Unlike the tribesman, he produced some food and goods for sale in local markets and purchased in turn exotic goods and specialized services. Through resident officials and priests, and by his contributions of labour and part of his crops to state revenue, he was linked to a political and economic system far larger than his village community. And he shared, in some degree, in the "high" culture produced in the urban centres by the priesthood, philosophers and writers.

The citizen enters the social order

The peasant, however, remained a subject with little involvement or interest in the larger forces that helped shape his life. He was both expendable and politically impotent, as was his unskilled compatriot in the city. But in the cities, and among the higher social strata, there emerged another new type of man, the citizen. The citizen was a person with legal rights and capacities defined in relation to the government. Although ancient law was generally more concerned with protecting the power of the state than safeguarding the rights of citizens, the latter enjoyed a measure of protection because of the simplicity and comparatively small scale of ancient bureaucracies.

The Great Wall of China was finished in 214 BC by the emperor Shih Huang Ti (259–210 BC), the first monarch of a united China under the Ch'in Dynasty. The wall, 2,400km (1,500 miles) long and 9m (30ft) high with watchtowers at regular intervals, protected China's northern boundary against the raids of nomadic tribes.

5

5 Codification of law was one result of the invention of writing. The most complete ancient law code is that of Hammurabi of Babylonia, which was devised in the eighteenth century BC.

6 The census was an important feature of ancient states. It determined government collection of revenue (usually taken in the form of labour or agricultural produce). The Domesday survey of 1086, shown here, ordered by William I, was England's first census.

6

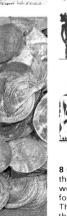

7

7 The cities of ancient civilizations were generally different in function as well as character from modern cities. Most were small, as even the so-called city states had predominantly rural populations. In early Egyptian and Central American societies, most conurbations seem to have been ceremonial centres with small resident populations of officials, priests and craftsmen. Since ancient societies were agrarian, their cities could seldom become manufacturing, commercial and fiscal centres of any size. Ancient cities such as Machu Picchu, the Inca city shown here, were primarily political and administrative centres.

8

9

8 Coins, such as these from Byzantium, were a helpful medium for taxation and trade. The Lydians were the first to use coins, in the 7th century BC.

9 This Egyptian wall painting of slaves and food being brought to the pharaoh reflects the fact that in ancient societies power was a source of wealth, not the other way about. Power and class were determined by political affiliation. Broadly, the population was divided into the governing class – the ruler, the royal family, state officials, landed aristocrats, priests, the military and merchants – and their dependants, who were mostly peasants.

Structure of societies

Sociology is the study of the way in which relationships between people are organized [Key]. The organization of social relationships produces social structures that range from the small group, such as the family, through larger groups, such as the peasant community or industrial corporation to national states and empires. The social structures of societies are characterized by mechanisms that integrate their members but they usually contain social divisions, conflicts of interest and inequalities of power. Social structures have characteristics that are independent of their members, but their workings must be understood in terms of the way these members experience them.

Understanding sociology

Although sociological theories and the results of sociological research are often complex and sophisticated, the basic principles of sociology are easily understood. This is because everyone who is aware of his membership of a human society is, in a sense, a sociologist. In order to cope with everyday social life we must all have some under-standing of the structure of our society, our social position within it, its rules, social divisions and hierarchies, and the way that other people will interpret social situations in which we are involved. It follows that sociology is as old as human society. Its emergence in Western Europe in the nineteenth century as a distinct academic discipline can be seen as a facet of the increasingly complex division of labour in developing Western capitalist society.

The pioneer work of the French sociologist Emile Durkheim (1858–1917) showed that explanations of social behaviour were to be found within the nature of society itself. Uniformities in the behaviour of individuals, together with constant variations in behaviour between societies, may be seen as products of the structure of societies rather than of the specific natures of their members.

Durkheim chose a highly individual act – suicide – and showed how variations in suicide rates between societies could be explained socially in terms of the strength of the societies' social bonds. It is the society of which a person is a member, together with the social groupings within that society to which he is attached, that determine most of his behaviour. His religious and political beliefs, his moral values, the type of house he lives in, the way it is furnished, the type of food he likes and the way he eats it, the clothes he wears – these are all products of the influence of society [1].

Social rules and social roles

Society influences and controls its members through their acceptance of its social rules (norms) and their occupancy of preordained social positions (roles) together with their willingness to act as expected in these positions. While some societies are more repressive than others, people are generally unaware of the extent to which their actions are socially controlled. In conforming to social rules and acting out the roles assigned to them they usually believe they are exercising freedom of choice.

The rules of society to which the individual member is required to conform include not only codified laws and written rules of organizations, but also a multitude of

1 The individual is presented by sociologists, at one extreme, as wholly determined by the society in which he lives. He is pressurized to conform to social norms through the variety of social institutions and organizations to which he is attached, such as the family, the educational system, the work place, the Church, or even the local sporting organization. This is an extreme characterization and represents only one side of his relationships.

2 While people are born into society, and both the norms of society and the roles into which they must fit are there before them, it is also true that social structures are initially created and continually changed by people's actions. Protest and other forms of social movement may bring about changes in social norms (eg campaigns for abortion law reform and homosexual law reform) and changes in social roles (eg campaigns to change the role of women in society).

3 Ferdinand Tönnies (1855–1936) claimed that industrialization and urbanization brought about major changes in social relationships. He depicted the trend as being away from personal, communal (*Gemeinschaft*) relationships characteristic of the pre-industrial rural community towards impersonal, formal, bureaucratic (*Gesellschaft*) relationships. *Gemeinschaft* relationships are based on shared values and sentiments; *Gesellschaft* ones are based on self-interest.

4 Uniformities in social behaviour usually imply the presence of social norms. Norms are known to be present only when sanctions ranging from social disapproval to more severe forms are applied to deviants. Often we become aware of social norms only when someone breaks them. In the cartoon all are aware that the central character has broken a norm by wearing the wrong clothes and they are applying their sanction by laughing at him.

unwritten rules regulating minute details of social encounters. These include knowing when to speak, how to address people (Sir, Mr Smith or Joe), when to laugh, when to stand up or sit down and when to leave.

The process of socialization

Through the process of becoming social beings (socialization) people learn the norms of their society and those of the diverse social groups in which they may participate during their lives – the family, the school, the factory shop floor, the sports club, the professional organization and so on.

People learn, through socialization, to accept social rules as their own standards of right and proper behaviour, so that they may be controlled as much through their own moral sense and avoidance of guilt feelings as by an external coercive force. But external pressures to conform to norms are also present in the form of sanctions applied to deviants, which may range from legal penalties of death or imprisonment and loss of job or position, to the social disapproval of one's fellows [4]. The severity of sanctions gener-

ally depends on how far common sentiments are offended by non-conformity.

People know their place in society through the variety of social roles that they occupy. Each person normally plays many roles in his daily life, his work, his family, his community and other social activities. Old roles are continually cast off and new ones taken on. Social roles are not so much descriptions of what people generally do in the positions they occupy, but rather what they are expected to do. While these expectations may include the performance of tasks, they are mainly requirements of behaviour in the social relationships that acting out the role involves. Some roles, such as that of husband, may involve a relationship with only one other person. Others, such as those of waiter [7] or social worker, may require relationships with several types of people. Role expectations are expectations of behaviour towards other people, but they come from the norms of society rather than from the opinions and desires of the participants. The structure of societies may be seen as based on organized sets of roles.

Auguste Comte (1798–1857) saw sociology as a science of social life that would discover the laws of social behaviour akin to the laws of the natural sciences. His belief that sociologists should be the priests of the new scientific social order has a contemporary relevance in that today critical sociologists are increasingly becoming the conscience-bearers of advanced Western societies. While Comte saw sociology as providing justification for the prevailing social order, modern sociologists are more likely to be moral critics of their societies and tend to reject Comte's belief in progress through the application of science.

5 Different roles will often bring out quite contradictory qualities in the individual. Effective performance of these different roles may be jeopardized if those involved with him in one such role are able to observe his performance of another. In the cartoon the CO's performance as CO might be undermined if his soldiers observed him in his role at home as a husband.

6 Georg Simmel (1858–1918) distinguished the formal structures of social groups from the content of social interaction within them. He saw sociology primarily as the study of forms of social groups: a variety of groups, such as those of officers and enlisted men, political leaders and their followers and parents and children, might share the form of superiority and subordination. He showed how the addition of a third person to a two-person group would radically change its form. The third person could act as a mediator between the original two or exploit their differences for his own ends.

7 While the individual usually plays several social roles, any one of these roles may involve relationships with a variety of different sorts of people. In a large restaurant the waiter may have to relate to customers, his supervisor, wine waiter, chef, service-pantry workers and other types of staff. When roles involve several relationships it is sometimes found that the demands of one relationship conflict with those of another. The customer wants the waiter to attend to his order promptly while the service-pantry worker does not want to be rushed with orders. The waiter must keep them both happy. Such roles require social skill and may be stressful.

263

Conflict, power and social inequality

The mechanisms by which society integrates and controls its members are never completely successful. All societies have deviants and criminals, and nearly all are subject at times in their histories to serious forms of social disorder, revolts and revolutions. Sociologists have shown that such occurrences are not just the result of failure of the social structure. Instead they believe that conflict and social division are inevitable in all social structures. Societies are unequal in their distribution of power, and of material and social rewards, and it is these inequalities that generate social divisions and conflicts of interest. If uncontrolled they may produce extreme forms of social strife.

The exercise of power

In all societies some social groups are able to exercise power over others [Key, 8]. Their power may result from their control over means of coercion – military or police forces – or control over material resources and social rewards. The German sociologist Max Weber (1864–1920) showed that the ability of groups to maintain their power depended on their success in persuading those subject to it that they had authority; in other words, that their exercise of power was legitimate. Acceptance of the existing power structure by the population may be based on tradition, as in the case of a traditional monarchy [1]; rational laws, as in a constitutional parliamentary system; or personal loyalty to a leader, as in the case of Benito Mussolini [2] or Adolf Hitler.

When the exercise of power is accepted as being legitimate it usually complies with prevailing social norms. A result of this is that the norms of society tend, in varying degrees, to serve the interests of powerful groups rather than the common interests of all. For example, in societies where the powerful groups are owners of private property, the norms tend to emphasize respect for private property. The exercise of power always tends to generate resistance on the part of the powerless, who challenge its legitimacy. However, just as people may be unaware of the extent to which their actions are controlled by the forces of society so they may not notice the power that others exercise over

them. They may have been socialized to want to act in ways that are in the interests of the powerful. Where the exercise of power is based on control over material resources and the means of production, as in nineteenth-century Western Europe, it is exercised by a dominant social class. Where power is based on control of the bureaucratic and military apparatus of the state, as in several contemporary African states, then that power lies in the hands of a ruling élite.

Social class structure

Social classes are groups of people sharing common social and economic interests. Where these interests are recognized, class consciousness develops, which may be the basis for social strife in forms ranging from strikes to full-scale revolutionary movements. Political parties often draw support on the basis of class membership, although parties may also be based on religious, regional, tribal or ethnic divisions. In nineteenth-century capitalist societies, the chief social class division was between property owners and the workers without property whom they

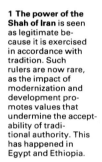

1 The power of the Shah of Iran is seen as legitimate because it is exercised in accordance with tradition. Such rulers are now rare, as the impact of modernization and development promotes values that undermine the acceptability of traditional authority. This has happened in Egypt and Ethiopia.

2 Charismatic leaders, such as Benito Mussolini (1883–1945), are seen as legitimate by their followers by virtue of the extraordinary personal qualities – their magnetism – which they are believed to possess. Such leaders often break with both established tradition and codified laws.

3 The policeman on traffic duty exercises rational-legal authority – power that is seen as legitimate because it is in accordance with rational laws. His authority is vested in the office, not in his person, and its scope is strictly defined by law. The authority of elected parliamentary governments is also rational-legal.

4 Tanks in the street, as here in Prague in 1968, show the exercise of power based solely on violence. If the power-holders lack legitimacy in the eyes of their subjects, they must rely on coercion. All forms of political domination use force to some degree, although they may try to achieve legitimacy.

employed. In modern advanced capitalist societies, this division has now been overshadowed by that between professionally qualified and managerial workers (often known as the middle class) and manual workers (the working class).

Social classes must be considered in relation to the historical development of each society, but there is some evidence that the dominance in advanced capitalist societies of the middle classes is spreading to a number of developing countries. The class structure may often be the key to the distribution of power, but social class membership can also help determine expectations. In societies where in theory there is equality of education, for example, educational opportunities for middle-class children have shown to be better than for working-class children.

Ruling élites and their power
Ruling élites are groups that actually control the instruments of power. They may act either in the general interest, or simply in their own interests. If their power is not sustained by a dominant class, then it is maintained by control of the police, armed forces and propaganda. The state bureaucracies of the Soviet Union and the military regimes of several Latin American countries are examples of such ruling élites, although some of these may be allied to particular social class interests.

The American sociologist C. Wright Mills (1916–62) argued that in the United States in the 1950s, power in matters of national importance rested in the hands of leading businessmen, top politicians and soldiers.

Power is often exercised by ruling élites in a democracy through their ability to choose the issues on which the majority may vote. If a new urban transport system is proposed, for example, information and debate on the advisability of having a system at all and the effects that it will have on the area may not be much in evidence. What will usually be presented are several alternative technological schemes. The élite, often businessmen or trade unions, have already made the most important decision, perhaps to their own advantage, and the majority are presented with a choice only on secondary issues.

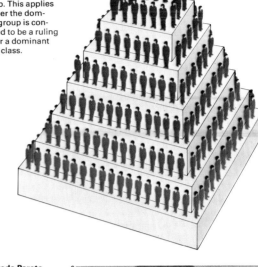

Society can be represented as a human pyramid in which power and socio-economic rewards are concentrated towards the top. This applies whether the dominant group is considered to be a ruling élite or a dominant social class.

6 Vilfredo Pareto (1848–1923), the Italian engineer, economist and sociologist, argued that people's actions are often based on sentiments rather than on rational thought. Political élites, he believed, maintain their power by manipulation and coercion requiring appeals to sentiment rather than to rational interests. Elites, he claimed, are liable to be overthrown unless they are also prepared to use force.

5 The Indian caste system is a unique form of social stratification representing the most extreme type of social hierarchy. People are born into caste membership and are segregated: for members of high castes to come into contact with low castes, such as these people collecting dung, involves ritual pollution. Although in modern India the caste system is not officially recognized, many features remain entrenched.

7 The British sociologist Basil Bernstein (1924–) has studied how people's use of language is related to their social class. He has distinguished the use of a simple "public" language from a more complex or "formal" language.

8 A person's position in society largely determines not only his level of material and social rewards, but also the power he has over others and the extent to which others have power over him. A member of a dominant group, in this case an employer, will have direct power over his employees. He also exerts more influence over local and national politics, the operation of the law and the mass media.

Employees

Law

Media

Politics: MP and local council

The individual in society

Classes and groups in society are not concrete entities: they exist only insofar as they are accepted as real by the individuals of which they are composed. Social structures cannot therefore be fully understood without examining their meaning for the participants.

Defining social situations

There are no objective definitions of a social situation independent of people's perception of them, rather people will act according to the way in which they define that situation. Different people may well define the same situation differently [4]: as an extreme example, whether or not the authorities open fire on an angry crowd will depend on whether they define the situation as a riot. The way people choose to define social situations is not a matter of personal whim but depends largely on social influences.

People may define situations by unconsciously imagining how other people who are significant in their lives would define them. These significant influences constitute the individual's reference groups. A person's reference groups may include his family, his neighbours, the people he works with, and his Church, political party, age group, social class and sex. He takes his moral standards from his reference groups and through them is influenced by social norms.

Reference groups also provide an individual with a basis for evaluating his social position. Although they are not actual social groups, but constructions in the mind of the individual using them, it is through them that the individual is subject to the influences of society. If, for example, a doctor's reference group is the medical profession it is his idea of his profession rather than the actual behaviour of its members that influences his professional actions.

The social self

The eminent American sociologist and social psychologist G. H. Mead (1863–1931) showed how the individual's very conception of himself, and of how he appears to others, is a product of social interaction [1]. In his work Mead stressed the constant tension in social life between the assertion of individuality and the pressures of social conformity. He distinguished two components of self, the "me" and the "I". The "me" (the social component) is the impression we think we make on other people whose opinion matters to us. We gather these impressions by putting ourselves in the place of these others. The individual's reference groups are therefore important for the development of the "me". The "I" (the individual component) consists of the actual response of the individual to the attitude of others. It is the creative element of self, spontaneous, uncertain and never fully predictable. People's actions are never fully socially determined.

Social roles are laid down by the norms of society, but this does not mean that people cannot express their individuality in the roles they play. The various devices that people use to express themselves and convey impressions of themselves have been explored by the American sociologist Erving Goffman (1922–) in a number of studies. He has shown that it is normal for people to distance themselves from the roles they are playing, so conveying to others that their real selves are not wholly involved in

1 The different impressions of a girl that are held by her employer, her mother, her boyfriend and herself are portrayed here [left to right]. These views all correspond to the different roles she plays – employee, daughter, girlfriend. In these various roles she is not seen by others merely as a player of a particular role but as a whole person. The impression the girl has of herself derives in part from her idea of the impressions that others have of her and of her actions in social encounters with these others. It is therefore a mistake to think of the individual as having a "true self" that is independent of the impressions that others have. At the same time the individual's self-image is no mere mechanical reflection of others' impressions. This girl can, for example, consciously influence these impressions. This may be done not only through choice of speech and action in social situations, but also through the clothes she wears, her hairstyle, body posture, facial expressions and gestures. Sometimes people try to convey to others impressions of themselves that they do not themselves believe in. This may take the form of presenting an idealized version of the self that the individual believes he falls short of – like the teacher who tries to create the impression that he is more knowledgeable than he believes himself to be. In the case of the confidence trickster a deliberately false impression of self is conveyed to others.

any one particular role. Goffman's work has shown how people's spontaneous individuality in social sitituations can itself be the subject of sociological study.

The sociology of everyday life often appears to deal with rather trivial things such as conversations in bars, but sociologists have shown how an understanding of social rules in everday situations may illuminate the processes governing the structuring of society. The routine, everyday activities of policemen (how they react to and treat people of various classes and races), for example, are relevant to the explanation of criminal behaviour.

Trends in sociology

In many major universities the establishment of sociology dates only from the 1940s, although much work of lasting importance had been carried out before then. The 1940s and 1950s were dominated by two types of American sociology. One of these was rather abstract theorizing in attempts to develop models applicable to all societies; the other was the large-scale collection of information about various aspects of social life through social surveys, questionnaires and statistical techniques of analysis. These surveys were characterized by a rigorously scientific approach in the technical sense. During this time it was the common stance of sociologists that they were disinterested scientists and that it was not for them to make policy recommendations or moral judgments on the societies they were studying. Since the end of the 1950s theory and research have been brought closer together. Sociologists have been less inclined to emulate the natural sciences in a narrow technical sense and have paid more attention to understanding people's experiences and the meaning of their lives.

Sociologists today are less prone to proclaim themselves disinterested academics and more willing to advocate policies and to criticize societies. In this they come closer to their founding fathers, whose aim was to produce better societies. By revealing the oppressive elements of social structures, sociologists hope to liberate man from them: once people know the social results of their actions, they have the chance to avoid them.

A theme of much sociological thought is the tension between the social and individual aspects of man. Man is a product of society but society itself is produced by men's actions. Man, represented by the cardboard cut-out figure, fits into the roles that society provides for him. But at the same time he continually asserts his individual identity in the various roles that he is called upon to play.

2 Alfred Schutz (1899–1959), philosopher and sociologist, believed in the importance of the "common sense" beliefs of the mundane world. He used this reality as the point from which to begin examining society.

3 The self is not a fixed entity. The images we have of ourselves are continually being modified according to our expectations and experiences in social encounters, as in this cartoon sequence.

4 Different people may have different views of the same social situation in which they are involved. In part at least these different perspectives are the result of their different reference groups. At this party one person will see it as romantic [A], another as empty and boring [B], a third as a drunken get-together [C] while a fourth may see it as a chance to renew old friendships [D].

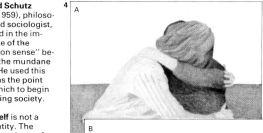

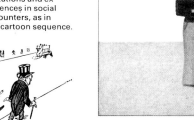
A

B

C

D

3

I am bidden to the War Office | I depart for it | I approach it

I enter | I am not observed | I am still not observed

I am observed | I am spoken to (and still live) | I continue to be spoken to

I am spoken to quite nicely | I am shaken hands with | I take my leave

267

Prejudice and group intolerance

Prejudice is a preconceived opinion, usually unfavourable, about a category of people or about individuals thought to belong in that category. It is likely to lead to discrimination – treating differently people who belong to particular social, ethnic or religious groups.

Aspects of prejudice

Prejudice has several different interlocking causes, but a major influence is the sharing of preconceived opinions by a group. People learn the prejudices of their parents, teachers and friends, and often take if for granted that their judgments are right. This sharing of opinions is one of the leading features of human cultures and because there are patterns of prejudice associated with different societies, but changing in the course of history, it is helpful first to distinguish the cultural aspect of prejudice.

While prejudice may be widely shared within a group, it is manifested in the speech and actions of individuals and has to be understood in relation to their personalities and to their positions in society. The word prejudice is frequently used to designate an attitude that is emotional and rigid. When someone tries to demonstrate to a person who is prejudiced about something that his opinion is false, the prejudiced person is apt to cling to his preconceived opinion and to twist the new evidence to make it fit. Prejudiced people are likely to behave differently from unprejudiced people in discriminating against members of a particular group, especially when they see them as economic or social threats.

Ethnocentrism and racialism

Under the heading of cultural prejudice it is important to separate feelings of superiority based on pride in cultural and social achievements and those based on supposed inherent, genetically fixed characteristics. The latter type of prejudice can be called racialism and is characterized by the feeling that members of some group are sub-human [Key, 3].

In every known society people tend to prefer those of "their own kind" as friends, workmates and relatives. They tend to see international affairs from the standpoint of their own society or nation and to put a lower value on the people and customs of other societies. This inclination to prefer peoples and things with which they identify themselves is called by sociologists "ethnocentrism" and necessarily implies a bias against those seen as different [4]. Ethnocentrism has been a constant factor in world history. But international contacts diminish it by making individuals of different groups aware of their common heritage.

Within a single society such feelings of preference and suspicion are reinforced by the way that people of like background tend to congregate together both for company and for protection in the demarcated residential zones that are a feature of towns and cities.

The history of prejudice shows no simple pattern. Although prejudice and discrimination have been evident in most societies, by and large in the ancient world the motivation was ethnocentric rather than racialist in type. This can be said even of the Hindu caste system that reflects the cultural split between the ancient, conquered Dravidian peoples and the invading Aryans. The kingdoms that are now Rwanda and Burundi in central

1 Few animals act as aggressively towards members of their own species as man does. And even when they do fight over territory or mates, for example, they have an automatic cut-off point in response to a submission signal. Man has similar submission signals, such as avoiding eye contact and hanging the head, but seems to be able to ignore them. This is made easier by the development of long-range weapons.

2 The fighting between young supporters of football teams in Britain suggests that these games are also often ceremonies in which opposing supporters stimulate each other to more violent expression. As members of groups emotionally committed to "different sides", people are capable of extreme behaviour. Parallels can be drawn with the feuding based on religion in Northern Ireland.

3 The government of ancient Rome had an effective way to divert popular anger when things went wrong. "If the Tiber floods or the Nile fails to flood, if the sky is darkened, if the earth trembles, if famine, war or plague occurs", then immediately one shout went up: "The Christians to the lions!" The Christians made a convenient scapegoat, just as did Jews in Nazi Germany.

4 Chinese ethnocentrism was evident in 1793 when Emperor Ch'ien Lung instructed the British envoy to tell King George III that he saw no reason to open diplomatic relations. "Even if your envoy were to acquire the rudiments of our civilization you could not possibly transplant our manners and customs to your alien soil . . . we possess all. . . ."

Africa formed a sharp contrast. Here the pre-eminence of the Tutsi (who made up only 15 per cent of the population) over the Hutu (83 per cent) and the pygmoid Twa (two per cent) was based largely on the Tutsi's physical characteristics.

One possible explanation for the lack of racialism in ancient societies is that they were much smaller and those who travelled did so in small groups, so contact was limited.

Slavery and prejudice
Colonization and the African slave trade across the Atlantic created a new set of relations between people and stimulated the growth of a new kind of prejudice. The chattel slavery of the New World [5] (in which the legal status of slaves resembled that of domestic animals and non-human property) was different from the personal slavery of ancient Rome and medieval Spain (in which the law accorded a slave a human status, the right to marry and some protection against abuse). To understand this difference it is necessary to consider the demand for labour. In the New World, land was abun-

dant. If a landowner imported a free white worker to labour on his estate, the workman was inclined to go off and establish his own farm. So the landowner found it best to bind labourers to indentures whereby they promised to work for him for seven years in order to pay off the cost of their passage. The move to outright slavery was an easy one to make. Europeans had for centuries regarded blackness unfavourably but when it became associated with slavery and, indeed, guilt about white maltreatment of blacks, it received a new emotional charge.

When, in the nineteenth century, scientists speculated about classifying mankind into types, just like flowers and fish, they were quick to describe Negroes as a distinct and inferior racial type. Up to the end of the nineteenth century the understanding of man's physical nature was confused and these early speculations about race were easily built up into pseudo-scientific theories of race. In later generations these have been used to justify people's prejudices and to serve their interest in keeping certain peoples beneath them in a separate category.

A Jewish youth is forced by the Nazis to parade through / the streets of Nuremberg carrying a placard that reads / "I shamed a Christian maiden" – a signal for discrimination.

6 British rule over India came to an end in 1947 and the subcontinent was divided into India and Pakistan. Both Muslims / and Hindus were attacked in areas in which they were minorities. This picture shows Muslims crowded on to the / roof and hanging on to the sides of a train in an attempt to escape from New Delhi to Pakistan. Shortly before, a / similar train had been attacked by Sikhs and estimates at the time were that 1,200 refugees were killed and 400 injured.

5 Slave families were often broken up when a slaveholder died and his estate was divided. In law slaves / were treated like farm animals and other possessions, as shown by this picture of a sale in the Rotunda, / New Orleans, where household effects are being auctioned at each side and the slaves in the centre.

7 An episode in one of the popular Fu Manchu films shows a diabolical-looking Oriental carrying off an innocent-looking Western heroine. Hollywood films of the 1920s and 30s developed popular stereotypes of black people as stupid and subservient and of yellow people as inscrutable and vilely dangerous.

8 An anti-American parade through the streets of Peking was a protest against US involvement in the Vietnam war. Uncle Sam is being pulled through the streets on a barrow. Governments use psychological warfare techniques not only against the enemy, but also often to persuade their own citizens to hate people on the other side.

269

Prejudice and personal choice

The frequently heard accusation that men are prejudiced against women provides an insight into the nature of prejudice in society. Obviously the accusation cannot be taken at its face value: it could scarcely be alleged that most men are against the existence of women. But many men are prejudiced against the upsetting of traditional roles by the claims of women to social equality [1]. Society has developed stereotyped notions of "femininity" and has attributed such qualities to women as weakness, mildness and passivity. These are then presented as being "natural" when in fact they are culturally determined insofar as they exist at all.

Social roles and prejudice
Prejudice thus often occurs in connection with particular social roles and expectations. When blacks are at the bottom of the social scale some whites wish to keep them there because they get emotional satisfaction from the belief that someone is beneath them, or because they obtain an economic advantage from the restraint upon competition from blacks, or because they have come to regard this as a natural state of affairs. Racial prejudice rises to the surface when members of the subordinated category appear to challenge the social pattern.

One way of examining patterns of prejudice is to measure degrees of social distance. In the traditional Hindu caste system social distance was translated into physical distance [2]. Europeans saw themselves as occupying the highest positions in India prior to independence in 1947, but Brahmins saw them differently. If they had to call on Europeans they would do so as early in the day as possible so that they could cleanse themselves and eat in a state of ritual purity.

Even when not expressed in linear terms the requirement of social distance can be very obtrusive. In some racially mixed societies whites display much more reluctance to accept blacks in certain relationships than in others. Marriage is a close relationship where resistance – and prejudice – is highest [3]. Accepting someone as a fellow-worker is easier because it is a much more distant relationship.

People's theoretical attitudes, however, often do not correspond to their actual behaviour. A survey in New York showed that of those white people who objected to black shop assistants, one in four did not even notice when he or she was served by a black assistant. When interviewed, the others said they did not mind black assistants in the department in which they had just been served (whether it was clothing or food) although they might not like it elsewhere. This underlines the irrationality of prejudice.

Attitudes and customs
People tend to accept the customs of their community and to adopt the attitudes that justify those customs. Attitudes and customs influence one another.

A classic study that points to this conclusion was reported in 1934 by a white American sociologist who with his wife and a Chinese couple took a trip in the western United States. Together they stopped at 184 restaurants and 66 hotels and were refused service only once. After returning to his university the sociologist sent questionnaires to the places he had visited, enquiring of each

1 The fight to achieve "rights for women" centred first on the struggle for the vote. In the USA women first obtained the vote in local elections and in the primary elections for the presidency. One by one state constitutions were amended and by 1918 women had acquired equal suffrage with men in 15 states. The Constitution was finally changed in 1920. The right to vote in national elections was given to women in New Zealand in 1893, Australia in 1902, Finland in 1906, Norway in 1913, the USSR in 1917, the UK in 1918 and China in 1947. There are still some Islamic countries where voting is not considered to be the business of women.

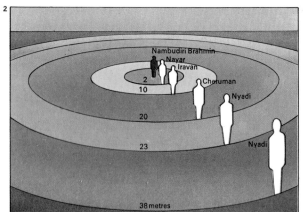

2 A study of Hindu caste in southern India published in 1947 stated that "a Nayar must keep 2.1m (7ft) from a Nambudiri Brahmin, an Iravan 9.75m (32ft), a Cheruman 19.5m (64ft) and a Nyadi from 22.6m (74ft) to 37.8m (124ft)." The Brahmin was polluted if a Nayar came closer than 2.1m (7ft) and the Nayar if an Iravan came within 7.6m (25ft) of him. Anyone who was polluted by the touch or proximity of someone of lesser purity was supposed to undergo a purification ritual. The Indian government is trying to prevent discrimination, but prejudice against intercaste marriage is strong.

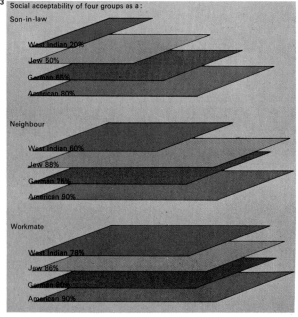

Social acceptability of four groups as a:

Son-in-law
West Indian 20%
Jew 50%
German 65%
American 80%

Neighbour
West Indian 60%
Jew 88%
German 76%
American 90%

Workmate
West Indian 78%
Jew 86%
German 80%
American 90%

3 One social study discerned a three-step pattern of social acceptability in Great Britain in the early 1960s. On the first step, nearly eight Britons out of ten were inclined to accept West Indians as fellow-workers. This figure compares favourably with ratings given to Jews (just over eight out of ten), Germans (eight out of ten) and Americans (nine out of ten). On the second step, Britons were more inclined to reject West Indians as neighbours than other groups because it suggested that a neighbourhood was of lower social status. As might be expected rejection was highest in respect of marriage because this involves close sexual and family relations.

proprietor if he would accept someone of the Chinese race as a guest in his establishment. Ninety-two per cent said they would not.

Other studies at children's camps have shown that when schoolboys are arranged in rival teams the competition evokes prejudice, but when they are in co-operative relationships prejudice is reduced [7]. People overcome their suspicions of one another when they have a common goal to work for, as, for example, in times of war and crisis. To an important extent, therefore, prejudice is a product of social organization.

Psychological origins of prejudice

Prejudice is often expressed in irrational behaviour that has a psychological origin. Research has shown that the people who express the strongest prejudices are hostile towards all strange groups that they consider socially inferior. They express hostility even towards fictitious groups. One of the great contributions of the psychoanalyst Sigmund Freud (1856–1939) was the interpretation of the dynamics of personality, which enables the student to understand why some people

need to display prejudice in order to make up for their own deficiencies. They show in extreme form a tendency present in everyone, namely, an inclination to relieve frustrations by displacing them. Like the ancient Jews, who loaded their sins on to a scapegoat and drove it out into the wilderness, people often have a psychological need to find a scapegoat. Complex and conventional societies impose many restrictions on their members, who consequently seek some occasion to release their emotional energy.

Prejudice can maintain itself because people often have little personal acquaintance with those they use as scapegoats. Beliefs that are oversimplified in content and unresponsive to the objective facts are called stereotypes. People who believe that all blacks are strongly sexed, all Jews are grasping, all Englishmen are snobbish, select the evidence to suit themselves. They avoid situations in which they might be forced to recognize their error. Their attitudes influence the way they participate in society and so the personal and social aspects of prejudice reinforce each other.

The members of this white anti-Civil Rights group in the United States chose the swastika as a symbol for their banner. This recalls the Nazi movement in Germany and its emphasis on the superiority of the white races. As such it would probably appeal mainly to aggressive extremists.

4 This diagram summarizes the answers given by white people in 20 cities in the USA in 1951 when asked about relations with black Americans. What people say in such circumstances is influenced by local customs and as these change so do attitudes. It may therefore be easier to reduce prejudice by introducing laws against discrimination than by promoting educational campaigns.

In large cities social acceptability is greater but more superficial.

Acceptable
Controversial
Unacceptable

4 Should blacks :

Try on dresses in white department store ?

Sit in same part of bus as whites ?

Use same public rest rooms (toilets) ?

Sit among whites in movies ?

Occupy hospital bed beside a white ?

Use same swimming pool ?

Be served in white restaurants ?

Belong to white Protestant Churches ?

Stay in white hotels ?

Use white barbershop ?

10%	20	30	40	50	60	70	80	90	100

5 Segregation in schools in the USA has been one of the most powerful factors encouraging racial prejudice. Since 1954 cities have been required to change the boundaries of school areas or mix the pupils by taking groups across town to other schools. In many cities this so-called busing policy has evoked antagonism. This poster seeks to increase hatred by appealing to sexual fears.

6 The struggle to enrol James Meredith at the University of Mississippi in 1962 was solved when federal law forced the university to accept him as the first Negro to attend classes.

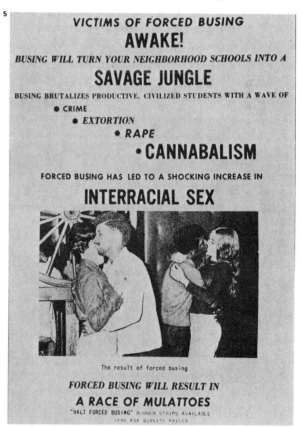

VICTIMS OF FORCED BUSING
AWAKE!
BUSING WILL TURN YOUR NEIGHBORHOOD SCHOOLS INTO A
SAVAGE JUNGLE
BUSING BRUTALIZES PRODUCTIVE. CIVILIZED STUDENTS WITH A WAVE OF
* CRIME
* EXTORTION
* RAPE
* CANNABALISM
FORCED BUSING HAS LED TO A SHOCKING INCREASE IN
INTERRACIAL SEX

The result of forced busing

FORCED BUSING WILL RESULT IN
A RACE OF MULATTOES
"HALT FORCED BUSING" BUMPER STRIPS AVAILABLE
SEND FOR QUANTITY PRICES

6

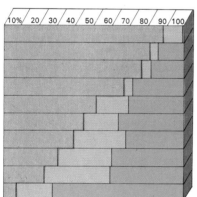

7

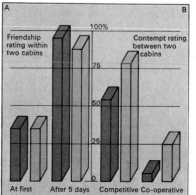

A — Friendship rating within two cabins; At first, After 5 days

B — Contempt rating between two cabins; Competitive, Co-operative

100%, 75, 50, 25

7 Several experiments with groups of boys in American summer camps have shown that although initially suspicious of each other [A], after only five days they had formed friendships in the same cabin. Two similar groups [B] meeting only in competitive situations regarded all members of the other cabin with contempt. After co-operative activities, however, the percentage who felt such contempt fell dramatically.

Fundamental political ideas

Political thought considers the nature and the validity of government. The perennial questions asked by political thinkers are: What purpose does government serve? Why and by what right do some men rule over others? Under what conditions and to what extent should one obey the state? To what degree should political opposition be tolerated?

The method of political thought ranges from generalizations and maxims derived from history to philosophical and theological analysis. It is wider than political science in that it has a strong ethical content – a concern not only with studying different forms of government but also with the kind of government that is best and the way in which it can be achieved. Above all, it is concerned with the question of political legitimacy and of the circumstances under which a person should obey the state or defy it [Key].

The Greek concern with justice
The Greeks believed that the city state of the Athenian type [1] arose from a quest for justice that their previous anarchic or tribal arrangements could not satisfy. Plato (c.

427–347 BC) in his *Republic* contrasts the view that justice is the rule of the strong over the weak (which being "natural" must therefore be right) with the rival Greek view that justice is the majority of the weak collectively imposing their rule upon the strong. He saw justice as a universal concept that consists of the right relationship of the individual parts to the harmony of the whole. Those he considered fit to rule were an intellectual élite, able to penetrate the nature of truth and reality. Such a view was far from the Athenian practice of democratic election which appeared to Plato and to Aristotle [2] as the rule of ignorance, likely to lead to strife.

The Roman concept of sovereignty
Roman thought was less speculative and more practical. The major theoretical contribution was the notion of *imperium*, or sovereign authority. When the yoke of the first alien Etruscan kings was thrown off, the Roman people declared that they alone had the right to rule themselves. Although the concept of sovereignty's residing in the people was not always followed in practice,

Rome prided itself on a balanced class system whereby patricians (the Roman aristocracy) had preponderant power and authority in the Senate and plebeians (the common people) had their own assembly and officers (called tribunes) as a necessary balance [3]. The Romans saw politics in terms of practical interests that needed protection. The Roman citizens' keen sense of legal rights developed into an elaborate legal system that included principles of law still used today.

Modern political thought has been haunted by the memory of the decline from the "golden age" of the Roman Republic to the decadence of imperial autocracy (rule by one man) backed by military might. This memory has added a strain of pessimism to the distinctively modern political idea, inherited from the eighteenth-century French Enlightenment, that human reason makes political progress inevitable.

Political order during feudal times was based on a political hierarchy of kings, vassals and serfs and a Church hierarchy of pope, bishops and priests. Inequality within these orders was generally accepted as the neces-

1 The Greek political heritage is dominated by the idea of democracy, exemplified by the Athenian state [A], a fluid and flexible system which embodied the principle that citizens should rule and be ruled in turn [arrows] by means of annual election to governmental office. But the philosopher Plato distrusted democracy and preferred the militaristic system of another Greek state, Sparta, with subordination of many tribes to one [B]. Cohesion of the governing élite in Sparta resulted in an inflexible system but Plato disapproved of the extreme libertarian constitution of Athens, believing it could lead to disorder and tyranny.

2 Aristotle (384–322 BC) disliked extreme Athenian democracy and held that justice meant giving virtue its due by electing the best to office. The best constitution lay somewhere between oligarchy (rule by the few) and democracy (rule by the many). Extremes of either kind were unjust and led to conflict. A liberal élitist, Aristotle advocated a balanced constitution and a strong middle class.

3 Attempts to base political order on a hierarchy have met with varying success. The Roman republican system [A] involved a class balance within a unified sovereign system. The two main classes, patricians and plebeians, cooperated in the activity of government. Believing in the dignity of leadership, the Romans gave patricians an influential role in the Senate with safeguards for the plebeians. A fundamental conflict was thus contained in a stable system. The medieval system [B] involved a conflict between two separate hierarchies, spiritual and temporal, whose powers could not easily be separated. Disputes arose concerning the extent and limits of each.

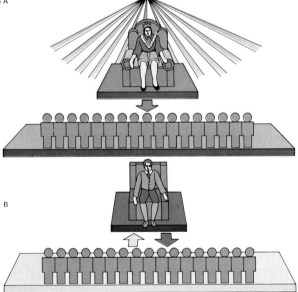

4 The issue of sovereignty came to a head at the beginning of the modern era when men questioned the ambivalence of the medieval theory that power came from God and was at the same time based on popular consent from below. Kings such as Louis XIV of France claimed that as power came from God, kings were responsible to God alone, not to their people. This theory of Divine Right [A] began to lose popular support when discontent grew over the autocratic mismanagement of public affairs. Opponents revived the Roman theory that sovereignty resides in the people and that governments must therefore hold themselves responsible to the people from whom their power to govern is derived [B].

272

sary hierarchical order of God. Chaos would result from any blasphemous attempt to challenge it. Monarchy based on election, acclaim or hereditary right was considered the best form of government as it was thought most likely to preserve unity, a prized ideal in an age of constant strife [7].

Government by consent

As nation states gradually claimed independence from the Church, kings also tried to claim sovereignty over their people by asserting that they were responsible to God alone for the affairs of the realm [4]. A protracted struggle over this point led finally to an acceptance that governing authority derived from the people and had to be exercised with their consent [6]. As feudal theory was replaced by an assumption of the natural equality of individuals, modern political thought became more secular and rational.

Despite a stress on individual rights, property owners were long considered to have the sole right to decide the general affairs of the whole community. The question soon emerged as to how men could be free and equal if they were ruled by others. In the view of Jean Jacques Rousseau (1712-78), the individual could be free only if he actively participated in formulating the laws by which he was governed. Socialist thought in the nineteenth century began to question the belief that the rich had a right to govern the poor and that the rights of individuals should be put before the welfare of the state as a whole. To Karl Marx (1818–83) all systems seemed class dictatorships based on exploitation. He believed political freedom was meaningful only if the economic system prevented some men from controlling others.

Marx's dictum "From each according to his abilities, to each according to his needs" also moved away from the concept that political justice is based on absolute equality. A generally optimistic theory of evolutionary progress in the 20th century has assumed that injustice will disappear eventually but there is no consensus on whether this will come by revolutionary conflict or by peaceful reform in an "open" society [9]. Despite this the "political" nature of political thought remains a perennial challenge.

Guerrilla warfare aimed at the overthrow of an established political order raises fundamental issues of political thought.

When do men have a right or duty to rebel against a government? Is revolution essential for progress and does it involve unnecessary violence and chaos? How can freedom and order be balanced and what is the origin and nature of legitimate governing authority?

5 The violence of the French Revolution brought a reaction against Locke's rational theories, which had been used by the 18th-century Enlightenment to attack the irrational basis of Church and state. Conservative thinkers such as Edmund Burke (1729-97) and Joseph de Maistre (1753-1821) concluded that society was based not on rationalism but on custom, prejudice and tradition, which preserved necessary harmony and mutual class respect. Religion was the bulwark of civilization. They believed that excessive social criticism resulted in a disturbance of society's equilibrium. This led to violence and ended with the guillotine.

6 The return of Juan Perón to Argentina in 1973 illustrates the idea that popular support (in this case for an exiled strongman) is the basis of political power. The answerability of government to the people was established by the English philosopher John Locke (1632–1704) who put more stress than Hobbes on man's natural rights and asserted that if a government set up to protect these rights betrayed its trust it could legitimately be removed, if necessary by revolution. Many of Locke's ideas were enshrined in the American Constitution under which men were entitled to "life, liberty and the pursuit of happiness".

7 Niccolò Machiavelli (1469-1527) was the most original political theorist of Renaissance Italy, advocating ruthless measures to maintain authority. Popularly misjudged as the incarnation of evil, he in fact admired the civic patriotism and the sense of duty of the Roman Republic. Yet in circumstances of corrupt politics, leaders were justified in doing what was necessary to uphold the strength and unity of their kingdom, even if it was contrary to private conscience.

8 Leviathan, written in 1651 by the English philosopher Thomas Hobbes (1588-1679), depicts the state as combining the wills of all men into one sovereign body with a single head to guide it.

Hobbes approved of autocracy but not one based on Divine Right. He thought men were too quarrelsome and self-centred to conduct their own affairs. To avoid constant strife and miserable insecurity

they must rationally agree to a social contract with a powerful coercive authority to make laws and keep the peace. Social freedom could thereby be maintained albeit at the expense of political freedom.

9 Freedom of political thought and expression, exemplified by the soap box orators at Speakers' Corner in London, is the idea of modern liberalism. Its most influential political theorist was John Stuart Mill (1806-73), who held that while

harmful actions should be curbed, no opinions should be, as an open society led to the emergence of truth. With Jeremy Bentham (1748-1832), Mill believed government should be based on the greatest happiness of the greatest number.

273

Political science

The science of politics seeks to acquire knowledge of the nature of politics and to reach general conclusions about it. Whereas historians often concentrate on particular events, political scientists work on the assumption that political phenomena have certain general causes and consequences. The theories of political scientists are usually statements about the connection between two or more aspects of politics. The statement that representative democracy is the most stable form of government is, for example, a theory whose truth a political scientist might try to refute or prove by comparing societies that are representative democracies with those that have different governmental systems.

Development of political science
Politicians (those who practise politics) seek to promote or balance sectional interests in finding solutions to political problems. Political scientists (students of politics), on the other hand, try to establish and analyse what the problems are, rather than attempting to solve them. But the knowledge gained by

political scientists may well have practical implications for those who rule. Equally, greater insight into the nature of politics by all the members of a society may enable all to participate in political decisions [Key].

The focus of interest among students has changed since political science was established as an independent discipline in universities at the beginning of this century. The various approaches to and conceptions of politics that exist today have been influenced by this development.

Initially, political scientists concentrated mainly on the study of constitutional problems. (A constitution is made up of basic laws and rules – written or unwritten - according to which other laws are made and a state is governed.) The assumption was that political life was carried on in accordance with a society's constitution. Typically, the political scientist was interested in such questions as: How can the constitution guarantee civil rights? Is a two-chamber parliament better than a one-chamber parliament? Thus the object of study was the various state institutions (the legislature, the executive, the

judiciary) and the laws regulating the relations between them [2]. This kind of political science, especially in Britain, was closely connected with organized politics. It was directed to those who framed the laws and gave them effect, the politicians, the civil servants and the judiciary, rather than to the people in general.

Constitutional problems [1] are still of considerable interest to political scientists but they are now only one branch of a wider study. The realization that political science should cover a broader field than constitutional questions first appeared in the writings of American political scientists who began to study other political fields in the 1920s.

A broader approach to political science
The mere study of state institutions is inadequate to political understanding because it throws little light on the way political decisions are taken or on attempts to influence these decisions by such pressure groups as trade unions, employer organizations and political parties. Second, because political science aims at producing generalizations

1 Devising new political systems is a difficult art, as British political experts found in Africa during the period of decolonization. Arrangements were made for 12 states to start out, as did Kenya on Independence Day, with two-party democratic political systems. Only two of these states still have such a system. This high failure rate was the result of the British seeing their own constitution as a model for the new states, despite considerable social and economic differences. It was soon realized that the problems involved in creating a new state went beyond those of working out a constitution.

2 Early political scientists focused on the relationships between the major institutions of the state – the executive [A], judiciary [B] and legislature [C]. In doing so they reflected the dominant ideas and preoccupations of the time. In the 18th and 19th centuries it was believed that man was a rational creature whose behaviour could be changed by altering laws. Consequently reform efforts, both in the USA and Europe, centred on a fight for written constitutions that could protect basic human rights. The emphasis was on ideal systems, not merely on description and explanation of politics.

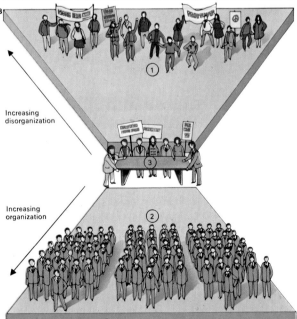

Increasing disorganization

Increasing organization

3 In modern political science, politics is often viewed as the resolution of a conflict between the inherent forces of violence and revolution in society [1] and the existing institutions – government, law and bureaucracy [2]. By means of consensus and compromise, or the use of power and force, a balance is achieved. The point along the spectrum at which this happens [3] governs the character of the state – increasingly disorganized or increasingly institutionalized.

4 A political system [1] can be seen as a machine that has to produce a certain number of goods; that is, decisions and actions [2]. The demands [3] made on the system are raw material the machine must process; political support for the system [4] is the fuel that powers the machine. Thus a political system needs a certain amount of support to cope with demands. A feedback mechanism [5] ensures that if the right decisions are made future support will follow.

about politics in all kinds of societies, including those that do not have the American or European kind of state institutions, a framework of analysis is needed that can be applied to various political systems.

Another general approach to political problems is systems analysis [4]. In this kind of analysis politics is defined not in terms of what goes on in specific institutions, but in terms of all behaviour connected with decisions that affect most members of society. Instead of talking about parliament, the cabinet and so on, systems analysts talk about the political system – by which they mean all the political interactions in a society. Their main interest is in how political systems persist under changing circumstances; how, for instance, a system will adapt to a situation in which certain groups inside or outside a society make greater demands than the existing system can immediately satisfy.

Sectional interests and practical politics

Many political scientists, rather than concerning themselves with the stability of any political system, view politics as dealing basically with struggles for power [3]. The powers of trade unions [6] can, for example, be studied in relation to the various political parties, as can the powers of trade union leaders in relation to the ordinary members [7]. Such studies examine not only who takes part in decisions but also the consequences of the decisions taken and whose interests are being taken care of. In Western industrial societies political scientists are at variance in their views or theories about power relations. Some contend that their societies consist of a range of groups whose powers balance each other, whereas others hold the view that specific élites or classes rule society [5].

Another important branch of political science is concerned with the motives behind political policies and the effect of these policies and of other factors on electoral support for particular politicians or parties. The analysis of voting trends and statistics is called psephology. Political scientists are often able to throw revealing light on the methods, performance and future prospects of candidates for political offices, and on the reasons behind electoral victories or defeats.

The people   Politicians

The political scientist

Problems studied by the political scientist [1] are considerably influenced by what the rest of society regards as important and relevant. The informed ideas of political scientists may have the practical consequence of increasing the power of the rulers [2] or may enable the ruled to exercise greater influence [3]. Whether it does one or the other depends largely on the goals of the society and the power relationships within it. In a democratic society political science tends to aim at wide participation in political life.

5   A   B   C

5 The power structure of Western societies has been interpreted in three different ways. Some political scientists hold the theory that in liberal societies such as the USA, UK, Sweden, France and West Germany, everybody has a say in the ruling of society. According to this pluralist theory [A] society is viewed as consisting of a number of pressure groups such as political parties, trade unions and business organizations which compete on an equal footing to promote the interests of their members. Some critics of the pluralist view claim that Western societies only appear to be ruled by the people and are, in reality, run by a limited number who form élite groups [B]. Who constitutes these élites is open to question but one of the most influential theories is that put forward by an American, C. Wright Mills. He identifies three élites that make important decisions together – politicians, military leaders and the owners of big industry. In the past priests and nobles often formed élites and today trade unions are beginning to form another. Finally there is the Marxist view that the state exists to serve the interest of a single ruling class which, in Western societies, is made up of capitalists or owners of capital [C]. In this view economic forces are decisive in shaping the political process and the apparent freedoms of democratic societies mask the real power structure.

6

7   A   C

B

6 Public demonstrations and strikes are among the powerful weapons available to modern trade unions in seeking to sway political decisions in the interests of their members. In Western industrialized societies the growing strength of unions and their ability to promote sectional interests is an important area of study for political scientists. Industrialization has brought an obvious shift in the power relationships within society in favour of the workers.

7 Internal power relations within trade unions have been an important field of study ever since these workers' organizations were first formed. A key question is whether trade union leaders [A] take care of the real interests of the workers [B] or whether they have other goals that are more in accord with interested groups outside the union [C]. Do trade union leaders identify with the workers or with political parties and outside pressures?

Types of political systems

Political systems can be classified in different ways, for example in terms of their political institutions. Western systems of representative government, for instance, can be differentiated according to the way in which the legislature or law-making body is elected, the main difference being between single-member systems such as Britain's and proportional electoral systems [1, 2].

The basis of elections

In single-member systems, each geographical area or constituency elects one representative on the basis of a simple majority within that area. Proportional electoral systems try to give greater weight to the proportion of votes given to each party, a method adopted by most continental European countries. Either districts elect several representatives on the basis of each party's percentage of the overall vote or, in other systems, where no one candidate has an absolute majority on the first ballot the second preferences of the voters are distributed or a second ballot is held between the leading candidates.

Yet another way of classifying political systems is in terms of how the executive arm of government is chosen. The two major Western systems are the parliamentary and presidential systems. In a parliamentary system [3] the head of government (the prime minister) is appointed on the basis of the distribution of power in parliament. The majority of the members of parliament must consent to the choice of the prime minister who then decides the composition of his government (his ministers). According to the "parliamentary principle" a government must resign if the majority of parliament votes against it on an important issue. In a presidential system [4] the head of government (the president) is elected directly by the people independently of the election of the legislative body. This means that the president and the government he chooses do not necessarily have a majority in the legislature.

Democracy and dictatorship

The term democracy usually refers to a political system in which people are involved in some way in the ruling of society. A dictatorship is a political system in which the few rule the many. The notion of equality is central to democracy in the sense that in an ideal democratic society all people are supposed to have an equal say in the making of important decisions. But in characterizing existing political systems there is considerable disagreement as to what is the most democratic type of rule. Western, liberal societies would contend that the most democratic system is the one with regular free elections for which any political party may stand. Socialist societies, on the other hand, claim that there can be no democracy unless all are economically equal and have equal say in determining the pattern of production of goods and services.

In view of this conflict of terms political scientists often use other categories to classify political systems. Three types of systems can be differentiated, for instance, according to the ways in which they try to solve conflicts of interest between the various groups in society [5]. Autocracy is a system in which one man or small group rules society and enforces his or its own interests without systematically consulting other members of society. Republican government attempts to bring together

1 The electoral principle of the single-member system [A] is that each geographical area must be represented by the candidate winning most votes. The principle of proportional representation [B] is that each political party should be represented according to its share of the total vote. Alternatively, minorities can be given a voice by some system of distributing their second preferences.

Percentage of total number of votes

2 The same voting support can produce different legislative representation under single-member and proportional representation systems. A country with five constituencies, each with 1,000 votes distributed between three parties might elect under a single-member system, 3 Red, 2 Yellow and no Green candidates while the same vote under proportional representation elected 2 Red, 2 Yellow and 1 Green. Single-member systems restrict smaller parties. As a practical example, the British Liberal Party won 19% of votes in February 1974, but gained only 14 seats, not the 125 possible under a proportional system.

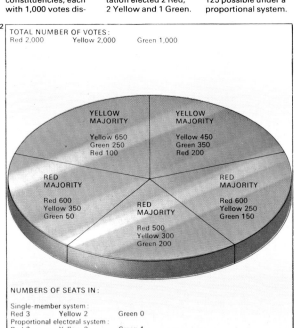

TOTAL NUMBER OF VOTES:
Red 2,000   Yellow 2,000   Green 1,000

YELLOW MAJORITY
Yellow 650
Green 250
Red 100

YELLOW MAJORITY
Yellow 450
Green 350
Red 200

RED MAJORITY
Red 600
Yellow 350
Green 50

RED MAJORITY
Red 600
Yellow 250
Green 150

RED MAJORITY
Red 500
Yellow 300
Green 200

NUMBERS OF SEATS IN:

Single-member system:
Red 3   Yellow 2   Green 0
Proportional electoral system:
Red 2   Yellow 2   Green 1

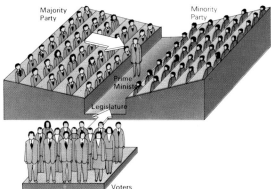

Majority Party
Minority Party
Prime Minister
Legislature
Voters

4 In a presidential system the president is elected independently of the legislative body and is the head of government as well as head of state. He has much more power than the prime minister, president or monarch in a parliamentary system and usually appoints his own prime minister and executive who are responsible directly to him rather than to parliament. Legislative control over the passing of laws, or other checks and balances, curb his power.

3 In a parliamentary system each voter casts one vote and the sum of these votes determines the composition of the parliament or legislature, which in turn determines which party (or coalition) makes up the government and chooses its head or prime minister. The head then selects an executive or cabinet which is collectively responsible for its acts. A head of state (monarch or president) has only nominal powers.

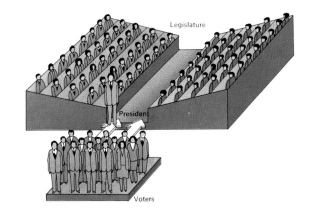

Legislature
President
Voters

the different interests by letting all citizens share in government or in the choosing of the government. Totalitarian government attempts to solve the problem of conflict by creating a society in which no major conflicts will arise; the creation of such a society is based upon a system of ideas (an ideology) that is supposed to guide the actions of the people and mobilize their support for the system. Each of these three categories can be subdivided – to distinguish different types of totalitarian government [6] for example.

Communism, socialism and liberalism

A common distinction between existing political systems is that between non-communist or Western, liberal countries and communist or socialist countries [7]. The difference between these two kinds of societies is basically economic. In the West, the economy is partly capitalist, with the means of production (factories, machinery and so on) largely in private ownership. In a socialist economy the means of production are publicly owned. The differences in economic systems are to some extent

reflected in political systems. Most political systems of the capitalist countries are of the republican type. The state interferes only to a limited extent in production and all political parties are allowed to compete in parliamentary and local elections.

The political systems of socialist countries are often described as totalitarian. They have emerged as a result of a revolution, through which the private ownership of the means of production has been abolished and the state, which is seen as representing the interests of the working people, controls production. Only one party, the Communist Party, is allowed to function; all other parties are considered to be undermining the interests of the working class. The system is based on Marxist ideology, according to which true equality can be achieved only in a society where production is controlled by the working people. Socialism, occurring in the period during which the state is supposed to rule on behalf of the people ("the dictatorship of the proletariat") is seen as a transition towards communism, the stage at which the state is assumed to have withered away.

People, with their capacity for mass loyalty or mass rebellion, are at the heart of any political system. The major differences between systems of government lie in the means by which the will of the people is transmitted to those who govern them and in the methods governments adopt to make decisions on their behalf. Debate about the means and methods of good government is almost as endless as debate about objectives that should be pursued.

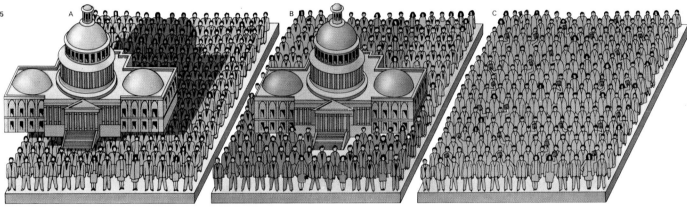

5 Democracies developed from authoritarian systems such as oligarchies [A] in which an elite rule. Modern Western democracy corresponds to republicanism [B]: power derives from the people and is given effect by an executive, legislature and judiciary through a bureaucracy. In the 20th century, a new political system has arisen [C], where only one party exists – it claims to represent all the people and thus to epitomise democracy.

6 Totalitarian systems governed Germany during the Nazi era of Adolf Hitler [A] and the Soviet Union, particularly under Joseph Stalin [B]. Both mobilized the people in support of the state. But the systems differed both in their economic bases and also in the ideologies used to mobilize the consent of the masses. Germany was economically capitalist and ideologically fascist. The Soviet Union remains economically socialist and ideologically Marxist.

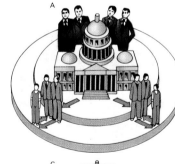
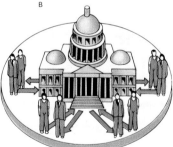
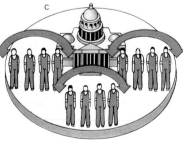

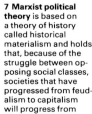
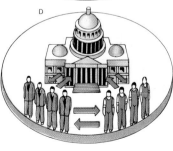

7 Marxist political theory is based on a theory of history called historical materialism and holds that, because of the struggle between opposing social classes, societies that have progressed from feudalism to capitalism will progress from capitalism [A] to socialism [B] and finally to communism [C]. Many Western scholars believe, however, that capitalism and socialism are becoming more and more like each other in the sense of being "mixed economies". Adherents of this "convergence thesis" [D] point to the fact that in many socialist countries, for instance the Soviet Union, some private production has been allowed, to encourage higher output. At the same time, in capitalist countries the state is increasingly interfering to regulate the economy. Therefore many hold that eventually the two kinds of society will become alike with a certain amount of private enterprise and a certain amount of state enterprise run by some kind of multi-party system.

Political participation

People can take part in politics directly by holding public office or by being active in political parties and pressure groups or, indirectly, by exercising the right to vote [3, 4,]. The constitution of a country establishes the institutions through which political power is exercised, the machinery for passing laws and administering policies, the qualifications for public office, the method of election or appointment and the composition of the electorate. Autocratic or oligarchic systems restrict participation to very few or to a minority [6], but in a democracy all adults are usually able to participate. Social and economic factors, however, as well as constitutional and legal ones, often determine how real a degree of political participation a society provides.

Parties and pressure groups
Most democracies have representative bodies [Key] whose members are responsible to those who elect them and act in their interests. The more control electors have over their representatives, the more chance they have of real political participation [1].

Frequent elections and the ability to recall representatives gives electors greater control. So do committee systems such as that operated by the US Senate which expose the policies and actions of officials to public inquiry. An alternative means of control is the establishment of political units small enough to allow citizens to participate directly [2] – a form of democracy that some believe could offset the tendency for government to become too remote and complex for the individual to understand or influence.

Organized political parties have developed to further the aims of those holding basic political beliefs in common. The parties provide a forum for discussion, machinery for political education and propaganda, and a method of achieving political goals by evolving a party policy and supporting candidates to implement it [9]. Participation in parties can range from passive membership, and occasional fundraising and electioneering, to membership of policy committees or the holding of public office. While parties further the interests of their members on a broad front and on a

permanent basis, pressure groups organize political participation on specific issues or promote the interests of a particular group. When a specific goal has been achieved such pressure groups often cease to function. Pressure groups cover the spectrum of political activities and may operate behind the scenes or by public campaign.

Assessing participation
In a democracy, the individual may take part in politics within a party or pressure group, or independently by taking an interest in public issues, voting in elections, watching his representatives' actions and perhaps even standing for political office himself. These forms of activity can take place only under certain conditions. The individual needs the ability and freedom to organize, discuss, publicize, obtain and disseminate information and criticize or question the existing rulers, policies and political institutions. Even when such conditions exist it is difficult to assess the amount of political participation. The percentage of the adult population who vote in an election [5], for example, is an indicator

1 **Ancient Greek city states** provided the first examples of direct democracies. In Athens, all citizens formed the legislature and participated directly in political affairs. They were actively engaged both in decision-making and details of administration. Slaves, foreigners and women were excluded.

2 **Direct participation by the people** in the making of political decision works in small-scale units such as this Chinese commune, or in a Swiss canton, or an Israeli kibbutz. It is, however, often impractical in societies organized on a larger scale where a system of elected representatives is much more efficient.

3 **Casting a vote in a plebiscite** or referendum, or at an election, is the most widespread form of political participation – often the climax of political discussions and propaganda in the media.

4 **Mechanized voting machines,** such as this American model, reduce the possibility of fraud and improve the speed and accuracy of obtaining election results, particularly when there is a complex voting system. Technical innovation can increase the

level of political participation; instant voting machines could reflect public opinion on many issues. But such a "populist" system could undermine representative democracy and the carrying through of wise but initially unpopular policies.

5 | Argentina | 80·9% |
Finland	83·0%
Hungary	97·2%
India	51·3%
Iran	66·7%
Italy	92·9%
Japan	71·1%
Netherlands	87·9%
New Zealand	90·5%
Nigeria	28·0%
South Africa	73·0%
Soviet Union	99·9%
Tunisia	96·7%
United Kingdom	77·1%
United States	58·7%
West Germany	86·8%
Yugoslavia	90·9%

Free, competitive elections · Arranged elections
Interference in elections · Voters as a % of the electorate

5 **Voting patterns** in a number of countries at recent elections indicate formal political participation. But variations in the number of adults who cast a vote do not necessarily measure political apathy or enthusiasm. They are often affected by a legal requirement to vote or limitations on freedom to do so.

of participation, but voting may simply be a formality or a compulsory obligation. A low poll might indicate widespread apathy or an absence of contentious issues, or it might be the result of an organized boycott demonstrating opposition to that particular poll. Membership strengths of political parties may be equally misleading if party membership is socially desirable or a consequence of trade union membership. Such indicators of political participation must therefore be treated with particular care.

Political frustrations

Democratic institutions of government and a free political system do not themselves guarantee wide and effective participation in political decisions. For the latter to occur, people need a certain level of political knowledge, leisure time, a consciousness of their political rights and a belief that their participation is worthwhile.

In modern industrialized societies the institutions of government are often complex, remote and inaccessible. Political issues that are not straightforward are often ob-

scured by jargon and presented as matters best left to "experts". Political parties, as governments themselves, tend to become bureaucratic and hinder new ideas. The business of government and the policies of public officials may be difficult for political representatives to understand or influence, let alone the public. In such circumstances individuals may be reluctant to participate in politics even if qualified to do so.

Political participation is lower among the illiterate and less well educated. People who believe that political power is the prerogative of superior groups or classes tend to be apathetic about politics. In some countries such apathy may be officially encouraged. This kind of situation has often led to a reaction, however. When existing institutions and channels are inadequate to the political needs of a community the people may resort to direct action [10]. Popular revolts, revolutions or other forms of mass political action [7] may lead to the establishment of more democratic forms of government and a higher level of genuine political participation by a better-educated community.

The British House of Commons, the "Mother of Parliaments", has been a model for many other legislative assemblies, providing a forum for the elected representatives of the people to frame new laws and to put the running of government to the test of open party political debate.

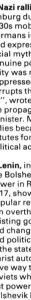

6 Nazi rallies at Nuremburg during the 1930s mobilized Germans in a ritualized expression of racial mythology, but genuine political activity was ruthlessly suppressed. "Politics corrupts the character", wrote Goebbels, the propaganda minister. Mass Nazi rallies became substitutes for real political activity.

7 Lenin, in leading the Bolshevik Party to power in Russia in 1917, showed how a popular revolution can overthrow an existing government and change the social and political nature of the state. The tsarist autocracy gave way to elected soviets which soon lost power to the Bolshevik Party.

8 Suffragettes in Britain fought successfully for the right of women to vote and participate in politics. The Women's Movement of the 1970s also seeks to extend participation in political affairs.

9 The Democratic Party Convention in Chicago in 1968 was an example of a party organization working through established channels to frame a policy for an impending election and to choose candidates for high office to implement it.

10 Political demonstrators who clashed with police outside the Democratic Convention in 1968 were dissatisfied with established channels of political participation. Direct action of this kind is often the result of frustration be- cause formal institutions are responding to an unpopular situation too slowly or not at all. The 1960s saw an eruption of worldwide protest movements in opposition to government policies and were met with varying state repression.

Machinery of government

In every political system there is a central body called the government which is responsible for the functions of the state. It must make provision for external defence and internal order, see that laws are administered and collect the revenue needed to finance these state activities.

The modern state

In the modern state governments are usually responsible for a much wider range of activities, including the provision of social services in education, health and housing. Even under a capitalist system based on private enterprise, the government regulates general economic trends, encourages export industries, helps low employment areas, aids research, controls international trade and adjusts the money supply. In a socialist state with a "command economy" government activity extends further to running all major industries as national state enterprises.

The structure of government usually reflects a division of responsibility both between national and regional government (or federal and state government) and between legislative, executive and judicial areas under the theory of "separation of powers" [Key]. However the government machinery is organized, a central body is needed to give coherence to government policy as a whole and to co-ordinate the activities of individual departments. Each department may be responsible for a particular government function, or for the supervision of a particular group of citizens, and will have its own administrative structure. Departmental activities may be co-ordinated through the central or cabinet office, or through a complex of inter-departmental committees.

Members of the political executive are each made responsible for a particular area of government activity by the head of the government. These members, often called ministers, become answerable for their department's policy and its shortcomings [1–21]. The machinery of government itself is staffed by civil servants who are employees of the state. Little decision-making will be required of civil servants engaged in the purely clerical or low-level administrative routine of government work. Similarly, much technical and scientific work may be merely routine testing, to enforce government standards. But at the highest levels civil servants work with political heads and suggest ways in which political decisions can be carried out and government policy implemented. The satisfactory functioning of government depends on the civil servants at this level.

High-level civil servants

Ministers, as politicians, take charge of government departments for given periods only but civil servants are often permanent officials and it is on their advice and administrative abilities that good government largely depends. These high-level civil servants may be recruited from the ranks of university graduates and selected for their general intellectual abilities; or they may have received their training at a college for administrators. Further training in public administration may be undertaken while they are in government service.

Management theories and familiarity with the quantitative techniques of economics and other relevant social sciences

1 A new minister of transport is appointed by the head of government to take charge of policy.

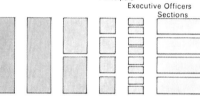

2 Minister
Private Secretary
Permanent Under-Secretary
Deputy Under-Secretary
Assistant Secretaries
Principals
Executive Officers
Sections

2 The ministry is made up of several sections responsible for different activities in different parts of the country, each with its head or principal. Section activities are co-ordinated by various officials who report to the minister.

3 In a meeting with his section heads the minister proposes a new motorway to help a region.

4 A departmental committee of heads of sections frames a report with three possible schemes.

5 Other ministries with related interests are consulted about the three schemes at an inter-departmental meeting.

6 Local government officials are consulted about the impact that the new road is likely to have on their own areas of responsibility. They give it their agreement in principle.

7 The minister receives a report on the merits of the schemes, chooses plan B and sends draft proposals to other ministries.

8 Treasury officials in the ministry of finance read the proposals with horror and advise their own minister to oppose the scheme as being too costly in the light of other government commitments and available funds.

9 A ministerial committee considers the plan outlined by the minister of transport and the objections of the minister of finance but adopts the plan as policy.

10 At a press conference the minister reveals the plan. Publicity is handled by his public relations section.

11 Some local residents, fearing that the road will spoil their neighbourhood, form an action committee to fight the scheme.

are the necessary background knowledge of the modern senior civil servant. Management and organization theory will be used not only to increase efficiency within a particular department but also on a larger scale to help plan the machinery of government itself.

As the political complexion of the executive changes, permanent civil servants will have new political masters and may have to administer new and even opposite policies to those that they have been following. A certain degree of political neutrality is therefore required, for as part of the government machine civil servants do, as their name suggests, serve the state. In some countries, notably the United States, key public service posts are more closely linked to political affiliations and may reflect electoral changes.

Control and accountability

Governments spend vast amounts of money, take decisions that provide large profits and make appointments that bestow power and prestige. It is therefore necessary to safeguard against corruption and dishonesty within the government machine. The task of ensuring honesty, efficiency and fairness may be undertaken both internally and externally. Internally the procedures for decision-making and administration are carefully designed to ensure consistency and secrecy where necessary; full records enable government processes to be traced. The machinery of government usually has its own unit to monitor efficiency and promote improvements in standards of administration and personnel. Occasionally special commissions may be appointed to survey the machinery of government, in part, or as a whole.

Members of the public may be protected against arbitrary or illegal administrative action through civil administrative courts. Many countries appoint an independent ombudsman who may investigate charges of maladministration and provide relief. Investigative committees of a legislature may bring to light misuse of the government machine. Similarly, public exposure of government processes by the mass media can play an important part in controlling excesses of bureaucratic behaviour and analysing the efficiency with which public funds are spent.

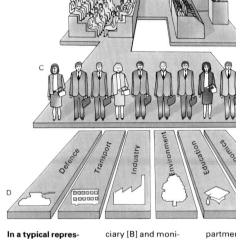

In a typical representative democracy the legislature [A] passes laws administered by the judiciary [B] and monitors the actions of the executive [C], which governs the country through departments [D]. The story below shows how these relationships work within a government machine.

12 Representatives of the action committee appeal to the courts, seeking an injunction to stop motorway work. Meanwhile, various sections within the ministry begin work to implement the plan, which has been endorsed by most local opinion including the press. The legal section uses powers under existing laws to secure purchase of the land required. Technical civil servants prepare details of construction plans. The finance and accounts sections then accept tenders and appoint contractors to undertake construction.

13 In court the ministry's actions are declared to be within the law.

14 Political pressure is put on the minister when the action committee organizes a campaign of adverse publicity based on the road's effect on the environment.

15 In parliament the minister, briefed by his civil servants, defends his action, outlines the complex consultations behind the final plan and shows the decision to be consistent with long-term government policies.

16 The minister opens the completed road.

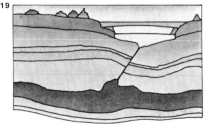

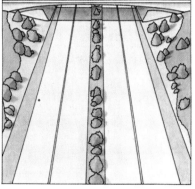

17 Serious subsidence is discovered some months later.

18 The minister orders a departmental enquiry.

19 Following intense press criticism the minister is asked to account for the subsidence. Addressing parliament on the basis of his departmental report he blames geological movements for causing the damage. But his performance under close questioning is unconvincing and raises doubts about his grasp of departmental work and his ministerial ability.

20 A local resident lodges a grievance which has to be heard before an administrative tribunal. The complaint is that he was inadequately compensated for loss of part of his property to the motorway and that he has been further affected by geological disturbance as a result of drilling during its construction. The tribunal examines the ministry files, exonerates the civil servants from any negligence, but recommends a review of the compensation paid to the complainant and of the method of geological survey.

21 A special commission is set up to hear expert evidence on the adequacy of government surveying techniques and to advise whether extra safeguards should be adopted to prevent a recurrence of the motorway subsidence. The recommendations of the commission lead to a restructuring of the transport department and a strengthening of its geological survey section. In a ministerial reshuffle, the minister is transferred to a less important department.

Rule of law

It is said that "the law and the courts are often in error, but they are never in doubt!" The law demands to be obeyed and its sanctions can be unpleasant: a fine, imprisonment or some other social disability. However, we are not usually conscious of the penalties the law can impose [1]: we accept the law because it upholds our chosen way of life.

Law and justice

There are many arguments about the source of legal authority. The eighteenth-century French philosopher Jean Jacques Rousseau (1712–78) felt that to merit obedience the law must have the status of a social contract freely agreed by free citizens. In direct contrast the English jurist John Austin (1790–1859) argued that laws are basically nothing more than a series of commands from the ruler to the ruled. At the same time in Germany Friedrich von Savigny (1779-1861) described law as a thing that grows naturally out of a nation's spirit, environment and history. And infact, every nations legal system has some characteristics which are distinctively its own.

Although laws vary from place to place certain concepts are basic to almost all legal systems. Perhaps the most important is the idea of justice: the desire to balance fairly the needs of the individual against the needs of society plus the desire to find a fair balance between the interests of one individual and those of another. The borderline between these two endeavours is, broadly speaking, the line of distinction between public law and private law.

One difficult problem in the search for justice was epitomized by William Blake (1757–1827): "One Law for the Lion and Ox is Oppression." The law that is fair to the lion may be unfair to the ox, and vice versa. But lawmakers cannot produce individual laws for each member of the community [3]. They have to legislate for the whole society. Many legal systems have felt the need for mechanisms to remedy such injustices as result. In medieval Europe the Church courts applied a system of equity to protect individuals from legal unfairness. And in imperial China judges were allowed to apply the law in a flexible way that took account of individual

circumstances. A simple everyday example of equity in action is that fire-engines and ambulances, for example, can break speed limits and other traffic laws in emergencies but not in other circumstances.

Law in ancient times

Legal systems seek certainty. Once early men had learned to write they tried to make their laws certain by writing them down. Later they constructed codes – systematic collections of legal rules – which had the advantage of making the community's laws clear and easy to refer to.

One of the earliest legal codes known to us is the Code of Hammurabi (c. 1792– c. 1750 BC), a king of Babylon. Its 300 laws deal in a matter-of-fact way with exactly the same kind of legal matters that exist in modern society, such as sale and purchase, inheritance, employment, marriage, theft and manslaughter.

A legal code of a different type is the one that – as the Bible recounts – Moses brought down from Mount Sinai in about 1200 BC as a law for the Israelites [2]. The Ten

1 Why do we obey the law? One powerful reason is the wish to avoid the sanctions by which law is enforced [A]. Another is that it is custom to obey [B]. For many people the law is also morally right [C].

2 Moses was given, by God on Mount Sinai, the tablets of law on which, according to the Bible, the Ten Commandments were inscribed. The law can be seen as based on fundamental moral principles.

3 Mr Bumble, in Charles Dickens's Oliver Twist (1837–9), declared, on being told that he was answerable for his wife's actions: "If the law supposes that, the law is a ass – a idiot." This statement has been echoed by many litigants and is the basis of the principle of equity which seeks to avoid wrongs resulting from strict adherence to the letter of the law. The process of law-making and judgment attempts to make the unfair fair.

4 Solon the Lawgiver, an Athenian statesman and poet (c. 640–c. 559 BC), tried to create a just society. Given power to change the law he reorganized the community, cancelled unfair debts and carried out many other reforms to improve the lot of the people. The dilemma of balancing the rights, duties and conflicting needs of all members of society is seen in his lack of success. Ultimately Solon managed to please few Athenians.

Commandments are essentially a body of principles. They enshrine ideas of morality that have subsequently helped to shape law in almost every part of the world.

The ancient Greeks tried to humanize law. They developed the idea that rules should be changed when they ceased to meet the needs of the community. This idea seems commonplace today but in early society laws were seen as God-given, fixed and immutable. The great thinkers of ancient Greece, including Socrates, Plato and Aristotle, also concerned themselves with the quality of law and its moral standards.

Some Greek ideas were adopted by the Romans. But the Roman genius was essentially practical. The lawmakers sought primarily for order and efficiency in the administration of their territories. Henry Maine (1822–88), writing of Roman law, said: "The most celebrated system of jurisprudence known to the world begins, as it ends, with a Code." He meant that it began with the rudimentary Law of the Twelve Tables in 450 BC and ended with the *Corpus Juris Civilis*, the complex collections of laws and doctrines made by the Emperor Justinian I in the sixth century AD. But Roman law did not really end there; much of its substance still exists in contemporary legal systems.

Modern legal systems

The modern world has hundreds of legal systems, but many of them have drawn principles and methods from the same sources and, for this reason, can be grouped together. The two largest groups are those with a major civil law component, and those with a major common law component [7].

Civil law systems utilize the experience and ideas of Roman law. They are found in most of western Europe, in South America, in parts of North America, Asia and North Africa, in South Africa and in the Soviet Union. Common law systems derive from the common law of England and are found in most English-speaking countries.

Contemporary systems draw from innumerable other sources, and are shaped by such influences as the teachings of Islam, political or economic theory and recent advances in jurisprudence and sociology.

Justice, with her sword and scales, represents the power and impartiality of the law. Often she is shown blindfolded to indicate that she is blind to prejudice. Legal systems may strive to give effect to the principles of reason and morality that constitute justice, but pure justice is unattainable.

5

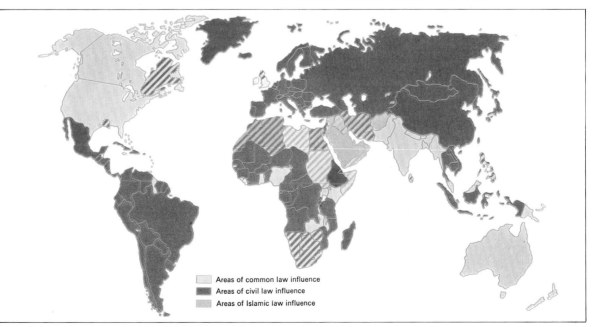

We the People

of the United States, in...

Article I

5 **"We the People"**, the famous opening words of the Constitution of the United States of America, were intended to indicate that the legal authority in the new country derived from its citizens, not from kings or other rulers. The Constitution was signed on 17 September 1797 and, after ratification, became the supreme law of the USA. It has been a model for the constitutions of many other countries. In most countries that have written constitutions, constitutional laws are specially protected. They can be changed only by procedures stricter than those employed for changing the other laws of the land.

6

6 **Napoleon I** (1769–1821), who is remembered chiefly as a military genius, ordered the compilation of the French code of law called the *Code Napoléon* or *Code Civil*. Promulgated in 1804, it was the first of the great modern legal codes. Based on existing French law and Roman law it contains the civil as opposed to criminal law of France and was a compromise between the customary law of the north and Roman traditions of the south. Pre-revolutionary elements co-exist with postrevolutionary innovations. The *Code* has strongly influenced the legal systems of many countries of Europe, Asia, North America, South America and Africa.

7 **Legal systems of various nations** show different influences. Civil law systems derive from Rome. Their pedigree runs through the *Code Napoléon*, the law schools of Bologna and other ancient universities and the Canon Law of the Christian Church to the *Corpus Juris Civilis*. In common law judges are bound by certain civilized assumptions, such as that all men were equal before the law. They also have regard to precedent – the recorded reasoning of their brother judges in similar cases. It is for these reasons that, over the years, judicial decisions were fused together to form a coherent system. There is much cross-pollination between the two systems.

7

Areas of common law influence
Areas of civil law influence
Areas of Islamic law influence

Sources and divisions of the law

In a democratic country, the legislature – the central law-making body, such as the National Assembly in France or the Knesset in Israel – has power to make almost any law it wishes, and to abolish or change any existing laws. The only major restriction on legislative power is where the fundamental laws of a country's constitution are specially protected. In federal countries the legislative power is shared between a number of law-making bodies. In the United States, for example, it is shared between the federal Congress and the legislatures of each state.

The lawmakers

In practice, the legislature is directly concerned in the making of only a small proportion of the laws by which people live. The life of a modern community is so complex that the legislature could not possibly find the time – even if it had the knowledge – to make all the rules needed for orderly existence. Usually, it has to confine itself to making laws, called statutes, on only the most important matters or on broad issues. To deal with the great bulk of day-to-day matters, it delegates some of its powers to government ministers and to other bodies.

An important example of this delegated legislation are the local laws made by local authorities in many countries relating to such activities as traffic control, education and the sale of food and drink. Similarly, water authorities may make laws about water supplies, electricity companies about electricity supplies and railway companies about safety and even about the behaviour of the public when using the railway. Taxation, social insurance, labour relations, air travel and agricultural marketing are among numerous other matters that come within the scope of delegated legislation in one country or another. The laws made by these various bodies have just the same force and authority as laws made by the legislature directly.

Many laws have their origin in judicial decisions – that is, in decisions made by judges when presiding over trials in court [1]. In countries where trial judges are obliged to follow precedents set by earlier judges, these decisions become part of the established law. But judge-made rules can, of course, be altered at any time by acts of the legislature.

Another important source of law is custom, the rules that have "always" been followed, but whose origins are obscure. As legal systems develop, customary rules tend to be incorporated in statute law or judge-made law. The writings of jurists – that is, legal scholars – are also an important source because they have an influence on judges.

Classification of laws

The ancient Romans divided their rules of law into public law and private law. This method of classification is still a convenient one today [Key]. Broadly speaking, public law includes all rules relating to matters in which the state or the community is directly involved. Private law – again broadly speaking – relates to private persons; that is, it deals with the relationship between one individual and another. There are occasions, such as in cases of contract and tort, when corporations (regarded as artificial persons) and even state institutions can be sued as if they were private individuals.

The major branches of public law are

CONNECTIONS

See also
282 Rule of law
286 Law in action
280 Machinery of government
274 Political science

1 **As society becomes more complex,** rules and regulations increase in number. To some critics it seems that modern man lives in a torrent of law-making. The rules that apply to any particular situation may come from one or more of many sources. They may belong to a written constitution as do many of the laws in the USA, or they may be part of a systematic legal code, such as the *Code Napoléon* in France. They may also be among the tens of thousands of laws in the country's statute book, or the even greater number of by-laws, orders or regulations made by delegate bodies such as local councils. They may be found in reports of judicial decisions, as in the English system of case law whereby a judge's decision, if it creates a precedent, may create a rule of law. Such rules may also be derived from legal writings and custom.

Constitution
Legal code
Statute law
Delegated law
Precedent
Legal writings
Custom

2 **Criminal and non-criminal acts** are distinguished by the legal procedures they set in motion. Some non-criminal acts are more harmful than others that society punishes as criminal. But, despite juristic arguments about the nature of crime, the position of the criminal has always been clear: most members of society dislike and fear him. Today, because of the pressure of living and increased regulation, traditional attitudes are modified. A large number of acts, such as minor traffic offences, that are treated as criminal, do not incur general social disapproval.

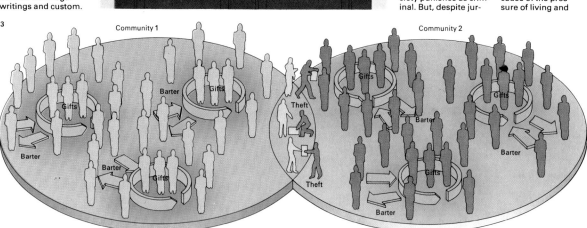

Community 1   Community 2

3 **Any human act** may have several levels of significance, each of which has to be comprehended by the law and custom and each of which requires its own rules. The transfer of property from one individual to another may be regarded as a gift when the relationship is close. At a greater remove it may be an exchange or a sale. Between two communities, where there is no mutual awareness of bond between the individuals, robbery or cheating may be involved.

constitutional law, administrative law and criminal law. A country's constitutional law consists of the fundamental principles underlying its life, such as the equality of all citizens before the law and the right of free speech and liberty of conscience. It also includes the laws that regulate the various organs of government. In this respect it overlaps to some degree with administrative law, which deals with the day-to-day administration of the country. Criminal law is concerned with actions (or failures to act) that are considered harmful to the community and that are punishable by law.

The rules of international law [5], dealing with relationships between countries, are also included in public law. So are the rules affecting the conflict of laws [4] that occurs when legal issues are affected by the laws of more than one country.

Private or civil law

Private law is sometimes called civil law, a term that is used in a different sense to indicate legal systems derived from Roman law. The chief branches of private law are the laws of property, of status, of contract and of delict. Contract and delict are sometimes grouped together as the law of obligations. The law of property deals with the ownership and possession of land or buildings and with related rights. It also deals with inheritance and with rights over such things as trademarks, patents or copyrights.

The laws of status regulate such relationships as marriage, parenthood and guardianship. They also determine the status of persons who have a special legal position, such as minors, lunatics and aliens. The law of contract is basic to modern society. It governs agreements entered into "for money or money's worth"; such agreements are legally enforceable. And the law of delict (called "tort" in some countries) deals with wrongful actions for which the injured party can claim money to compensate for personal injury. Examples of delict are defamation of character and physical injury caused by another person's negligence. In substance, there is often little or no difference between the type of injurious action that is treated as a delict and the type treated as a crime.

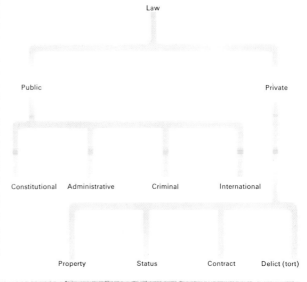

KEY

The law can be roughly divided into two parts – public and private. Public law deals with matters in which the state or community has an interest, while private law regulates the actions of individuals towards each other.

Law

Public — Private

Constitutional — Administrative — Criminal — International

Property — Status — Contract — Delict (tort)

4 A conflict of laws arises whenever an occurrence subject to litigation is affected by the laws of more than one country. An example is a collision at sea between ships of different nationalities, possibly within the territorial waters of a third country. Most of the larger countries have reached some kind of international agreement about responsibility in such incidents.

5 International law is a flimsy and piecemeal structure painfully inadequate to deal with the circumstances of the modern world. Ease of travel and increasing contact between peoples have brought new problems. Not least is that of hijacking. One such incident, in September 1970, ended with Palestinian hijackers blowing up three planes in Amman, Jordan.

6 The breakdown of social cohesion in a country leading to continuing street violence and a state of "undeclared civil war" can put strain on the legal system. One such example was the student rioting in the streets of Paris in 1968.

7 Space-age travel has added a new dimension to legal imponderables: who owns space? The US astronauts who landed on the Moon in July 1969 planted their country's flag on the surface. It can be assumed that the astronauts were not asserting American ownership of the Moon or even the portion of the surface on which their spacecraft landed. However, sooner or later some regulation of space and of the reachable planets will be needed. Areas of sovereignty and "spheres of influence" may be vaguely defined. But the real test will come when some country feels obliged to demonstrate its ability to protect the rights to which it lays claim, possibly by force.

Law in action

The French writer Voltaire (1694–1778) complained that the law comes to life only when something goes wrong. The ultimate test and expression of the law is a trial in court – a contest between two parties, only one of whom can win [Key]. The trial may be a criminal one involving a prosecutor who acts on behalf of the community, and an alleged offender who faces punishment if found guilty. Or it may be a civil trial between two private persons, one of whom claims compensation for injury said to have been caused directly or indirectly by the other.

Finding the facts

Every trial has two essential elements. The court has first to establish the facts of the case and then find the legal rule that applies. If the court consists of a jury as well as a judge, the jury decides questions of fact and the judge decides questions of law.

The most difficult problem is establishing the facts – discovering what actually occurred or what was omitted that should, legally, have been done [1]. The court depends on evidence given by the parties to the case and by witnesses who have information concerning the points at issue. A distinction is made between direct evidence, the testimony of an eyewitness, and circumstantial evidence – evidence of fact from which the facts in dispute may be inferred [6]. A witness who says: "I saw the accused throw a bottle through the window" is giving direct evidence. A fingerprint expert who says "I found the accused's fingerprints on the bottle that was thrown through the window" is giving circumstantial evidence. Often, many pieces of circumstantial evidence have to be fitted together to prove a single point.

Some people feel that the value of direct evidence is greater than that of circumstantial evidence. But witnesses, even the most convinced witnesses, are notoriously inaccurate in identifying people they have seen and in describing events they have observed [7]. Moreover, unknown to the court, an eyewitness may be drawing entirely false conclusions from mistaken ideas. One such witness, a bystander at a traffic accident, stated positively that she had seen the accused drive his car through a crowded street at 50 miles an hour. No amount of persuasive questioning could shake her testimony. But eventually it occurred to the accused's lawyer to ask: "How fast do you walk?" To which the witness replied with complete assurance: "About 20 miles an hour." Such inaccurate evidence can be hard to disprove.

Rules regarding evidence

Most legal systems have rules governing the types of evidence that may be given in court and the way in which evidence is to be presented. In criminal cases, for example, precautions are taken to ensure that a confession by the accused is voluntary and not the result of threats or promises of leniency. A witness may not be allowed to express an opinion – as distinct from explaining a fact of which he or she has knowledge – unless it is an expert opinion, such as a doctor's on medical facts. If a document is used its authenticity must be established as evidence before its contents are admissible as evidence. These rules are more strict in common-law systems than in civil-law systems, which often allow judges more scope for using their discretion.

CONNECTIONS

See also
296 Changing patterns of punishment
282 Rule of law
284 Sources and divisions of the law
288 Nature and causes of criminality
290 Changing patterns of crime

1 Establishing the real facts of a case with certainty is a problem that has baffled people for thousands of years. In the past, torture [A] was sometimes used to extract confessions from accused persons. Sometimes, champions [B] fought as representatives of the parties to a trial; it was hoped that God would not permit the just to lose. The contemporary method is to put a witness on his solemn oath or affirmation to tell the "whole truth" [C]. And the decision may be left to a jury [D] on the assumption that ordinary citizens should recognize the facts when they hear them. In technical cases, expert assessors [E] may be called in to assist the judge and jury in arriving at their conclusion.

2 The advocate's task is to put his client's case in the most favourable way possible within the law: as the client would put it for himself were he familiar with the law's technicalities. He also seeks – by cross-examination, for example – to show weaknesses in the opposing side's case. In a criminal case, the advocate need not consider whether his client is innocent or guilty. That is the function of the court. But he should not argue the innocence of a client who has privately admitted his guilt.

3 A commission of the US Senate met in the early 1970s to investigate charges of wrongful conduct brought against President Nixon during the Watergate scandal. Such commissions are usually appointed in circumstances involving grave public issues. Their purpose is purely exploratory. They often follow trial procedures, but have wider discretion than an ordinary court. Unlike courts they may be televised.

4 Judges discharge their duties in ways that vary from country to country. In common-law countries the judge's role is more that of an umpire seeing fair play between parties [A]. But he may question witnesses. In civil-law countries he acts as an impartial inquisitor [B], whose task is to uncover the truth. Thus he has to take a more active part in presenting and sifting evidence.

There are also rules regulating procedure in court whose purpose is to ensure that both parties are fairly treated and meet on equal terms. Again, judges in civil-law systems are generally allowed more freedom than common-law judges, but the purpose is the same: each side is to be allowed to state its case in its own way – within the rules – and to have a chance to answer the argument put forward by the other side.

Finding the rules that apply

Because of the volume and complexity of modern law, finding the correct legal rules to apply is sometimes only slightly less difficult than finding the facts. The statute book – the record of legislation – in any developed country today contains tens of thousands of rules; and to these must be added the mass of indirect legislation emanating from such delegate bodies as local authorities. Even when the relevant rules have been identified, their precise significance may still have to be established, because few statutes are indisputably clear in their meaning when applied to human situations of the kind

that occupy the attention of the lawcourts. "Interpreting" legislation – deciding its meaning – is one of the functions of the judge in a trial. He will look to earlier legislation on the same topic and will seek to discover what the legislators had in mind in modifying the law. He will also have to study their particular choice of words with great care.

The judge also has a creative role when he finds "a gap in the statute", that is, when there is no existing rule relevant to the points in dispute. He then has to devise a rule that is in accord with the principles of the law. In common-law countries, the grounds on which he decides a case – called the *ratio decidendi* – form a precedent that is binding, in certain circumstances, on judges trying similar cases in the future. In civil-law countries precedents are not binding, but judges often tend to follow them for practical reasons. One strong reason is the desire for achieving certainty and consistency in the law. Another is the reluctance of a judge in a lower court to insist on formulating a principle that he suspects may have a chance of being rejected by other judges on appeal.

KEY

The dramatis personae of a trial include the judge, who presides; the clerk; the recorder; the plaintiff – the person bringing the action – and his advocate; the defendant and his advocate; the jury to decide the facts; the witnesses; and the press and public.

5 Most countries have a hierarchy of courts. These range from local courts dealing with minor disputes and offences to a supreme court whose function is to decide legal questions of the highest importance. In some systems all allegations of crime have first to be examined in a lower court. The more serious cases are then passed on to a higher court for further examination.

Criminal courts   Civil courts

1 Pickpocket
2 Shoplifting
3 Drunken driving
4 Assault
5 Murder
6 Bank robbery
7 Large insurance claim
8 Libel action
9 Fraud
10 Hire purchase debt
11 Dog bites postman

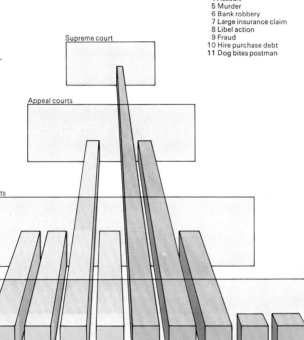

Supreme court

Appeal courts

High courts

Local courts

1 2 3 4 5 6 7 8 9 10 11

6 The police and courts rely on two types of evidence to establish the actual facts of a case – the testimony of an eyewitness and circumstantial evidence. In different countries and legal systems different emphasis is placed on these two elements. In a smash-and-grab raid prosecution, for example, an eyewitness's claim to have recognized the criminal, although seemingly reliable, may be distorted, or countered by a strong alibi for the time of the crime. Circumstantial evidence – the fingerprints of the accused on the shop window, matching blood types, the stolen property being found in the accused's accommodation and so on – often form a stronger case.

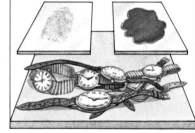

7 In a dispute about the facts the testimony of eyewitnesses can be seriously misleading. Eyewitnesses frequently "see" what they expect to see. Few people will clearly observe or make mental notes of the details. A simple road accident [A], for example, will often produce radically conflicting evidence from the various eyewitnesses [B, C, D], although each believes his version is true.

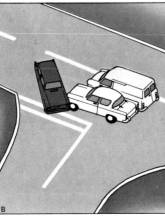

Nature and causes of criminality

Criminal behaviour takes many forms – from murder to petty theft, from treason to misappropriation of funds. There can be no single explanation for such a wide variety of human actions, and attempts to produce a general theory have been discredited.

Theories of crime must explain two quite different facets of the crime problem: first, why certain forms of behaviour are defined by society as crimes and second what causes, entices or compels certain persons to adopt such forms of behaviour.

Crime and its causes

Theories about what constitutes criminal behaviour have varied from the ancient belief that the criminal code represents the embodiment of God's law and is independent of the will of man, to the modern radical idea that criminal law is simply the instrument by which the ruling class entrenches its power.

Explanations of the causes of crime are also varied. There is the theory that assumes that an individual chooses immoral (or criminal) conduct of his own free will and is therefore responsible for his actions [Key]. Other theories portray the criminal more or less as the helpless pawn of biological, psychological or social forces beyond his reason, understanding or control [2, 3].

Traditional explanations of criminal behaviour have tended to focus upon the individual and on particular attributes of his personality that supposedly cause his criminal behaviour. Moralistic explanations emphasize his "evil" qualities such as greed, envy, corruption, revenge or the like and assume that the individual "gives in" to such feelings of his own free will – that he deliberately chooses to behave criminally. But, at the same time, it is recognized that in some cases criminal acts are performed by the insane, who therefore cannot be held responsible for their actions. This tradition has been incorporated into modern legal systems by the acceptance that psychological stress or disturbance can diminish the guilty party's responsibility for his actions. Included are states caused by brain damage, psychotic states such as schizophrenia and manic depression, neuroses such as hysteria and various psychopathic personality disorders.

Modern theories of crime rely more on statistics than on firmly established causal relations. Many studies have established, for example, that burglary and assault are most associated with the poorer areas of cities [1].

Crime, class and circumstance

Not all crime can be accounted for by unfavourable socio-economic factors. "White-collar crime" – crimes such as fraud, embezzlement and tax evasion – committed by people of respectability and high social status [5] in the course of their jobs, is on the increase. Murders of passion and jealousy occur within all social classes and in rural as well as urban areas. Vandalism, drunken driving, forgery and counterfeiting are evenly spread over cities and rural areas.

Often, in crimes of violence, consideration must be given to the role of the victim. There is ample evidence that a victim may provoke or even seek out his own murder. That there are cases in which the role of the offender and victim are reversed and the victim causes the crime – the American criminologist Hans von Hentig's claim – is

y

1 Statistical studies to determine which areas have the highest crime rates in cities, and to identify the conditions of life that are most typical of those areas, have been conducted for over a century in Europe and for many years in the USA. The findings of these studies and the picture that they draw are remarkably and depressingly consistent. Over and over again it is seen that burglary, robbery and serious assaults occur most frequently in those areas with low income, buildings that are tumbling down or in very bad repair, and concentrations of several different races and cultures. The houses are often extremely overcrowded, with few facilities. Owner-occupation is rare and population density high. The people generally have little or no formal education and are mostly unskilled labourers. There is a high proportion of single males and unemployed persons. Families are often broken up, with the parents either divorced or separated. Consequently, the mothers have to go out to work and leave their children behind. Health is generally poor, with high rates of tuberculosis and infant mortality. It is not surprising that such conditions, still common even in the most affluent countries, are associated with crime, although to what extent removing these conditions would eradicate crime is open to question. Criminologists have put forward several explanations for the fact that once such an underprivileged area is established with a criminal subculture it readily encourages crime. By a process called "dissociation" the potential criminal becomes part of an already established group and adopts their often antisocial values and standards.

now well established. Sexual crimes may be the result of economic necessity, as with some prostitution, or they may be associated with mental abnormality, as in the case of the exhibitionist or the child molester. Drunkenness is one of the most common causes of criminal behaviour, for excessive alcohol produces faulty judgment and impulsive behaviour – significant factors in many crimes. Drug addicts often steal in order to obtain money to buy narcotics.

The "training" of the criminal

Crime is one of a large variety of nonconforming patterns of behaviour. As such, it cannot be explained solely by the failure of social control over man's innate urges, although many theorists investigating the causes of crime believe this. Rather, as observed by the American Robert K. Merton, society and its values exert a definite pressure upon some people to be nonconforming rather than conforming. In this view deviant or criminal behaviour is regarded as a symptom of a gap between the aspirations that society encourages and socially accept-

able ways of realizing those aspirations. Many individuals, in response to pressures or frustrations or because of discrepancies between ambitions and possibilities, break the laws of society and commit crimes.

Yet certain forms of crime represent a special type of professional career. In common with members of other, more acceptable, professions, the professional criminal requires special training and recognition by others in his profession. In this way criminal behaviour is learned behaviour, acquired through social interaction.

Criminal behaviour may, on the other hand, be a successful conformity to the values of a delinquent sub-culture. Such a subculture group becomes a collective enterprise in crime, opposed to the institutions and values of the greater society [4].

It is clear that criminal behaviour, as all behaviour, is caused by, or correlated with, a vast number of social, psychological, economic, political, legal and moral factors. It is as futile to search for the one cause of crime as it is to account for all law-abiding behaviour with a single explanation.

The Devil was traditionally seen as an almost physical force as shown in this 14th century French manuscript of the temptations of a nun. Criminal behaviour was thought to be the result of a deliberate choice between the individual's good and evil instincts. Modern theories stress various forces that excuse criminal acts. Whether society can afford to recognize such mitigating circumstances is still hotly debated.

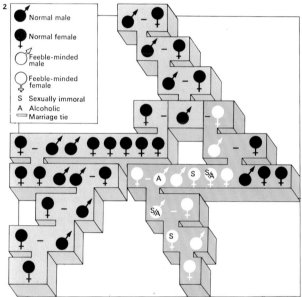

S Sexually immoral
A Alcoholic
— Marriage tie

Normal male
Normal female
Feeble-minded male
Feeble-minded female

3 Various psychological theories have been advanced to explain why some people are prone to crime and others not. One such theory is that the criminal is the victim of a "crime neurosis" [A]. He suffers from great internal conflict and in committing the crime he is acting out that conflict. One American psychologist has put forward the thesis that juvenile delinquency is the result of the offender acting out the forbidden wishes of his parents [B]. In Freud's theory of personality [C] the individual follows the dictates of his id (his fundamental drives or desires) if his superego or conscience is not properly developed.

2 The idea that criminals are born degenerate and that "bad blood" is caused by "bad genes" was popular at the beginning of this century. One of the most famous case studies, published in 1912, was of the American Kallikak family, descended from a Revolutionary soldier. The "good" clan [blue] resulted from his marriage to a Quaker, the "bad" [green] from his union with a feeble-minded girl. Their son, "Old Horror", had five "bad" children and, among the hundreds of their descendants traced, a remarkably high proportion had criminal or antisocial tendencies. Studies in prisons and mental hospitals have linked male criminal behaviour with an extra Y chromosome but the results are inconclusive.

4 Antisocial behaviour can be defined as the breaking of social rules or norms. It can also be the strict adherence to the norms of a delinquent sub-culture. These Hell's Angels, as a group, solve their individual problems of adaptation.

5 White-collar crime generally involves a betrayal of trust. Horatio Bottomley (1860–1933), British journalist, MP and financier, is typical. During a career that included starting nearly 50 companies with a total capital of £20 million, and raising £900,000 for various enterprises during World War I, he faced three charges of fraudulent conversion and was served with nearly 70 bankruptcy petitions and writs. Such professional people do not see themselves as ordinary criminals.

Changing patterns of crime

Every known community has found it necessary to declare certain actions criminal – and to enforce the ban by punishing anyone who performs them. Down the ages there has been a remarkable consistency in the areas of behaviour regulated in this way – killing, extramarital sexual activity, property ownership, keeping the peace and so on. Yet in detail the definition of what is regarded as criminal and how seriously it is treated varies so widely that it is difficult to find a single absolute universal prohibition.

Crime and the law

Crimes can be classified in various ways. Perhaps the most widely used divisions are crimes against people (rape, kidnapping, murder, assault); against property [2] (theft, arson, forgery, embezzlement, vandalism); and against public order or morality (drunkenness, gambling, prostitution).

Attitudes and hence laws change, thus what once was criminal may cease to be so and vice versa. Attempted suicide was a crime in Britain until 1961; the USA once permitted practices, now declared illegal,

that encouraged racial separateness, practices that are still legally enforced in South Africa. Crime is not fixed, but something that evolves and changes with each society.

Although most people regard certain activities – murder for example – as invariably criminal, it is the definition of such activities as crimes by the law that is crucial. There is argument, for example, about whether the law should concern itself only with antisocial behaviour and leave private "immoral" activities alone. Debates about prostitution, homosexuality, pornography and obscenity have highlighted this problem. Most communities control or ban the use of violence, robbery and antisocial sexual activities and have also developed an increasing corpus of law to regulate driving, drug use, the use of credit and business transactions. These have been the "growth points" of modern crime.

The statistics of crime

Despite the existence of highly organized crime syndicates, especially in the USA, most people arrested and convicted as criminals

have committed relatively minor offences such as small-scale thefts. Successful criminals are either never arrested or never convicted because of skilful defence lawyers.

Discovering hard facts about trends in crime is extremely difficult [1, 5]. Reliable statistics have only relatively recently been compiled. In the USA, the FBI has been keeping national statistics only since 1930. There are many types of crime that are often not reported at all to the police, notably rape and blackmail which are embarrassing to the victim, and those in which organized crime is involved. The latter tends to concentrate on illegal service activities – narcotics, prostitution, loan sharking (charging excessively high rates of interest) and violations of drinking laws. In such crimes the injured party is usually a willing participant. Organized crime also tends to put pressure on witnesses not to testify and can suppress investigations through bribery or violence.

Statistics are therefore confused, but certainly the number of crimes committed rises inexorably every year (serious crime rose 144 per cent in the USA from 1960 to 1970) but

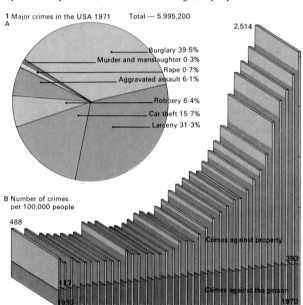

1 Major crimes in the USA 1971   Total — 5,995,200
A

Burglary 39·5%
Murder and manslaughter 0·3%
Rape 0·7%
Aggravated assault 6·1%
Robbery 6·4%
Car theft 15·7%
Larceny 31·3%

2,514

B Number of crimes per 100,000 people

488

Crimes against property

393

117

Crimes against the person

1935   1970

1 Statistics about crime have only recently been regarded as reliable. Figures from the USA suggest that the level of crime per capita depends in part on the ratio of young to old, since 45% of all major crimes are committed by persons under 18. Crimes against property make up a large proportion of all serious crimes [A] and the trend since 1935 [B] has been for numbers of these crimes to rise faster than crimes against people.

2 Acts of some groups, such as football supporters, are branded as "mindless destruction by a delinquent sub-culture". These actions may, however, be due to unemployment and racial frustration.

3 The "great train robbery" in England in 1963 was a classic, well-planned crime in which over one million used notes were stolen by a large gang after months of planning. Most of the people involved were soon arrested – the leaders being sent to prison for 30 years – but an alleged "master mind" was never found. In fact probably only a fraction of crimes committed ever come to the notice of the police. In one American study almost 90% of students who were asked to tick a list of serious crimes admitted committing at least one, yet none had been caught. The chances of a crime's being solved vary considerably. Ninety per cent of murder cases, for example, are solved whereas small thefts are particularly hard to trace.

4 Coldblooded violence has been a feature of 20th-century crime. Terror was widely used to enable gangs to extort protection from potential victims and to scare off rivals from their territory. This was the motive for the 1929 St Valentine's Day Massacre in which seven men of the Bugs Moran gang were gunned down in a Chicago garage by Al Capone's men. Capone (1899-1947) controlled the Chicago underworld in the 1920s but retired after being convicted of tax evasion in 1931 and spending eight years in prison. Crime syndicates grew initially in response to demand for liquor during Prohibition (1920–33). After World War II they concentrated on vice, gambling, loan sharking and drugs.

this may be partly due to more efficient police reporting procedures. Crimes of violence [8], although they attract most attention and outrage, are not rising faster than other crimes, nor do they account for the majority of arrests. "Minor" offences such as drunkenness, disorderly conduct and vagrancy account for up to a third of all arrests. In the ten years 1955–65, crimes against property rose 100 per cent in France, the UK, Italy and Norway; almost 200 per cent in the USA; and more than 200 per cent in Finland, The Netherlands, Sweden and West Germany.

Levels of crime seem to be linked with the proportion of city dwellers to country people, the degree of racial, social and cultural tension and the proportion of young to old. Whether people feel they have a chance to "better" themselves may also affect them.

Contemporary criminal trends

It was confidently believed that with the expanding affluence of the postwar years crimes associated with poverty and deprivation would disappear. But in fact all advanced societies experienced an upsurge, particu-

larly of violent crimes, notably among youths such as the Teddy Boys in Britain and *stilyagi* in Russia. Later, in the 1960s, there was violence on university campuses and generally a greater willingness to use force for political ends. By the 1970s urban guerrillas, whether hijacking planes, bombing strategic targets in cities or holding hostages for ransom, became almost commonplace [7].

While most offenders seen by the courts in most countries are still the routine petty traffic and property offenders, certain changes can be discerned. Young offenders are more prevalent and more violent in cities. Fashions play their part in crime as they do everywhere else in modern society, aided by the impact of the mass media, which bring these acts into the home. It is thus sometimes difficult to say when a "real" change of behaviour is taking place and when imitation is producing a temporary wave after some notorious event. A feature of recent crime, especially urban guerrilla acts and terrorism, has been the increasing prominence of women [9], who have not formerly been regarded as a threat to social stability.

Most of us distinguish "right" from "wrong" and teach our children to do so. We recognize the idea of "sin". We also recognize nonconforming or deviant behaviour, although this may be more a matter of taste or practicality. At different times and in different countries criminal law covers parts of both these categories and omits parts. The "sin" of adultery, for instance, may or may not be recognized as a crime and be legally punishable.

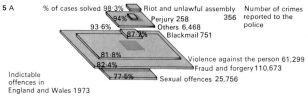
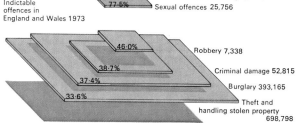

Indictable offences in England and Wales 1973

5 A

% of cases solved 98·3% — Riot and unlawful assembly 356
94% — Perjury 258
93·6% — Others 6,468
87·7% — Blackmail 751
81·8%
82·4% — Violence against the person 61,299
77·5% — Fraud and forgery 110,673
— Sexual offences 25,756

Number of crimes reported to the police

46·0% — Robbery 7,338
38·7% — Criminal damage 52,815
37·4% — Burglary 393,165
33·6% — Theft and handling stolen property 698,798

5 The image of industrialized societies as being submerged in a wave of serious crime is not borne out by statistics in the UK. These show that minor offences make up a substantial proportion of all crimes committed [A]. Different countries have differing patterns of violence [B]. In the USA there are nearly four times as many murders per million as in Canada or Hong Kong, and almost ten times as many as in England and Wales.

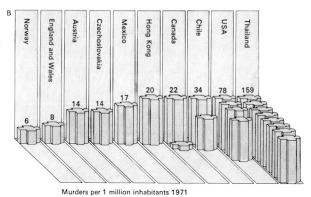

B
Norway 6
England and Wales 8
Austria 14
Czechoslovakia 14
Mexico 17
Hong Kong 20
Canada 22
Chile 34
USA 78
Thailand 159

Murders per 1 million inhabitants 1971

6 Drug taking is seen as a modern menace, although the taking of heroin or marijuana is an ancient practice. What has changed is the group that now uses these drugs – young people in the West. Drug-linked crimes in which ad- dicts rob to pay for their supplies have increased. Most countries ban drug production and sale but these laws are flouted and illicit trade continues despite a measure of international co-operation between law enforcement agencies.

7

8 The brutal murder of Sharon Tate and others in Los Angeles in 1969 by Charles Manson (seen here) and his "family" stresses the problems of predicting and preventing crimes by people who not only work outside society's rules but reject its whole basis.

8

7 Crime, like beauty, is in the eye of the beholder. What most people regard as a senseless outrage – such as this bomb attack that killed 12 people on a bus carrying army families – is to a few a blow for freedom. Political crimes are an increasingly disturbing modern phenomenon, if only because they demand unacceptable counter-measures. The crimes include robbery, kidnapping and various forms of terrorism.

9 Leila Khaled became something of a heroine when she was involved in an unsuccessful hijacking of an El Al aircraft in 1970 during which her accomplice was killed. Women have been prominent and ruthless members in several modern liberation fronts and have featured in political crimes. This contrasts with their traditional criminal image as "bad" girls with hearts of gold or as gangsters' molls who never initiated or participated in crime themselves.

9

Crime prevention

Crime prevention, once the responsibility of Jonathan Wild, Eugène Vidocq, and other "thief takers" who did their business out of sight in a *demi-monde* of felons and miscreants, is now the daily work of millions of people throughout the world. In its efforts to modernise, it has found itself the centre of controversy with its needs and prerogatives often in direct conflict with the rights of the individuals it is supposed to protect.

History of police power
Some nations, such as France, where the paid informants and *agents provocateurs* of Louis XIV (1638–1715) were to be found in the back alleys and boudoirs of Paris, created their police forces from the centre outwards. But it was not until 1829 that an Act of Parliament established England's first paid, uniformed police force. The "bobbies" of London (at first called "Bobby Peelers" after their founder Robert Peel) were armed only with wooden truncheons and functioned in an atmosphere hostile to police power.

In the United States law enforcement grew up even more haphazardly, with order being maintained by vigilante groups in the west and police forces under the control of big city political machines in the urban centres of the east. Police were not only corrupt and incompetent but were also hindered by jurisdictional conflicts that allowed criminals to evade arrest by the simple expedient of crossing city, county and state lines.

This situation encouraged Allan Pinkerton (1819–84) to leave his job with the Chicago police force in 1850 and open his own investigation business. Using a wide open eye (hence the term "private eye") for his symbol, Pinkerton soon filled the country with a para-police force of detectives functioning as national bounty hunters [Key] and strike breakers. Pinkerton helped to create a permanent role for the private investigator in law enforcement practice.

While the problems of policing a modern metropolis such as Rome or Tokyo are not the same as those in Micronesia, the spread of industrialization has produced the same law enforcement problems and common solutions. All national police forces (those of most Western nations are members of Interpol, originally founded in 1923 to facilitate international investigation of crimes and criminals) tend to look to technology to solve their problems. After the civil disturbances of the 1960s, for example, there was emphasis on the development of non-lethal weapons for crowd control and riot suppression [6].

Most major metropolitan police forces now have helicopters to aid in tactical work, and special weapons units and speciality teams trained to deal with specific crises such as bomb alerts or kidnapping. Computers store information and speed its recall and transmission between cities and nations [3].

Detective agencies
The growth of crime has increased the number of detective firms, which were in something of an eclipse after national police forces such as the Federal Bureau of Investigation (FBI, created in 1908) took over most of the semi-official crime-fighting functions once held by Pinkerton's and its imitators. Private detectives still have many investigative functions – negligence cases concerning property and cars, insurance

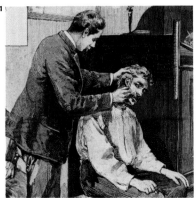

1 Alphonse Bertillon (1853–1914), of the Paris police headquarters, devised a system of identifying suspects in 1882 known as anthropometry or Bertillonage. By a series of careful measurements of the head, body and limbs of the criminal, he acquired a detailed portrait for his filing system. This system was superseded 11 years later by the discovery that every individual's fingerprints are unique.

2 Identikit is a method whereby the police can communicate a description of a wanted man to the general public and other police forces. From an eyewitness, an image can be built up by selecting facial characteristics from an existing bank of material by placing transparent films of a head shape, nose, eyes and mouth over each other. The Identikit system has 550 variables and from these 62 million million combinations can be produced. The witness can also construct a picture by briefing an artist. A third method is the Photo-fit (shown here) with a combination of photographs of typical characteristics.

3 The arrest of Hawley Crippen (1862–1910) on board the liner *Montrose* at Quebec marked the occasion on which wireless telegraphy first played a role in crime detection. Crippen had poisoned his wife and fled with his mistress. A search of his house revealed his wife's remains and the news was telegraphed across the Atlantic. He was tried, convicted and hanged. The police now use a variety of communication services such as radio, television, teletypewriters, computers and picture transmitters. Two modern developments are computer terminals in police cars, and communications centres that process all calls from the public, monitor police resources over a wide area and dispatch police.

4 Interrogation still plays a central part in any investigation but poses problems of assessing the truth, accuracy and perspective of the witness's memory. The polygraph or lie detector is an attempt to assist this evaluation scientifically by measuring body changes said to be caused by a person's emotional state. The first was constructed by John Larson in the USA in 1921 and it recorded blood pressure, pulse and respiration rate. Later versions recorded "psychogalvanic skin reflex", the flow of current between two different parts of the body. Under stress a witness sweats more and so conductivity rises. Lie-detector evidence is not admissible in many courts.

fraud, missing persons and the investigative groundwork for civil and criminal defendants that attorneys once performed themselves. But the primary thrust of private investigative firms is now security and guard work [7].

Big business and invasion of privacy
By 1975 individuals and corporations were spending some $4,000 million a year on private security and crime prevention in the USA alone, with another $1,000 million going into closed-circuit television monitors, burglar alarms and other security equipment. Crime prevention had become big business, creating the need for what amounts to an immense private police force.

There are areas of potential friction between private detective firms and public police agencies. Existing licensing procedures do little to lessen police suspicion of private investigators' qualifications since they tend to certify only that an investigator has a minimum level of professional competence. Other countries are carefully watching the Swedish example of requiring police clearance for private detective firms investi-

gating crimes such as industrial espionage.

The growth of public and private police bureaucracies has surrounded law enforcement with new problems. Many private investigators, for instance, have adapted to their trade the advanced electronic surveillance equipment developed and used by the intelligence services in the postwar era [5]. When used in sensitive areas such as industrial espionage and counter-espionage, this equipment has caused controversy. But the debate over it use in these areas has certainly not been as heated as that concerning its invasion of individual privacy.

Yet the surveillance practised by individual private investigators is minimal compared with the data collected by clandestine organizations such as the Central Intelligence Agency (CIA) and national police forces all over the world. In their attempt to grapple with the complexity of modern life and its institutions law enforcement agencies have sought constantly to increase and centralize their own powers. This has given a new urgency to an old question: *Quis custodiet ipsos custodes*? Who guards the guardians?

KEY

The James gang was founded by Jesse (1847–82) and Frank (1843–1915) in 1867 after Jesse, a Southerner, was declared an outlaw at the end of the American Civil War (1861–5). They plundered and murdered, specializing in bank and train hold-ups. Law enforcement at that time depended largely on posting rewards to attract bounty hunters and fellow gang members who bore a grudge or who wanted to retire. Jesse was shot in the back of the head by Robert Ford, a gang member, for the bounty. Frank surrendered and was tried but twice acquitted, spending the rest of his life on a farm. Their exploits have been romanticized in films and song.

REWARD

$15,000 REWARD
FRANK JAMES
DEAD or ALIVE

$25,000 REWARD FOR JESSE JAMES
$5000 Reward for any Known Member of the James Band
SIGNED
ST. LOUIS MIDLAND RAILROAD

5 Technology has enormously increased the risks and opportunities of industrial espionage. There is a large choice of devices – cameras, microphones, transmitters and tape recorders – that are widely available. They can be readily hidden under clothes and their range is prodigious: most microphones can pick sound up at 6m (20ft). Weapons are also more sophisticated. The Swiss produce a pen gun that writes as well as firing a bullet from the cap.

1 Briefcase with tape recorder, switch, microphone and volume control
2 Miniature tape recorder for holster or pocket
3 Pen microphone connected to tape recorder
4 Pen transmitter with built-in microphone
5 Transmitter with built-in microphone to fit in a cigarette case
6, 7 Silent motorized camera that can be worn as a wrist watch
8 Automatic 16mm camera
9 Ashtray with microphone and transmitter
10 Transmitter
11 Microphone watch
12 Telephone transmitter
13 Cigarette lighter with transmitter and microphone in base
14 Tiny transmitters
15 Subminiature microphone and amplifier

6 Civil disturbances of the 1960s have caused many police forces to equip themselves with new protective clothing as well as non-lethal weapons for crowd control. Research contractors have de-

veloped irritants such as Mace and "pepper fog" (a dense concentration of tear gas from a mobile unit); tranquillizer dart guns similar to those used by big-game veterinarians and "stun guns" to shoot rubber

bullets at high velocity; chemical dye to be sprayed on demonstrators to facilitate later identification; and a "banana skin" foam that makes it impossible for a crowd to maintain its footing.

7 Private security firms have increased in number and size because the hard-pressed police are unable to provide adequate protection for industrial premises or the transportation of valuables or money. Such firms are also hired

to prevent espionage between companies and pilfering of goods by customers and employees. The relationship between security firms and police forces poses certain problems. Firms can be infiltrated by criminals, staff wear uniforms

that are misleadingly like those of the official police and their activities in such areas as checking the credit-worthiness of potential customers are liable to abuse. In the UK security men may carry weapons not available to the police themselves.

Crime detection

Whenever a serious crime is committed the police can call on the assistance of a variety of experts to bring the offenders to justice. These are the police surgeons, pathologists, toxicologists, chemists, biologists, ballistics experts and other highly trained specialists who provide a service in forensic science or criminalistics – the application of medical, scientific and technological skills to the investigation of crime.

The most widely used of these behind-the-scenes services is the fingerprint collection. Fingerprints [8] were first used to solve crime in the late nineteenth century. To aid the speed and accuracy of detection, fingerprints from the scene of the crime are now compared with central records by computer.

Introduction of forensics

Forensic medicine has its roots far back in history. Among the earliest evidence of the use of medical knowledge to combat crime is a thirteenth-century Chinese treatise, *The Washing Away of Wrongs*, which includes helpful hints on how to tell if a drowned person has been strangled. More recently,

there has been a tendency for forensics to be divided into two: the examination of the living (sometimes both victims and suspects) and the pathology of the dead.

Forensic pathology in particular is the cornerstone of homicide investigation. Here the doctor called to examine a body found in suspicious circumstances uses his knowledge of the patterns of violent or unexpected death to decide first whether the case is one of murder, suicide or accident. The actual cause of death is established at an autopsy.

Forensic science as such is a much more recent innovation. Before the turn of the century laboratory science played very little part in the detection of crime. Even in a case of murder it was left to the policeman to make his routine enquiries and to the doctor to examine the body and make a few simple tests on the medical side. It was not until the early 1900s that forensic science, a product of police work and medicine, began to take on a separate identity and to emerge as a new and vital service to the police and the courts.

Today the forensic science laboratory offers a range of specialities that includes

chemistry, physics, biology, pathology, metallurgy, ballistics and document examination. Much of the daily routine is taken up with blood-alcohol analysis in "drink-and-driving" cases and with drug offences. But the laboratory can also call on almost any branch of medicine, science or technology in the fight against crime.

The workload of individual laboratories varies with geographical location. In the USA, where there are many more shootings than in Britain, the examination of firearms is a major preoccupation. Clearly, too, there are more warehouse raids requiring elaborate tool-fit comparisons in cities than in the country. In general, the three types of offence most often calling for examination of contact traces [14] (clues left at the scene, on the person or in vehicles) are break-ins, assaults and hit-and-run cases.

Features of scientific investigation

Detectives are taught during their training that every contact leaves a trace – that the criminal leaves some clue, however microscopic, at the scene of the crime [3–13]. At

CONNECTIONS

See also
60 Skin and hair
288 Nature and causes of criminality
290 Changing patterns of crime
292 Crime prevention
296 Changing patterns of punishment

1 In this imaginary crime a casual intruder has broken into a house to steal. After wandering from room to room (the black line traces his route), he then disturbs a woman who has been asleep upstairs and strangles her. Unlike the average homicide – most are so-called "domestic" murders usually solved within two or three days – this could mean lengthy investigations because the killer has no previous connec-

tion with the victim. As well as the police, various specialists are called in: a police photographer to record details before anything is disturbed; a police surgeon to certify death; fingerprint experts; a forensic pathologist who examines the body before it is taken away for post-mortem; and a forensic scientist (although collection of evidence is often the job of specially trained policemen).

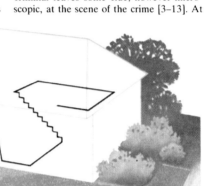

Intruder's route

2 The experienced investigator gradually pieces together his own interpretation of events from the physical evidence found at the scene. To preserve an accurate record, the scene of the crime must be photographed and sketched in minute detail immediately the police arrive. The victim was apparently strangled with a knotted nylon scarf. On the floor detectives find blood stains and the dressmaking scissors used by the victim to stab wildly at her assailant. The bedroom is in disarray, indicating a possible line of struggle from the bed to the dressing-table and then across to the body's final resting place. The forensic pathologist has the difficult task of estimating time of death, a calculation based mainly on the rate of cooling of the body and the degree of rigor mortis. The position of the bed-clothes and the overturned furniture indicate the violence of the struggle during which the victim tried to defend herself with scissors. The rumpled carpet suggests that the body was dragged away from the area of maximum disturbance, the space between the window and the door. Clues to the identity of the attacker include one or two hairs on the pillow, partial fingerprints on the dressing-table and door knob and, leading out to the bathroom, traces of blood from a wound caused by the scissors.

the start of an investigation the police look to the scientists for what is sometimes called inceptive evidence – information on which to base the search for the offender. As the case progresses corroborative evidence may emerge linking a suspect with the crime.

A feature of some scientific evidence is what is known as "inferred uniqueness" – an overwhelming statistical probability that the item in question incriminates the suspect. Fingerprints are a case in point. Since nobody has ever fingerprinted the entire world population, there is no way of knowing for certain that no two people have identical prints. But because, in all the many millions of prints on record, no two sets ever correspond, it is reasonable to assume that each person's prints are unique.

More satisfactory, however, is "proof positive", and here the best example is the "jigsaw fit". This is the matching of fragments of the same article found at the scene and elsewhere. Cloth and paper are most likely to feature in a jigsaw fit, although it may also involve wood, glass and metals. It is by no means rare for a housebreaker to be

caught by the forensic matching of a piece of metal left at the scene of the crime with a broken tool edge found in his possession [5]. Or, reassembling a headlight may prove conclusive in a case of hit-and-run.

Importance of forensic science

Ultimately the value of forensic science lies in providing objective physical evidence – hard facts which, unlike some witness testimony, will withstand close scrutiny in a court of law. It is for this reason that crime laboratories bristle with increasingly elaborate hardware for microscopic examination and analysis of the most minute particles of trace material.

A major attribute of forensic work, therefore, is its inevitable mystique – not least because the average criminal can never be sure just what the experts can do. For the future, there is, in theory, no limit to the possible application of science to the detection of crime. This prospect is particularly reassuring because criminologists claim that the most powerful deterrent to criminal behaviour is not so much the fear of punishment as the certainty of discovery.

KEY

The pathologist's case, the so-called murder bag, contains medical paraphernalia and other items needed for forensics. Particularly vital are

outline diagrams to record details of the scene and the position of the wounds on the body; plastic bags in which to protect the victim's head

and hands; a special thermometer used to take rectal temperature for estimating the time of death; and containers to hold evidence samples.

3 When a window is broken from outside about two-thirds of the glass goes inside. The rest flies backwards and may lodge in an intruder's clothes, to be matched for type at a later stage.

4 Tyre marks are photographed and casts taken. Tread pattern may indicate the make of car, and worn tread could lead to identification. There are laboratories with indexed collections of tyre treads.

5 Entry damage to woodwork may leave a clear impression, identifying the tool by its pattern of scratch marks. The tool and a section of wood are scanned together using a comparison microscope.

6 Often an intruder will eat on the premises; some cook a meal. As few people have perfectly regular teeth, impressions left in food such as cheese, chocolate or apples can be compared, by casting

or photography, with a suspect's bite. Bite marks on victims' bodies are also matched with their assailants' teeth. Forensic dentistry (odontology) is a valuable resource in identifying the dead.

7 Shoe impressions reveal the sole pattern and size. Characteristic wear and defects may pinpoint an actual shoe. Prints can be brought up on carpet by scattering plastic beads which adhere to tread marks.

8 Fingerprints may occasionally be left by indentation on soft surfaces, or by transfer of materials such as blood or grease. More common are latent prints – faint impressions left by transfer of

sweat or natural oil. These are dusted with powder and photographed. Barring tissue damage, fingerprints remain the same throughout life and no two people have exactly the same fingerprints.

11 There is some transfer of fibres in practically any indoor crime. These strands of cloth caught on a splinter will be examined microscopically for fibre type, colour and weave.

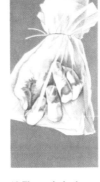

9 When blood falls vertically it produces a star-shaped splash; but if a wounded man is moving the tailed splash reveals the direction he took. This information is useful in dismissing the false confessions that often follow a murder.

10 The microscope can reveal whether hair is human or animal and whether from the head or elsewhere but it cannot identify the owner. In the future, metallic deposits derived from diet and environment may help to individualize hair.

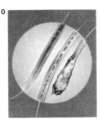

12 The pathologist protects the head and hands of the victim with plastic bags to preserve any clues that may help identify the assailant. Blood or skin fragments, for example, may be trapped under the fingernails.

13 The ligature is always cut off well away from the knot in cases of strangulation, then labelled and tied off so that the cut ends do not unravel. It is important that the knot, which may be of special significance, is kept intact.

14 Random murder is often difficult to solve because the killer, usually mentally disturbed, is an unknown quantity, someone who has no links with his victim. In this case the police already have a rich haul of clues from

the scene. In addition, valuable trace materials such as the glass from the window of entry, will have transferred to the killer's clothing. When an arrest is made the two sets of clues will be matched up to establish a watertight case.

295

Changing patterns of punishment

In simple, small-scale societies the obvious way of righting a wrong is for the victim (or his relatives and friends) to exact vengeance or seize compensation. As societies grow and become more complex such a procedure threatens their stability. Offences come to be seen as wrongs against the community rather than against the individual and the right to punish is taken over by the state. Punishment is imposed as a retribution for an offence, to deter would-be offenders and as a means of reforming those who have broken the legal or moral code. The balance between these objectives has varied in different societies and at different periods, as have the methods adopted to achieve them. These fall into four categories: compensation, humiliation, mutilation and elimination.

Methods of punishment

The usual punishment for minor offenders was either compensation (including various forms of fine) or public humiliation [1-4], the latter being abandoned in most countries during the eighteenth and early nineteenth centuries. For more serious or persistent

criminals, mutilation – such as cutting off a limb, slitting the nose, cropping ears or branding [5] – was used, along with flogging or birching. Torture was largely abolished as a criminal punishment by the middle of the eighteenth century in Europe.

The law's final sanction against offenders, however, was elimination. This could be exile, as when Britain and France, for example, transported prisoners [Key] during the nineteenth century to distant penal colonies, such as Australia and Devil's Island. But the most drastic method was execution, or, as it is called today, capital punishment.

As a means of enforcing the law capital punishment presents problems, particularly if, as in former times, it is imposed for a wide range of offences. (In eighteenth-century England death was decreed for several hundred offences, mostly against property.) Offenders are not likely to be deterred if they are to suffer the same penalty for large and small, violent and non-violent crimes. Legislators tried to find a solution by making the execution more terrible for graver offences.

Hanging, drawing and quartering, for

instance, was introduced in the Middle Ages. In many European countries breaking on the wheel [9] and crucifixion were used until the seventeenth century as similar aggravated forms of the death penalty.

The development of the prison system

A milder form of elimination is imprisonment. Early prisons, such as the Houses of Correction established in England in 1533 and the Rasp Huis in Amsterdam in 1595, were generally squalid and intended solely as retributive deterrents. The notion of reformatory imprisonment grew chiefly through the efforts of John Howard (1726–90) and other penal reformers. Their belief in the beneficial effects of hard labour [10] and strict rules is reflected in the regimes of the Auburn prison in New York (1825) and Cherry Hill prison, Pennsylvania (1829). The first was based on complete separation of prisoners, the latter on prevention of communication through a rigid silence regulation. Both ideas established considerable reputations and later greatly influenced European practice.

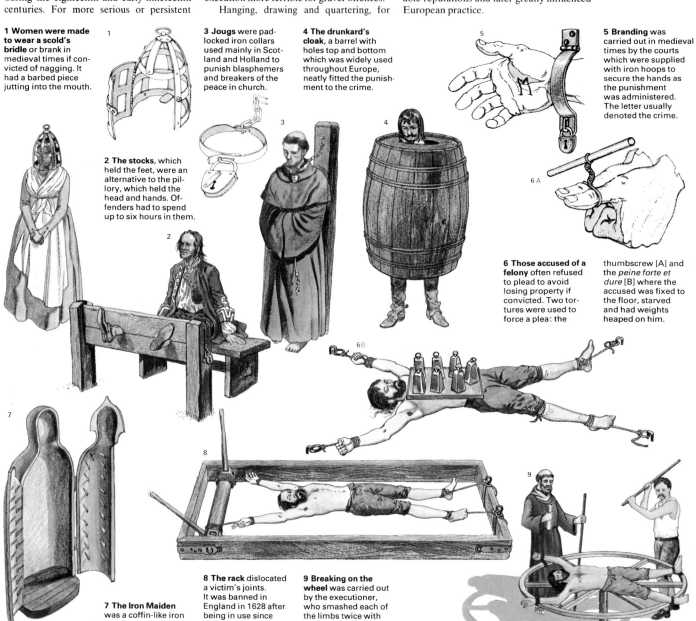

1 **Women were made to wear a scold's bridle** or brank in medieval times if convicted of nagging. It had a barbed piece jutting into the mouth.

2 **The stocks**, which held the feet, were an alternative to the pillory, which held the head and hands. Offenders had to spend up to six hours in them.

3 **Jougs** were padlocked iron collars used mainly in Scotland and Holland to punish blasphemers and breakers of the peace in church.

4 **The drunkard's cloak**, a barrel with holes top and bottom which was widely used throughout Europe, neatly fitted the punishment to the crime.

5 **Branding** was carried out in medieval times by the courts which were supplied with iron hoops to secure the hands as the punishment was administered. The letter usually denoted the crime.

6 **Those accused of a felony** often refused to plead to avoid losing property if convicted. Two tortures were used to force a plea: the thumbscrew [A] and the *peine forte et dure* [B] where the accused was fixed to the floor, starved and had weights heaped on him.

7 **The Iron Maiden** was a coffin-like iron chamber with spiked doors that impaled the victim when closed.

8 **The rack** dislocated a victim's joints. It was banned in England in 1628 after being in use since 1447. It was also favoured by the Spanish Inquisition.

9 **Breaking on the wheel** was carried out by the executioner, who smashed each of the limbs twice with an iron bar, before killing with a blow to the heart.

During the past hundred years, ideas of how best to reform prisoners have become more liberal. These have led to the introduction of less brutalizing labour, more emphasis on occupational training and education generally, better living conditions and the possibility of earning remission of sentence and release on parole through good conduct. But limited resources, lack of trained personnel and a generally unsympathetic public attitude towards "softer" treatment have held back reforms in many prisons.

The effectiveness of imprisonment
Prisons, although differing in their degrees of security, visiting privileges, recreation and so on, are generally repressive with total regulation of the prisoners' lives and few opportunities for constructive, meaningful activity. The result is that for the inmate a prison sentence means loss of material goods, loss of heterosexual relationships, loss of individuality, permanent association with people who do not abide by society's rules and withdrawal of support by society.

As a result, in recent years many people have begun to doubt that prison experience does prevent individuals from returning to crime on release. Although more than two-thirds of those imprisoned for the first time do not return to prison, there is little evidence to suggest that being behind bars has contributed to their "going straight". Experience suggests that young offenders and those imprisoned more than once are particularly likely to repeat their offences.

In some countries, therefore, the trend is away from imprisonment and towards supervision in the home and community setting, where it is thought that many of the pressures resulting in criminal behaviour arise. In the United States, for example, two-thirds of all offenders are under supervision outside prisons. Probation, parole and suspended sentences conditional on good behaviour, particularly for juvenile and first offenders, are increasingly being adopted because they are both cheaper and more effective than imprisonment. The disadvantages include shortage of trained personnel, few job opportunities and difficulties in the selection of offenders for this approach.

Ships or hulks were used as prisons in the 18th and 19th centuries. Felons were kept in them to await transportation to penal colonies. If the sentence was short it would all be served aboard in cramped conditions. Hulk prisoners were employed in dock construction, river dredging and other public works which required the use of heavy manual labour.

10 The tread-wheel was adopted widely in the 19th century to solve some of the prison labour problems. For it meant that a fixed quantity of punitive work could be extracted from each prisoner individually. A period of 15 to 20 minutes, hard effort was followed by a short rest. The mental and physical effects were hotly debated. In some prisons the wheels were used to grind corn or raise water. Elsewhere the prisoner's effort was deliberately not harnessed so that its very futility would act as a deterrent. The tread-wheel was eventually abandoned because it tended to brutalize rather than deter or reform.

11 The gas chamber was used until recently in ten states of the USA. This one is in San Quentin prison, in California. The condemned person was strapped into a chair and globes of cyanide dropped into acid held in the cylinder, bottom right. This released the lethal hydrocyanic acid gas. Unconsciousness was almost instantaneous and death usually occurred within five minutes. Like most modern forms of capital punishment, it was meant to minimize the suffering which such penalties necessarily involve. This is in clear contrast to the ancient methods of execution where intense, protracted pain before death was deliberately inflicted.

12 Prisoners lined up in their striped uniforms in the exercise yard at Sing-Sing prison, Ossining, New York (c. 1895). Even today they may spend more than 22 hours a day in their cells. Good behaviour is rewarded with privileges and greater freedom. Parole can be applied for after serving only a third of their sentence.

13 Open prisons, like this one at Leyhill, Gloucestershire, UK, may be the answer to the problems of prison life. Established between the wars, they are minimum security institutions without bars, where inmates work and live in relaxed conditions and are involved as a matter of policy in the life of the local community.

Work: curse and pleasure

In societies with slaves, such as ancient Greece, work was a necessary material evil that the élite avoided [2]. The Hebrews considered work a painful necessity, but added the belief that it was a product of original sin. Protestantism, and particularly the Puritan followers of John Calvin (1509–64), established work in the modern mind as "the base and key of life": the best way to serve God was to do most perfectly the work of one's profession. Conversely, idleness began to be seen as sinful. In his humanistic book *Utopia*, Thomas More (*c.* 1478–1535) envisaged a working day of six hours with all men taking their turn at all kinds of work.

Why people work

Utopian socialism of the ninteenth century was also influenced by religious ideas about the dignity of labour and the value of work. Karl Marx (1818–83) believed that production should be carried on for use rather than for profit so that men would "live to work" rather than "work to live" – thus acknowledging that satisfaction could be found in labour, apart from any economic reward.

In the nineteenth century and the early part of the twentieth, when it was customary to seek wholly biological explanations of human behaviour, psychologists tended to think in terms of instincts or drives to explain why people might "live to work". Sigmund Freud (1856–1939) implied that hidden sexual satisfactions must lie behind any pleasure derived from working. Other psychologists believed that people worked out of an instinct for self-preservation or that an "instinct of mastery" accounted for man's striving for control of his environment.

Later theories insisted that man was so largely a social animal that the question of why and how people work must be investigated from a cultural and social viewpoint. Studies of what work means to people in modern societies suggest that a complex mixture of economic, social and psychological factors is involved [Key, 1]. People may work primarily for money, prestige or power. They may be motivated by a sense of social purpose or simply by the need to establish personal relationships with a wider circle of people. Their satisfaction may be intellectual

or purely physical or a mixture of the two.

The degree of emphasis now placed on the material rewards of work is a comparatively recent development. Initially, the central driving force of Western industrial society was the Protestant work ethic. But as the religious justification for hard work waned and, as mass consumption of the goods produced by work became economically important, monetary rewards were increasingly offered as an incentive. One result of this was that the interest of those who worked began to turn from production to consumption; in other words, work became for many a tedious means by which to accomplish a material end.

Job satisfaction

Evidence of widespread job dissatisfaction has led in recent years to several hundred investigative studies covering many occupations (although mostly factory and office work). Satisfaction with work for its own sake is often found among skilled factory workers and craftsmen, especially when the job involves completion of a whole project [4].

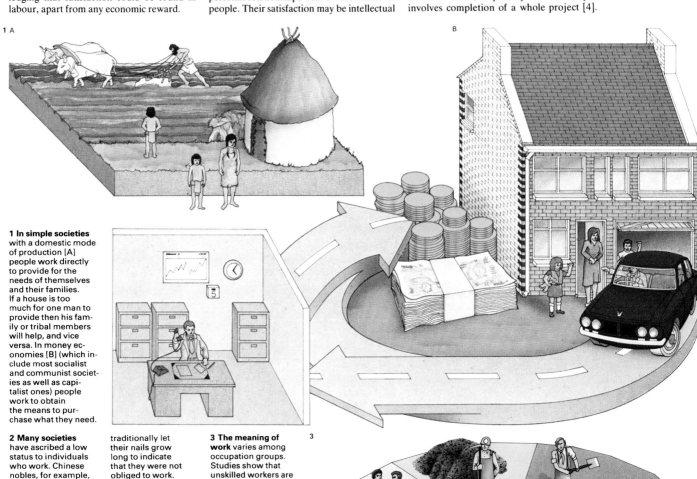

1 A

B

1 In simple societies with a domestic mode of production [A] people work directly to provide for the needs of themselves and their families. If a house is too much for one man to provide then his family or tribal members will help, and vice versa. In money economies [B] (which include most socialist and communist societies as well as capitalist ones) people work to obtain the means to purchase what they need.

2 Many societies have ascribed a low status to individuals who work. Chinese nobles, for example, traditionally let their nails grow long to indicate that they were not obliged to work.

2

3 The meaning of work varies among occupation groups. Studies show that unskilled workers are likely to see their work as having no other meaning than that of earning money. Skilled craftsmen emphasize work as a source of self-respect. Physicians stress the public service aspect of their jobs. Salespeople value friendship and sociability. Coalminers often have a personal sense of struggling against their environment and having to conquer it.

3

Assembly line workers like to be able to control, to some extent, the pace and methods of their work. Variety is important to both factory and office workers, particularly for women, as is the friendliness of the other members of the working group.

Of other factors that influence satisfaction, social relationships seem to be most important, although freedom to make independent decisions and take responsibilities is valued. Satisfaction is also linked with tactful and flexible supervision and leadership and with being consulted in advance about changes in work methods. In general, jobs that involve dealing with people are seen as more satisfying than those which do not.

People's attitude to work may also be influenced by their age, sex and personality. Women, for example, are generally more satisfied with their work than men, even when their jobs give them less authority, status or income. The relatively low importance women have traditionally placed on career success may partially account for this. People are usually happier at work as they grow older, although there is a tendency for

them to find satisfaction in the social surroundings of a job while young people are more attracted by the nature of the job itself. People from the higher social classes tend to have more satisfying work. But well educated people may be less contented with the kind of job they have. Menial workers are sometimes less dissatisfied because they have invested less of their self-image in their work.

Working conditions

Studies of job satisfaction [5] need to be treated with care. Dissatisfaction at work is often not admitted because of a widespread feeling that this is tantamount to an admission of personal failure in life. The diversity of variables known to influence work behaviour is also a warning against over simplifying the factors that make for satisfaction. Like most activities, work is influenced by the environment in which it takes place. But instead of better physical conditions (or more pay), dissatisfied workers may actually need a more human environment – one that makes them feel less impersonal and estranged from decisions, and the people who make them.

People work for a variety of reasons that are not always easy to separate. First and foremost, perhaps, they work to provide the necessities for themselves and their families, including a measure of security. They may work to increase their standard of living and to buy luxuries. They may also work because they enjoy what they do in itself. Work allows them to realize their potential and gain recognition as individuals. Finally they may work to achieve power and status.

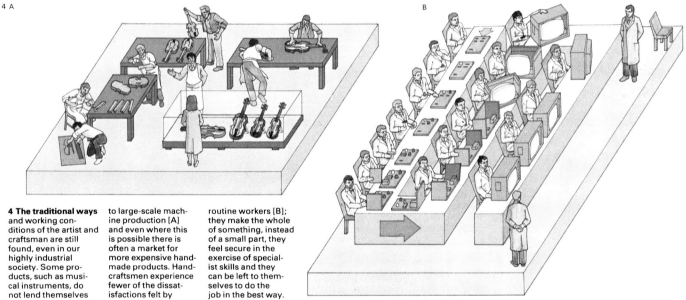

4 The traditional ways and working conditions of the artist and craftsman are still found, even in our highly industrial society. Some products, such as musical instruments, do not lend themselves to large-scale machine production [A] and even where this is possible there is often a market for more expensive handmade products. Handcraftsmen experience fewer of the dissatisfactions felt by routine workers [B]; they make the whole of something, instead of a small part, they feel secure in the exercise of specialist skills and they can be left to themselves to do the job in the best way.

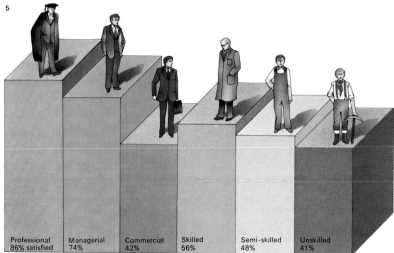

| Professional 86% satisfied | Managerial 74% | Commercial 42% | Skilled 56% | Semi-skilled 48% | Unskilled 41% |

5 One American study of job satisfaction indicated that levels of satisfaction differed widely in both white-collar and blue-collar workers. In an equivalent UK study nine out of ten people claimed to be either very or fairly satisfied with their work. This proportion, which appears unrealistically high, may be the result of a reluctance to admit dissatisfaction because it would appear to be a confession of failure. It could also be that "satisfaction" is no more than acceptance of the status quo.

6 Occupation confers status in every society. In the industrial USA [blue] the relationship between job and status is less ritualized than in the agrarian Swat region of Pakistan [yellow], although ranking in both cases is roughly equivalent, as is occupation.

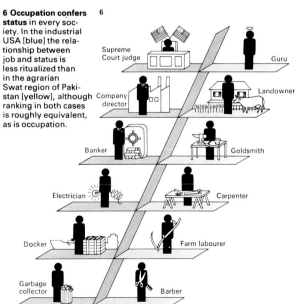

Supreme Court judge — Guru
Company director — Landowner
Banker — Goldsmith
Electrician — Carpenter
Docker — Farm labourer
Garbage collector — Barber

Work: motivation and performance

The first efforts to study work behaviour scientifically were focused on work merely as an input in the process of production. The criterion used in those early studies, which began at about the turn of the century, was purely and simply that of mechanical efficiency. In America, Frederick W. Taylor (1856–1915), pioneered what has become known as time-and-motion study [1], and another early investigator, Frank B. Gilbreth (1868–1924), coined the term "therblig" to describe the smallest time-motion units into which any job can be divided.

Workers as machines

The central aim of these early studies was to make the worker as much like a machine as possible. While Taylor and his followers were not chiefly concerned with comfort or fatigue they were aware that a tired worker is an inefficient producer. Thus a by-product of Taylorism has been some reduction in the discomforts of manual labour. Workers and their representatives, however, have always tended to look askance at the figure of the efficiency engineer with his stopwatch.

Subsequent studies of work performance have been more concerned with the most efficient deployment of men and machines and the organization of the relations between them. Consequently there is now as much interest in adapting the machine to the worker as in adapting the worker to the machine [2, 3]. The attitude and involvement of the worker in what he is doing are now seen as significant.

Apart from the importance of avoiding a poor physical environment – noise, temperature, overcrowding, and so on [6] – work efficiency is also affected by the cohesiveness of the group. In one study it was found that absenteeism was highest where people had a low opinion of other members of their group. Close-knit groups have about a quarter of the average absenteeism rate. Cohesiveness is affected by the amount of communication that exists between workers. Some jobs demand frequent interactions while the technology of others either restricts communication to sign language or else makes it unnecessary or even impossible.

More lighthearted but nonetheless valu-

able and perceptive approaches to work performance are those of C. N. Parkinson (1909–) and Laurence Peter. *Parkinson's Law* (1965) expands on the basic proposition that work expands to fill the time available for its completion. Parkinson also suggested that the total number of people employed in bureaucracies rises whether the volume of work increases or diminishes. An official wants to multiply subordinates, not rivals, and officials make work for each other.

The Peter Principle (1969) states that in a hierarchy every employee tends to rise to his level of incompetence. The actual work is accomplished by those employees who have not yet reached their level of incompetence.

Aspects of supervision

An important element in the quality of work performance is the amount and the style of supervision. A foreman or first-level supervisor's job differs from that of other managers because the group he works with is different. He is seen in one of five basic roles: as a key man in management, sitting astride the chains of authority and communication and

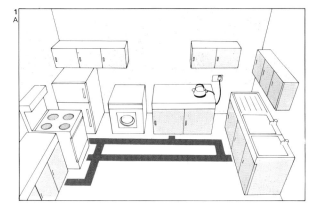
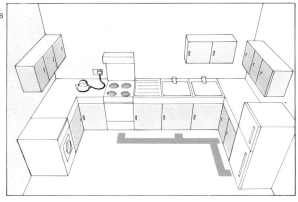

1 **A time-and-motion study** seeks to minimize wasted time and effort. Its principles can be as easily applied to the home as to the factory. Kitchens [A] are often badly laid out so that every job entails a large amount of unnecessary movement [red] between stove, storage, refrigerator, sink and work surface. In remodelling [B], these units can be regrouped so as to minimize effort [green].

2 **In the traditional assembly line** method of producing cars the bodies move along a conveyor belt and each worker performs a specific operation on each car. The job cycle usually has to be completed in less than two minutes, and individuals have to keep up with the speed of the line. Although theoretically efficient, the resulting physical and mental strain may promote inefficiency over a prolonged period.

3 **Using modern planning methods** some car firms have organized a system of small work groups. The members of each group are responsible for the entire car, except for some pre-assembly work. During the assembly work there is no mechanical control of flow. The maximum job cycle time is increased to 30 minutes. The system makes the work more interesting and vulnerability to disturbances is drastically reduced. The disadvantages are that more floor space is needed and more material is tied up, but work efficiency apparently reaches the same level as on traditional machine-paced lines.

4 **Efficient communication** is vital to the success of an organization. In the traditional pyramid model [A] nearly all contacts take the form of orders, with each subordinate receiving instructions from one boss. In this way an individual has contact only with the supervisor above him and the subordinates below. The link-pin concept [B] emphasizes social contact and co-operation between groups. At least one member of each group is also a member of the unit immediately above.

with the power to block anything going upwards or downwards; as a man in the middle, pressed between opposing forces of management and workers; as a marginal man, left out of or on the margin of the main relationships between management and workers; as another worker who is regarded as lacking authority because real decisions are made by others; and as a human relations specialist, handling the human problems of those with whom he works closely.

Work potential

The role of worker is only one of several roles for people in modern industrial society and this can lead to difficulties in changing from one role to another. The same man may have to be compliant at work, the "father figure" at home and outgoing with his friends.

In an effort to fit square pegs into square holes various methods of assessing work potential have been developed [7]. Four of them are: mental testing, which consists basically of intelligence tests; job analysis, which concentrates more on the type of work than on the personality of the worker; the work-

sample approach, which incorporates both of these other methods by setting up simulated industrial operations; and the situational approach, which also involves simulation but is based more on work behaviour.

One way of training people is the system of apprenticeship. In the Middle Ages this consisted of a paternalistic relationship between the craftsman and the trainee and formed an integral part of the guild system. Today the paternalism has largely vanished but many of the associated customs remain. Apprenticeships are served, mostly for periods of four or five years, from the ages of 15 or 16 and are a mixture of on-the-job and back-to-school training.

New methods of training differ from the earlier methods in several ways. They try to operate on an emotional as well as an intellectual level; they concentrate on the group rather than just on the individual; they attempt to make people more aware of the needs of others; and they conduct the training programme on a long-term basis so that eventually the whole organization will share these new values.

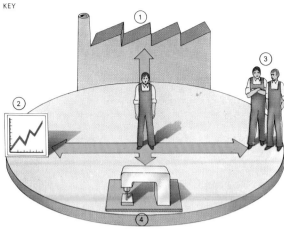

Workers relate to their job environment on four different levels. First there are the broad rules, regulations and purposes of the organization [1]. Next there are the goals of the organization, those of the worker's specific job and his actual achievement [2]. At the same time the worker has a set of informal relations with his colleagues [3] through which he gains status from his personal characteristics. Finally there is the worker's direct skill at the job itself [4]. The well-adjusted worker operates on all four levels.

5 A worker's level of responsibility can be measured by the amount of time between successive reviews of his performance. Elliott Jaques suggested in 1961 that the time-span during which a member of an organization makes decisions at his own discretion defines his position. For a production worker it may be an hour, for the managing director as long as a year.

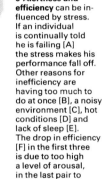

| 1 hour | 1 day | 1 week | 1 month | 1 year |

6 Alertness and efficiency can be influenced by stress. If an individual is continually told he is failing [A] the stress makes his performance fall off. Other reasons for inefficiency are having too much to do at once [B], a noisy environment [C], hot conditions [D] and lack of sleep [E]. The drop in efficiency [F] in the first three is due to too high a level of arousal, in the last pair to too low a level.

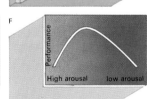

7 The traditional interview, that relies on the interviewer's intuition, can be supplemented by techniques ranging from computer analysis of a questionnaire to hand-writing analysis. Common methods include the measurement of motor ability by the Five Choice Task [A] and evaluation of educational skills by written tests [B]. Social ability can be gauged by video-taping the interview [C], while the stress interview [D] indicates how a person reacts in difficult circumstances such as when table tennis balls are being thrown at him.

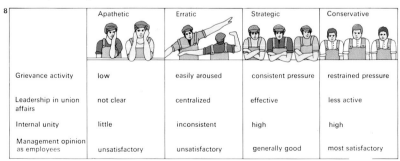

	Apathetic	Erratic	Strategic	Conservative
Grievance activity	low	easily aroused	consistent pressure	restrained pressure
Leadership in union affairs	not clear	centralized	effective	less active
Internal unity	little	inconsistent	high	high
Management opinion as employees	unsatisfactory	unsatisfactory	generally good	most satisfactory

8 Work groups vary according to the level and type of their complaints, their union leadership, their own internal unity and the opinion management has of them as employees. The diagram shows the characteristics of the four types of work group distinguished by Leonard Sayles (1926–): apathetic (many unskilled workers), erratic (automobile assembly line workers), strategic (welders) and conservative (garment cutters and toolroom personnel).

9 Many types of workers could technically do their jobs from home seated in front of their television sets. The Canadian sociologist Marshall McLuhan (1911–) proposes that companies feed all their information into central computers, which can be linked to individual terminals in the home. All the information that employees are likely to want could be instantly available and their reactions and decisions quickly relayed back. People would have to retrain for computer work but they would save time, money and the nervous energy involved in travelling to and from work in overcrowded conditions. However, the loss of social contacts could in many cases prove to be counter-productive.

Work: management and organization

In modern societies most work is done within the framework of an organization. In addition to allowing specialization by allocating tasks and resources, organizations satisfy the needs of various groups. They provide employees with jobs and wages, shareholders with a return on capital and the public with the products they want. Analysis of how an organization functions can be based on two contrasting ideas of how society works: as a system which has naturally evolved or as a sphere of rational action and choice.

Consensus and individualism
The "system" view of organizations regards them as having a life of their own and existing above and beyond the people who comprise them. It assumes that each part of an organization contributes to and receives something from the whole; that organizations are governed by needs, which they must satisfy if they are to survive; and that actions can be attributed both to organizations as a whole and also to their individual members.

The alternative or "individual" view maintains that organizations are the ever-changing product of the self-interest of individual members. It also maintains that the policies of organizations are determined more by power conflicts than by agreement on common goals.

Two major aspects of the controversy over organization theory are bureaucracy and the relationship between the individual and the organization. To most people bureaucracy means red-tape and procrastination but the word also has a technical meaning – the administration of organizations. The German sociologist Max Weber (1864–1920) distinguished in bureaucracy a number of classic features, including the distribution of official duties, an authority structure based on graded ranks and formally established rules and regulations. Despite its shortcomings, an effective bureaucracy may be preferable to other methods of control – for example, administration dominated by a single leader.

The question of the relationship between the individual and the organization sometimes takes the form of asking what type of organization will best meet human needs.

The American sociologist Chris Argyris considers an organization good if it encourages loyalty among its members and enables them to be creative, flexible and productive.

Organizations differ in the types of power on which they are based. Coercive power is the use of threats and sanctions to force co-operation. It often alienates participants, who see the organization as an oppressor. Examples are prisons and some mental hospitals. Remunerative power means that people are paid to co-operate; they exchange effort for money. Co-operative power is achieved when people work willingly because they value the goals of the organization.

The nature of management
Management is an important feature of an organization [4] and can be looked at in three ways: as a unit whose significance is mainly economic; as an élite social group, prepared and perpetuated by selective education and entry; and as a system of authority within which individuals pursue personal goals.

The increasing scale of economic operations and technical complexity during this

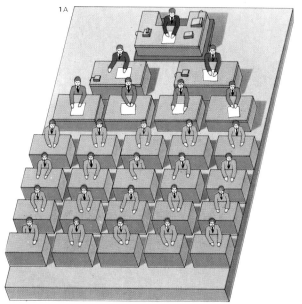

1 In management theory the "span of management" refers to the number of persons a manager controls directly. Many factors determine the number he can effectively manage. A manager with too many subordinates risks wasting time dealing with their problems. But with too few, he may do too much lower-level work. His own skill, that of his subordinates, the stability of operations, and the type of work managed are other factors influencing the span of management. Organizations with a small span of management [A] have closer co-ordination and control because each manager works with fewer people; but the communication chain is long. In a larger span [B] there is a shorter communication chain, which employees often seem to prefer, although each manager controls more people.

2 Managerial styles vary considerably. An autocratic manager [A] centralizes power and decision-making in himself. He structures the complete work situation for his employees. They do what they are told. He takes full authority and assumes full responsibility. Participative management shares managerial power [B]. The participative manager's decisions are not unilateral because they arise from consultation with followers and participation by them. The manager and the group act as a unit. An extension of this approach is the free-rein manager who avoids power. He depends largely upon the group to establish its own goals and work out its problems for itself.

century has meant that a higher level of administrative and technical competence is required of managers. Management has become an élite group, that is increasingly differentiated both from other employees and from business owners.

Management attitudes

Although the term "management" is often used to denote a group of people who share common interests and a common social identity, the assumption is open to challenge. Research into managerial attitudes and actions demonstrates that managers diverge among themselves. Within any one company, for example, research personnel and marketing managers have been found to be more willing to take risks and retain open minds about the solutions to problems than are financial and quality control managers.

In popular usage the term "organization man" implies a lack of individuality and independence. But the need to conform has been exaggerated. Struggles for power form as essential a part of behaviour in organizations as the planning and execution of work

itself. In one study older managers without further promotion prospects were found to have formed protective "cliques" to counter prevailing values of the organization that were not to their advantage. Younger managers, in contrast, identified with these values and formed cliques of their own in order to promote their interests.

Underlying management techniques are two basic approaches [2]. In the traditional, autocratic approach to management those in command have the power to demand that an employee follows orders. The other approach is more democratic. Those in command depend on leadership instead of power. Management provides a climate that helps each employee in the organization accomplish as much as he can with the talents he has. The leader assumes that workers will take responsibility and be motivated to contribute to the organization and to improve themselves if management gives them the chance. In other words, the approach is designed to support the employee's performance and to diminish the sense of estrangement from the goals of production [3].

KEY

One source of conflict between workers and management arises from their different appreciation of the role of work. To management the worker is just one among many inputs [A] (often troublesome and unreliable at that). Buildings, machinery, materials and services are all interwoven parts of the single process of production. For the worker [B] a personal investment of time and effort on his part requires a respect for his individuality, which management may ignore.

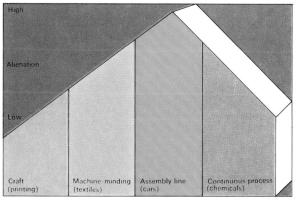

3 If a person feels powerless and useless in his job he tends to withdraw from its realities. The term describing this reaction is "alienation" and it is more typical of some types of work than others.

Research shows that in craft industries such as printing, which have no standardized product, alienation is usually low. It is higher in machine-minding industries such as textiles and much

higher in assembly lines, where work is fragmented and products standardized. In continuous-process industries, where workers control a whole process, alienation is reduced to near the craft level.

5 The rewards of work have risen remarkably during the last 100 years as a result of changing social attitudes and growing union power. Traditionally the worker was given a fixed wage in return for a certain number of hours worked. The owner and shareholders (private and institutional) reaped all the benefits of the organization's profit. Increasingly, workers have shared in the profits, seeing theirs as the most significant (and praiseworthy) contribution. Paid holidays, pension and insurance schemes, profit bonus and share options are part of this pattern. A parallel development has been worker participation in management, with joint committees and worker directors.

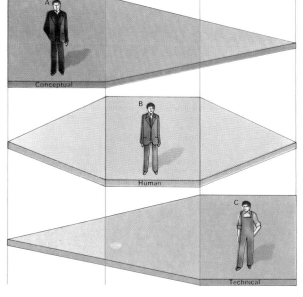

4 A leader in an organization uses three separate skills; conceptual (green), human (brown) and technical (blue). Different levels of management are called on to use different combinations of these skills. For top-level managers [A], conceptual skills are crucial, enabling them to deal with abstractions, to set up models and to devise plans. In middle management [B], human skills – the ability to interact effectively with people – are of paramount importance and conceptual skills less needed. Finally, a foreman [C] relies heavily on his technical skills. For he must know, and be a master of, all the processes that he has to supervise.

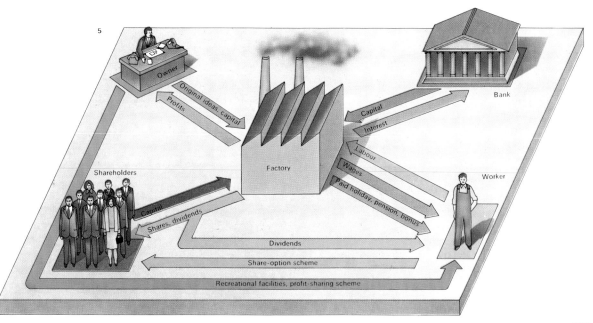

303

Play and sports

Play is an almost universal activity of higher animals but one that is hard to define. In essence it consists of two main ingredients – activity for its own sake and experimentation with types of behaviour that will be subsequently carried out in earnest (such as hunting, fighting and hiding).

Animals at play
In animals both these kinds of play are carried out by the young and are clearly a part of the learning process. Experimental activity familiarizes the young animal with its capabilities. Leaf-chasing by kittens prepares them for hunting, for instance, and tussling by fox cubs establishes a dominance order in the family. Among adult animals, by contrast, it is the first or exuberant type of play that predominates, particularly in social and highly active species such as the dolphin and otter. At times an adult will revert to infant behaviour (old cats sometimes chase leaves) but, in adults, play is normally transformed into ritual and serves a social function (as in courtship). It is debatable how far the concept of "play" can be extended to include

ritual social behaviour and whether behaviour such as grooming in primates (which increases group cohesion) can be described as play.

Human play clearly originates in the same two classes of behaviour, but is enormously extended and complicated by human self-awareness and sociability. The earliest child play, as in animals, takes the form of exploration and experimenting with bodily skills. Then interaction with others becomes important, involving role playing and fantasy enactment. As in monkey play, for example, the "pretend" element in play sometimes involves learning what roles and values the individual is expected to adopt.

Fantasy enactment, the use of play to explore dreams, may be uniquely human or not (we cannot know, because we are unable to converse with apes) but it is important. Where the physical side of play (running, jumping, throwing and so on) develops physical co-ordination, fantasy play serves much the same purpose for emotional development. Play violence, as in cowboys and Indians [Key], far from encouraging violent,

aggressive instincts, may teach the child that his aggressive fantasies do not destroy others.

The real difference between animal and human play is the persistent importance of play in the life of the human adult. Organized sports and games are an important point of social contact for many people and provide relaxation or escape from boredom.

The psychology of games
There are several psychological theories about games. It has been suggested that through them the players experience release from their aggressions in a socially acceptable way. The fact that retired sportsmen sometimes suffer from a variety of personality disorders suggests that the absence of this means of release may be part of the problem. It has also been found that some people (from gamblers to racing drivers) deliberately put themselves at risk both for the intense emotional arousal that it gives and for the relief afterwards [8]. They may become dependent on this pain-pleasure combination. Games are also a natural expression of human competitiveness and of the need to

CONNECTIONS

See also
156 Social development
306 Uses of leisure

1 Sailing is now a sport practised at various levels, from dinghy racing to single-handed circum-navigation of the globe. But like most sports based on the idea of transport, its origins lie in work skills. Its rise as a sport, boosted by modern technology and such new materials as fibre-glass, occurred as its practical function declined.

2 Angling, like the other field sports of hunting, shooting and hare coursing, evolved from the necessary search for food in early societies. Although pleasurable, it was nevertheless work until technological change supplanted its essential role.

3 Archery is a good example of play with a military origin. Although of no practical use in industrial societies, it was once of prime importance; Richard II banned football because it distracted the English people from their archery.

4 Japanese wrestling has arguable origins in combat. Like many of the Eastern martial arts such as ju-jitsu and kendo, it shows a feature prominent in most games – extreme ritualization and a tightly formulated set of rules.

5 Basketball, like most team sports, reflects the values of the mass society that supports it and is involved in its organization; its popularity is second only to soccer. But unlike most team sports, it was an invention – in 1891 by James Naismith (1861–1939) rather than a development from an older sport.

push both body and mind as near to their limits as possible.

As far as spectators are concerned, sport is one way of achieving a sense of belonging to a group, but whether they also burn off their aggressions vicariously, in a purging of pent-up emotion, is open to question. Violence sparked by soccer matches suggests that the reverse may be true.

Another form of play that pervades adult life is the play-use of sexuality, which is greatly extended compared to the rather summary mating of primates. Sexuality has a wide range of functions connected with pair-bonding, gender learning, dominance and fantasy enactment.

Puritan communities have sometimes opposed play in its dramatic, sexual and non-serious aspects, severally distrusting the whole notion of pleasure – the emotion generated by effective play. Hindu philosophy by contrast makes playfulness a part of reality in that the real world reflects the playful activity of the gods. For modern psychiatry, free and spontaneous playfulness is a good index of an unanxious person who

has no difficulty in giving and accepting pleasure, valuing the experience of the body and interacting creatively with others.

Games and society

Play, especially in the form of sports and games, is also important for what it can tell us about society. Forms of sport reflect the nature of society [10] (just as forms of art, theatre and literature do). In pre-industrial society sports often developed from necessary work and other activities of contemporary life. Jousting, archery and other martial sports [3, 4] once provided useful practice for the real thing. Hunting, rowing, sailing [1] and skiing evolved naturally from the demands for food and means of transport.

In industrial society two main influences on sports can be seen. Technological advances have led to the invention of new sports such as motor racing and free-fall parachuting [7] and the organization of games such as soccer [6]. Team sports, working as a group and group, competitiveness within clearly defined rules – reflect and reinforce the structure and values of mass society.

Child play, far from being inconsequential or trivial, serves many useful functions. It is in play that children develop their personalities, as well as their physical skills. They also learn many of the written and unwritten rules of their society, for the various games that both boys and girls play clearly reinforce their social and cultural roles. Surprisingly perhaps, studies have shown that play violence does not promote real violence.

6 Association football or soccer, an 11-a-side game, is the most common form of football and the most popular world sport. The game is an ancient one but modern codes are a post-industrial development. Massed spectators and modern communications are essential for its current organization. Equally, the size of its following and the financial investment in the game have made it crucial to the leisure of some societies.

7 Free-fall parachuting is another sport with military origins. Forty nations now compete in the biennial world championships. The ceiling for jumps is 3,657m (12,000ft) with a free fall of 2896m (9,500ft).

8 Motor racing is an example of man's apparent need to compete and to put himself at risk. The latest in a long line of "wheeled" sports, motor racing has taken place on road, field and track since the car's invention in the 19th century.

9 Man's natural competitiveness makes racing, in person or by proxy, a feature of all ages and all cultures. He has raced frogs, snails, dogs, pigeons, balloons, aircraft – anything that moves. Horse-racing, the so-called sport of kings, has adherents of all classes and illustrates well the competitive instinct. Numbers directly involved in the sport, competing, organizing or watching at the track, are minute compared with those who "compete" by the placing of a bet.

10 Tennis is perhaps the best example of sport as a barometer of social change. Five centuries ago, real or royal tennis was an indoor sport for the aristocracy. In the 19th century it was taken out of doors and, although in a radically different form, was played on grass, primarily by the upper middle class. Even today, the most famous tennis clubs are found in the wealthier suburbs of large cities, although the spread of professionalism has done much to democratize tennis.

Uses of leisure

Leisure is not so much a specific activity as a state of mind. As such it can be experienced while lying on a beach, while working or while vigorously exercising. It describes a person's attitude towards what he or she is doing and the quality of the time spent doing it. The advantage of such a broad and unspecific definition is seen when two people take the 24 hours in a day and try to subtract from them periods they do not consider to be leisure, such as working, sleeping and eating [Key] – no two people are found to agree fully about what should be taken out and what should be left in.

However defined, leisure is certainly important both in the life of the individual and for the society of which he is a part. For the individual it may provide relaxation from daily pressures and routines, creative experience such as education or voluntary work that helps to liberate and develop the personality, or entertainment as an antidote to boredom or drudgery. This last element is reflected in the fact that many employers provide leisure facilities such as football pitches, piped music, games rooms and holiday centres, because they believe that these will produce fitter and happier workers. Equally, in many countries government agencies provide facilities for certain types of leisure for the "good" of people generally.

Leisure and work
In many ways leisure is bound up with work. It is easy, although misleading, to think of leisure as the opposite of work, or to define it as time left over after work. But not all human societies make the distinction. Rural life has always involved an integration of the two. The tradition of the artist and the craftsman is also one in which there is little division between work and play.

In assessing our lives we may give priority to work or to leisure, or equality to both. Those who say that work is the most important thing in their lives are not necessarily saying that their lives are devoid of leisure-like experiences: perhaps they obtain from work some of the satisfaction others get from leisure. Those who value leisure highly may be in arduous, dull or otherwise unsatisfying jobs. It is a commentary on the material values of our society that most people prefer additional paid work to more leisure.

Leisure activities and expenditure
People today generally have more free time than they did a few decades ago, although the actual gain (bearing in mind increased time spent travelling to work and so on) is sometimes overstated. Equally, there are now more diverse ways of spending leisure: more facilities, indoor and outdoor, provided by both private enterprise and public authorities. Television, sports centres and holidays abroad (other than for the wealthy) were unknown a few years ago. Leisure is now big business and all too often a business that sells people something that they do not really want or need as part of a convenient, impersonal package.

It is impossible to estimate accurately the total annual expenditure on leisure because of the difficulty of knowing what to include. In recent years leisure spending has increased, although it is difficult to know how much this is the result of inflation.

One exception is money spent on the

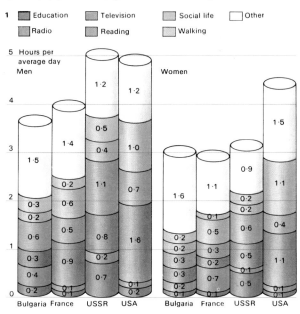

1
- Education
- Radio
- Television
- Reading
- Social life
- Walking
- Other

5 Hours per average day

1 A multi-national survey (1972) compared the amount of time men and women have for leisure in different countries and the ways in which they spent it. Men in France and Bulgaria had less leisure than men in the USA and USSR and women had most leisure in the USA. The share of the leisure devoted to television was greatest in the USA and least in Bulgaria. Differences in time spent on reading were generally small, except in the USSR where more time was spent in this way. Most time was spent on social life in the USA. In the "other leisure" category, a higher percentage of time was spent by the French and Bulgarians on "resting".

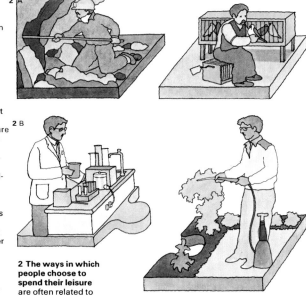

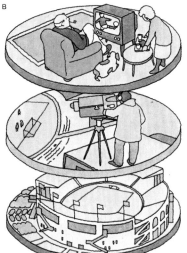

2 The ways in which people choose to spend their leisure are often related to the type of work they are engaged in. Research suggests that there are three main patterns of work and leisure. There may be said to be opposition between the two [A] when leisure is deliberately different from work – as when a miner trains and races pigeons or greyhounds and the factory worker races a motor cycle. On the other hand, work can extend into leisure [B] when the leisure activity bears some similarity to the work done. A chemist, for example, may experiment with fungicides in his garden or the professional social worker may take part in a voluntary project in his spare time. Finally, there is neutrality where passive work is matched by passive leisure [C] and when the person concerned is a spectator rather than an active participant. An example is the factory worker who plays bingo or watches television for most of the time.

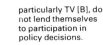

3 People can take an active part in the way their leisure is used, or they can simply "consume" something produced by others. A sportsman [A] may also be on the committee that organizes the sport. The mass leisure industries, and particularly TV [B], do not lend themselves to participation in policy decisions.

purchase and maintenance of motor vehicles, which has been increasing faster than most other expenditure.

Despite the growth of commercialized leisure, there are still many forms of recreation enjoyed in the home (although adolescents are often an exception to this). There are conflicting views on the effect of television on family life, but these do suggest that we need to distinguish different stages in the growth of television viewing. Until the immediate postwar years, recreation and entertainment outside the home had been steadily gaining ground, resulting in less time being spent in the family circle. In the 1950s television brought families together again. But in the 1960s in the United States, and in the 1970s in Western Europe, families began to have two or more sets, with consequent separation of family members and fragmentation of the family unit again.

Leisure and national resources
Many ways of spending leisure involve the use of scarce resources – land for parks and golf courses, water for swimming and boating, and so on. Planning is necessary to make the best use of these resources, although there is a basic paradox in planning for the freedom that leisure ought to mean. Some leisure activities – usually the more artistic ones – are catered for by public or local authorities and require subsidies to make them pay their way. Traditional live arts such as opera, ballet and classical music would have great difficulty surviving without some form of patronage. Against this, it is argued, people ought to be allowed to decide for themselves which leisure interests they want and are prepared to support, instead of having this decision taken for them.

There are two important questions about the future of leisure – what form will it take and how much time will be devoted to it? The confident predictions that by the end of this century the average working week will be less than 30 hours and the average annual holiday several months rather than weeks seem less justified in the slumping 1970s than in the booming 1960s. And if the United States is an indicator, "anti-leisure" – a competitive, frenetic approach to leisure – may spread.

KEY

Leisure for some people is a completely separate part of life [A] – the only part that makes life worth living. They see work as a means to enjoy leisure. For others, leisure is more integrated with the rest of life [B]. They make no rigid separation between work and leisure.

4 Types of holiday fall into a number of different groups and may be classified according to what people want from them. Four kinds of organized holidays are: sociable but mainly passive (holiday camps) [A]; sociable and more active (package tours) [B]; solitary or family holidays (hotels) [C]; or more active pursuits such as sailing schools [D]. Less organized holidays may be classified in the same way: sociable but mainly passive (staying with relatives) [E]; sociable and more active (holiday camps based on a particular interest) [F]; and holidays that are more solitary such as caravanning [G] or quiet camping [H].

5 Retirement presents a major leisure problem since work has filled so much of a person's time. How people adjust to retirement depends on how interesting their work has been and how much they have been involved in it. People who have been highly involved with creative work will probably seek and be able to do something similar in their "retirement" [A]. Those who have been accustomed to divide their lives into separate compartments of work and leisure may be able to expand the leisure portion to fill most of their time [B]. Perhaps the biggest problems face those who simply leave off routine work, since they have fewer inner resources to fill their new-found leisure [C].

Money and capital

Modern man uses money in a wide variety of forms [Key, 1], offering various degrees of liquidity, risk and return. The cash in our purses and wallets is used to cover a diminishing number of our needs, most are now met by cheques or, increasingly, by credit cards.

Where our grandfathers kept a gold sovereign or two to tide them over rainy days, we use the savings bank, premium bonds, national savings certificates and bank deposit accounts as our first line of reserve. These are all "liquid assets", so called because they can quickly be changed into cash. Insurance policies, partly paid mortgages, the shares of public companies all can, be turned into cash with varying notice and cost. In this way the concept of money gradually shades into other forms of financial assets.

The banking system
A modern banking and monetary system has certain constituent parts. Large banks with many local branches (clearing banks) are in touch with the general public, holding their deposits and dealing with their everyday transactions. In continental Europe the most common medium for everyday transactions is the giro method of monetary transfer based upon the postal system. Above the clearing banks stands the central bank – a government institution that is a banker's bank and "lender of last resort". The certainty that the central bank will ordinarily support any subsidiary bank should it get into difficulties has removed the crises of confidence and "runs on the bank" that so disrupted eighteenth- and nineteenth-century business.

Managing the economy
The relationship between the central bank and the clearing banks is a crucial part of economic management as practised by modern governments. They use the banking system to influence the balance between monetary demand and supply. They may when necessary require the clearing banks to increase the proportion of their liquid funds deposited with the central bank, which forces the clearing banks to call in some overdrafts and loans so that their liquid assets are still able to cover any normal claims on them.

The central bank can also influence the monetary situation by changing its "base rate" or "minimum lending rate". This is the rate of interest at which subsidiary banks can borrow money from the central bank if they need to and it forms a logical base for all other interest rates. So, if this bank rate is raised or lowered, other interest rates follow. The central bank or other appropriate authority can influence the money supply by "open market operations" in which it buys or sells government bonds and so competes in the money market for available funds.

Banks are only part of the complex interlocking system of financial institutions that characterizes the modern monetary economy. Peculiar to the City of London is the discount market – a small group of firms that deal in government and commercial short-term paper and form a buffer between the central bank and other financial institutions. There are also the merchant banks, which do not accept deposits from the general public but provide specialized banking services to business. They engage in the financing of trade (particularly of commodities) and are active in company develop-

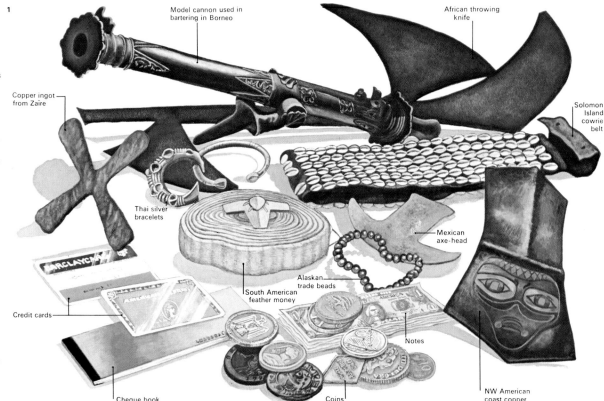

1 **Money** is, at root, really a matter of confidence. It is anything that a society generally accepts as having value. Once that confidence is destroyed money reverts to its intrinsic value. Throughout history money has assumed many forms. Apart from those shown here, precious stones, fish hooks, nails, compressed tea (the original "cash"), livestock, special stones and, in prison camps, cigarettes and tinned food have all been used. Man needs money as a store of wealth, a medium of exchange and a unit against which to value other goods. To be adopted as money an article needs to be relatively scarce, durable, easily stored and portable. It is for this reason that gold and silver have formed the basis of many coinages. That confidence is vital is seen by the gradual development of paper money, the intrinsic value of which is negligible, despite its face value.

Model cannon used in bartering in Borneo

African throwing knife

Copper ingot from Zaire

Solomon Island cowrie belt

Thai silver bracelets

Mexican axe-head

Credit cards

Alaskan trade beads

South American feather money

Notes

Cheque book

Coins

NW American coast copper

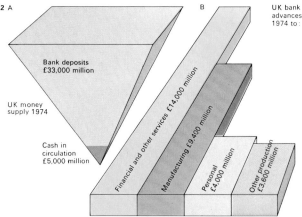

2 A B

Bank deposits £33,000 million

UK money supply 1974

Cash in circulation £5,000 million

Financial and other services £14,000 million

Manufacturing £9,400 million

Personal £4,000 million

Other production £3,600 million

UK bank advances 1974 to:

2 Cash makes up only a small part of the total money supply [A], 13% of the 1974 UK total of £38,000 million. Credit [B] in the form of bank advances totalled £31,000 million of which only 13% went to private individuals.

3 Currencies such as the dollar and pound are "reserve currencies". Other countries hold part of their reserves in New York or London as deposits or investments in government securities for easy encashment.

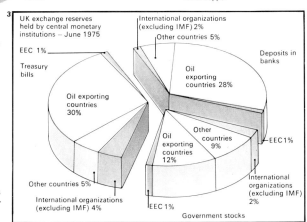

3

UK exchange reserves held by central monetary institutions – June 1975

International organizations (excluding IMF) 2%

Other countries 5%

EEC 1%

Treasury bills

Deposits in banks

Oil exporting countries 28%

Oil exporting countries 30%

Oil exporting countries 12%

Other countries 9%

EEC 1%

Other countries 5%

International organizations (excluding IMF) 4%

EEC 1%

International organizations (excluding IMF) 2%

Government stocks

ment and floatation as well as in organizing the issue of new companies' shares to the general public (as issuing houses) [5, 6].

Significance of the stock exchange
Another key financial institution in the money market is the bourse or stock exchange, where bonds and shares, which represent the physical wealth of the community, can be traded. The stock exchange makes a market through stockbrokers – firms that buy and sell shares on behalf of investors [7].

In London the market is made by stock jobbers ("specialists" in the USA) who are wholesalers of stocks and shares. They hold a "float" of shares at any time and it is the marking up and down of prices as they balance shares on offer with shares demanded that sets the daily prices. Transactions on a stock exchange are undertaken within a complex network of law, government regulations and house rules that are laid down by the governing bodies.

One striking characteristic of stock markets throughout the world has been their growing institutionalization since 1945.

Although small savings have expanded greatly, they have in general been used to boost the great growth in personal insurance and pension provision. Hence the main supply of funds to the stock market both for existing shares and for taking up new ones is now through insurance companies and pension funds which invest on behalf of their millions of customers.

Another important development has been the rise of unit trusts, or mutuals as they are called in the USA. These buy groups of shares so that the investor, by buying units in the trust, can spread his risks over a wide range of companies even though he is investing a relatively small amount. Another constituent of the modern financial scene is some form of institution to finance house building and purchase, which in Britain is carried on by the building societies and some insurance companies. Since the 1920s, there has been an increase in other institutions specializing in the provision of short- and medium-term credit, both to individuals and also to industry in the form of hire purchase facilities and leasing contracts.

KEY

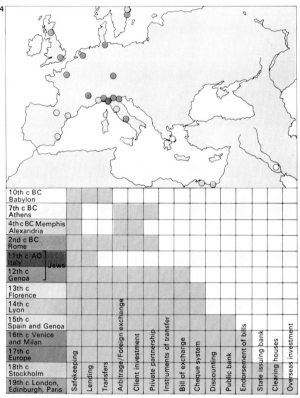

4 Banking first appeared about 1000 BC in Babylon in the form of safekeeping, lending and transfers. Modern banking began with Italian merchants and London goldsmiths who gave credit to depositors. The formation of the Bank of England in 1694 marked the realization that a central bank was needed. In the time chart, below left, brown squares indicate services available.

5 A business [1] can raise capital either through a bank [2] or by selling a new issue of shares on the stock market [3] through a guarantor, the merchant bank [4]. The public [5] and institutions [6] can then invest.

	Capital
	Insurance and pensions
	Interest
	Dividends
	Stocks and shares

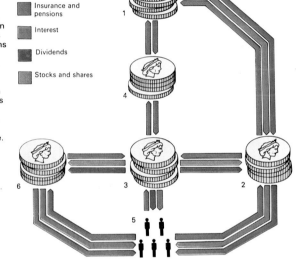

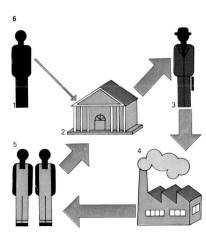

6 The public [1] deposits savings with the banking system [2] which in turn advances money to entrepreneurs [3]. This money is spent on plant [4], materials and wages and returns to the bank via the recipients' accounts [5]. Since the public does not call on all the money at the same time, the banking system can safely expand the money supply in the form of loans by lending about five times the amount of its cash reserves to hand.

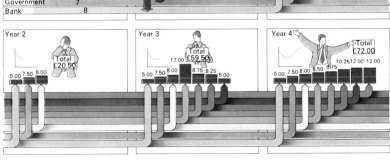

7 Different types of shares carry different rates of interest and degrees of risk. The illustration shows the fortunes of a man who divides £700 equally between a bank deposit, government stock and shares in a company. The red boxes show the interest he receives over a four-year period. Bank deposit, government bonds and debentures are the most reliable. Preference shares have first call on dividends over common shares, and both carry the chance of an increase in unit value.

Man as an economic being

Choice is one of the fundamental ideas in economic thought and scarcity is another. Man is seen as rationally choosing between alternatives – between different goods, between material benefits and leisure, between consumption now and in the future and between alternative uses for the scarce or limited production resources such as land, labour, skills and capital that he uses to achieve his economic standard of living.

What is economic man?

Economic man is a term used to describe a series of generalizations and abstractions that have been developed over two centuries of economic theory to determine the essence of "economic life". Economic theory is concerned with three areas: the way in which man's demands are generated; the behaviour of organizations that supply those demands; and the behaviour of groups within the economy and of national economies interrelated in the world economy. On this substructure rest more detailed theories about the monetary system, taxation, monopolistic bodies such as trade unions, cartels and multinational firms and so forth.

Economic man is a maximizer. As a consumer, according to his own preferences and the prices that confront him, he adjusts his expenditure so as to make the best use of it [2]. This balancing act is performed by adjusting personal consumption of various goods until a given amount spent on any one of them will yield equal satisfaction to him [1]. As a producer, economic man works within the constraint of market demand and his supply of resources in order to make the most profit. Resources are put into production to the point where a given amount spent on any one of them yields equal profit. The "marginalist" character of economic theory derives from the law of diminishing returns – the general rule that as more money is spent on goods or on a resource, the return (utility or profit) to the purchaser from each successive expenditure decreases.

The standard of living

Although economic man is a maximizer, he may not always choose to maximize his consumption of material goods. Workers in unpleasant jobs frequently react to a rise in pay by working less and absenteeism rises. The mid-twentieth century has seen a small but significant minority opting out of the competitive pressures and the full material benefits of industrial society [6]. If, as living standards rise in the world, there should be any marked shift towards a preference for leisure and peace of mind and away from material goods, this could have the most fundamental effect on economic life.

As a general rule the standard of living in any country is fixed by the average output of each person [7]. The opulent few in a community have virtually no impact on the standard of living of the masses of their countrymen. That standard is determined by the efficiency with which the community as a whole works. This is as true in the "advanced" countries as in peasant economies, even though the complex organization of the former and their access to capital tends to confuse the issue. Since the eighteenth century many countries in the world have achieved a rapid rise in output per man and hence in their standard of living.

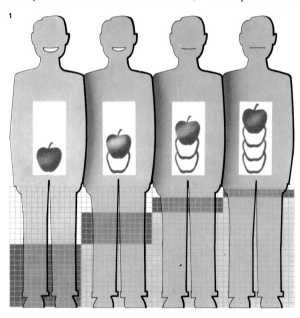

1 "Ravenous, hungry, peckish, full-up" illustrates a basic law of economics – diminishing marginal utility. As we fill our most urgent needs, additional consumption gives less satisfaction or "utility" (dark area on graph) until the point is reached when we gain no additional satisfaction from anything more consumed.

2 Indifference curves [A] and budget lines [B] show how a man will get the most use from his consumption with a given income and relative prices. The optimum occurs where line A touches line B at X. Here the marginal utility of the last unit of income spent on each item becomes equal.

Quantity bought at original price

Quantity bought after price falls

3 Demand is said to be elastic [1] when a given percentage fall in price [yellow columns] produces a higher percentage rise in demand. It is unity when the two balance [2] and inelastic when a fall in price produces a smaller percentage rise in demand [3].

4 In a simple market demand [D] changes inversely with prices [P] so that as prices rise, demand decreases [1]. Supply [S] directly increases with prices [2] so that when prices rise more goods are produced. An excess of demand over supply leaves unsatisfied customers [3] and an excess of supply over demand leaves unsold goods [4]. Equilibrium is reached when demand equals supply at the current prices [5, 6].

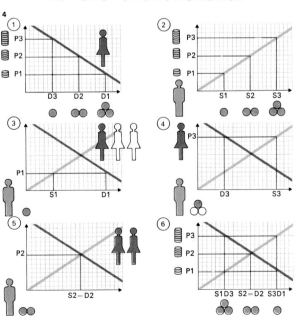

Welfare economics [8] grew out of a realization in the early 1900s that competitive economic individualism did not necessarily bring about, in a phrase used by Jeremy Bentham (1748–1832), "the greatest happiness of the greatest number". Poor parents, to relieve their misery, might seek the solace of gin rather than feed milk to their children. Exploitation of resources led to large-scale destruction of the landscape. Some people felt that a degree of state intervention was preferable to the largely untrammelled competition experienced in Western Europe and Russia in the period 1850–1900. From this has developed the concept of "cost-benefit" analysis whereby the benefits and costs of an irrigation scheme, chemical factory, food subsidy, indirect tax or whatever are calculated for the community as a whole.

Statistical sampling of economic man

Knowledge is more useful and complete when it can be quantified. Therefore population, trade, output and money are counted and compared over time. Because it is usually physically impossible or impractically expensive to make a total count of whatever is being studied, statistical sampling techniques are used. However, the periodic censuses of population are an attempt at total counting and the results contain only a small element of estimation. Once the total population has been counted and classified, "sampling" can be used to collect data about a few thousand representative people (selected in relation to age, social class or income level) and the results can be grossed up by reference to the total population.

Statistical indexes are a particularly effective way of describing changes over time. They also give meaning to such ideas as "changes in the price level" and "the volume of industrial production" which, because they are composed of myriad changes relating to a variety of products, are virtually impossible to discuss without indexes.

"Cost of living" indexes [9] are familiar enough; a representative quantitative collection of goods is priced in the base years. Developments in the successive years are shown as a percentage change and thus indexes usually show the base year as 100.

KEY

Advanced countries generate their high standard of living through high-pressure selling methods that are here epitomized by the Ginza, the hub of Tokyo's commercial life.

5

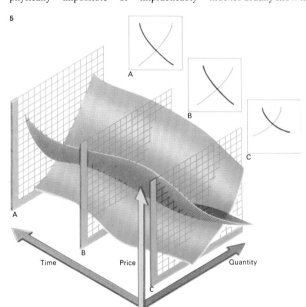

A
B
C

A
B
C

Time   Price   Quantity

5 Economists use simplified models of actual economic situations, which may be either dynamic (tracing changes over time) or static (presenting a single situation). The diagram illustrates both models. A, B, C show the static relationship between supply [green curve] and demand [pink curve], and are three points charting the change from A to C through time.

6

6 Drop-out communities are a luxury afforded only by high-pressure wealthy economies. Those disenchanted with stressful urban life seek simpler ways – but their "poverty" may include such things as cars, TV, alcohol, modern medicine and social security benefits.

7

150   Productivity index 1970 = 100
Japan
130
West Germany
USA
110   USSR
UK
Bolivia
90
Productivity trends
70
50
1964   1966   1968   1970   1972   1974

7 Productivity depends on many factors – efficiency of both management and workforce, hours worked, investment levels and degree of industrialization. Comparing figures between countries is difficult because of special local conditions. But it is possible to monitor trends in productivity within a country. Here the output per person in 1970 is taken as 100 and productivity before and after compared to that. The steeper the line, the faster the standard of living can rise.

8   British government spending 1875–1972

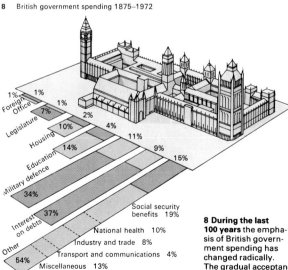

1%   1%
Foreign Office   1%
Legislature   7%   2%
10%   4%
Housing   11%
Education   14%
9%
Military defence   15%
34%
Interest on debts   37%
Social security benefits   19%
National health   10%
Other   Industry and trade   8%
54%   Transport and communications   4%
Miscellaneous   13%

1875 Total £73 million
1972 Total £27,144 million

8 During the last 100 years the emphasis of British government spending has changed radically. The gradual acceptance, since 1875, of the principles of welfare economics has meant a shift from defence expenditure to welfare.

9
A

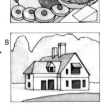

B

C

D

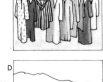

9 One important way of gauging a country's economic health is to measure shifts in the cost of buying a standard range of common goods. This so-called cost-of-living index is based, in the UK, on the average price of some 348 individual items ranging from a pound of cheese to drawing up a will. The items are listed under four main headings.

	Number of items	% of total cost
A Food, drink and tobacco	156	41·6
C Consumer durables	113	14·7
B Housing and heating	12	18·1
D Services and transport	67	25·6
Total	348	100·0

Industry and economics

Economic analysis is applied both to the individual and to group or aggregate situations. Micro-economics deals with the problems of the individual consumer, household or firm and uses supply and demand as its basic model, whereas macro-economics studies communities or countries and uses the overall flow of money as its basis. The division is not clear-cut but largely one of convenience. They are simply different aspects of the same picture.

Behaviour of organizations

The behaviour of firms is analysed generally in terms of perfect competition and monopoly. These two categories probably do not exist in their extreme forms, and real-life situations lie on a spectrum between them. Perfect competition assumes a world in which everybody knows what is happening and in which production freely responds to changes in demand. Above all it requires that individual firms be small in relation to the total market so that they cannot affect price levels by changing their volume of supply, but equally they can always sell as much as they want at the current market price [1].

In a monopolistic situation, a firm is large enough to affect the market price offered for its wares by withholding or increasing supplies. Under conditions of monopoly [2], the greatest profit is achieved when the volume of sales or output is such that the additional revenue resulting from selling one more unit equals the cost of producing it – that is, when the marginal revenue equals the marginal cost. If a monopolist could be forced to sell that volume of output at which price just equals his average cost per unit, he would sell a much larger volume and the consumer would pay a much lower price.

Control over company activities

Monopoly, oligopoly and imperfect competition are names for situations in which, because of relative market size or because of product branding and publicity, a firm can to some extent control the price at which it sells. In the last 100 years governments have exercised more and more control over the behaviour of firms. Company law is the chief means of control. It determines the way in which firms are established; how they are controlled by their shareholders; how they are financed; and it makes them accountable to the community by forcing them to publish certain financial information.

A wide range of other laws covering safety, health, location of plant and pollution also impinge on the businessman's life. Among the most important are laws, such as the American anti-trust laws, that restrict or break up monopolistic firms. Much of this state concern is a reaction to the concentration of economic and social power within industrial and financial corporations where the four or five largest in an industry may control over 60 per cent of the total assets.

The boom/slump cycles of the 1920s [7] and the slump of 1931–3 spawned the theoretical work of the British economist John Maynard Keynes (1883–1946), who argued that governments could and should take counter-cyclical action (deflation and reflation) to regulate the level of economic activity. Since those days and particularly since 1945, governments throughout the world have been expected by their citizens to

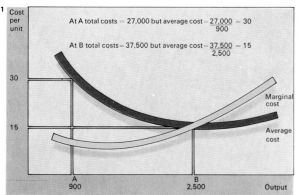

1 **Unit costs** depend on volume of production. At its designed capacity a plant achieves minimum costs but if output is pushed higher costs tend to rise through machine breakdown, congestion and so on. Under perfect competition, firms enter or leave an industry until market price and marginal cost coincide with the minimum average cost. Below this price firms are forced out of production, above it excess profits attract competitors.

2 **A dominant supplier or monopolist** can exploit its customers. By restricting output, for example, it can force the price up and increase profit. The consumer pays a higher price but consumes less.

3 **Primary industry** [A] produces food, oil, wood, steel, aluminium and energy. These materials are worked into higher forms of manufacture by secondary industry [B] which produces chemicals, coats, cans, cars and so on. Shops, warehouses, banking facilities, etc form tertiary industry. An advanced economy has a growing sector of these "service" trades [C], such as credit card companies and the professions.

4 **By acting in concert,** the main oil producers of the world in OPEC were able to triple the world price of oil in 1973–4. This was monopoly in action. However cartels are not new and they have proved unstable in the past.

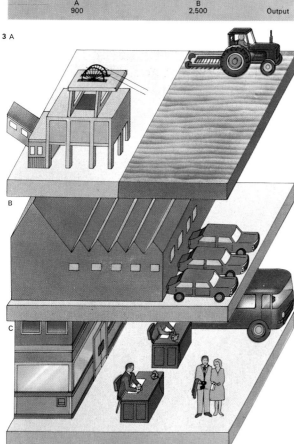

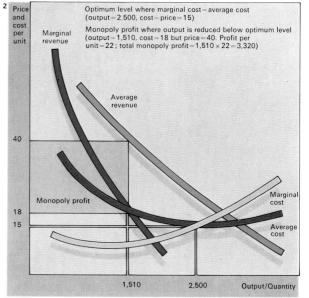

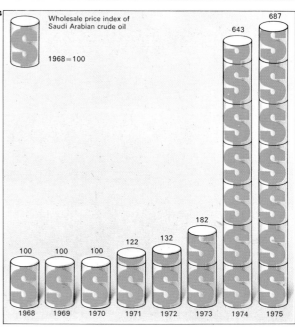

follow full employment policies. Gradually it has become clear that at first creeping, and latterly galloping, inflation limits even the most powerful government's ability to maintain a high level of employment without occasional recessions and resultant increases in the numbers of unemployed.

Inflation and modern society

Inflation is a baffling phenomenon. It shows itself usually as a persistent rise in prices or fall in the value of money and has long been simply summed up as "too much money chasing too few goods". But inflation is an extremely complex process; its results are obvious but they are reached by myriad paths [6]. Prices have been rising persistently and with relatively few interruptions throughout recorded human history – and nobody knows why. It has required a major catastrophe such as the collapse of a civilization or a holocaust like the plague in fourteenth-century Europe to produce a serious check to rising prices.

Many causes are suggested for inflation; governments spending more than they have the courage to collect in taxes; individual

optimistic expectations produced by decades of full employment; growing trade union or industrial monopoly power; international cartels such as those that typify both the demand and supply side of the oil industry [4]; the exhaustion of world raw materials; the inexorable growth of world population; the side-effects of mass media; and so on.

Modern industrial society is promoting great unease in men's minds. Giant firms, apparently out of the control of their nominal owners, giant trade unions sometimes in the hands of criminals, sliding ethical standards and so forth, have produced a political demand for more and more intervention by national governments. The most fundamental issue is probably a question of whether the earth with its finite resources can support perpetually growing industrial output. Man's magnificent technology may have lightened his physical labours (at least for a minority) but in subtle ways it may have enslaved him to the giant institutions, complex administrative systems and mass media manipulation that are necessary to utilize this technology at the large scale it demands.

KEY
Production requires the assembling and co-ordination of capital [A], labour [B] and materials [C]. These are divided into fixed or overhead costs (rent, rates) and

variable costs (materials, overtime) which relate directly to the level of output. The greater the level of output, the lower are fixed costs per unit. These

costs tend to fall at first as work is sub-divided and material waste reduced, but they rise later when additional costs do not produce an equivalent rise in output.

A

B

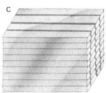

C

5 Spiral of demand

Resources of the economy

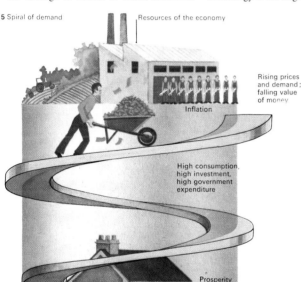

Inflation

High consumption, high investment, high government expenditure

Prosperity

Full employment

Recession

High unemployment

Low consumption, low investment, low government investment

Rising prices and demand; falling value of money

5 "Demand-pull" inflation is used to explain price changes. Inflation is seen as a result of rises in demand pulling on the limited resources of an economy and forcing prices up a self-perpetuating spiral.

Stable prices and demand

Levels of consumption, investment or government expenditure can cause changes in demand. Any increases have an inflationary effect as demand increases, whereas a fall in their levels causes a downward movement on the spiral.

Low prices and demand; high value of money

At the bottom of the spiral, in a recession, demand for resources is low, production and employment fall and with them income and expenditure (demand). Thus the downward movement develops its own momentum.

6 Increased costs of any or all the three factors of production is the basis of "cost-push" inflation. Raw materials [A], capital [B] and labour [C] may all increase in price, thus pushing upwards the

cost of goods produced. World shortages, trade union wage increases that exceed productivity, or monopoly profits are examples of cost push. In a modern interdependent economy, price rises in

any one area lead to increases in many other sectors, the multiple effect of a particular increase developing its own self-perpetuating momentum, as in the case of a large increase in oil prices.

6 A

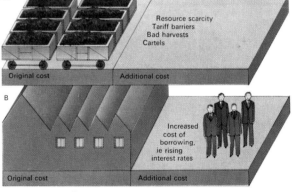

Resource scarcity
Tariff barriers
Bad harvests
Cartels

Original cost   Additional cost

B

Increased cost of borrowing, ie rising interest rates

Original cost   Additional cost

C

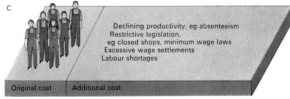

Declining productivity, eg absenteeism
Restrictive legislation, eg closed shops, minimum wage laws
Excessive wage settlements
Labour shortages

Original cost   Additional cost

7

1923
1922 147,000   126,000,000,000,000
1921 3,500
1920 1,400
1919   800
1918   245

7 Runaway price inflation was experienced by several European countries immediately after World War I but Germany was hardest hit. Between 1918 and 1925 wholesale prices rose a million million times. German wholesale prices tripled in 1919, doubled in 1920, re-doubled in 1921, and finally took off in 1922 with a 40-fold increase. In 1923 came the explosion. Prices and money wages grew so high that a wheelbarrow was needed to carry a pay packet.

313

International trade and finance

Trade, or the exchange of goods between parties, is one of the major wealth-creating activities of mankind [Key]. This is so even if the people of a country or region are the most efficient producers in every conceivable field of production: they are still better off if they exchange goods with countries whose pattern of relative efficiencies is different from their own. This principle is called the theory of comparative advantage [1].

Free trade and tariff barriers
It is easy to see the sense in manufacturing producers and primary producers exchanging their goods, but it is less easy to see why the USA, Germany, Japan and the UK are able to profit by importing and exporting similar products among themselves. Yet it is a fact of world trade that the biggest and most rapidly growing markets for manufacturing countries are not the primary producers but other industrial nations. Trade between low-wage and high-wage countries provides an area for controversy. While Americans do not suggest putting a duty on shirts imported from low-wage Tennessee into high-wage Detroit, they

may think it reasonable to put a duty on textiles coming from Korea.

The theory of comparative advantage requires that goods be able to move freely between the trading areas. This raises few problems within a country but it may be a different matter if boundaries of politics, language and culture have to be crossed. Equally, free trade is a wonderful idea if everybody plays the same game, but for various political reasons tariff quotas, levies and similar restrictions may be more popular than totally free-trade policies.

Governments impede trade in many ways, usually by tariffs but sometimes by non tariff barriers [2] and quotas. In the years 1950 to 1970 tariffs were substantially reduced throughout the world with resultant benefits to trade. Today, other barriers have become more important as limitations on international trade.

A country's trading position vis-à-vis the rest of the world is summed up in its balance of payments [4]. The net result of all transactions on both current (basic flow of goods and services) and capital accounts

(loans or debts to other countries) will be seen as a change in the country's international reserves (gold, dollars, sterling or other convertible currency). If a country has a surplus on its current account but lends more than this abroad, its reserves will fall.

There is a close link between national economic policy and the state of a country's balance of payments. Good-neighbour behaviour in world trade dictates that no country should run a persistent surplus or deficit in its balance of payments. After a year or two of surplus a country should stimulate demand for imports by reflation (for example, lowering taxes or undertaking public works); equally it should deflate its economy after a year or two of deficit [5].

Exchange rates of currency
An exchange rate is simply the price of one currency in terms of another. This is determined on foreign exchange markets, which exist in major financial centres such as London, New York, Frankfurt and Tokyo. When the gold standard was used, the currencies of participating countries were all

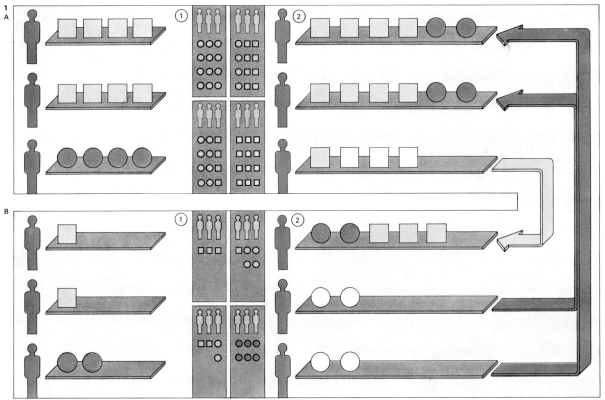

1 **The theory of comparative advantage** explains how benefits arise from trade. In the pre-trade situation in country A, two men produce four crates of lemons each and one man produces four bags of corn. In country B two men each produce one crate of lemons and one man produces two bags of corn. Each country can produce various combinations of lemons and corn by transferring labour between the activities [1]. However, when trade takes place each country specializes in producing the commodity in which it is relatively more efficient. (Of the various possible combinations [2] the best is in colour.) A produces only lemons; B only corn, and they trade four bags of corn for three crates of lemons. Each country now consumes an extra crate of lemons so that both have gained from specialization and trade.

2 **Stringent checks on incoming products**, food hygiene laws, labelling requirements, electrical safety regulations or tendering procedures that handicap foreigners can be greater barriers to trade than tariffs.

3 **Floating exchange rates** gradually replaced fixed exchange rates so that by the early 1970s most major currencies were responding daily to foreign exchange market movements. The most spectacular unshackling was the dollar price of gold.

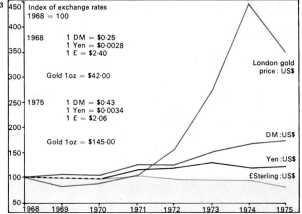

Index of exchange rates
1968 = 100

1968
1 DM = $0·25
1 Yen = $0·0028
1 £ = $2·40

Gold 1oz = $42·00

1975
1 DM = $0·43
1 Yen = $0·0034
1 £ = $2·06

Gold 1oz = $145·00

London gold price : US$
DM :US$
Yen :US$
£Sterling :US$

tied to one another by their gold content. The gold standard was largely abandoned during the 1920s because it was thought to be too inflexible and too limiting. Thereafter the major trading currencies became international standards of value.

In the decades after World War II this system came under pressures that resulted partly from structural weaknesses in the world economy, partly from side-effects of other policies such as the USA's commitment to massive economic and military aid, and partly from institutional rigidities such as fixed exchange rates and the fixed value of the US dollar in terms of gold. Both of these were maintained somewhat pointlessly for many years. The result was a flurry of foreign exchange speculation precipitating devaluations and revaluations of dollars, marks, pounds, yen and finally gold, that led in 1970–71 to a position where there was no international standard of value against which currencies could be expressed.

This situation produced the "Smithsonian parities" in which each major currency was priced against a "basket" of other currencies

weighted in relation to their importance in world trade, and thereby provided a conventional standard of value against which movements in exchange rates could be judged.

Floating systems in world currency
The world has moved slowly from a regime of fixed exchange rates to one in which most currencies are "floating"; that is, their value in other currencies is allowed to fluctuate from day to day on the foreign exchange markets [3]. Systems have been introduced that allow a currency to float within certain fixed limits so that most major currencies now have two fixed values instead of one – the upper and lower limits of the "float" – at which points the monetary authorities will intervene to sell or buy the currency. A further modification has been the introduction of systems in which short-term and relatively minor market fluctuations are permitted in the value of the currency but in which the fixed limits are sporadically moved up and down. In this manner the nominal value of a currency is adapted to long-term movements and economic forces.

KEY

Trade within a country and trade between countries (typified by thriving ports) have similar economic causes and results. Differences in climate and resources produce different regional patterns of efficiency in the production of goods, hence making trade worthwhile.

Current account

Capital account

Foreign exchange reserves

Visible trade

Invisible trade

Short-term capital

Long-term capital

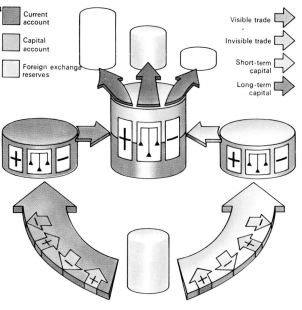

4 The balance of payments is the combined net surplus or deficit of current and capital accounts and records the flow into [+] and out of [−] a country. Current account covers visible and invisible (shipping, insurance, etc) trade; capital account includes long- and short-term capital. A surplus will add to the reserves of foreign exchange, a deficit reduces them.

5 Deficits are corrected by deflation, devaluation or direct controls, whose purpose is to decrease imports [A/D], increase exports [A/C], attract short-term capital [B/C] and reduce capital outflow [B/D]. All three policies add to reserves.

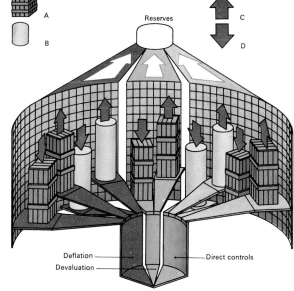

A

B

Reserves

C

D

Deflation

Devaluation

Direct controls

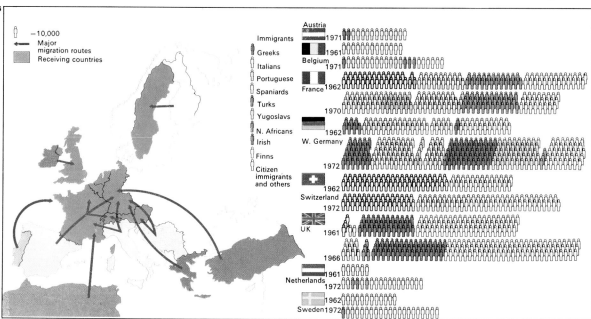

= 10,000

Major migration routes

Receiving countries

Immigrants

Greeks
Italians
Portuguese
Spaniards
Turks
Yugoslavs
N. Africans
Irish
Finns
Citizen immigrants and others

Austria 1971
1961
Belgium 1971
France 1962
1970
W. Germany 1962
1972
Switzerland 1962
1972
UK 1961
1966
Netherlands 1961
1972
Sweden 1962 1972

6 Labour migration has been an important stimulus to world economic developments. The opening up of the USA, Canada, Australia, New Zealand and Argentina in the 19th century drew millions of migrants from Europe. In the 1960s and 1970s the booming heartland of Europe drew migrant workers as shown from the less economically active peripheral countries. The flow would have been greater but for restrictions imposed on migration; for example from East to West European countries. An important difference between this latest intra-European migration and earlier movements is that recent migrants have only temporary work permits.

International co-operation and development

The booms and slumps that scarred the interwar years clearly revealed the need for international economic co-operation. One result has been the establishment of two key institutions to finance trade and economic development. The International Monetary Fund (IMF) provides temporary help to member countries that are having balance of payments problems, tries to stabilize exchange rates and provides an adequate monetary base for trade. The International Bank for Reconstruction and Development (the World Bank) provides long-term loans and expertise to aid economic development.

Other world bodies [Key] include the Food and Agriculture Organization (FAO), which was set up to improve world standards of nutrition by promoting agricultural development. The General Agreement on Tariffs and Trade (GATT) provides a framework for reductions in trade barriers.

Co-operation in regional groupings

Regional groupings of countries, because of their close cultural, political and geographical links, have achieved much practical economic co-operation. Undoubtedly the most fruitful so far has been the European Economic Community closely followed by COMECON (the economic organization to which Russia and the East European countries belong). Other successful regional, economic organizations include the European Free Trade Area (EFTA), the Organization for European Economic Co-operation and Development (OECD) and the Latin American Free Trade Area (LAFTA).

Co-operation in international finance has concentrated on arrangements to finance balance of payments deficits and surpluses and to regulate international liquidity. Areas of controversy have been whether the IMF should have more power to deal with countries that persistently run surpluses or deficits, thus upsetting the world's financial equilibrium, and the role of gold, its relationships with the dollar and the wisdom of replacing it with some form of "paper gold" such as the IMF Special Drawing Rights (SDRs).

After successive rounds of tariff-cutting negotiated by GATT, worldwide tariffs, in the early 1970s, are at an historic low. The USA, for example, has moved from tariffs averaging 45 per cent to tariffs of less than 10 per cent in 1974. Now the international community is beginning to turn its attention to non-tariff barriers to trade – such as national food hygiene laws, labelling requirements, or weights and measures rules – that are sometimes unfairly invoked to exclude foreign goods from the domestic market.

Economic growth

Geographical, cultural, psychological and religious phenomena play a part in explaining the past growth and decline of great civilizations but a comprehensive theory explaining how growth starts has yet to be formulated. Nevertheless, modern economic theory has greatly extended our understanding of the inter-related parts of the economic growth process once it has started. The world has experienced an exceptionally large growth of wealth since 1950 but the gap between rich and poor countries has if anything widened. Helped by the World Bank and national aid programmes the poor have become better off

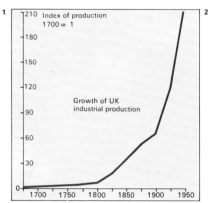

1 For centuries, until the mid-18th century, human population and wealth grew at only a very modest pace. Then the application of power to industrial processes increased productivity. This produced capital surpluses which, when applied to farming, mining and transport, diffused the rise in productivity. Population increases followed improved hygiene and living standards.

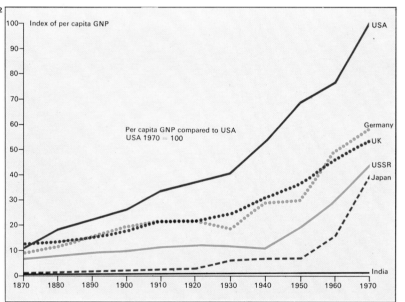

2 Rates of growth vary widely between nations. A mature, relatively wealthy economy may not show the same rate of growth as a poor but emergent one, but in absolute terms its growth may be much greater. Since 1950 the additional income available each year to the USA has been greater than the total national income of India. In the 20th century growth rates have ranged from the slow climb of India and Brazil to the spectacular increases in Germany, Japan and the USSR. However, the hardships imposed on the peoples of totalitarian states such as the USSR to attain their industrial growth would be unacceptable in democratic societies.

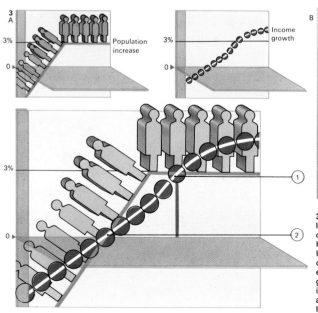

3 Poor countries with low levels of per capita income and high rates of population growth face difficulties in development. Population growth [A] rises as incomes increase above subsistence level [2] up to an average high point of 3% where it levels off. In order to close the widening gap between population increase and the per capita income curve, growth in total income has to be accelerated to point 1. This can be achieved by a massive saving and investment programme. Below point 1 population growth is faster than income growth and per capita income will fall back to subsistence. Beyond point 1 there can be a continuing increase in per capita income. To move to a higher rate of growth a country must make the decision to divert its resources [B] so as to produce more investment goods [3] at the cost of consumption goods [4]. The effect of technology [C] is to increase the output achieved per head [5] at each level of investment per head [6].

– but less so than have the richer countries giving the aid [2]. It is estimated that even now less than ten per cent of world income accrues to the poorer half of the world.

The attempt to create higher economic growth in low-income areas involves the transfer of both capital and technology – the one helping the other [4]. In this process errors have been made. For prestige reasons a country may want an airline or a steelworks that confers no trading advantage on it. With the best of intentions, an advanced country giving aid tends to offer its highest technology with the result that the underdeveloped world is dotted with large-scale plants that are difficult to link with the general economic development. More "intermediate" technology that can be grafted into less developed economies should be applied [7]. But this is not necessarily available, the techniques having perhaps been abandoned by the more advanced countries 50 years or more ago. Despite these problems, per capita incomes have been rising at three or four per cent per annum in the less developed world with the expectation

that, as more is learnt about the social and economic conditions for successful growth, a higher rate will be achieved.

Growth in population and production

Since the mid-eighteenth century the world has experienced growth in both population and production. This has been due principally to a better technology [1]. At the beginning of the nineteenth century, current ideas, especially those put forward by the English economists Thomas Malthus (1766–1834) and David Ricardo (1772–1823), suggested that the world was destined to become a stationary economy with population growth limited by disease and malnutrition among poorer peoples, and capital investment limited by falling profit rates and an upward surge of rents. These views overlooked technology, which has raised the productivity of labour and expanded resources of usable land and available minerals. However, since natural resources are limited, perpetual exponential growth is impossible and technology has only postponed rather than removed the spectre of Malthus's ideas [8].

Economic co-operation has flourished since the end of World War II. The setting up of international organizations such as those whose symbols are shown here provides its institutional basis.

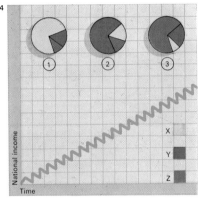

4 As a country's national income grows steadily over time the economy passes through three stages. In underdeveloped societies [1] agriculture [X] is the dominant activity and source of income. As the economy develops [2] manufacturing industry [Y] grows. Finally [3] the service industries [Z] become more important, concerned with areas such as entertainment, social welfare, transport and commerce.

National income / Time / X / Y / Z

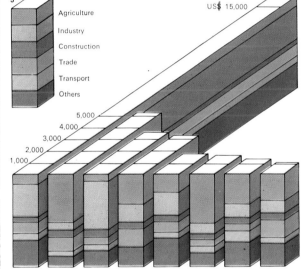

Agriculture / Industry / Construction / Trade / Transport / Others

US$ 15,000 / 5,000 / 4,000 / 3,000 / 2,000 / 1,000

South Africa / Nigeria / Libya / Morocco / Zaïre / Ethiopia / Kenya / Senegal

5 The output of goods and services produced internally by a country (the gross national product) is indicative of that country's economic state. Shown here is the composition of the gross national product for eight representative African countries and the value of their GNP expressed in millions of dollars. The countries have been chosen to represent different areas of the continent: oil-rich Libya and less-developed Morocco in North Africa; Nigeria and Senegal in West Africa; Zaïre in Central Africa; and Ethiopia and Kenya in East Africa. These are contrasted with South Africa.

6 Abu Dhabi shows the ironic contrast between a sterile desert surface and the fabulous wealth of buried oil that has transformed the minute population of herdsmen in the oil sheikhdoms into some of the wealthiest citizens of the world. The question now is whether these islands of technology can be developed to provide an adequate living for their citizens after the oil has been exhausted. Will they become within 50 years a series of Middle East Switzerlands, earning their living by providing services? Or is it possible that these countries will become nations of comfortable rentiers living on the huge investments in other people's industry that they will have amassed by 2050.

7 The need in developing countries is for technology that is appropriate to local conditions. An example of such "intermediate" technology is this pedal-operated cassava grinder in Nigeria made from old bicycle parts. Although less prestigious than expensive imported machinery it is ideal for countries where labour is plentiful and spare parts, trained personnel and foreign exchange are scarce.

8 The Malthusian theory of population attempts to explain the relationship between the size of population [horizontal axis] and the level of output [vertical axis]. As output rises [1, 2] above subsistence [light blue], so does the population. But as more marginal land is brought into production and fertility declines, output falls [4] below subsistence [dark blue]. Finally equilibrium between output and population is reached [3]. The yellow boxes represent the total goods to be shared amongst the population.

317

INDEX

Astronomy
  and astrology, 200, 201
  map, *207*
Asvattha, *208*
Aterian culture, *31*
Athene, *206*
Athens, political system, 272, *272, 278*
Atherosclerosis, 90, *90, 91*
Athlete's foot, *98*
Atlas (bone), *56*
Atman, 220
Atomism, 46
Atrium, *63*
Atropa belladonna, 86
Atropos, *185*
Atum. *See* Re-Atum
Auburn prison, New York, 296
Auditory canal, 50
Auditory nerve, *51*
Auditory ossicles, 50, *50*
Audumulla, *208*
Augustine, St, *232–3*
Aum, *195, 220*
Aura and non-medical healing, 122
Aurelius Antoninus, Marcus, *232–3*
Aurignacian culture, *27, 31*
Austin, John, 282
Australasia, hominid migration, 30–1, *31*
Australia
  mythology, *209*
  race, *34*
Australopithecines, 22–3, *22–3*, 24, 28, *31*
Australopithecus, 21–3
  africanus, 22–5, *22–3*
  boisei, 23, *23*
  habilis, 24–5
  robustus, 22–3, *22–3*
Austria
  hominid remains, 27
  medical facilities, *108*
Authority
  in moral issues, *161*
  in organizations, *264*
Autocracy, 272, 276, 278
Automatic writing, *197*
Autonomic nervous system, 38, *38–9,* 65
Auto-suggestion, 196
Averrhöes, 226, *227, 232–3*
Aversion therapy, 104, *146,* 147
Avicenna, 226, *227, 232–3*
Axis (bone), *56*
Axon, 38, *38*
Aztec mythology, *212–14*

Baby
  development, *162,* 163
  See also Child
Babylon
  astrology, 200
  mythology, *209*
Bacchus, *211*
Bachofen, J. J., 250, *251*
Bacillus anthracis, 85
Bacterial and infectious disease, 84–5, *85, 111*
  digestive system, 93, *93*
  food, *110*
  genitals, *103*
  leprosy, *107*
  pneumonia, 88
  prevention, 112–13, *113*
  respiratory system, *88*
  skin, 98–9, *98–9*

tetanus, 96, *96*
tonsilitis, 88
urinary system, 102
Third World, 106–7, *106–7*
See also specific diseases
Bacteriology, 114
  antibiotics, *119*
  nutrition, *71*
Bacteriostatic, 120
Balance, *50,* 51
Balance of payments, 314, *315*
Baldness, 99
Ball-and-socket joint, *56*
  disorder, 95, *95*
Bandura, Arthur, *159*
Banking system, 308, *309*
Bank of England, *309*
Bantu, *35, 255*
Baptism, *205*
Barber Surgeon's Hall, anatomy class, *85*
Barbiturate, 105, 121, 144
Bar Mitzvah, 165, *165*
Barnard, Christiaan, *182*
Barium sulphate
  use in radiology, *93,* 124, *124*
Basilar membrane, 50, *50*
Basketball, *304*
Basque language, 238, 240
Becquerel, Antoine, 124
Bedlam, *137*
Beethoven, Ludwig van, 174
Behaviour
  animal, *238*
  and brain, 40
  in communication, 236, *237*
  disorder, 135, *135*
  social, 262–3, *262–3,* 266–7, *267*
  therapy, 146–7, *146–7*
Behaviourist psychology, 146–7, *146–7,* 150–1, 187
Belladonna, 119
Bengali language, *240*
Bentham, Jeremy, *232–3, 235, 273,* 311
Benzodiazepine, 144
Bergson, Henri, 232, *232–3*
Beri-beri, 84–5
Beringia, 30
Berkeley, George, *232–3*
Bernstein, Basil, *265*
Bertillon, Alphonse, *292*
Beta rays in medicine, 124, *125*
Bethlem Royal Hospital, *137*
Bhagavad Gita, 222
Bible,
  John i:3, 222
  John xii:24, 219
  Matt x:39, 219
  Matt xiii:12, 223
  Matt xxii:37–40, 218
Biceps brachii, *58*
Biceps femoris, *59*
Bigrammatic playfair system, 243
Bile, 69, *69*
Bilharzia, 107, *107*
Bilharz, Theodor, *107*
Bilirubin, *92, 93,* 93
Binet, Alfred, *143, 153*
Biochemistry, 37
Bioelectricity
  brain, *40*
  nervous system, 38–9, *39*
  sight, 49
Biofeedback
  controlling self-image, 192
  and healing, 122–3
  machines, 193, *193*

medical use, 193
Bipedalism, 20, *20, 21,* 23
Birth
  breech, 79, *79*
  Caesarian, 79, *79,* 126
  date calculation, 76
  stages, 78, *78*
  techniques, 79
Birth chart, 200, *200*
Birth control
  development, 80
  family planning, 81, *81*
  methods, 74, 80, *80–1*
  morality, 183
Birth of Tragedy, The, 235
Birth-rate, world, *81*
Black, Davidson, *24*
Black letter script, 247, *247*
Bladder
  male, and reproductive system, *74*
  muscle structure, 59
  tumour, 103
  in urinary system, 102, *102–3*
Blake, William, 282
Blastocyst, 75, *75*
Blavatsky, Helena, 197
Blind spot, 49, *49*
Blood
  and absorption, 69
  acidity, 67
  bank, *126,* 128
  calcium, *64, 101*
  carbon monoxide, 86
  cell, 57, *62, 63,* 91, *91, 93, 114*
  circulation, 62
  composition, 62–3, *62*
  defence system, *114,* 115
  disorders, 90, 91, *91*
  function, 62
  group classification, *62*
  protein, *102*
  respiration, 66, *66–7*
  tests, 76, *94,* 128
  transfusion, 127
  ABO system, 34, *34–5,* 62
  Duffy system, *34,* 35
  Rhesus system, 34, 76, 87
Boar, *25*
Body and mind, 36–7, *36–7*
  See also individual organs and systems
Body functions, 36–7, *36*
  ages 30–90 yrs (chart), *178*
  and alcohol, *104*
  control of involuntary, 192–3
Body image, 192–3, *192–3*
  changing, 192, *192–3*
  importance, 193
Body language, 193
Body temperature
  diurnal rhythm, *135*
  regulation, *52*
  in surgery, 127
Bolivia, productivity index, *311*
Bolus, 68, *68*
Bomb, atomic, morality of, 182, *182*
Bone
  first aid, *133*
  flat, 56
  fracture, *115*
  formation, 56
  growth, 56, *57*
  irregular, 56
  and joints, *56,* 57, 94–5, *94–5*
  long, 37, 56
  marrow, 57

short, 56
structure, 56, *57*
surgery, 95, *95*
tissue, 56
Bonpland, Aime, *249*
Book of Kells, 223
Book of Mencius, 235
Book of the Dead, 212, 214, 222
Bosch, Hieronymus, *217*
Bos primigenius, 25
Botticelli, *219*
Bottomley, Horatio, *289*
Bowman's capsule, *69*
Brace, Loring, 23
Brachioradialis, *58*
Brahman, 220, *220*
Braille, *243*
Brain, 36–7, 38, 40–1, *40–1*
  capacity, *43*
  hominid, 24, *24,* 26–7, *26, 31*
  primate, 20
  cell, 40, *43*
  damage, 96–7, *96–7,* 104, 134, 139
  development, *163*
  language control areas, *239*
  and memory, *44*
  and mind, 42–3, *42–3*
  scanning, 124–5, *125*
  and senses, 52–5, *52–5*
  stem, 40–1, *40*
  surgery, 145, *145*
  tumour, 96, *96*
  wave patterns (EEG), *194,* 194–5
Branding, *296*
Breaking on the wheel, *296*
Breast
  cancer, 103, 113
  in pregnancy, 76, *77*
  structure and function, *61*
Breathing, 66–7, *66*
  disease, 88–9, *88–9*
Breech birth, 79, *79*
Breughel. *See* Bruegel
Broca's area of cerebral cortex, *40*
Bronchial asthma, 89, 142, *142*
Bronchitis, 88, *88–9*
Bronchus, 66, *66*
  cancer, 89
  disease, 88
Bronze Age, *31*
Bronzino, *193*
Brucellosis, *110*
Bruegel, Pieter, the Elder, *169, 191*
Buber, Martin, 218
Bubonic plague, 106, *106*
Buccinator, *59*
Buddha, 221, *221*
Buddhism, 221–2, *221*
  meditation, 194, *195*
  mythology, *207*
  Eightfold Path, *221*
  Tibetan, *221*
  Zen, *221*
Bulgaria, leisure statistics, *306*
Bunyan, John, *187*
Bureaucracy, 280–1, *280–1,* 302
Burial rites,
  Bronze Age, *181*
  Gravettian culture, *28*
  Neanderthal man, *26*
  Suttee, *181*
Burke, Edmund, *273*
Burmese script, 245, *245*
Burns, skin
  first aid, *133*

Management sciences, 302-3, *302–3*
  in civil service, 280–1
  employee training, 301
  industrial research, 300, *300*
Mandala, 194, *194–5*
Mandarin Chinese language, *240*
Mandible, *56–7*
Mandus, Brother, 122
Mania, 138–9
  treatment, 144
Manic-depressive psychosis, 135, 138
  treatment, 144
Manichean, *221*
Man-made remedy, 120–1, *120–1*
Mantra, 194, *195*
Manu, 213, *213*
Manuscript, 245–7, *247*
Manville, Tommy, *169*
Mao, Chairman, *123*
MAO (monoamine oxidase) inhibitor, 144, *145*
Maori, greeting ritual, *202*
Marconi, 242
Marihuana, 105
Marital therapy, *149*
Marriage, *169*
  arranged, *168*
  early years, 170–1
  homosexual, *169*
  later years, 172–3
  lower age limit for, *165*
  romantic idea of, 168
  sexual behaviour and, 172
  social class and, 168
  in society, *250*, 251
  trial, 169
Mars, *200*
Marx, Karl, *232–3*, 273, *275*, 298
Marxism, 277, *277*
Massage, 73, *73*
Masseter, *59*
Mastectomy, 128
Mastication, *59*
Masturbation, 166
Materialism, 186, *187, 227, 233*
Mathematics
  and logic, 228, *228*
  and philosophy, 233
Mating behaviour, 251
Matthew, St, gospel
  x:39, 219
  xiii:12, 223
  xxii:37–40, 218
Maturity
  in adulthood, 30–40, 172–3
  in personal crises, *180*, 181
Maui, *209*, 214
Mauveine, 120
Maxilla, *57*
Maya
  civilization, *252*
  script, 244, *244*
May, M. A, 160, *161*
Mead, G.H, 266
Meadow sweet, *118*
Measles, *85*, 97, 112, 143
Mechanoreceptor, 52, *52*
Medial pterygoid, *59*
Medial rectus, *48*
Medical check-up, 82–3, 113, *113*
Medicine
  community, 110–11, *110–11*
  dentistry, 130–1, *130–1*
  diagnosis, *93*, 113, *113*, 124–5, *124–5*
  forensic, 294
  history

anatomy, *85*, 126
astrology, *82, 201*
curative, 82–3, *82–3*, 116–19, *116–19*
law, 117
mental illness, 136–7, *136–7*
pharmacology, *118–19*, 119
  Arab, 136
  Chinese, 117, 123, *123*
  Egyptian, *83*
  Greek, 136
morality, 182–3, *182–3*
preventive, 112–13, *112–13*
radiology, 124–5, *124–5*
rehabilitation, *90, 97*, 144–9, *144–9*
spiritualism, 224
surgery, 126–9, *126–9*, 145, *145*
unorthodox, 122–3, *122–3*
WHO, 108–9, *109*
Medieval scholastic philosophy, *230, 233*
Meditation, 194, *194–5*, 225
  and crime, *195*
  in non-medical healing, 122
  physiological changes, 194–5, *194–5*
  in religion, 216, 222, *222*
  research, 224
  transcendental, 195, *195*, 225
Mediterranean, 32
Medium Coeli, *200*
Medulla, *40*
  renal, *69*
Medusa, *206, 210*
Meduse, 186
Meissner's corpuscles, *52*
Melanesians, 31, 33, *34*
Melanin, 34, *35*, 61, *61*
Memory, 44–5, *44–5*
  disorders, 134–5
  lapse, 104
Mencius, *232–3, 235*
Meninges, *38*
Meningitis, 97, 143
Menisectomy, 128
Menopause, *101*, 174
Menorah, 218
Menstruation, 75, *75*, 164, *164*
Mental illness. *See* Diseases, mental
Mental retardation
  behavioural technique, *143*
  causes, 143
  and IQ, 142–3
  prevention, 143
Meperdine hydrochloride, *119*
Merchant bank, 308–9
Mercury (astrology), *200*
Merkel's discs, *52*
Mermaid, *206*
Merton, Robert K, 289
Merton, Thomas, *219*
Meru, Mount, *216*
Mescaline, 104, *105*, 144
Mesmer, Franz, *83*, 136
Mesmerism, *83*, 136, 137
Mesolithic man, 27, 29, *29, 31*
Mesopotamia
  astrology, 200–1
  mythology, 213–14
Message
  concealed, 242–3, *242–3*
  open, *242–3*
  pictorial, *244*
Metabolism, *40*

control, *71*
disorder, *96*
meditation, 195
and skin disease, 99
Metallurgy, origins, 258
Metaphysics, 226, 232, *233*
Metatarsals, *57*
Methionine, 70
Metoscopy, *196*
Microcosm, man as, *216*
Micro-economics, 312
Micronesians, 31, *34*
Microsporum andoni, 98
Mid-brain, *40*
Middle age, 174–5, *174–5*
Middle Ages
  apprenticeship, 301
  punitive devices, 296, *296*
Middle ear, 50
Mid-life crisis, 174
Migration
  hominid, 28–31, *31*
  of labour, *315*
Milgram, Stanley, *161*
Militarism, 272
Military one-time pad code, 242–3
Military service, lower age limit for, 165
Milk
  digestion, 71
  hygiene, *110*
  nutritional value, *70*
  pasteurization, *111*
  production, human, *77*, 78
Milk teeth, *131*
Mill, John Stuart, 228, *232–3, 235, 273*
Mills, C. Wright, 265, *275*
Mind
  and body, 36–7, *36–7*
  and brain, 42–3, *42–3*
  and consciousness, 224
  control by meditation, *195*
  and ego, *187*
  health of, 134–5, *134–5*
  illness of, 138–43, *138–43*
  memory, 44–5, *44–5*
  and philosophy, 229
  potential, 46–7, *46–7*
  treatment, 136–7, *136–7*, 144–9, *144–9*
Mindel-Riss interglaciation, hominid remains, 26
Mind patterns, 44
Mineral
  in diet, 71, *71*
  springs, *83*
Ministry, government, 280–1
Minoan script, 246, *246*
Minos, King of Crete, 210
Minotaur, *206*, 210
Minuscule cursive script, *247*
Mithraism, *212*
Mithras, *212*
Mitral valve, *63*
Mnemonic system, 45, *45*
Modelling, in economics, *311*
Mogadon, 144
Mohammed, 220
Moirai, *185*
Molars, *131*
Monarchy
  in feudal times, 273
  power, 264
Monetary system, 308, *308–9*
Mongolism, 84, *84*, 112, 143, *143*
Mongol race, 32–3, *33–5*, 35

early, 31, *31*
  society, *252*
Monism, *233*
Moniz, Egas, 145, *145*
Monoamine oxidase (MAO) inhibitor, 144
Monocyte, *62*
Monogamy, 168–9
Monopoly, 312, *312*
Monotheism, 220
Mons pubis (veneris), 74
Montessori, Maria, 153
Monumental capital script, *247*
Moon
  astrology, 200, *200*
  mythology, 212
Moore, G.E, 232, *233*
Moral development, 160–1, *160–3*
Morality, 160–1, *161*, 184
  absolutist, 182
  and children, *161*
  Christian, 161
  death, 182–3, *182–3*
  and individual, 183
  in law code, 282, *282*
  life, 182–3, *182–3*
  philosophy, 234–5, *234–5*
  roots of, 183
  and science, 182–3, *182*
  and sex, 169
  situationalist, 182
  utilitarian, 182
Moral reasoning, 160–1, *160*
More, Thomas, 298
Moreno, Jacon, 149, *149*
Morgan, J.J.B, 158
Morgan, Lewis H, 248, 250, *253*
Morgan and Watson theory on emotions, 158
Morita therapy, 148–9
Morocco
  GNP, *317*
  hominid remains, 26
Morphine, 104, *105, 119*
Morphology, 239–40
Morse, Samuel, *242*
Morse code, *242*
Mortality, 177, *184, 184*
  See also Death-rate
Morton, William, 126
Morula, *75*
Moses, 219, 282, *282*
Moslem. *See* Islam
Mothers
  surrogate, *158*
  of young children, 170
Motor cell, *37*
Motor disorder, 135
Motor nerve, 38
Motor racing, *305*
Mo Tzu, *232–3*
Mourning, Chinese, *181*
Mousterian culture, *22, 27, 31*
Moustier, Le, hominid remains, 27
Mouth tumour, 131
Movement, illusions of, *47*
Mtessa, King, *169*
Mucus, 54–5, 67–8, 88
Mudra, 194, *194*
Muhamed. *See* Mohammed
Multiple sclerosis, 97
Mummy, first, *207*
Mumps, *85, 101*, 143
Mumtaz, Mahal, *184*
Munch, Edvard, *139*
Munchausen, Baron von, *141*
Munchausen syndrome, *141*

Picture Credits

335

Artwork Credits

Art Editors
Angela Downing; George Glaze; James Marks; Mel Peterson; Ruth Prentice; Bob Scott

Visualizers
David Aston; Javed Badar; Allison Blythe; Angela Braithwaite; Alan Brown; Michael Burke; Alistair Campbell; Terry Collins; Mary Ellis; Judith Escreet; Albert Jackson; Barry Jackson; Ted Kindsey; Kevin Maddison; Erika Mathlow; Paul Mundon; Peter Nielson; Patrick O'Callaghan; John Ridgeway; Peter Saag; Malcolme Smythe; John Stanyon; John Stewart; Justin Todd; Linda Wheeler

Artists
Stephen Adams; Geoffrey Alger; Terry Allen; Jeremy Alsford; Frederick Andenson; John Arnold; Peter Arnold; David Ashby; Michael Badrock; William Baker; John Barber; Norman Barber; Arthur Barvoso; John Batchelor; John Bavosi; David Baxter; Stephen Bernette; John Blagovitch; Michael Blore; Christopher Blow; Roger Bourne; Alistair Bowtell; Robert Brett; Gordon Briggs; Linda Broad; Lee Brooks; Rupert Brown; Marilyn Bruce; Anthony Bryant; Paul Buckle; Sergio Burelli; Dino Bussetti; Patricia Casey; Giovanni Casselli; Nigel Chapman; Chensie Chen; David Chisholm; David Cockcroft; Michael Codd; Michael Cole; Gerry Collins; Peter Connelly; Roy Coombs; David Cox; Patrick Cox; Brian Cracker; Gordon Cramp; Gino D'Achille; Terrence Daley; John Davies; Gordon C. Davis; David Day; Graham Dean; Brian Delf; Kevin Diaper; Madeleine Dinkel; Hugh Dixon; Paul Draper; David Dupe; Howard Dyke; Jennifer Eachus; Bill Easter; Peter Edwards; Michael Ellis; Jennifer Embleton; Ronald Embleton; Ian Evans; Ann Evens; Lyn Evens; Peter Fitzjohn; Eugene Flurey; Alexander Forbes; David Carl Forbes; Chris Fosey; John Francis; Linda Francis; Sally Frend; Brian Froud; Gay Galfworthy; Ian Garrard; Jean George; Victoria Goaman; David Godfrey; Miriam Golochoy; Anthea Gray; Harold Green; Penelope Greensmith; Vanna Haggerty; Nicholas Hall; Horgrave Hans; David Hardy; Douglas Harker; Richard Hartwell; Jill Havergale; Peter Hayman; Ron Haywood; Peter Henville; Trevor Hill; Garry Hinks; Peter Hutton; Faith Jacques; Robin Jacques; Lancelot Jones; Anthony Joyce; Pierre Junod; Patrick Kaley; Sarah Kensington; Don Kidman; Harold King; Martin Lambourne; Ivan Lapper; Gordon Lawson; Malcolm Lee-Andrews; Peter Levaffeur; Richard Lewington; Brian Lewis; Ken Lewis; Richard Lewis; Kenneth Lilly; Michael Little; David Lock; Garry Long; John Vernon Lord; Vanessa Luff; John Mac; Lesley MacIntyre; Thomas McArthur; Michael McGuinness; Ed McKenzie; Alan Male; Ben Manchipp; Neville Mardell; Olive Marony; Bob Martin; Gordon Miles; Sean Milne; Peter Mortar; Robert Morton; Trevor Muse; Anthony Nelthorpe; Michael Neugebauer; William Nickless; Eric Norman; Peter North; Michael O'Rourke; Richard Orr; Nigel Osborne; Patrick Oxenham; John Painter; David Palmer; Geoffrey Parr; Allan Penny; David Penny; Charles Pickard; John Pinder; Maurice Pledger; Judith Legh Pope; Michael Pope; Andrew Popkiewicz; Brian Price-Thomas; Josephine Rankin; Collin Rattray; Charles Raymond; Alan Rees; Ellsie Rigley; John Ringnall; Christine Robbins; Ellie Robertson; James Robins; John Ronayne; Collin Roso; Peter Sarson; Michael Saunders; Ann Savage; Dennis Scott; Edward Scott-Jones; Rodney Shackell; Chris Simmonds; Gwendolyn Simson; Cathleen Smith; Lesley Smith; Stanley Smith; Michael Soundels; Wolf Spoel; Ronald Steiner; Ralph Stobart; Celia Stothard; Peter Sumpter; Rod Sutterby; Allan Suttie; Tony Swift; Michael Terry; John Thirsk; Eric Thomas; George Thompson; Kenneth Thompson; David Thorpe; Harry Titcombe; Peter Town; Michael Trangenza; Joyce Tuhill; Glenn Tutssel; Carol Vaucher; Edward Wade; Geoffrey Wadsley; Mary Waldron; Michael Walker; Dick Ward; Brian Watson; David Watson; Peter Weavers; David Wilkinson; Ted Williams; John Wilson; Roy Wiltshire; Terrence Wingworth; Anne Winterbotham; Albany Wiseman; Vanessa Wiseman; John Wood; Michael Woods; Owen Woods; Sidney Woods; Raymond Woodward; Harold Wright; Julia Wright

Studios
Add Make-up; Alard Design; Anyart; Arka Graphics; Artec; Art Liaison; Art Workshop; Bateson Graphics; Broadway Artists; Dateline Graphics; David Cox Associates; David Levin Photographic; Eric Jewel Associates; George Miller Associates; Gilcrist Studios; Hatton Studio; Jackson Day; Lock Pettersen Ltd; Mitchell Beazley Studio; Negs Photographic; Paul Hemus Associates; Product Support Graphics; Q.E.D. [Campbell Kindsley]; Stobart and Sutterby; Studio Briggs; Technical Graphics; The Diagram Group; Tri Art; Typographics; Venner Artists

Agents
Artist Partners; Freelance Presentations; Garden Studio; Linden Artists; N.E. Middletons; Portman Artists; Saxon Artists; Thompson Artists